Theory and Applications of Fractional Differential Equations

Theory and Applications of Fractional Differential Equations

Editor: Bernard Gibson

NY RESEARCH PRESS
New York

Published by NY Research Press
118-35 Queens Blvd., Suite 400,
Forest Hills, NY 11375, USA
www.nyresearchpress.com

Theory and Applications of Fractional Differential Equations
Edited by Bernard Gibson

International Standard Book Number: 978-1-64725-440-7 (Hardback)

Cataloging-in-Publication Data

Theory and applications of fractional differential equations / edited by Bernard Gibson.
p. cm.
Includes bibliographical references and index.
ISBN 978-1-64725-440-7
1. Fractional differential equations. 2. Differential equations. 3. Fractional calculus. I. Gibson, Bernard.
QA372 .T44 2023
515.35--dc23

Contents

Preface

Fractional differential equations or fractional order differential equations are generalized and non-integer order differential equations, which can be obtained in time and space with a power law memory kernel of the non-local relationships. They are also known as extraordinary differential equations. They are used to describe the functioning of various complex and non-local systems with memory. Fractional differential equations are applied extensively in various fields including medicine, mechanics, control theory, signal and image processing, environmental science, mathematics, physics, chemistry, and biology. Some of its prominent applications include time-space fractional diffusion equation models, acoustic wave equations for complex media and electrochemical analysis. This book covers the latest researches on the theory and applications of fractional differential equations. It strives to provide a fair idea about these equations and to help develop a better understanding of the latest advances in their study. The book is an invaluable asset for researchers working in the areas of pure mathematics, applied mathematics, statistics, and engineering.

The information contained in this book is the result of intensive hard work done by researchers in this field. All due efforts have been made to make this book serve as a complete guiding source for students and researchers. The topics in this book have been comprehensively explained to help readers understand the growing trends in the field.

I would like to thank the entire group of writers who made sincere efforts in this book and my family who supported me in my efforts of working on this book. I take this opportunity to thank all those who have been a guiding force throughout my life.

Editor

1

A Coupled System of Fractional Difference Equations with Nonlocal Fractional Sum Boundary Conditions on the Discrete Half-Line

Jarunee Soontharanon [1], Saowaluck Chasreechai [1,*] and Thanin Sitthiwirattham [2,*]

1. Department of Mathematics, Faculty of Applied Science, King Mongkut's University of Technology North Bangkok, Bangkok 10800, Thailand; jarunee.s@sci.kmutnb.ac.th
2. Mathematics Department, Faculty of Science and Technology, Suan Dusit University, Bangkok 10700, Thailand

* Correspondence: saowaluck.c@sci.kmutnb.ac.th (S.C.); thanin_sit@dusit.ac.th (T.S.)

Abstract: In this article, we propose a coupled system of fractional difference equations with nonlocal fractional sum boundary conditions on the discrete half-line and study its existence result by using Schauder's fixed point theorem. An example is provided to illustrate the results.

Keywords: existence; coupled system of fractional difference equations; fractional sum; discrete half-line

1. Introduction

Recently, many mathematicians and researchers have extensively studied fractional difference calculus since this subject can be used for describing many problems of real-world phenomena such as mechanical, control systems, flow in porous media, and electrical networks (see [1,2] and the references therein). The basic definitions and properties of fractional difference calculus are given in the book [3]. The applications and developments of the theory can be found in [4–47] and the references cited therein. For example, Ferreira [20] studied the fractional difference equation of order less than one. Goodrich [22] presented the fractional difference equation of order $1 < \alpha \leq 2$ with a constant boundary condition. Chen et al. [28] proposed the initial value problem of order less than one. Chen and Zhou [29] studied the antiperiodic boundary value problem of order $1 < \alpha \leq 2$. Sitthiwirattham et al. [38] initiated the study of the fractional sum boundary value problem of order $1 < \alpha \leq 2$. Sitthiwirattham [40] proposed the sequential fractional difference equation with the fractional sum boundary condition. We observe that these research works are fractional problems containing only one equation.

The study of coupled systems of fractional differential equations is an important topic in this area (see [48–53] and the references cited therein), and a recent example of the application of systems of fractional difference equations is [54].

For the boundary value problems for systems of discrete fractional equations, there are some studies in this area (see [55–60] and the references cited therein).

Pan et al. [55] proposed the system of discrete fractional difference equations as given by:

$$\begin{aligned} -\Delta^{\nu} y_1(t) &= f(y_1(t+\nu), y_2(t+\mu-1)), \\ -\Delta^{\mu} y_2(t) &= g(y_1(t+\nu), y_2(t+\mu-1)), \end{aligned} \tag{1}$$

for $t \in \mathbb{N}_{0,b+1} := \{0, 1, 2, ..., b+1\}$, with the difference boundary conditions:

$$\begin{aligned} y_1(\nu - 2) &= \Delta y_1(\nu + b) = 0, \\ y_2(\mu - 2) &= \Delta y_2(\mu + b) = 0, \end{aligned} \tag{2}$$

where $b \in \mathbb{N}_0 := \mathbb{N} \cup \{0\}$; $1 < \mu, \nu \leq 2$; $0 < \beta \leq 1$; and $f, g : \mathbb{R} \times \mathbb{R} \to \mathbb{R}$ are continuous functions. Δ^ν and Δ^μ are fractional difference operator of order ν and μ, respectively.

In 2015, Goodrich [58] discussed the coupled system of discrete fractional difference equations:

$$\begin{aligned} -\Delta^{-\nu} x(t) &= \lambda_1 f(t + \nu - 1, y(t + \mu - 1)), \quad t \in \mathbb{N}_{0,b+1}, \\ -\Delta^{-\mu} y(t) &= \lambda_2 g(t + \mu - 1, y(t + \nu - 1)), \end{aligned} \tag{3}$$

with the nonlinearities satisfying no growth conditions:

$$\begin{aligned} x(\nu - 2) &= H_1 \left(\textstyle\sum_{i=1}^{n} a_i y(\xi_i)\right), & x(\nu + b + 1) &= 0, \\ y(\mu - 2) &= H_2 \left(\textstyle\sum_{j=1}^{m} b_j x(\zeta_i)\right), & y(\mu + b + 1) &= 0, \end{aligned} \tag{4}$$

where $1 < \nu \leq 2$; $1 < \mu \leq 2$; $\lambda_1, \lambda_2 > 0$; $\{a_i\}_{i=1}^{n}, \{b_j\}_{j=1}^{m} \subseteq (0, \infty)$; and $H_1, H_2 : [0, \infty) \to [0, \infty)$ are continuous functions.

In this paper, we considered the coupled system of fractional difference equations:

$$\begin{cases} \Delta^{\alpha_1} u_1(t) = F_1(t + \alpha_1 - 1, t + \alpha_2 - 1, \Delta^{\beta_1} u_1(t + \alpha_1 - \beta_1), u_2(t + \alpha_2 - 1)), \\ \Delta^{\alpha_2} u_2(t) = F_2(t + \alpha_1 - 1, t + \alpha_2 - 1, \Delta^{\beta_2} u_2(t + \alpha_2 - \beta_1), u_1(t + \alpha_1 - 1)), \end{cases} \tag{5}$$

for $t \in \mathbb{N}_0$, subject to the nonlocal fractional sum boundary conditions on the discrete half-line $\mathbb{N}_0$:

$$\begin{cases} u_1(\alpha_1 - 2) = \phi_1(u_1, u_2), \\ u_2(\alpha_2 - 2) = \phi_2(u_1, u_2), \\ \lim_{t \to \infty} u_1(t + \alpha_1 - 2) = \lambda_2 \Delta^{-\theta_2} g_2(\eta_2 + \theta_2) u_2(\eta_2 + \theta_2), \\ \lim_{t \to \infty} u_2(t + \alpha_2 - 2) = \lambda_1 \Delta^{-\theta_1} g_1(\eta_1 + \theta_1) u_1(\eta_1 + \theta_1). \end{cases} \tag{6}$$

For $i = 1, 2$, $\alpha_i \in (1, 2]$; $\nu_i, \gamma_i, \theta_i \in (0, 1]$; $\beta_i \in (\alpha_i - 1, \alpha_i)$; $\lambda_1, \lambda_2 > 0$, and $\eta_i \in \mathbb{N}_{\alpha_i - 1, T + \alpha_i - 1}$ are given constants; $F_i \in C\left(\mathbb{N}_{\alpha_1 - 2} \times \mathbb{N}_{\alpha_2 - 2} \times \mathbb{R}^2, \mathbb{R}\right)$ and $g_i \in C\left(\mathbb{N}_{\alpha_i - 2, T + \alpha_i}, \mathbb{R}^+\right)$ are given functions; $\phi_i(u_1, u_2)$ are given functionals; and $\Delta^{-\theta_i}$ are fractional sums of order θ_i.

The goal of this study is to show the existence of solutions of the governing problems (5) and (6). The paper is structured as follows. Some definitions and basic lemmas are recalled in Section 2. In Section 3, we prove the existence of solutions of the boundary value problem (5) by employing Schauder's fixed point theorem. Finally, we present an example to illustrate our result in the last section.

2. Preliminaries

In what follows, the notation, definitions, and lemmas used in the main results are given.

Definition 1. *The generalized falling function is defined by* $t^{\underline{\alpha}} := \dfrac{\Gamma(t+1)}{\Gamma(t+1-\alpha)}$, *for any* t *and* α *for which the right-hand side is defined. If* $t + 1 - \alpha$ *is a pole of the Gamma function and* $t + 1$ *is not a pole, then* $t^{\underline{\alpha}} = 0$.

Lemma 1. *[4] Assume the falling factorial functions are well defined. If* $t \leq r$, *then* $t^{\underline{\alpha}} \leq r^{\underline{\alpha}}$ *for any* $\alpha > 0$.

Definition 2. *For $\alpha > 0$ and f defined on $\mathbb{N}_a$, the α-order fractional sum of f is defined by:*

$$\Delta^{-\alpha} f(t) := \frac{1}{\Gamma(\alpha)} \sum_{s=a}^{t-\alpha} (t-\sigma(s))^{\underline{\alpha-1}} f(s),$$

where $t \in \mathbb{N}_{a+\alpha}$ and $\sigma(s) = s+1$.

Definition 3. *For $\alpha > 0$ and f defined on $\mathbb{N}_a$, the α-order Riemann–Liouville fractional difference of f is defined by:*

$$\Delta^{\alpha} f(t) := \Delta^N \Delta^{-(N-\alpha)} f(t) = \frac{1}{\Gamma(-\alpha)} \sum_{s=a}^{t+\alpha} (t-\sigma(s))^{\underline{-\alpha-1}} f(s),$$

where $t \in \mathbb{N}_{a+N-\alpha}$ and $N \in \mathbb{N}$ are chosen so that $0 \leq N-1 < \alpha \leq N$.

Lemma 2. *[4] Let $0 \leq N-1 < \alpha \leq N$. Then,*

$$\Delta^{-\alpha}\Delta^{\alpha} y(t) = y(t) + C_1 t^{\underline{\alpha-1}} + C_2 t^{\underline{\alpha-2}} + \ldots + C_N t^{\underline{\alpha-N}},$$

for some $C_i \in \mathbb{R}$, with $1 \leq i \leq N$.

The following lemma deals with the linear variant of the boundary value problems (5) and (6) and gives a representation of the solution.

Lemma 3. *Let $\alpha_i \in (1,2]$, $\theta_i \in (0,1]$, $\lambda_1, \lambda_2 > 0$ and $\eta_i \in \mathbb{N}_{\alpha_i-1, T+\alpha_i-1}$ be given constants, $k_i \in C(\mathbb{N}_{\alpha_i-2}, \mathbb{R})$ and $g_i \in C(\mathbb{N}_{\alpha_i-2, T+\alpha_i}, \mathbb{R}^+)$ given functions, and $\phi_i(u_1, u_2)$ given functionals. For each $i, j \in \{1,2\}$ and $i \neq j$, then the problems:*

$$\Delta^{\alpha_i} u_i(t) = k_i(t+\alpha_i-1), \quad t \in \mathbb{N}_0, \tag{7}$$

$$u_i(\alpha_i - 2) = \phi_i(u_1, u_2), \tag{8}$$

$$\lim_{t\to\infty} u_i(t+\alpha_i) = \lambda_j \Delta^{-\theta_j} g_j(\eta_j+\theta_j) u_j(\eta_j+\theta_j). \tag{9}$$

have the unique solutions:

$$\begin{aligned} u_1(t_1) &= t_1^{\underline{\alpha_1-1}} \Bigg\{ \frac{\lambda_1}{\Lambda\Gamma(\theta_1)} \sum_{s=\alpha_1-2}^{\eta_1} (\eta_1+\theta_1-\sigma(s))^{\underline{\theta_1-1}} g_1(s) s^{\underline{\alpha_1-1}} \mathcal{P}(k_1,k_2) \\ &\quad - \frac{\lambda_2}{\Lambda\Gamma(\theta_2)} \sum_{s=\alpha_2-2}^{\eta_2} (\eta_2+\theta_2-\sigma(s))^{\underline{\theta_2-1}} g_2(s) s^{\underline{\alpha_2-1}} \mathcal{Q}(k_1,k_2) \Bigg\} \\ &\quad + \frac{t_1^{\underline{\alpha_1-2}} \phi_1(u_1,u_2)}{\Gamma(\alpha_1)} + \frac{1}{\Gamma(\alpha_1)} \sum_{s=0}^{t_1-\alpha_1} (t_1-\sigma(s))^{\underline{\alpha_1-1}} k_1(s+\alpha_1-1), \ t_1 \in \mathbb{N}_{\alpha_1-2}, \end{aligned} \tag{10}$$

$$\begin{aligned} u_2(t_2) &= t_2^{\underline{\alpha_2-1}} \Bigg\{ \frac{\lim_{t_2\to\infty} t_2^{\underline{\alpha_2-1}}}{\Lambda} \mathcal{P}(k_1,k_2) - \frac{\lim_{t_1\to\infty} t_1^{\underline{\alpha_1-1}}}{\Lambda} \mathcal{Q}(k_1,k_2) \Bigg\} \\ &\quad + \frac{t_2^{\underline{\alpha_2-2}} \phi_2(u_1,u_2)}{\Gamma(\alpha_2)} + \frac{1}{\Gamma(\alpha_2)} \sum_{s=0}^{t_2-\alpha_2} (t_2-\sigma(s))^{\underline{\alpha_2-1}} k_2(s+\alpha_2-1), \ t_2 \in \mathbb{N}_{\alpha_2-2}, \end{aligned} \tag{11}$$

provided that both $u_1(t_1), u_2(t_2)$ are uniformly bounded on $\mathbb{N}_{\alpha_1-2}$ and $\mathbb{N}_{\alpha_2-2}$, respectively, and:

$$\Lambda = \frac{\lambda_2 \lim_{t_2\to\infty} t_2^{\underline{\alpha_2-1}}}{\Gamma(\alpha_2)} \sum_{s=\alpha_2-1}^{\eta_2} (\eta_2+\theta_2-\sigma(s))^{\underline{\theta_2-1}} g_2(s)\, s^{\underline{\alpha_2-1}} \tag{12}$$

$$-\frac{\lambda_1 \lim_{t_1\to\infty} t_1^{\underline{\alpha_1-1}}}{\Gamma(\alpha_1)} \sum_{s=\alpha_1-1}^{\eta_1} (\eta_1+\theta_1-\sigma(s))^{\underline{\theta_1-1}} g_1(s)\, s^{\underline{\alpha_1-1}}, \quad t_i \in \mathbb{N}_{\alpha_i-2},$$

$$\begin{aligned} \mathcal{P}(k_1,k_2) &= \frac{\lim_{t_1\to\infty} t_1^{\underline{\alpha_1-2}} \phi_1(u_1,u_2)}{\Gamma(\alpha_1)} - \frac{\lambda_2\phi_2(u_1,u_2)}{\Gamma(\alpha_2)\Gamma(\theta_2)} \sum_{s=\alpha_2-2}^{\eta_2} (\eta_2+\theta_2-\sigma(s))^{\underline{\theta_2-1}} g_2(s)\, s^{\underline{\alpha_2-2}} \\ &\quad + \frac{1}{\Gamma(\alpha_1)} \lim_{t_1\to\infty} \sum_{s=0}^{t_1-\alpha_1} (t_1-\sigma(s))^{\underline{\alpha_1-1}} k_1(s+\alpha_1-1) - \frac{\lambda_2}{\Gamma(\alpha_2)\Gamma(\theta_2)} \times \\ &\quad \sum_{\xi=\alpha_2}^{\eta_2} \sum_{s=0}^{\xi-\alpha_2} (\eta_2+\theta_2-\sigma(\xi))^{\underline{\theta_2-1}} (\xi-\sigma(s))^{\underline{\alpha_2-1}} g_2(s+\alpha_2-1) k_2(s+\alpha_2-1), \end{aligned} \tag{13}$$

$$\begin{aligned} \mathcal{Q}(k_1,k_2) &= \frac{\lim_{t_2\to\infty} t_2^{\underline{\alpha_2-2}} \phi_2(u_1,u_2)}{\Gamma(\alpha_2)} - \frac{\lambda_1\phi_1(u_1,u_2)}{\Gamma(\alpha_1)\Gamma(\theta_1)} \sum_{s=\alpha_1-2}^{\eta_1} (\eta_1+\theta_1-\sigma(s))^{\underline{\theta_1-1}} g_1(s)\, s^{\underline{\alpha_1-2}} \\ &\quad + \frac{1}{\Gamma(\alpha_2)} \lim_{t_2\to\infty} \sum_{s=0}^{t_2-\alpha_2} (t_2-\sigma(s))^{\underline{\alpha_2-1}} k_2(s+\alpha_2-1) - \frac{\lambda_1}{\Gamma(\alpha_1)\Gamma(\theta_1)} \times \\ &\quad \sum_{\xi=\alpha_1}^{\eta_1} \sum_{s=0}^{\xi-\alpha_1} (\eta_1+\theta_1-\sigma(\xi))^{\underline{\theta_1-1}} (\xi-\sigma(s))^{\underline{\alpha_1-1}} g_1(s+\alpha_1-1) k_1(s+\alpha_1-1). \end{aligned} \tag{14}$$

Proof. For each $i, j \in \{1,2\}$ and $i \neq j$, using Lemma 2 and the fractional sum of order $\alpha \in (1,2]$ for (7), we obtain:

$$u_i(t_i) = C_{1i} t_i^{\underline{\alpha_i-1}} + C_{2i} t_i^{\underline{\alpha_i-2}} + \frac{1}{\Gamma(\alpha_i)} \sum_{s=0}^{t_i-\alpha_i} (t_i-\sigma(s))^{\underline{\alpha_i-1}} k_i(s+\alpha_i-1), \tag{15}$$

for $t_i \in \mathbb{N}_{\alpha_i-2}$.

By using the boundary condition (8), we find that:

$$C_{2i} = \frac{\phi_i(u_1,u_2)}{\Gamma(\alpha_i)}. \tag{16}$$

Then, for $t_i \in \mathbb{N}_{\alpha_i-2}$, we have:

$$\begin{aligned} u_i(t_i) &= C_{1i} t_i^{\underline{\alpha_i-1}} + \frac{\phi_i(u_1,u_2)}{\Gamma(\alpha_i)} t_i^{\underline{\alpha_i-2}} \\ &\quad + \frac{1}{\Gamma(\alpha_i)} \sum_{s=0}^{t_i-\alpha_i} (t_i-\sigma(s))^{\underline{\alpha_i-1}} k_i(s+\alpha_i-1). \end{aligned} \tag{17}$$

Taking the fractional sum of order $0 < \theta_i \leq 1$ for (17), we obtain:

$$\begin{aligned} &\Delta^{-\theta_i} u_i(t_i) \\ = &\frac{C_{1i}}{\Gamma(\theta_i)} \sum_{s=\alpha_i-2}^{t_i} (t_i+\theta-\sigma(s))^{\underline{\theta_i-1}} g_i(s)\, s^{\underline{\alpha_i-1}} + \frac{\phi_i(u_1,u_2)}{\Gamma(\alpha_i)} \times \\ &\sum_{s=\alpha_i-2}^{t_i} (t_i+\theta_i-\sigma(s))^{\underline{\theta_i-1}} g_i(s)\, s^{\underline{\alpha_i-2}} + \frac{1}{\Gamma(\theta_i)\Gamma(\alpha_i)} \sum_{\xi=\alpha_i}^{t_i} \sum_{s=0}^{\xi-\alpha_i} (t_i+\theta_i-\sigma(\xi))^{\underline{\theta_i-1}} \times \\ &(\xi-\sigma(s))^{\underline{\alpha_i-1}} g_i(s+\alpha_i-1)\, k_i(s+\alpha_i-1), \end{aligned} \tag{18}$$

for $t_i \in \mathbb{N}_{\alpha_i-2}$.

Employing the boundary condition (9), this implies that:

$$\begin{aligned} &C_{11}\lim_{t_1\to\infty} t_1^{\underline{\alpha_1-1}} + \frac{\phi_1(u_1,u_2)}{\Gamma(\alpha_1)}\lim_{t_1\to\infty} t_1^{\underline{\alpha_1-2}} + \frac{\lim_{t_1\to\infty}}{\Gamma(\alpha_1)}\sum_{s=0}^{t_1-\alpha_1}(t_i-\sigma(s))^{\underline{\alpha_1-1}}k_1(s+\alpha_1-1) \\ &= \frac{\lambda_2 C_{12}}{\Gamma(\theta_2)}\sum_{s=\alpha_2-1}^{\eta_2}(\eta_2+\theta_2-\sigma(s))^{\underline{\theta_2-1}}g_2(s)\,s^{\underline{\alpha_2-1}} \\ &+ \frac{\lambda_2\phi_2(u_1,u_2)}{\Gamma(\alpha_2)\Gamma(\theta_2)}\sum_{s=\alpha_2-2}^{\eta_2}(\eta_2+\theta_2-\sigma(s))^{\underline{\theta_2-1}}g_2(s)\,s^{\underline{\alpha_2-2}} \\ &+ \frac{\lambda_2}{\Gamma(\alpha_2)\Gamma(\theta_2)}\sum_{\xi=\alpha_2}^{\eta_2}\sum_{s=0}^{\xi-\alpha_2}(\eta_2+\theta_2-\sigma(\xi))^{\underline{\theta_2-1}}(\xi-\sigma(s))^{\underline{\alpha_2-1}}g_2(s+\alpha_2-1)\,k_2(s+\alpha_2-1), \end{aligned} \tag{19}$$

and:

$$\begin{aligned} &C_{12}\lim_{t_2\to\infty} t_2^{\underline{\alpha_2-1}} + \frac{\phi_2(u_1,u_2)}{\Gamma(\alpha_2)}\lim_{t_2\to\infty} t_2^{\underline{\alpha_2-2}} + \frac{1}{\Gamma(\alpha_2)}\lim_{t_2\to\infty}\sum_{s=0}^{t_2-\alpha_2}(t_2-\sigma(s))^{\underline{\alpha_2-1}}k_2(s+\alpha_2-1) \\ &= \frac{\lambda_1 C_{11}}{\Gamma(\theta_1)}\sum_{s=\alpha_1-1}^{\eta_1}(\eta_1+\theta_1-\sigma(s))^{\underline{\theta_1-1}}g_1(s)\,s^{\underline{\alpha_1-1}} \\ &+ \frac{\lambda_1\phi_1(u_1,u_2)}{\Gamma(\alpha_1)\Gamma(\theta_1)}\sum_{s=\alpha_1-2}^{\eta_1}(\eta_1+\theta_1-\sigma(s))^{\underline{\theta_1-1}}g_1(s)\,s^{\underline{\alpha_1-2}} \\ &+ \frac{\lambda_1}{\Gamma(\alpha_1)\Gamma(\theta_1)}\sum_{\xi=\alpha_1}^{\eta_1}\sum_{s=0}^{\xi-\alpha_1}(\eta_1+\theta_1-\sigma(\xi))^{\underline{\theta_1-1}}(\xi-\sigma(s))^{\underline{\alpha_1-1}}g_1(s+\alpha_1-1)\,k_1(s+\alpha_1-1). \end{aligned} \tag{20}$$

After solving the system of Equations (19) and (20), we obtain:

$$\begin{aligned} C_{11} &= \frac{\lambda_1}{\Lambda\Gamma(\theta_1)}\sum_{s=\alpha_1-2}^{\eta_1}(\eta_1+\theta_1-\sigma(s))^{\underline{\theta_1-1}}g_1(s)s^{\underline{\alpha_1-1}}\mathcal{P}(k_1,k_2) \\ &\quad - \frac{\lambda_2}{\Lambda\Gamma(\theta_2)}\sum_{s=\alpha_2-2}^{\eta_2}(\eta_2+\theta_2-\sigma(s))^{\underline{\theta_2-1}}g_2(s)s^{\underline{\alpha_2-1}}\mathcal{Q}(k_1,k_2), \end{aligned} \tag{21}$$

and:

$$C_{12} = \frac{\lim\limits_{t_2\to\infty} t_2^{\underline{\alpha_2-1}}}{\Lambda}\mathcal{P}(k_1,k_2) - \frac{\lim\limits_{t_1\to\infty} t_1^{\underline{\alpha_1-1}}}{\Lambda}\mathcal{Q}(k_1,k_2), \tag{22}$$

where $\Lambda, \mathcal{P}(k_1,k_2)$ and $\mathcal{Q}(k_1,k_2)$ are defined as (12)–(14), respectively. □

The following lemma deals with the solutions $u_i(t_i)$, $i = 1,2$ of the problems (7)–(9), and $\Delta^{\beta_i}u_i(t_i-\beta_i+1)$ are uniformly bounded on $\mathbb{N}_{\alpha_i-2}$, $\beta_i \in (\alpha_i-1,\alpha_i)$.

Lemma 4. *For each $i,j\in\{1,2\}$ and $i\neq j$, let $k_i \in C(\mathbb{N}_{\alpha_i-2},\mathbb{R})$ and $g_i\in C(\mathbb{N}_{\alpha_i-2},\mathbb{R}^+)$ be given functions, $\phi_i(u_1,u_2)$ be given functionals, $\rho_i > \max\{\beta_i-\alpha_i\}$, $\beta_i\in(\alpha_i-1,\alpha_i)$, and $0<g_i\le g_i(s_i)\le G_i$, for each $s_i\in\mathbb{N}_{\alpha_i-2,T+\alpha_i}$.*

The solution $u_i(t_i)$ of the problems (7)–(9) and $\Delta^{\beta_i}u_i(t_i-\beta_i+1)$ are uniformly bounded on $\mathbb{N}_{\alpha_i-2}$, if and only if $u_i(t_i)$ and $\Delta^{\beta_i}u_i(t_i-\beta_i+1)$ satisfy the following properties:

(A_1) There exist constants $M_1, N_1, m_1, n_1 > 0$ such that, for u_1 and $\Delta^{\beta_1}u_1$,

$$|k_i(t_i)| \le M_1\,e^{-m_1(2t_1+t_2)},$$

$$|\phi_i(u_1,u_2)| \le N_1(t_1+\rho_1)^{\underline{\rho_1}} e^{-n_1(t_1+1)}.$$

(A_2) *There exist constants* $M_2, N_2, m_2, n_2 > 0$ *such that, for* u_2 *and* $\Delta^{\beta_2}u_2$,

$$|k_i(t_i)| \le M_2\,(t_2+\rho_2)^{\underline{\rho_2}} e^{-m_2(t_1+2t_2)},$$
$$|\phi_i(u_1,u_2)| \le N_2(t_1+\rho_1)^{\underline{\rho_1}}\left[(t_2+\rho_2)^{\underline{\rho_2}}\right]^2 e^{-n_2(t_2+1)}.$$

(A_3) *There exist constants* $\Omega_i > 0,\ i=1,2$ *such that,*

$$\left\{\frac{(t_1-\alpha_1+1)^{\underline{2-\alpha_1}}(t_2-\alpha_2+1)^{\underline{2-\alpha_2}}}{1+(t_1+\rho_1)^{\underline{\rho_1}}(t_2+\rho_2)^{\underline{\rho_2}}}\right\}|u_i(t_i)| < \Omega_i,$$
$$\left\{\frac{(t_1-\alpha_1+1)^{\underline{2-\alpha_1}}(t_2-\alpha_2+1)^{\underline{2-\alpha_2}}}{1+(t_1+\rho_1)^{\underline{\rho_1}}(t_2+\rho_2)^{\underline{\rho_2}}}\right\}|\Delta^{\beta_i}u_i(t_i)| < \Omega_i.$$

Proof. Firstly, taking the fractional difference of order $\alpha_i-1<\beta_i<\alpha_i,\ i=1,2$ for (10) and (11), we obtain:

$$\begin{aligned}
&\Delta^{\beta_1}u_1(t_1)\\
&= \frac{1}{\Gamma(-\beta_1)}\sum_{s=\alpha_1-1}^{t_1+1}(t_1-\beta_1+1-\sigma(s))^{\underline{-\beta_1-1}}s^{\underline{\alpha_1-1}}\times\Bigg\{\frac{\lambda_1}{\Lambda\Gamma(\theta_1)}\times\\
&\sum_{s=\alpha_1-2}^{\eta_1}(\eta_1+\theta_1-\sigma(s))^{\underline{\theta_1-1}}g_1(s)s^{\underline{\alpha_1-1}}\mathcal{P}(k_1,k_2)-\frac{\lambda_2}{\Lambda\Gamma(\theta_2)}\times\\
&\sum_{s=\alpha_2-2}^{\eta_2}(\eta_2+\theta_2-\sigma(s))^{\underline{\theta_2-1}}g_2(s)s^{\underline{\alpha_2-1}}\mathcal{Q}(k_1,k_2)\Bigg\} \qquad (23)\\
&+\frac{\phi_1(u_1,u_2)}{\Gamma(-\beta_1)\Gamma(\alpha_1)}\sum_{s=\alpha_1-2}^{t_1+1}(t_1-\beta_1+1-\sigma(s))^{\underline{-\beta_1-1}}s^{\underline{\alpha_1-2}}\\
&+\frac{1}{\Gamma(-\beta_1)\Gamma(\alpha_1)}\sum_{\xi=\alpha_1}^{t_1+1}\sum_{s=0}^{\xi-\alpha_1}(t_1-\beta_1+1-\sigma(s))^{\underline{-\beta_1-1}}(\xi-\sigma(s))^{\underline{\alpha_1-1}}k_1(s+\alpha_1-1),
\end{aligned}$$

and:

$$\begin{aligned}
&\Delta^{\beta_2}u_2(t_2)\\
&= \frac{1}{\Gamma(-\beta_2)}\sum_{s=\alpha_2-1}^{t_2+1}(t_2-\beta_2+1-\sigma(s))^{\underline{-\beta_2-1}}s^{\underline{\alpha_2-1}}\times\Bigg\{\frac{\lim_{t_2\to\infty}t_2^{\alpha_2-1}}{\Lambda}\mathcal{P}(k_1,k_2) \qquad (24)\\
&-\frac{\lim_{t_1\to\infty}t_1^{\alpha_1-1}}{\Lambda}\mathcal{Q}(k_1,k_2)\Bigg\}+\frac{\phi_2(u_1,u_2)}{\Gamma(-\beta_2)\Gamma(\alpha_2)}\sum_{s=\alpha_2-2}^{t_2+1}(t_2-\beta_1+1-\sigma(s))^{\underline{-\beta_2-1}}s^{\underline{\alpha_2-2}}\\
&+\frac{1}{\Gamma(-\beta_2)\Gamma(\alpha_2)}\sum_{\xi=\alpha_2}^{t_2+1}\sum_{s=0}^{\xi-\alpha_2}(t_2-\beta_2+1-\sigma(s))^{\underline{-\beta_2-1}}(\xi-\sigma(s))^{\underline{\alpha_2-1}}k_2(s+\alpha_2-1).
\end{aligned}$$

If $u_i(t_i)$ and $\Delta^{\beta_i}u_i(t_i-\beta_i+1)$ are uniformly bounded on $\mathbb{N}_{\alpha_i-2}$, we have:

$$|\Lambda| \quad \le \quad \left|\frac{\lambda_2\lim_{t_2\to\infty}t_2^{\alpha_2-1}}{\Gamma(\alpha_2)}\sum_{s=\alpha_2-1}^{\eta_2}(\eta_2+\theta_2-\sigma(s))^{\underline{\theta_2-1}}g_2(s)\,s^{\underline{\alpha_2-1}}\right.$$

$$-\frac{\lambda_1 \lim_{t_1\to\infty} t_1^{\underline{\alpha_1-1}}}{\Gamma(\alpha_1)} \sum_{s=\alpha_1-1}^{\eta_1} (\eta_1+\theta_1-\sigma(s))^{\underline{\theta_1-1}} g_1(s)\, s^{\underline{\alpha_1-1}}\Bigg| \qquad (25)$$

$$\leq \max\left\{\left|\lim_{t_2\to\infty} t_2^{\underline{\alpha_2-1}}(\eta_2+\theta_2-\alpha_2)^{\underline{\theta_2-1}} G_2\lambda_2\mathcal{A}_2\right|, \left|\lim_{t_1\to\infty} t_1^{\underline{\alpha_1-1}}(\eta_1+\theta_1-\alpha_1)^{\underline{\theta_1-1}} G_1\lambda_1\mathcal{A}_1\right|\right\}.$$

Furthermore, considering $u_1(t_i)$ and $\Delta^{\beta_1}u_i(t_i)$, we obtain:

$$|k_i(t_i)| < \begin{cases} M_1\, e^{1-(t_1+t_2)} & ,\text{for } u_1(t_1) \\ M_1\, e^{-(2t_1+t_2)} & ,\text{for } \Delta^{\beta_1}u_1(t_1-\beta_1+1) \\ M_2\,(t_2+\rho_2)^{\underline{\rho_2}}\, e^{-(t_1+t_2)} & ,\text{for } u_2(t_2) \\ M_2\,(t_2+\rho_2)^{\underline{\rho_2}}\, e^{-(t_1+2t_2)} & ,\text{for } \Delta^{\beta_2}u_2(t_2-\beta_2+1) \end{cases} \qquad (26)$$

and:

$$|\phi_i(t_1,t_2)| < \begin{cases} N_1\,(t_1+\rho_1+1)^{\underline{\rho_1}} & ,\text{for } u_1(t_1) \\ N_1\,(t_1+\rho_1)^{\underline{\rho_1}}\, e^{-(t_1+1)} & ,\text{for } \Delta^{\beta_1}u_1(t_1-\beta_1+1) \\ N_2(t_1+\rho_1)^{\underline{\rho_1}}\left[(t_2+\rho_2+1)^{\underline{\rho_2}}\right]^2 & ,\text{for } u_2(t_2) \\ N_2(t_1+\rho_1)^{\underline{\rho_1}}\left[(t_2+\rho_2)^{\underline{\rho_2}}\right]^2 e^{-(t_2+1)} & ,\text{for } \Delta^{\beta_2}u_2(t_2-\beta_2+1) \end{cases} \qquad (27)$$

where:

$$M_1 = \min\left\{\frac{\lambda_1 G_1(\eta_1-\alpha_1+\theta_1)^{\underline{\theta_1-1}}\mathcal{A}_1}{\lambda_2 g_2\Gamma(\alpha_2)\mathcal{A}_2}, \frac{G_1(\eta_1-\alpha_1+\theta_1)^{\underline{\theta_1-1}}}{g_1}, \Gamma(\alpha_1), \frac{G_1(\eta_1-\alpha_1+\theta_1)^{\underline{\theta_1-1}}}{\lambda_2 g_1 g_2\Gamma(\alpha_1)\mathcal{C}_2}, \frac{G_1(\eta_1-\alpha_1+\theta_1)^{\underline{\theta_1-1}}\mathcal{A}_1}{\lambda_2 g_1 g_2\Gamma(\alpha_2)\mathcal{A}_2\mathcal{C}_1}\right\}, \qquad (28)$$

$$M_2 = \min\left\{\Gamma(\alpha_2), \lambda_2 G_2(\eta_2-\alpha_2+\theta_2)^{\underline{\theta_2-1}}\mathcal{A}_2, G_1(\eta_1-\alpha_1+\theta_1)^{\underline{\theta_1-1}}\mathcal{A}_1, \frac{G_1(\eta_1-\alpha_1+\theta_1)^{\underline{\theta_1-1}}\mathcal{A}_1}{g_1\mathcal{C}_1}, \frac{G_2(\eta_2-\alpha_2+\theta_2)^{\underline{\theta_2-1}}\mathcal{A}_2}{g_2\mathcal{C}_2}\right\}, \qquad (29)$$

$$N_1 = \min\left\{\Gamma(\theta_1), \Gamma(\theta_2), \Gamma(\alpha_1), \frac{\Gamma(\theta_1)}{\lambda_2 g_2\Gamma(\alpha_1)\mathcal{B}_2}, \frac{G_1(\eta_1-\alpha_1+\theta_1)^{\underline{\theta_1-1}}\mathcal{A}_1}{\lambda_2 g_1 g_2\Gamma(\alpha_1-1)\mathcal{B}_1\mathcal{A}_2}\right\}, \qquad (30)$$

$$N_2 = \min\left\{\Gamma(\alpha_2), \lambda_2 G_2\Gamma(\alpha_1)(\eta_2-\alpha_2+\theta_2)^{\underline{\theta_2-1}}\mathcal{A}_2, \lambda_1 G_1\Gamma(\alpha_2)(\eta_1-\alpha_1+\theta_1)^{\underline{\theta_1-1}}\mathcal{A}_1, \frac{G_2(\eta_2-\alpha_2+\theta_2)^{\underline{\theta_2-1}}\mathcal{A}_2}{g_2\mathcal{B}_2}, \frac{G_1(\eta_1-\alpha_1+\theta_1)^{\underline{\theta_1-1}}\mathcal{A}_1}{g_1\mathcal{B}_1}\right\}. \qquad (31)$$

Consequently, the conditions $(A1)$ and $(A2)$ hold.
We next show that the condition $(A3)$ holds. By using the conditions $(A1)$ and $(A2)$, we obtain:

$$\left|u_i(t_i)\right| \leq t_1^{\underline{\alpha_1}} t_2^{\underline{\alpha_2}}\,\Omega_i < t_1^{\underline{\alpha_1+\rho_1-1}} t_2^{\underline{\alpha_2+\rho_2-1}}\,\Omega_i < \left[\frac{1+t_1^{\underline{\rho_1+2}} t_2^{\underline{\rho_2+2}}}{(t_1-\alpha_1+1)^{\underline{2-\alpha_1}}(t_2-\alpha_2+1)^{\underline{2-\alpha_2}}}\right]\Omega_i$$

and:

$$
\begin{aligned}
\left|\Delta^{\beta_i}u_i(t_i-\beta_i+1)\right| &\le (t_i-\alpha_i-\beta_i+1)^{\underline{-\beta_i}}\,t_j^{\underline{\alpha_j}}\,\Omega_i < \ t_i^{\underline{\alpha_i}}\,t_j^{\underline{\alpha_j}}\,\Omega_i \\
&< \left[\frac{1+t_1^{\rho_1+2}t_2^{\rho_2+2}}{(t_1-\alpha_1+1)^{\underline{2-\alpha_1}}(t_2-\alpha_2+1)^{\underline{2-\alpha_2}}}\right]\Omega_i,\ \ i\neq j=1,2
\end{aligned}
$$

where:

$$
\begin{aligned}
\Omega_1 = \max&\left\{\frac{N_1}{\Gamma(\theta_1)}+M_1+\frac{\lambda_2G_2\Gamma(\alpha_1)(\eta_2-\alpha_2+\theta_2)^{\underline{\theta_2-1}}}{\Gamma(\theta_2)}\times\right.\\
&\left(N_2\mathcal{B}_2+\frac{M_2\mathcal{C}_2}{\lambda_1G_1(\eta_1-\alpha_1+\theta_1)^{\underline{\theta_1-1}}\mathcal{A}_1}\right),\\
&\frac{N_2}{\Gamma(\theta_2)}+M_2+\frac{\lambda_1G_1\Gamma(\alpha_2)(\eta_1-\alpha_1+\theta_1)^{\underline{\theta_1-1}}}{\Gamma(\theta_1)}\times\\
&\left.\left(N_1\mathcal{B}_1+\frac{M_1\mathcal{C}_1}{\lambda_2G_2(\eta_2-\alpha_2+\theta_2)^{\underline{\theta_2-1}}\mathcal{A}_2}\right)\right\}\\
&+\frac{1}{\Gamma(\alpha_1)}(M_1+N_1),
\end{aligned}
\tag{32}
$$

$$
\begin{aligned}
\Omega_2 = \max&\left\{\frac{1}{G_2(\eta_2-\alpha_2+\theta_2)^{\underline{\theta_2-1}}\mathcal{A}_2}\left[\frac{M_1+N_1}{\Gamma(\alpha_1)\lambda_2}+\frac{1}{\Gamma(\theta_2)\mathcal{A}_2}\left(N_2\mathcal{B}_2+M_2\mathcal{C}_2\right)\right],\right.\\
&\left.\frac{1}{G_1(\eta_1-\alpha_1+\theta_1)^{\underline{\theta_1-1}}\mathcal{A}_1}\left[\frac{M_2+N_2}{\Gamma(\alpha_2)\lambda_1}+\frac{1}{\Gamma(\theta_1)\mathcal{A}_1}\left(N_1\mathcal{B}_1+M_1\mathcal{C}_1\right)\right]\right\}\\
&+\frac{1}{\Gamma(\alpha_2)}(M_2+N_2),
\end{aligned}
\tag{33}
$$

with

$$\mathcal{A}_i = {}_2F_1\left(\alpha_i,\alpha_i-\eta_i-1;\alpha_i-\eta_i-\theta_i;1\right) \tag{34}$$
$$\mathcal{B}_i = {}_2F_1\left(\alpha_i-1,\alpha_i-\eta_i-1;\alpha_i-\eta_i-\theta_i-1;1\right) \tag{35}$$
$$\mathcal{C}_i = {}_2F_1\left(\alpha_i+1,\alpha_i-\eta_i;\alpha_i-\eta_i-\theta_i+1;1\right). \tag{36}$$

Therefore, the condition $(A3)$ holds.

Finally, if the conditions $(A1)$–$(A3)$ hold, it is clear that $u_i(t_i)$ and $\Delta^{\beta_i}u_i(t_i-\beta_i+1)$ are uniformly bounded on $\mathbb{N}_{\alpha_i-2}$. Our proof is complete. □

We next provide the following theorems used for proving the existence result for the problems (5) and (6).

Theorem 1. *(Arzelá–Ascoli theorem [61])*

A set of functions in $C[a,b]$ *with the sup norm is relatively compact if and only if it is uniformly bounded and equicontinuous on* $[a,b]$.

Theorem 2. *[61] If a set is closed and relatively compact, then it is compact.*

Theorem 3. *(Schauder's fixed point theorem [61])*

If S is a convex compact subset of a normed space, every continuous mapping of S into itself has a fixed point.

3. Main Result

In this section, we aim to establish the existence result for the problems (5) and (6). To accomplish this, we let $\mathcal{C}_i = C(\mathbb{N}_{\alpha_i-2}, \mathbb{R})$ be a Banach space of all functions on $\mathbb{N}_{\alpha_i-2}$, for each $i, j \in \{1,2\}$ and $i \neq j$. Obviously, the product spaces:

$$\mathcal{U}_i = \Big\{(u_1,u_2) \in \mathcal{C}_1 \times \mathcal{C}_2 : \ \Delta^{\beta_i} u_i(t_i - \beta_i + 1) \in \mathcal{C}_i \text{ and } \chi|u_j(t_j)|, \\ \chi|\Delta^{\beta_i} u_i(t_i - \beta_i + 1)| \text{ are bounded on } \mathbb{N}_{\alpha_j-2}, \mathbb{N}_{\alpha_i-2}, \text{respectively,}\Big\}$$

is also the Banach space endowed with the norm defined by:

$$\|(u_1,u_2)\|_{\mathcal{U}_i} = \|\Delta^{\beta_i} u_i\|_{\mathcal{C}_i} + \|u_j\|_{\mathcal{C}_j},$$

where:

$$\|\Delta^{\beta_i} u_i\|_{\mathcal{C}_i} = \max_{t_i \in \mathbb{N}_{\alpha_i-2}} \chi\Big|\Delta^{\beta_i} u_i(t_i - \beta_i + 1, t_j)\Big| \text{ and } \|u_j\|_{\mathcal{C}_j} = \max_{t_j \in \mathbb{N}_{\alpha_j-2}} \chi|u_j(t_i,t_j)|,$$

with for $\rho_i > \max\{\beta_i - \alpha_i\}$ and $\beta_i \in (\alpha_i - 1, \alpha_i)$,

$$\chi = \frac{(t_1 - \alpha_1 + 1)^{\underline{2-\alpha_1}}(t_2 - \alpha_2 + 1)^{\underline{2-\alpha_2}}}{1 + t_1^{\underline{\rho_1+2}} t_2^{\underline{\rho_2+2}}}. \tag{37}$$

Let $\mathcal{U} = \mathcal{U}_1 \cap \mathcal{U}_2$; clearly, the space $\Big(\mathcal{U}, \|(u_1,u_2)\|_{\mathcal{U}}\Big)$ is the Banach space with the norm:

$$\|(u_1,u_2)\|_{\mathcal{U}} = \max\left\{\|(u_1,u_2)\|_{\mathcal{U}_1}, \ \|(u_1,u_2)\|_{\mathcal{U}_2}\right\}.$$

Next, we define the operator $\mathcal{F} : \mathcal{U} \to \mathcal{U}$ by:

$$(\mathcal{F}(u_1,u_2))(t_1,t_2) = \Big((\mathcal{F}_1(u_1,u_2))(t_1,t_2), (\mathcal{F}_2(u_1,u_2))(t_1,t_2)\Big), \tag{38}$$

and:

$$\begin{aligned}(\mathcal{F}_1(u_1,u_2))(t_1,t_2) = {} & \frac{t_1^{\underline{\alpha_1-1}}}{\Lambda}\left\{\frac{\lambda_1}{\Gamma(\theta_1)}\sum_{s=\alpha_1-2}^{\eta_1}(\eta_1+\theta_1-\sigma(s))^{\underline{\theta_1-1}} g_1(s) s^{\underline{\alpha_1-1}}\mathcal{P}(F_1,F_2)\right. \\ & \left. - \frac{\lambda_2}{\Gamma(\theta_2)}\sum_{s=\alpha_2-2}^{\eta_2}(\eta_2+\theta_2-\sigma(s))^{\underline{\theta_2-1}} g_2(s) s^{\underline{\alpha_2-1}}\mathcal{Q}(F_1,F_2)\right\} \\ & + \frac{t_1^{\underline{\alpha_1-2}}\phi_1(u_1,u_2)}{\Gamma(\alpha_1)} + \frac{1}{\Gamma(\alpha_1)}\sum_{s=\alpha_1-1}^{t_1-1}(t_1+\alpha_1-1-\sigma(s))^{\underline{\alpha_1-1}} \times \\ & F_1\big(s,t_2,\Delta^{\beta_1}u_1(s-\beta_1+1),u_2(t_2)\big), \ \ t_i \in \mathbb{N}_{\alpha_i-2},\end{aligned} \tag{39}$$

$$\begin{aligned}(\mathcal{F}_2(u_1,u_2))(t_1,t_2) = {} & \frac{t_2^{\underline{\alpha_2-1}}}{\Lambda}\left\{\lim_{t_2\to\infty} t_2^{\underline{\alpha_2-1}}\mathcal{P}(F_1,F_2) - \lim_{t_1\to\infty} t_1^{\underline{\alpha_1-1}}\mathcal{Q}(F_1,F_2)\right\} \\ & + \frac{t_2^{\underline{\alpha_2-2}}\phi_2(u_1,u_2)}{\Gamma(\alpha_2)} + \frac{1}{\Gamma(\alpha_2)}\sum_{s=\alpha_2-1}^{t_2-1}(t_2+\alpha_2-1-\sigma(s))^{\underline{\alpha_2-1}} \times \\ & F_2\big(t_1,s,u_1(t_1),\Delta^{\beta_2}u_2(s-\beta_2+1)\big), \ \ t_i \in \mathbb{N}_{\alpha_i-2},\end{aligned} \tag{40}$$

where Λ is defined as (12), and:

$$\mathcal{P}(F_1,F_2) = \frac{\lim_{t_1\to\infty} t_1^{\underline{\alpha_1-2}}\phi_1(u_1,u_2)}{\Gamma(\alpha_1)} - \frac{\lambda_2\phi_2(u_1,u_2)}{\Gamma(\alpha_2)\Gamma(\theta_2)}\sum_{s=\alpha_2-2}^{\eta_2}(\eta_2+\theta_2-\sigma(s))^{\underline{\theta_2-1}}g_2(s)\,s^{\underline{\alpha_2-2}}$$
$$+\frac{1}{\Gamma(\alpha_1)}\lim_{t_1\to\infty}\sum_{s=\alpha_1-1}^{t_1-1}(t_1+\alpha_1-1-\sigma(s))^{\underline{\alpha_1-1}}F_1(s,t_2,\Delta^{\beta_1}u_1(s-\beta_1+1),u_2(t_2))$$
$$-\frac{\lambda_2}{\Gamma(\alpha_2)\Gamma(\theta_2)}\sum_{\xi=\alpha_2}^{\eta_2}\sum_{s=\alpha_2-1}^{\xi-1}(\eta_2+\theta_2-\sigma(\xi))^{\underline{\theta_2-1}}(\xi+\alpha_2-1-\sigma(s))^{\underline{\alpha_2-1}}\times$$
$$g_2(s)F_2(t_1,s,u_1(t_1),\Delta^{\beta_2}u_2(s-\beta_2+1)), \tag{41}$$

$$\mathcal{Q}(F_1,F_2) = \frac{\lim_{t_2\to\infty} t_2^{\underline{\alpha_2-2}}\phi_2(u_1,u_2)}{\Gamma(\alpha_2)} - \frac{\lambda_1\phi_1(u_1,u_2)}{\Gamma(\alpha_1)\Gamma(\theta_1)}\sum_{s=\alpha_1-2}^{\eta_1}(\eta_1+\theta_1-\sigma(s))^{\underline{\theta_1-1}}g_1(s)\,s^{\underline{\alpha_1-2}}$$
$$+\frac{1}{\Gamma(\alpha_2)}\lim_{t_2\to\infty}\sum_{s=\alpha_2-1}^{t_2-1}(t_2+\alpha_2-1-\sigma(s))^{\underline{\alpha_2-1}}F_2(t_1,s,u_1(t_1),\Delta^{\beta_2}u_2(s-\beta_2+1))$$
$$-\frac{\lambda_1}{\Gamma(\alpha_1)\Gamma(\theta_1)}\sum_{\xi=\alpha_1}^{\eta_1}\sum_{s=\alpha_1-1}^{\xi-1}(\eta_1+\theta_1-\sigma(\xi))^{\underline{\theta_1-1}}(\xi+\alpha_1-1-\sigma(s))^{\underline{\alpha_1-1}}\times$$
$$g_1(s)F_1(s,t_2,\Delta^{\beta_1}u_1(s-\beta_1+1),u_2(t_2)). \tag{42}$$

We next make the following assumptions:

(H_1) There exist positive numbers ${}_{ip}\rho_2 \in (-1,\rho_2)$ and $M_{ip}, m_{ip} > 0$ ($i = 1,2$ and $p = 1,2,3$) such that, for each $t_i \in \mathbb{N}_{\alpha_i-2}$ and $v_i \in \mathbb{R}$,

$$\left|F_i\left(t_1,t_2,\frac{1}{\chi}v_1,\frac{1}{\chi}v_2\right) - M_{i1}\,(t_2+{}_{i1}\rho_2)^{{}_{i1}\rho_2}\,e^{-m_{i1}(t_1+t_2)}\right|$$
$$\le M_{i2}\,(t_2+{}_{i2}\rho_2)^{{}_{i2}\rho_2}\,e^{-m_{i2}(t_1+t_2)}\,|v_1| + M_{i3}\,(t_2+{}_{i3}\rho_2)^{{}_{i3}\rho_2}\,e^{-m_{i3}(t_1+t_2)}\,|v_2|.$$

(H_2) There exist positive numbers ${}_{ip}\tilde{\rho}_i \in (-1,\rho_i)$ and $N_{ip}, n_{ip} > 0$ ($i = 1,2$ and $p = 1,2,3$) such that, for $v_i \in \mathcal{C}_i$,

$$\left|\phi_i\left(\frac{1}{\chi}v_1,\frac{1}{\chi}v_2\right) - N_{i1}\,(t_1+{}_{i1}\tilde{\rho}_1)^{{}_{i1}\tilde{\rho}_1}\left[(t_2+{}_{i1}\tilde{\rho}_2)^{{}_{i1}\tilde{\rho}_2}\right]^2 e^{-n_{i1}(t_1+t_2)}\right|$$
$$\le N_{i2}\,(t_1+{}_{i2}\tilde{\rho}_1)^{{}_{i2}\tilde{\rho}_1}\left[(t_2+{}_{i2}\tilde{\rho}_2)^{{}_{i2}\tilde{\rho}_2}\right]^2 e^{-n_{i2}(t_1+t_2)}\,\|v_1\| + N_{i3}\,(t_1+{}_{i3}\tilde{\rho}_1)^{{}_{i3}\tilde{\rho}_1}\left[(t_2+{}_{i3}\tilde{\rho}_2)^{{}_{i3}\tilde{\rho}_2}\right]^2 e^{-n_{i3}(t_1+t_2)}\,\|v_2\|.$$

(H_3) $g_i \le g_i(\eta_i)$ for all $\eta_i \in \mathbb{N}_{\alpha_i-1,T+\alpha_i-1}$.

Lemma 5. *Suppose that (H_1)–(H_3) hold. Then, the fixed point of $\mathcal{F}$ coincides with the solution of the problems (5) and (6), and $\mathcal{F}:\mathcal{U}\to\mathcal{U}$ is completely continuous.*

Proof. Let $(u_1,u_2)\in\mathcal{U}$, for each $i,j\in\{1,2\}$ and $i\neq j$. By the above assumptions (H_1) and (H_2), it follows that:

$$\left|F_i\left(t_1,t_2,\Delta^{\beta_i}u_i(t_i-\beta_i+1),u_j(t_j)\right)\right|$$

$$\begin{aligned} &= \left| F_i \left(t_1, t_2, \frac{1}{\chi} \left[\chi \Delta^{\beta_i} u_i(t_i - \beta_i + 1) \right], \frac{1}{\chi} \left[\chi u_j(t_j) \right] \right) \right| \\ &\leq \; M_{i1} \left(t_2 + {}_{i1}\rho_2 \right)^{i1\rho_2} e^{-m_{i1}(t_1+t_2)} + M_{i2} \left(t_2 + {}_{i2}\rho_2 \right)^{i2\rho_2} e^{-m_{i2}(t_1+t_2)} \|\Delta^{\beta_i} u_i\|_{C_i} \\ &\quad + M_{i3} \left(t_2 + {}_{i3}\rho_2 \right)^{i3\rho_2} e^{-m_{i3}(t_1+t_2)} \|u_j\|_{C_j}, \end{aligned} \tag{43}$$

and

$$\begin{aligned} |\phi_i(u_1, u_2)| &= \left| \phi_i \left(\frac{1}{\chi} [\chi u_1], \frac{1}{\chi} [\chi u_2] \right) \right| \\ &\leq \; N_{i1} \left(t_1 + {}_{i1}\tilde{\rho}_1 \right)^{i1\tilde{\rho}_1} \left[\left(t_2 + {}_{i1}\tilde{\rho}_2 \right)^{i1\tilde{\rho}_2} \right]^2 e^{-n_{i1}(t_1+t_2)} \\ &\quad + N_{i2} \left(t_1 + {}_{i2}\tilde{\rho}_1 \right)^{i2\tilde{\rho}_1} \left[\left(t_2 + {}_{i2}\tilde{\rho}_2 \right)^{i2\tilde{\rho}_2} \right]^2 e^{-n_{i2}(t_1+t_2)} \|u_1\|_{C_1} \\ &\quad + N_{i3} \left(t_1 + {}_{i3}\tilde{\rho}_1 \right)^{i3\tilde{\rho}_1} \left[\left(t_2 + {}_{i3}\tilde{\rho}_2 \right)^{i3\tilde{\rho}_2} \right]^2 e^{-n_{i3}(t_1+t_2)} \|u_2\|_{C_2}. \end{aligned} \tag{44}$$

The rest of the proof follows from Lemmas 3 and 4. This implies that the fixed point of $\mathcal{F}$ coincides with the solution of the problems (5) and (6).

To show that $\mathcal{F}$ is completely continuous, we organize the proof as the following four steps.

Step I. $\mathcal{F}$ is well defined and maps bounded sets into bounded sets.

Let $B_R = \{(u_1, u_2) \in \mathcal{U} : \|(u_1, u_2)\|_{\mathcal{U}} \leq R\}$, then for $(u_1, u_2) \in \mathcal{U}$:

$$\begin{aligned} R \;&\geq\; \max\left\{ \|(u_1, u_2)\|_{\mathcal{U}_1}, \|(u_1, u_2)\|_{\mathcal{U}_2} \right\} \\ &= \; \max\left\{ \chi \Big[|\Delta^{\beta_1} u_1(t_1 - \beta_1 + 1)| + |u_2(t_2)| \Big], \right. \\ &\qquad \left. \chi \Big[|u_1(t_1)| + |\Delta^{\beta_2} u_2(t_2 - \beta_2 + 1)| \Big] \right\}. \end{aligned} \tag{45}$$

By the definition of $\mathcal{F}$, we get $\mathcal{F}_i(u_1, u_2), \Delta^{\beta_i}\mathcal{F}_i(u_1, u_2) \in \mathcal{U}$. Therefore, (43) and (44) imply that:

$$\begin{aligned} &\left| F_i(t_1, t_2, \Delta^{\beta_i} u_i(t_i - \beta_i + 1), u_j(t_j)) \right| \\ &\quad = \left| F_i \left(t_1, t_2, \frac{1}{\chi} \left[\chi \Delta^{\beta_i} u_i(t_i - \beta_i + 1) \right], \frac{1}{\chi} \left[\chi u_j(t_j) \right] \right) \right| \\ &\quad \leq M_{i1} \left(t_2 + {}_{i1}\rho_2 \right)^{i1\rho_2} e^{-m_{i1}(t_1+t_2)} + R M_{i2} \left(t_2 + {}_{i2}\rho_2 \right)^{i2\rho_2} \times \\ &\qquad e^{-m_{i2}(t_1+t_2)} + R M_{i3} \left(t_2 + {}_{i3}\rho_2 \right)^{i3\rho_2} e^{-m_{i3}(t_1+t_2)}, \end{aligned} \tag{46}$$

and

$$\begin{aligned} |\phi_i(u_1, u_2)| &= \left| \phi_i \left(\frac{1}{\chi} [\chi u_1], \frac{1}{\chi} [\chi u_2] \right) \right| \\ &\leq N_{i1} \left(t_1 + {}_{i1}\tilde{\rho}_1 \right)^{i1\tilde{\rho}_1} \left[\left(t_2 + {}_{i1}\tilde{\rho}_2 \right)^{i1\tilde{\rho}_2} \right]^2 e^{-n_{i1}(t_1+t_2)} \\ &\quad + R N_{i2} \left(t_1 + {}_{i2}\tilde{\rho}_1 \right)^{i2\tilde{\rho}_1} \left[\left(t_2 + {}_{i2}\tilde{\rho}_2 \right)^{i2\tilde{\rho}_2} \right]^2 e^{-n_{i2}(t_1+t_2)} \\ &\quad + R N_{i3} \left(t_1 + {}_{i3}\tilde{\rho}_1 \right)^{i3\tilde{\rho}_1} \left[\left(t_2 + {}_{i3}\tilde{\rho}_2 \right)^{i3\tilde{\rho}_2} \right]^2 e^{-n_{i3}(t_1+t_2)}. \end{aligned} \tag{47}$$

Let

$$\begin{aligned} \tilde{\Omega}_1 =: \; &(N_{11} + N_{12}R + N_{13}R)\tilde{\Omega}_{11} + (N_{21} + N_{22}R + N_{23}R)\tilde{\Omega}_{12} \\ &+ (M_{11} + M_{12}R + M_{13}R)\tilde{\Omega}_{13} + (M_{21} + M_{22}R + M_{23}R)\tilde{\Omega}_{14} \\ \tilde{\Omega}_2 =: \; &(N_{11} + N_{12}R + N_{13}R)\tilde{\Omega}_{21} + (N_{21} + N_{22}R + N_{23}R)\tilde{\Omega}_{22} \\ &+ (M_{11} + M_{12}R + M_{13}R)\tilde{\Omega}_{23} + (M_{21} + M_{22}R + M_{23}R)\tilde{\Omega}_{24} \end{aligned}$$

where:

$$\begin{aligned}
\tilde{\Omega}_{11} &= \left[\max\left\{\frac{1}{\Gamma(\theta_1)}, \frac{\lambda_1 G_1 \Gamma(\alpha_2)}{\Gamma(\theta_1)}(\eta_1-\alpha_1+\theta_1)^{\underline{\theta_1-1}}\mathcal{B}_1\right\} + \frac{1}{\Gamma(\alpha_1)}\right],\\
\tilde{\Omega}_{12} &= \max\left\{\frac{1}{\Gamma(\theta_2)}, \frac{\lambda_2 G_2 \Gamma(\alpha_1)}{\Gamma(\theta_2)}(\eta_2-\alpha_2+\theta_2)^{\underline{\theta_2-1}}\mathcal{B}_2\right\},\\
\tilde{\Omega}_{13} &= \left[\max\left\{1, \frac{\lambda_1 G_1 \Gamma(\alpha_2)}{\Gamma(\theta_1)}(\eta_1-\alpha_1+\theta_1-1)^{\underline{\theta_1-1}}\mathcal{C}_1\right\} + \frac{1}{\Gamma(\alpha_1)}\right],\\
\tilde{\Omega}_{14} &= \max\left\{1, \frac{\lambda_2 G_2 \Gamma(\alpha_1)}{\Gamma(\theta_2)}(\eta_2-\alpha_2+\theta_2-1)^{\underline{\theta_2-1}}\mathcal{C}_2\right\},\\
\tilde{\Omega}_{21} &= \max\left\{\frac{1}{G_2\lambda_2\Gamma(\alpha_1)(\eta_2-\alpha_2+\theta_2)^{\underline{\theta_2-1}}\mathcal{A}_2}, \frac{\mathcal{B}_1}{\Gamma(\theta_1)\mathcal{A}_1}\right\},\\
\tilde{\Omega}_{22} &= \left[\max\left\{\frac{1}{G_1\lambda_1\Gamma(\alpha_2)(\eta_1-\alpha_1+\theta_1)^{\underline{\theta_1-1}}\mathcal{A}_1}, \frac{\mathcal{B}_2}{\Gamma(\theta_2)\mathcal{A}_2}\right\} + \frac{1}{\Gamma(\alpha_2)}\right],\\
\tilde{\Omega}_{23} &= \max\left\{\frac{1}{G_2\lambda_2\Gamma(\alpha_1)(\eta_2-\alpha_2+\theta_2)^{\underline{\theta_2-1}}\mathcal{A}_2}, \frac{\mathcal{C}_1}{\Gamma(\theta_1)\mathcal{A}_1}\right\},\\
\tilde{\Omega}_{24} &= \left[\max\left\{\frac{1}{G_1\lambda_1\Gamma(\alpha_2)(\eta_1-\alpha_1+\theta_1)^{\underline{\theta_1-1}}\mathcal{A}_1}, \frac{\mathcal{C}_2}{\Gamma(\theta_2)\mathcal{A}_2}\right\} + \frac{1}{\Gamma(\alpha_2)}\right].
\end{aligned}$$

Hence, we obtain:

$$\begin{aligned}
&\chi\left|\left(\mathcal{F}_1(u_1,u_2)\right)(t_1,t_2)\right|\\
&\le (N_{11}+N_{12}R+N_{13}R)\left[\max\left\{\frac{1}{\Gamma(\theta_1)}, \frac{\lambda_1 G_1 \Gamma(\alpha_2)}{\Gamma(\theta_1)}(\eta_1-\alpha_1+\theta_1)^{\underline{\theta_1-1}}\mathcal{B}_1\right\} + \frac{1}{\Gamma(\alpha_1)}\right]\\
&\quad + (N_{21}+N_{22}R+N_{23}R)\max\left\{\frac{1}{\Gamma(\theta_2)}, \frac{\lambda_2 G_2 \Gamma(\alpha_1)}{\Gamma(\theta_2)}(\eta_2-\alpha_2+\theta_2)^{\underline{\theta_2-1}}\mathcal{B}_2\right\}\\
&\quad + (M_{11}+M_{12}R+M_{13}R)\left[\max\left\{1, \frac{\lambda_1 G_1 \Gamma(\alpha_2)}{\Gamma(\theta_1)}(\eta_1-\alpha_1+\theta_1-1)^{\underline{\theta_1-1}}\mathcal{C}_1\right\} + \frac{1}{\Gamma(\alpha_1)}\right]\\
&\quad + (M_{21}+M_{22}R+M_{23}R)\max\left\{1, \frac{\lambda_2 G_2 \Gamma(\alpha_1)}{\Gamma(\theta_2)}(\eta_2-\alpha_2+\theta_2-1)^{\underline{\theta_2-1}}\mathcal{C}_2\right\}\\
&= \tilde{\Omega}_1,
\end{aligned} \tag{48}$$

and:

$$\begin{aligned}
&\chi\left|\left(\mathcal{F}_2(u_1,u_2)\right)(t_1,t_2)\right|\\
&\le (N_{11}+N_{12}R+N_{13}R)\max\left\{\frac{1}{G_2\lambda_2\Gamma(\alpha_1)(\eta_2-\alpha_2+\theta_2)^{\underline{\theta_2-1}}\mathcal{A}_2}, \frac{\mathcal{B}_1}{\Gamma(\theta_1)\mathcal{A}_1}\right\}\\
&\quad + (N_{21}+N_{22}R+N_{23}R)\times\\
&\quad \left[\max\left\{\frac{1}{G_1\lambda_1\Gamma(\alpha_2)(\eta_1-\alpha_1+\theta_1)^{\underline{\theta_1-1}}\mathcal{A}_1}, \frac{\mathcal{B}_2}{\Gamma(\theta_2)\mathcal{A}_2}\right\} + \frac{1}{\Gamma(\alpha_2)}\right]\\
&\quad + (M_{11}+M_{12}R+M_{13}R)\max\left\{\frac{1}{G_2\lambda_2\Gamma(\alpha_1)(\eta_2-\alpha_2+\theta_2)^{\underline{\theta_2-1}}\mathcal{A}_2}, \frac{\mathcal{C}_1}{\Gamma(\theta_1)\mathcal{A}_1}\right\}\\
&\quad + (M_{21}+M_{22}R+M_{23}R)\times\\
&\quad \left[\max\left\{\frac{1}{G_1\lambda_1\Gamma(\alpha_2)(\eta_1-\alpha_1+\theta_1)^{\underline{\theta_1-1}}\mathcal{A}_1}, \frac{\mathcal{C}_2}{\Gamma(\theta_2)\mathcal{A}_2}\right\} + \frac{1}{\Gamma(\alpha_2)}\right]
\end{aligned}$$

$$= \tilde{\Omega}_2. \tag{49}$$

Similarly, we have:

$$\chi\left|\Delta^{\beta_1}\left(\mathcal{F}_1(u_1,u_2)\right)(t_1-\beta_1+1,t_2)\right| < \tilde{\Omega}_1, \tag{50}$$

$$\chi\left|\Delta^{\beta_2}\left(\mathcal{F}_2(u_1,u_2)\right)(t_1,t_2-\beta_2+1)\right| < \tilde{\Omega}_2. \tag{51}$$

Therefore, $\mathcal{F}_i(u_1,u_2) \in \mathcal{U}$. This implies that $\mathcal{F}:\mathcal{U} \longrightarrow \mathcal{U}$ is well defined.

Furthermore, we obtain:

$$\begin{aligned} \|\mathcal{F}(u_1,u_2)\|_{\mathcal{U}_i} = \max\Big\{&\chi|\Delta^{\beta_i}(\mathcal{F}_i(u_1,u_2))(t_i-\beta_i+1,t_j)| \\ &+\chi\left|(\mathcal{F}_j(u_1,u_2))(t_i,t_j)\right| \text{ for } i,j\in\{1,2\},\ i\neq j\Big\}. \end{aligned} \tag{52}$$

Hence,

$$\|\mathcal{F}(u_1,u_2)\|_{\mathcal{U}} \quad = \quad \max\left\{\|\mathcal{F}(u_1,u_2)\|_{\mathcal{U}_1}, \|\mathcal{F}(u_1,u_2)\|_{\mathcal{U}_2}\right\} < \tilde{\Omega}_1+\tilde{\Omega}_2. \tag{53}$$

Thus, $\mathcal{F}$ maps bounded sets into bounded sets.

Step II. $\mathcal{F}$ is continuous.

Let $\epsilon > 0$ be given. Since F_i and ϕ_i are continuous, then F_i and ϕ_i are uniformly continuous. Therefore, there exists $\delta = \min\{\delta_i,\hat{\delta}_i\} > 0$ such that, for each $t_i \in \mathcal{N}_{\alpha_i-2}$, $u_i,v_i \in \mathcal{C}_i$ with $\max\left\{\chi|\Delta^{\beta_i}u_i(t_i-\beta_i+1) - \Delta^{\beta_i}v_i(t_i)| + \chi|u_j(t_i)-v_j(t_j)|\right\} < \delta_i$,

$$\begin{aligned} &\left|F_i(t_1,t_2,\Delta^{\beta_i}u_i(t_i-\beta_i+1),u_j(t_j)) - F_i\left(t_1,t_2,\Delta^{\beta_i}v_i(t_i-\beta_i+1),v_j(t_j)\right)\right| \\ &= \left|F_i\left(t_1,t_2,\frac{1}{\chi}\left[\chi\Delta^{\beta_i}u_i(t_i-\beta_i+1)\right],\frac{1}{\chi}\left[\chi u_j(t_j)\right]\right)\right. \\ &\quad \left. - F_i\left(t_1,t_2,\frac{1}{\chi}\left[\chi\Delta^{\beta_i}v_i(t_i-\beta_i+1)\right],\frac{1}{\chi}\left[\chi v_j(t_j)\right]\right)\right| \\ &< 2M_{i1}+2M_{i2}R+2M_{i3}R < \frac{\epsilon}{4\Omega_i}. \end{aligned} \tag{54}$$

For each $u_i,v_i \in \mathcal{C}_i$ with $|u_i-v_i| < \hat{\delta}_i$,

$$\begin{aligned} |\phi_i(u_1,u_2)-\phi_i(v_1,v_2)| &= \left|\phi_i\left(\frac{1}{\chi}[\chi u_1],\frac{1}{\chi}[\chi u_2]\right) - \phi_i\left(\frac{1}{\chi}[\chi v_1],\frac{1}{\chi}[\chi v_2]\right)\right| \\ &< 2N_{i1}+2N_{i2}R+2N_{i3}R < \frac{\epsilon}{4\Omega_i}. \end{aligned} \tag{55}$$

Similar to Step I, we obtain:

$$\chi\left|(\mathcal{F}_i(u_1,u_2)) - (\mathcal{F}_i(v_1,v_2))\right| < 2\tilde{\Omega}_i < \tfrac{\epsilon}{2}$$

and $\quad \chi\left|\Delta^{\beta_i}(\mathcal{F}_i(u_1,u_2)) - \Delta^{\beta_i}(\mathcal{F}_i(v_1,v_2))\right| < 2\tilde{\Omega}_i < \tfrac{\epsilon}{2}.$

Thus, we have:

$$\begin{aligned} &\|\mathcal{F}_i(u_1,u_2)-\mathcal{F}_i(v_1,v_2)\|_{\mathcal{U}_i} \\ =\ &\|\Delta^{\beta_i}\mathcal{F}_i(u_1,u_2)-\Delta^{\beta_i}\mathcal{F}_i(v_1,v_2)\|_{\mathcal{C}_i} + \|\mathcal{F}_j(u_1,u_2)-\mathcal{F}_j(v_1,v_2)\|_{\mathcal{C}_j} \\ <\ &2(\tilde{\Omega}_1+\tilde{\Omega}_2) < \epsilon. \end{aligned} \tag{56}$$

This means that each $\mathcal{F}_i$, $i = 1,2$ is continuous. This shows $\mathcal{F}$ is continuous.

In order to prove that $\mathcal{F}$ maps bounded sets of $\mathcal{U} \subset \mathcal{C}_1 \times \mathcal{C}_2$ to relatively compact sets of $\mathcal{U} \subset \mathcal{C}_1 \times \mathcal{C}_2$, it suffices to show that both $\mathcal{F}_1$ and $\mathcal{F}_2$ map bounded sets to relatively compact sets. Let $\Theta_i \subset \mathcal{C}_i$, $i = 1,2$ be bounded sets and $\Theta_1 \times \Theta_2 \subset \mathcal{U}$. Recall that Θ_i are relatively compact if:

- both Θ_i are bounded,
- both $\chi\Theta_i$ are equicontinuous on any closed subintervals of $\mathbb{N}_{\alpha_i-2}$,
- both $\chi\Theta_i$ are equiconvergent as $t_i \to \infty$.

It has been shown from in Step I that both $\mathcal{F}_i$ are uniformly bounded. Now, we show that $\mathcal{F}_i$ maps bounded sets into equicontinuous sets of $\mathcal{U}$.

Step III. Both $\mathcal{F}_i : \Theta_1 \times \Theta_2 \to \mathcal{U}$ are equicontinuous on $\left([a_1,b_1] \cap \mathbb{N}_{\alpha_1-2}\right) \times \left([a_2,b_2] \cap \mathbb{N}_{\alpha_2-2}\right) := \mathcal{D}$.

For any $\epsilon > 0$, there exists $\delta > 0$ such that, for each $t_{i1}, t_{i2} \in \mathbb{N}_{\alpha_i-2} \cap [a_i, b_i]$,

$$\left|(t_{11}+\rho_1)^{\underline{\rho_1}}(t_{21}+\rho_2)^{\underline{\rho_2}} - (t_{12}+\rho_1)^{\underline{\rho_1}}(t_{22}+\rho_2)^{\underline{\rho_2}}\right| \le \frac{\epsilon}{2\max\{\tilde{\Omega}_1^* + \tilde{\Omega}_2^*\}} = \delta, \tag{57}$$

where:

$$\begin{aligned}
\tilde{\Omega}_1^* &= \Big([N_{11}+N_{12}R+N_{13}R]+[M_{11}+M_{12}R+M_{13}R]\Big)\left[\frac{1}{\Gamma(\theta_1)}+\frac{1}{\Gamma(\alpha_1)}\right] \\
&\quad +\Big([N_{11}+N_{12}R+N_{13}R]\mathcal{B}_1+[M_{11}+M_{12}R+M_{13}R]\mathcal{C}_1\Big)\times \\
&\quad \frac{\lambda_1 G_1\Gamma(\alpha_2)(\eta_1-\alpha_1+\theta_1)^{\underline{\theta_1-1}}}{\Gamma(\theta_1)\Gamma(\theta_2)} \\
&\quad +\Big([N_{21}+N_{22}R+N_{23}R]+[M_{21}+M_{22}R+M_{23}R]\Big)\frac{1}{\Gamma(\theta_2)} \\
&\quad +\Big([N_{21}+N_{22}R+N_{23}R]\mathcal{B}_2+[M_{21}+M_{22}R+M_{23}R]\mathcal{C}_2\Big)\times \\
&\quad \frac{\lambda_2 G_2\Gamma(\alpha_1)(\eta_2-\alpha_2+\theta_2)^{\underline{\theta_2-1}}}{\Gamma(\theta_1)\Gamma(\theta_2)},
\end{aligned} \tag{58}$$

$$\begin{aligned}
\tilde{\Omega}_2^* &= [N_{11}+N_{12}R+N_{13}R]\left[\frac{1}{\lambda_2 G_2(\eta_2-\alpha_2+\theta_2)^{\underline{\theta_2-1}}\mathcal{A}_2}+\frac{1}{\Gamma(\alpha_2)}\right] \\
&\quad +[N_{21}+N_{22}R+N_{23}R]\frac{1}{\lambda_1 G_1(\eta_1-\alpha_1+\theta_1)^{\underline{\theta_1-1}}\mathcal{A}_1} \\
&\quad +[M_{11}+M_{12}R+M_{13}R]\frac{\mathcal{B}_1}{\Gamma(\theta_1)\mathcal{A}_1}+[M_{21}+M_{22}R+M_{23}R]\frac{\mathcal{B}_2}{\Gamma(\theta_2)\mathcal{A}_2}.
\end{aligned} \tag{59}$$

Hence, for each $t_{i1}, t_{i2} \in \mathbb{N}_{\alpha_i-2} \cap [a_i, b_i]$, and $u_i \in \Theta_i$, we have:

$$\begin{aligned}
&\left|\chi(\mathcal{F}_1u_1)(t_{11},t_{21}) - \chi(\mathcal{F}_1u_1)(t_{12},t_{22})\right| \\
\le\ & \chi\left|t_{11}^{\underline{\alpha_1}}\left\{\left[\frac{|\phi_1(u_1,u_2)|}{\Gamma(\theta_1)}-\frac{|\phi_2(u_1,u_2)|}{\Gamma(\theta_2)}+\frac{|\phi_1(u_1,u_2)|}{\Gamma(\alpha_1)}\right]+\frac{1}{\Gamma(\theta_1)\Gamma(\theta_2)}\times\right.\right. \\
&\left[\frac{\lambda_1\Gamma(\alpha_2)\mathcal{B}_1|\phi_1(u_1,u_2)|}{t_{21}^{\underline{\alpha_2-1}}\Gamma(\alpha_1)}+\frac{\lambda_2\Gamma(\alpha_1)\mathcal{B}_2|\phi_2(u_1,u_2)|}{t_{11}^{\underline{\alpha_1-1}}\Gamma(\alpha_2)}\right]+\frac{1}{\Gamma(\theta_1)\Gamma(\theta_2)}\times \\
&+\left[\frac{\lambda_1\Gamma(\alpha_2)\mathcal{B}_1|F_1(t_1,t_2,\Delta^{\beta_1}u_1,u_2)|}{t_{22}^{\underline{\alpha_2-1}}\Gamma(\alpha_1)}+\frac{\lambda_2\Gamma(\alpha_1)\mathcal{B}_2|F_2(t_1,t_2,u_1,\Delta^{\beta_2}u_2)|}{t_{12}^{\underline{\alpha_1-1}}\Gamma(\alpha_2)}\right] \\
&\left.\left[\frac{|F_1(t_1,t_2,\Delta^{\beta_1}u_1,u_2)|}{\Gamma(\theta_1)}-\frac{|F_2(t_1,t_2,u_1,\Delta^{\beta_2}u_2)|}{\Gamma(\theta_2)}+\frac{|F_1(t_1,t_2,\Delta^{\beta_1}u_1,u_2)|}{\Gamma(\alpha_1)}\right]\right\}
\end{aligned}$$

$$
\begin{aligned}
& -t_{12}^{\alpha_1}\Bigg\{\left[\frac{|\phi_1(u_1,u_2)|}{\Gamma(\theta_1)}-\frac{|\phi_2(u_1,u_2)|}{\Gamma(\theta_2)}+\frac{|\phi_1(u_1,u_2)|}{\Gamma(\alpha_1)}\right]+\frac{1}{\Gamma(\theta_1)\Gamma(\theta_2)}\times \\
& \left[\frac{\lambda_1\Gamma(\alpha_2)\mathcal{B}_1|\phi_1(u_1,u_2)|}{t_{21}^{\alpha_2-1}\Gamma(\alpha_1)}+\frac{\lambda_2\Gamma(\alpha_1)\mathcal{B}_2|\phi_2(u_1,u_2)|}{t_{11}^{\alpha_1-1}\Gamma(\alpha_2)}\right]+\frac{1}{\Gamma(\theta_1)\Gamma(\theta_2)}\times \\
& +\left[\frac{\lambda_1\Gamma(\alpha_2)\mathcal{B}_1|F_1(t_1,t_2,\Delta^{\beta_1}u_1,u_2)|}{t_{22}^{\alpha_2-1}\Gamma(\alpha_1)}+\frac{\lambda_2\Gamma(\alpha_1)\mathcal{B}_2|F_2(t_1,t_2,u_1,\Delta^{\beta_2}u_2)|}{t_{12}^{\alpha_1-1}\Gamma(\alpha_2)}\right] \\
& \left[\frac{|F_1(t_1,t_2,\Delta^{\beta_1}u_1,u_2)|}{\Gamma(\theta_1)}-\frac{|F_2(t_1,t_2,u_1,\Delta^{\beta_2}u_2)|}{\Gamma(\theta_2)}+\frac{|F_1(t_1,t_2,\Delta^{\beta_1}u_1,u_2)|}{\Gamma(\alpha_1)}\right]\Bigg\}\Bigg| \\
< & \left|(t_{11}+\rho_1)^{\underline{\rho_1}}(t_{21}+\rho_2)^{\underline{\rho_2}}-(t_{12}+\rho_1)^{\underline{\rho_1}}(t_{22}+\rho_2)^{\underline{\rho_2}}\right|\tilde{\Omega}_1^* \\
< & \frac{\epsilon}{2},
\end{aligned}
\tag{60}
$$

and:

$$
\begin{aligned}
& \left|\chi\left(\mathcal{F}_2u_2\right)(t_{11},t_{21})-\chi\left(\mathcal{F}_2u_2\right)(t_{12},t_{22})\right| \\
\le\ & \chi\Bigg|t_{21}^{\alpha_2}\Bigg\{\left[\frac{t_{11}^{\alpha_1-1}|\phi_1(u_1,u_2)|}{\lambda_2G_2(\eta_2-\alpha_2+\theta_2)^{\underline{\theta_2-1}}\mathcal{A}_2}-\frac{t_{21}^{\alpha_2-1}|\phi_2(u_1,u_2)|}{\lambda_1G_1(\eta_1-\alpha_1+\theta_1)^{\underline{\theta_1-1}}\mathcal{A}_1}+\frac{|\phi_2(u_1,u_2)|}{\Gamma(\alpha_2)}\right] \\
& +t_{21}^{\alpha_2-1}\left[\frac{\mathcal{B}_1|\phi_1(u_1,u_2)|}{\Gamma(\theta_1)\mathcal{A}_1}-\frac{\mathcal{B}_2|\phi_2(u_1,u_2)|}{\Gamma(\theta_2)\mathcal{A}_2}\right]+\left[\frac{t_{11}^{\alpha_1-1}|F_1(t_1,t_2,\Delta^{\beta_1}u_1,u_2)|}{\lambda_2G_2(\eta_2-\alpha_2+\theta_2)^{\underline{\theta_2-1}}\mathcal{A}_2}\right. \\
& \left.-\frac{t_{21}^{\alpha_2-1}|F_2(t_1,t_2,u_1,\Delta^{\beta_2}u_2)|}{\lambda_1G_1(\eta_1-\alpha_1+\theta_1)^{\underline{\theta_1-1}}\mathcal{A}_1}+\frac{|F_2(t_1,t_2,u_1,\Delta^{\beta_2}u_2)|}{\Gamma(\alpha_2)}\right] \\
& +t_{21}^{\alpha_2-1}\left[\frac{t_{11}^{\alpha_1}\mathcal{C}_1|F_1(t_1,t_2,\Delta^{\beta_1}u_1,u_2)|}{\Gamma(\theta_1)\mathcal{A}_1}-\frac{t_{21}^{\alpha_2}\mathcal{C}_2|F_2(t_1,t_2,u_1,\Delta^{\beta_2}u_2)|}{\Gamma(\theta_2)\mathcal{A}_2}\right]\Bigg\} \\
& -t_{22}^{\alpha_2}\Bigg\{\left[\frac{t_{11}^{\alpha_1-1}|\phi_1(u_1,u_2)|}{\lambda_2G_2(\eta_2-\alpha_2+\theta_2)^{\underline{\theta_2-1}}\mathcal{A}_2}-\frac{t_{21}^{\alpha_2-1}|\phi_2(u_1,u_2)|}{\lambda_1G_1(\eta_1-\alpha_1+\theta_1)^{\underline{\theta_1-1}}\mathcal{A}_1}+\frac{|\phi_2(u_1,u_2)|}{\Gamma(\alpha_2)}\right] \\
& +t_{21}^{\alpha_2-1}\left[\frac{\mathcal{B}_1|\phi_1(u_1,u_2)|}{\Gamma(\theta_1)\mathcal{A}_1}-\frac{\mathcal{B}_2|\phi_2(u_1,u_2)|}{\Gamma(\theta_2)\mathcal{A}_2}\right]+\left[\frac{t_{11}^{\alpha_1-1}|F_1(t_1,t_2,\Delta^{\beta_1}u_1,u_2)|}{\lambda_2G_2(\eta_2-\alpha_2+\theta_2)^{\underline{\theta_2-1}}\mathcal{A}_2}\right. \\
& \left.-\frac{t_{21}^{\alpha_2-1}|F_2(t_1,t_2,u_1,\Delta^{\beta_2}u_2)|}{\lambda_1G_1(\eta_1-\alpha_1+\theta_1)^{\underline{\theta_1-1}}\mathcal{A}_1}+\frac{|F_2(t_1,t_2,u_1,\Delta^{\beta_2}u_2)|}{\Gamma(\alpha_2)}\right] \\
& +t_{21}^{\alpha_2-1}\left[\frac{t_{11}^{\alpha_1}\mathcal{C}_1|F_1(t_1,t_2,\Delta^{\beta_1}u_1,u_2)|}{\Gamma(\theta_1)\mathcal{A}_1}-\frac{t_{21}^{\alpha_2}\mathcal{C}_2|F_2(t_1,t_2,u_1,\Delta^{\beta_2}u_2)|}{\Gamma(\theta_2)\mathcal{A}_2}\right]\Bigg\} \\
< & \left|(t_{11}+\rho_1)^{\underline{\rho_1}}(t_{21}+\rho_2)^{\underline{\rho_2}}-(t_{12}+\rho_1)^{\underline{\rho_1}}(t_{22}+\rho_2)^{\underline{\rho_2}}\right|\tilde{\Omega}_2^* \\
< & \frac{\epsilon}{2}.
\end{aligned}
\tag{61}
$$

Similarly, for each $i,j\in\{1,2\}$ and $j\neq i$, we obtain:

$$
\left|\Delta^{\beta_i}\left(\mathcal{F}_i(u_1,u_2)\right)(t_{i1}-\beta_i+1,t_{j1})-\Delta^{\beta_i}\left(\mathcal{F}_i(u_1,u_2)\right)(t_{i2}-\beta_i+1,t_{j2})\right|<\frac{\epsilon}{2}. \tag{62}
$$

Hence:

$$
\begin{aligned}
& \|\mathcal{F}_i(u_1,u_2)(t_{11},t_{21})-\mathcal{F}_i(u_1,u_2)(t_{12},t_{22})\|_{\mathcal{U}_i} \\
= & \|\Delta^{\beta_i}\mathcal{F}_i(u_1,u_2)(t_{i1}-\beta_i+1,t_{j1})-\Delta^{\beta_i}\mathcal{F}_i(u_1,u_2)(t_{i2}-\beta_i+1,t_{j2})\|_{\mathcal{C}_i}
\end{aligned}
$$

$$+\|\mathcal{F}_j(u_1,u_2)(t_{11},t_{21})-\mathcal{F}_j(u_1,u_2)(t_{12},t_{22})\|_{\mathcal{C}_j}$$
$$< \frac{\epsilon}{2}+\frac{\epsilon}{2}=\epsilon. \tag{63}$$

This implies that both $\mathcal{F}_1$ and $\mathcal{F}_2$ are equicontinuous on $\mathcal{D}$, which shows that $\mathcal{F}$ is equicontinuous on $\mathcal{D}$. Therefore, by the Arzelá–Ascoli theorem and Theorem 2, we can conclude that $\mathcal{F}$ is completely continuous.

Step IV. Both $\mathcal{F}_i:\Theta_1\times\Theta_2\to\mathcal{U}$ are equiconvergent as $t_1,t_2\to\infty$.

By the assumption $(H1)-(H2)$, we obtain:

$$\chi|\left(\mathcal{F}_1(u_1,u_2)\right)(t_1,t_2)| < \frac{\tilde{\Omega}_1^*}{t_1^{\alpha_1}t_2^{\alpha_2}} \to 0 \ \text{ uniformly in } \Theta_1\times\Theta_2 \text{ as } t_1,t_2\to\infty,$$
$$\chi|\left(\mathcal{F}_2(u_1,u_2)\right)(t_1,t_2)| < \frac{\tilde{\Omega}_2^*}{t_1^{\alpha_1}t_2^{\alpha_2}} \to 0 \ \text{ uniformly in } \Theta_1\times\Theta_2 \text{ as } t_1,t_2\to\infty,$$

where $\tilde{\Omega}_1^*,\tilde{\Omega}_2^*$ are defined as (58) and (59).

Furthermore, we have:

$$\chi|\Delta^{\beta_1}\left(\mathcal{F}_1(u_1,u_2)\right)(t_1-\beta_1+1,t_2)| < \frac{\tilde{\Omega}_1^*}{t_1^{\alpha_1}t_2^{\alpha_2}} \to 0 \ \text{ uniformly in } \Theta_1\times\Theta_2 \text{ as } t_1,t_2\to\infty,$$
$$\chi|\Delta^{\beta_2}\left(\mathcal{F}_2(u_1,u_2)\right)(t_1,t_2-\beta_2+1)| < \frac{\tilde{\Omega}_2^*}{t_1^{\alpha_1}t_2^{\alpha_2}} \to 0 \ \text{ uniformly in } \Theta_1\times\Theta_2 \text{ as } t_1,t_2\to\infty.$$

Hence, both $\mathcal{F}_i$ are equiconvergent as $t_1,t_2\to\infty$.

Consequently, from Step I–Step IV, we conclude that $\mathcal{F}$ is completely continuous. This complete the proof. □

Finally, we present the main result of the article. For the sake of convenience, we set:

$$\Psi_1 = (N_{12}+N_{13})\tilde{\Omega}_{11}+(N_{22}+N_{23})\tilde{\Omega}_{12}+(M_{12}+M_{13})\tilde{\Omega}_{13} + (M_{22}+M_{23})\tilde{\Omega}_{14}, \tag{64}$$

and (65)

$$\Psi_2 = (N_{12}+N_{13})\tilde{\Omega}_{21}+(N_{22}+N_{23})\tilde{\Omega}_{22}+(M_{12}+M_{13})\tilde{\Omega}_{23} + (M_{22}+M_{23})\tilde{\Omega}_{24}, \tag{66}$$

where $\tilde{\Omega}_{1p},\tilde{\Omega}_{2p},\ p=1,2,3,4$ are defined as (48) and (49).

Theorem 4. *Suppose that (H_1)–(H_2) hold. Then, the problems (5) and (6) has at least one solution if:*

$$\Psi_1+\Psi_2 < 1. \tag{67}$$

Proof. Under the Banach space $\mathcal{U}$ equipped with the norm $\|\cdot\|_{\mathcal{U}}$, we let:

$$\omega_1(t_1,t_2)$$
$$= \frac{t_1^{\underline{\alpha_1-1}}}{\Lambda}\left\{\frac{\lambda_1 G_1\Gamma(\alpha_1)(\eta_1-\alpha_1+\theta_1)^{\underline{\theta_1-1}}\mathcal{A}_1}{\Gamma(\theta_1)}\left[(t_1+\rho_1)^{\underline{\rho_1}}e^{-n_1(t_1+1)}\times\right.\right.$$

$$\left(\frac{N_{11}t_1^{\underline{\alpha_1-2}}}{\Gamma(\alpha_1)} - \frac{N_{21}\lambda_2 G_2(\eta_2-\alpha_2+\theta_2)^{\underline{\theta_2-1}}\mathcal{B}_2}{\Gamma(\theta_2)}\right) + \frac{M_{11}e^{-m_1t_2}}{\Gamma(\alpha_1)}\times$$
$$\sum_{s=0}^{t_1-\alpha_1}(t_1-\sigma(s))^{\underline{\alpha_1-1}}e^{-2m_1s} - \frac{M_{21}G_2(\eta_2-\alpha_2+\theta_2-1)^{\underline{\theta_2-1}}e^{-m_1(2t_1+t_2)}}{\Gamma(\theta_2)}\Bigg]$$
$$+\frac{\lambda_2 G_2\Gamma(\alpha_2)(\eta_2-\alpha_2+\theta_2)^{\underline{\theta_2-1}}\mathcal{A}_2}{\Gamma(\theta_2)}\Bigg[(t_1+\rho_1)^{\underline{\rho_1}}e^{-n_1(t_1+1)}\times$$
$$\left(\frac{N_{21}t_2^{\underline{\alpha_2-2}}}{\Gamma(\alpha_2)} - \frac{N_{11}\lambda_1 G_1(\eta_1-\alpha_1+\theta_1)^{\underline{\theta_1-1}}\mathcal{B}_1}{\Gamma(\theta_1)}\right) + \frac{M_{21}e^{-2m_1t_1}}{\Gamma(\alpha_2)}\times$$
$$\sum_{s=0}^{t_2-\alpha_2}(t_2-\sigma(s))^{\underline{\alpha_2-1}}e^{-m_1s} - \frac{M_{11}G_1(\eta_1-\alpha_1+\theta_1-1)^{\underline{\theta_1-1}}e^{-m_1(2t_1+t_2)}}{\Gamma(\theta_1)}\Bigg]\Bigg\}$$
$$+\frac{N_{11}}{\Gamma(\alpha_1)}(t_1+\rho_1)^{\underline{\rho_1}}e^{-n_1(t_1+1)} + \frac{M_{11}e^{-m_1t_2}}{\Gamma(\alpha_1)}\sum_{s=0}^{t_1-\alpha_1}(t_1-\sigma(s))^{\underline{\alpha_1-1}}e^{-2m_1s}, \tag{68}$$

and

$$\omega_2(t_1,t_2)$$
$$= \frac{t_2^{\underline{\alpha_2-1}}}{\Lambda}\Bigg\{t_2^{\underline{\alpha_2-1}}\Bigg[(t_1+\rho_1)^{\underline{\rho_1}}\left[(t_2+\rho_2)^{\underline{\rho_2}}\right]^2e^{-n_2(t_2+1)}\Bigg(\frac{N_{12}t_1^{\underline{\alpha_1-2}}}{\Gamma(\alpha_1)}$$
$$-\frac{N_{22}\lambda_2G_2(\eta_2-\alpha_2+\theta_2)^{\underline{\theta_2-1}}\mathcal{B}_2}{\Gamma(\theta_2)}\Bigg) + \frac{M_{12}e^{-m_2t_2}}{\Gamma(\alpha_1)}(t_2+\rho_2)^{\underline{\rho_2}}\sum_{s=0}^{t_1-\alpha_1}(t_1-\sigma(s))^{\underline{\alpha_1-1}}\times$$
$$e^{-2m_2s} - \frac{M_{22}G_2(\eta_2-\alpha_2+\theta_2-1)^{\underline{\theta_2-1}}e^{-m_2(t_1+2t_2)}(t_2+\rho_2)^{\underline{\rho_2}}}{\Gamma(\theta_2)}\Bigg] + t_1^{\underline{\alpha_1-1}}\times$$
$$\Bigg[(t_1+\rho_1)^{\underline{\rho_1}}\left[(t_2+\rho_2)^{\underline{\rho_2}}\right]^2e^{-n_2(t_2+1)}\left(\frac{N_{22}t_2^{\underline{\alpha_2-2}}}{\Gamma(\alpha_2)} - \frac{N_{12}\lambda_1G_1(\eta_1-\alpha_1+\theta_1)^{\underline{\theta_1-1}}\mathcal{B}_1}{\Gamma(\theta_1)}\right)$$
$$+\frac{M_{22}e^{-m_2t_1}}{\Gamma(\alpha_2)}\sum_{s=0}^{t_2-\alpha_2}(t_2-\sigma(s))^{\underline{\alpha_2-1}}e^{-2m_2s}(t_2+\rho_2)^{\underline{\rho_2}}$$
$$-\frac{M_{22}G_1(\eta_1-\alpha_1+\theta_1-1)^{\underline{\theta_1-1}}e^{-m_2(t_2+2t_2)}}{\Gamma(\theta_1)}\Bigg] + \frac{N_{22}}{\Gamma(\alpha_2)}\left[(t_2+\rho_2)^{\underline{\rho_2}}\right]^2e^{-n_2(t_2+1)}t_2^{\underline{\alpha_2-2}}\Bigg\}$$
$$+\frac{M_{22}e^{-m_2t_1}}{\Gamma(\alpha_2)}\sum_{s=0}^{t_2-\alpha_2}(t_2-\sigma(s))^{\underline{\alpha_2-1}}e^{-2m_2s}(t_2+\rho_2)^{\underline{\rho_2}}. \tag{69}$$

It is clear that $(\omega_1,\omega_2)\in\mathcal{U}$. For $\ell>0$, we define:

$$\Xi_\ell = \{(u_1,u_2)\in\mathcal{U} : \|(u_1,u_2)-(\omega_1,\omega_2)\|_{\mathcal{U}} \le \ell\}. \tag{70}$$

For $(u_1,u_2)\in\Xi_\ell$, we have:

$$\|(u_1,u_2)\|_{\mathcal{U}} \le \|(u_1,u_2)-(\omega_1,\omega_2)\|_{\mathcal{U}} + \|(\omega_1,\omega_2)\|_{\mathcal{U}} \le \ell + \|(\omega_1,\omega_2)\|_{\mathcal{U}}, \tag{71}$$
$$\|(u_1,u_2)\|_{\mathcal{U}} = \max\{\|(u_1,u_2)\|_{\mathcal{U}_1}, \|(u_1,u_2)\|_{\mathcal{U}_2}\} \le \ell + \|(\omega_1,\omega_2)\|_{\mathcal{U}}. \tag{72}$$

Using the conditions (H_1)–(H_2), together with the procedure employed in Lemma 5, we have:

$$\left|F_i\left(t_1,t_2,\Delta^{\beta_i}u_i,u_j\right) - M_{i1}\left(t_2+{}_{i1}\rho_2\right)^{\underline{i1\rho_2}}e^{-m_{i1}(t_1+t_2)}\right|$$
$$\le \Big\{M_{i2}\left(t_2+{}_{i2}\rho_2\right)^{\underline{i2\rho_2}}e^{-m_{i2}(t_1+t_2)}$$

$$+ M_{i3}\left(t_2 + {}_{i3}\rho_2\right)^{\underline{i3\rho_2}} e^{-m_{i3}(t_1+t_2)} \Big\} \|(u_1,u_2)\|_{\mathcal{U}}, \tag{73}$$

and:

$$\begin{aligned}
&\left|\phi_i(u_1,u_2) - N_{i1}\left(t_1 + {}_{i1}\tilde{\rho}_1\right)^{\underline{i1\tilde{\rho}_1}}\left[\left(t_2 + {}_{i1}\tilde{\rho}_2\right)^{\underline{i1\tilde{\rho}_2}}\right]^2 e^{-n_{i1}(t_1+t_2)}\right| \\
&= \Big\{ N_{i2}\left(t_1 + {}_{i2}\tilde{\rho}_1\right)^{\underline{i2\tilde{\rho}_1}}\left[\left(t_2 + {}_{i2}\tilde{\rho}_2\right)^{\underline{i2\tilde{\rho}_2}}\right]^2 e^{-n_{i2}(t_1+t_2)} \\
&+ N_{i3}\left(t_1 + {}_{i3}\tilde{\rho}_1\right)^{\underline{i3\tilde{\rho}_1}}\left[\left(t_2 + {}_{i3}\tilde{\rho}_2\right)^{\underline{i3\tilde{\rho}_2}}\right]^2 e^{-n_{i3}(t_1+t_2)} \Big\} \|(u_1,u_2)\|_{\mathcal{U}}.
\end{aligned} \tag{74}$$

Therefore, we obtain:

$$\chi\left|\left(\mathcal{F}_i(u_i,u_j)\right)(t_1,t_2) - \omega_i(t_1,t_2)\right| \le \Big(\ell + \|(\omega_1,\omega_2)\|_{\mathcal{U}}\Big)\Psi_i. \tag{75}$$

Furthermore, we have:

$$\begin{aligned}
&\chi\left|\Delta^{\beta_i}\left(\mathcal{F}_i(u_i,u_j)\right)(t_i - \beta_i + 1, t_j) - \Delta^{\beta_i}\omega_i(t_i - \beta_i + 1, t_j)\right| \\
&\le \Big(\ell + \|(\omega_1,\omega_2)\|_{\mathcal{U}}\Big)\Psi_i.
\end{aligned} \tag{76}$$

Hence, it follows that:

$$\left\|\left(\mathcal{F}_i(u_1,u_2)\right) - \omega_i\right\|_{\mathcal{U}_i} \le \Big(\ell + \|(\omega_1,\omega_2)\|_{\mathcal{U}}\Big)2\Psi_i. \tag{77}$$

Therefore,

$$\left\|\left(\mathcal{F}(u_1,u_2)\right) - (\omega_1,\omega_2)\right\|_{\mathcal{U}} \le \Big(\ell + \|(\omega_1,\omega_2)\|_{\mathcal{U}}\Big)\, 2\max\{\Psi_1,\Psi_2\}. \tag{78}$$

Choosing:

$$\ell \ge \frac{\left\|(\omega_1,\omega_2)\right\|_{\mathcal{U}} 2\max\{\Psi_1,\Psi_2\}}{1 - 2\max\{\Psi_1,\Psi_2\}}, \tag{79}$$

and for $(u_1,u_2) \in \Xi_\ell$, we consequently obtain:

$$\left\|\left(\mathcal{F}(u_1,u_2)\right) - (\omega_1,\omega_2)\right\|_{\mathcal{U}} \le \ell. \tag{80}$$

From the Schauder fixed point theorem, this implies that $\mathcal{F}$ has a fixed point $(u_1,u_2) \in \Xi_\ell$, which is a bounded solution of the problems (5) and (6). The proof is complete. □

4. Example

In order to illustrate our result, we consider the following fractional sum boundary value problem:

$$\begin{aligned}
\Delta^{\frac{3}{2}} u_1(t) &= \frac{2}{3}\left(t + \frac{4}{3}\right)e^{-(12t+5)} + \frac{\left(t+\frac{5}{3}\right)^{\frac{4}{3}} e^{-\left(25t+\frac{31}{3}\right)} u_2\left(t+\frac{1}{3}\right)}{4000\left(t+\frac{301}{3}\right)^2\left(1+\cos^2 u_2\pi\right)} \\
&+ \frac{\left(t+\frac{11}{6}\right)^{\frac{3}{2}} e^{-\left[(12t+5)+\left(t+\frac{1}{2}\right)\pi\right]} \Delta^{\frac{1}{3}} u_1\left(t+\frac{4}{3}\right)}{5000e + 10\cos^2\left(t+\frac{1}{2}\right)\pi}, \quad t \in \mathbb{N}_0
\end{aligned}$$

$$\begin{aligned}
\Delta^{\frac{4}{3}}u_2(t) &= \frac{1}{2}\left(t+\frac{4}{3}\right)e^{-(12t+5)}+\frac{\left(t+\frac{5}{3}\right)^{\frac{4}{3}}e^{-\left(25t+\frac{21}{3}\right)}u_1\left(t+\frac{1}{2}\right)}{1000\left(e^{\left(t+\frac{1}{2}\right)}+10\right)^2}\\
&\quad+\frac{\left(t+\frac{11}{6}\right)^{\frac{3}{2}}e^{-(12t+5)}\arctan\left(\cos^2\left(t+\frac{1}{3}\right)\pi\right)\Delta^{\frac{3}{4}}u_2\left(t_2+\frac{7}{12}\right)}{1000\pi\left(t+\frac{10}{3}\right)^2},\quad t\in\mathbb{N}_0\\
u_1\Big(-\frac{1}{2}\Big) &= \phi_1(u_1,u_2)=\frac{|u_1|}{2000e^3}\cos^2|\pi u_1|+\frac{|u_2||u_2^2+2|^{1-|u_2^2+2|}}{4000\pi^2(u_2^2+e)},\\
u_2\Big(-\frac{2}{3}\Big) &= \phi_2(u_1,u_2)=\frac{|u_2|}{5000e^2}\sin^2|\pi u_2|+\frac{|u_1||u_1^2+3|^{1-|u_1^2+3|}}{2000\pi(u_1^2+\pi)},\\
\lim_{t\longrightarrow\infty}u_1\Big(t-\frac{1}{2}\Big) &= \frac{1}{2}\Delta^{-\frac{1}{4}}\Big(12e+\cos(4)\Big)^2u_2(4)\\
\lim_{t\longrightarrow\infty}u_2\Big(t-\frac{2}{3}\Big) &= \frac{3}{4}\Delta^{-\frac{2}{3}}\left(10e-\sin\left(\frac{15}{4}\right)\right)^3u_1\Big(\frac{15}{4}\Big).
\end{aligned}\tag{81}$$

Here, $\alpha_1=\frac{3}{2}$, $\alpha_2=\frac{4}{3}$, $\beta_1=\frac{1}{3}$, $\beta_2=\frac{3}{4}$, $\gamma_1=\frac{3}{4}$, $\gamma_2=\frac{5}{6}$, $\theta_1=\frac{1}{4}$, $\theta_2=\frac{2}{3}$, $\eta_1=\frac{7}{2}$, $\eta_2=\frac{10}{3}$, $\lambda_1=\frac{1}{2}$, $\lambda_2=\frac{3}{4}$, $T=4$, $g_1(t_1)=(10e-\sin t_1)^3$, $g_2(t_2)=(12e+\cos t_2)^2$, and:

$$\begin{aligned}
F_1\left(t_1,t_2,\Delta^{\frac{1}{3}}u_1\left(t_1+\tfrac{2}{3}\right),u_2(t_2)\right) &= \tfrac{2}{3}(t_2+1)e^{-6(t_1+t_2)}+\frac{\left(t_2+\frac{4}{3}\right)^{\frac{4}{3}}e^{-[12(t_1+t_2)+t_2]}u_2(t_2)}{4000(t_2+10)^2(1+\cos^2u_2\pi)}\\
&\quad+\frac{\left(t_2+\frac{3}{2}\right)^{\frac{3}{2}}e^{-[6(t_1+t_2)+t_1\pi]}\Delta^{\frac{1}{3}}u_1\left(t_1+\frac{2}{3}\right)}{5000e+10\cos^2t_1\pi}\\
F_2\left(t_1,t_2,\Delta^{\frac{3}{4}}u_2\left(t_2+\tfrac{1}{4}\right),u_1(t_1)\right) &= \tfrac{1}{2}(t_2+1)e^{-6(t_1+t_2)}+\frac{\left(t_2+\frac{4}{3}\right)^{\frac{4}{3}}e^{-[12(t_1+t_2)+t_1]}u_1(t_1)}{1000(e^{t_1}+10)^2}\\
&\quad+\frac{\left(t_2+\frac{3}{2}\right)^{\frac{3}{2}}e^{-6(t_1+t_2)}\arctan\left(\cos^2t_2\pi\right)\Delta^{\frac{3}{4}}u_2\left(t_2+\frac{1}{4}\right)}{1000\pi(t_2+3)^2}.
\end{aligned}$$

Choose $\rho_1=1$, $\rho_2=2$, ${}_{i1}\rho_1={}_{i1}\tilde{\rho}_1=\frac{1}{2}$, ${}_{i2}\rho_1={}_{i2}\tilde{\rho}_1=\frac{2}{3}$, ${}_{i3}\rho_1={}_{i3}\tilde{\rho}_1=\frac{3}{4}$ ${}_{i1}\rho_2=1$, ${}_{i2}\rho_2=\frac{3}{2}$, ${}_{i3}\rho_2=\frac{4}{3}$, $m_{i1}=n_{i1}=6$, $m_{i2}=n_{i2}=6$, $m_{i3}=n_{i3}=12$, where $\rho_i>\max\{\beta_1-\alpha_1,\beta_2-\alpha_2\}$, ${}_{ip}\rho_1\in(-1,1)$, ${}_{ip}\rho_2\in(-1,2)$ for $i=1,2$ and $p=1,2,3$.

Let $t_1\in\mathbb{N}_{-\frac{1}{2},\frac{11}{2}}$, $t_2\in\mathbb{N}_{-\frac{2}{3},\frac{16}{3}}$ and $\chi=\frac{\left(t_1-\frac{1}{2}\right)^{1/2}\left(t_2-\frac{1}{3}\right)^{2/3}}{1+t_1^{\frac{3}{4}}t_2^{\frac{4}{4}}}$. Since:

$$\begin{aligned}
&\left|F_1\left(t_1,t_2,\frac{1}{\chi}\Delta^{\frac{1}{3}}u_1,\frac{1}{\chi}u_2\right)-\frac{2}{3}(t_2+1)e^{-6(t_1+t_2)}\right|\\
\leq\ &\frac{1}{361000}\left(t_2+\frac{4}{3}\right)^{\frac{4}{3}}e^{-12(t_1+t_2)}|u_2|+\frac{1}{5000e^{10}+10}\left(t_2+\frac{3}{2}\right)^{\frac{3}{2}}e^{-6(t_1+t_2)}\left|\Delta^{\frac{1}{3}}u_1\right|,\\
&\left|F_2\left(t_1,t_2,\frac{1}{\chi}\Delta^{\frac{3}{4}}u_2,\frac{1}{\chi}u_1\right)-\frac{1}{2}(t_2+1)e^{-6(t_1+t_2)}\right|\\
\leq\ &\frac{1}{121000}\left(t_2+\frac{4}{3}\right)^{\frac{4}{3}}e^{-12(t_1+t_2)}|u_2|+\frac{9}{196000}\left(t_2+\frac{3}{2}\right)^{\frac{3}{2}}e^{-6(t_1+t_2)}\left|\Delta^{\frac{3}{4}}u_2\right|,
\end{aligned}$$

we find that (H_1) holds with $M_{11}=0.666$, $M_{12}=9.080$, $M_{13}=2.770$ and $M_{21}=0.500$, $M_{22}=0.000046$, $M_{23}=0.0000083$.

Furthermore, we obtain:

$$\left|\phi_1\left(\frac{1}{\chi}u_1,\frac{1}{\chi}u_2\right)-\frac{2}{5}\left(t_1+\frac{1}{2}\right)^{\frac{1}{2}}\left(t_2+\frac{1}{2}\right)e^{-6(t_1+t_2)}\right|$$

$$\leq \frac{1}{2000e^3}\left(t_1+\frac{2}{3}\right)^{\frac{2}{3}}\left(t_2+\frac{2}{3}\right)^{\frac{4}{3}}e^{-6(t_1+t_2)}\|u_1\| + \frac{1}{4000\pi^2}\left(t_1+\frac{3}{4}\right)^{\frac{3}{4}}\left(t_2+\frac{3}{4}\right)^{\frac{3}{2}}\times$$
$$e^{-3(t_1+t_2)}\|u_2\|,$$
$$\left|\phi_2\left(\frac{1}{\chi}u_1,\frac{1}{\chi}u_2\right)-\frac{5}{6}\left(t_1+\frac{1}{2}\right)^{\frac{1}{2}}\left(t_2+\frac{1}{2}\right)e^{-6(t_1+t_2)}\right|$$
$$\leq \frac{1}{2000\pi}\left(t_1+\frac{2}{3}\right)^{\frac{2}{3}}\left(t_2+\frac{2}{3}\right)^{\frac{4}{3}}e^{-6(t_1+t_2)}\|u_1\| + \frac{1}{500e^2}\left(t_1+\frac{3}{4}\right)^{\frac{3}{4}}\left(t_2+\frac{3}{4}\right)^{\frac{3}{2}}\times$$
$$e^{-3(t_1+t_2)}\|u_2\|,$$

and $(10e-1)^3 < g_1(t_1) < (10e+1)^3$ and $(12e-1)^2 < g_2(t_2) < (12e+1)^2$.

Thus, $(H_2), (H3)$ hold with $N_{11} = 0.4$, $N_{12} = 0.0000249$, $N_{13} = 0.0000253$, $N_{21} = 0.833$, $N_{22} = 0.000159$, $N_{23} = 0.000271$, $g_1 = 17949.37$, $g_2 = 999.79$, $G_1 = 22384.80$, and $G_2 = 1130.26$.

Finally, we find that:

$$\tilde{\Omega}_{11} = 2313.238,\ \tilde{\Omega}_{12} = 319.647,\ \tilde{\Omega}_{13} = 27053.522,\ \tilde{\Omega}_{14} = 2597.063,$$
$$\tilde{\Omega}_{21} = 0.0212,\ \tilde{\Omega}_{22} = 1.1403,\ \tilde{\Omega}_{23} = 0.0198,\ \tilde{\Omega}_{24} = 1.5093.$$

Therefore, we have:

$$\Psi_1 + \Psi_2 = 0.00057 + 0.4692 = 0.4698 < 1.$$

Hence, by Theorem 4, this boundary value problem has at least one solution. □

Author Contributions: These authors contributed equally to this work.

Acknowledgments: The last author of this research was supported by Suan Dusit University.

References

1. Wu, G.C.; Baleanu, D. Discrete fractional logistic map and its chaos. *Nonlinear Dyn.* **2014**, *75*, 283–287. [CrossRef]
2. Wu, G.C.; Baleanu, D. Chaos synchronization of the discrete fractional logistic map. *Signal Process.* **2014**, *102*, 96–99. [CrossRef]
3. Goodrich, C.S.; Peterson, A.C. *Discrete Fractional Calculus*; Springer: New York, NY, USA, 2015.
4. Atici, F.M.; Eloe, P.W. A transform method in discrete fractional calculus. *Int. J. Differ. Equ.* **2007**, *2*, 165–176.
5. Atici, F.M.; Eloe, P.W. Initial value problems in discrete fractional calculus. *Proc. Am. Math. Soc.* **2009**, *137*, 981–989. [CrossRef]
6. Atici, F.M.; Eloe, P.W. Two-point boundary value problems for finite fractional difference equations. *J. Differ. Equ. Appl.* **2011**, *17*, 445–456. [CrossRef]
7. Abdeljawad, T. On Riemann and Caputo fractional differences. *Comput. Math. Appl.* **2011**, *62*, 1602–1611. [CrossRef]
8. Abdeljawad, T. Dual identities in fractional difference calculus within Riemann. *Adv. Differ. Equ.* **2013**, *2013*, 36. [CrossRef]
9. Abdeljawad, T. On delta and nabla Caputo fractional differences and dual identities. *Discret. Dyn. Nat. Soc.* **2013**, *2013*, 406910. [CrossRef]
10. Abdeljawad, T.; Baleanu, D. Fractional differences and integration by parts. *J. Comput. Anal. Appl.* **2011**, *13*, 574–582.
11. Holm, M. Sum and difference compositions in discrete fractional calculus. *Cubo* **2011**, *13*, 153–184. [CrossRef]
12. Anastassiou, G. Foundations of nabla fractional calculus on time scales and inequalities. *Comput. Math. Appl.* **2010**, *59*, 3750–3762. [CrossRef]

13. Jia, B.; Erbe, L.; Peterson, A. Two monotonicity results for nabla and delta fractional differences. *Arch. Math.* **2015**, *104*, 589–597. [CrossRef]
14. Jia, B.; Erbe, L.; Peterson, A. Convexity for nabla and delta fractional differences. *J. Differ. Equ. Appl.* **2015**, *21*, 360–373.
15. Cermák, J.; Kisela, T.; Nechvátal, L. Stability and asymptotic properties of a linear fractional difference equation. *Adv. Differ. Equ.* **2012**, *2012*, 122. [CrossRef]
16. Jarad, F.; Abdeljawad, T.; Baleanu, D.; Biçen, K. On the stability of some discrete fractional nonautonomous systems. *Abstr. Appl. Anal.* **2012**, *2012*, 476581. [CrossRef]
17. Mozyrska, D.; Wyrwas, M. The Z-transform method and delta type fractional difference operators. *Discret. Dyn. Nat. Soc.* **2015**, *2015*, 12. [CrossRef]
18. Mozyrska, D.; Wyrwas, M. Explicit criteria for stability of fractional h-difference two-dimensional systems. *Int. J. Dyn. Control* **2017**, *5*, 4–9. [CrossRef]
19. Ferreira, R.A.C.; Torres, D.F.M. Fractional h-difference equations arising from the calculus of variations. *Appl. Anal. Discret. Math.* **2011**, *5*, 110–121. [CrossRef]
20. Ferreira, R.A.C. Existence and uniqueness of solution to some discrete fractional boundary value problems of order less than one. *J. Differ. Equ. Appl.* **2013**, *19*, 712–718. [CrossRef]
21. Ferreira, R.A.C.; Goodrich, C.S. Positive solution for a discrete fractional periodic boundary value problem. *Dyn. Contin. Discret. Impuls. Syst. Ser. A Math. Anal.* **2012**, *19*, 545–557.
22. Goodrich, C.S. Existence and uniqueness of solutions to a fractional difference equation with nonlocal conditions. *Comput. Math. Appl.* **2011**, *61*, 191–202. [CrossRef]
23. Goodrich, C.S. On a discrete fractional three-point boundary value problem. *J. Differ. Equ. Appl.* **2012**, *18*, 397–415. [CrossRef]
24. Goodrich, C.S. A convexity result for fractional differences. *Appl. Math. Lett.* **2014**, *35*, 58–62. [CrossRef]
25. Goodrich, C.S. The relationship between sequential fractional difference and convexity. *Appl. Anal. Discret. Math.* **2016**, *10*, 345–365. [CrossRef]
26. Dahal, R.; Goodrich, C.S. A monotonicity result for discrete fractional difference operators. *Arch. Math.* **2014**, *102*, 293–299. [CrossRef]
27. Erbe, L.; Goodrich, C.S.; Jia, B.; Peterson, A. Survey of the qualitative properties of fractional difference operators: Monotonicity, convexity, and asymptotic behavior of solutions. *Adv. Differ. Equ.* **2016**, *2016*, 43. [CrossRef]
28. Chen, F.; Luo, X.; Zhou, Y. Existence results for nonlinear fractional difference equation. *Adv. Differ. Equ.* **2011**, *2011*, 713201. [CrossRef]
29. Chen, F.; Zhou, Y. Existence and Ulam stability of solutions for discrete fractional boundary value problem. *Discret. Dyn. Nat. Soc.* **2013**, *2013*, 459161. [CrossRef]
30. Chen, Y.; Tang, X. Thee difference between a class of discrete fractional and integer order boundary value problems. *Commun. Nonlinear Sci.* **2014**, *19*, 4057–4067. [CrossRef]
31. Lv, W. Solvability for discrete fractional boundary value problems with a p-laplacian operator. *Discret. Dyn. Nat. Soc.* **2013**, *2013*, 679290. [CrossRef]
32. Lv, W. Solvability for a discrete fractional three-point boundary value problem at resonance. *Abstr. Appl. Anal.* **2014**, *2014*, 601092. [CrossRef]
33. Lv, W.; Feng, J. Nonlinear discrete fractional mixed type sum-difference equation boundary value problems in Banach spaces. *Adv. Differ. Equ.* **2014**, *2014*, 184. [CrossRef]
34. Chen, H.Q.; Cui, Y.Q.; Zhao, X.L. Multiple solutions to fractional difference boundary value problems. *Abstr. Appl. Anal.* **2014**, *2014*, 879380. [CrossRef]
35. Chen, H.Q.; Jin, Z.; Kang, S.G. Existence of positive solutions for Caputo fractional difference equation. *Adv. Differ. Equ.* **2015**, *2015*, 44. [CrossRef]
36. Kang, S.G.; Li, Y.; Chen, H.Q. Positive solutions to boundary value problems of fractional difference equations with nonlocal conditions. *Adv. Differ. Equ.* **2014**, *2014*, 7. [CrossRef]
37. Dong, W.; Xu, J.; Regan, D.O. Solutions for a fractional difference boundary value problem. *Adv. Differ. Equ.* **2013**, *2013*, 319. [CrossRef]
38. Sitthiwirattham, T.; Tariboon, J.; Ntouyas, S.K. Existence Results for fractional difference equations with three-point fractional sum boundary conditions. *Discret. Dyn. Nat. Soc.* **2013**, *2013*, 104276. [CrossRef]
39. Sitthiwirattham, T.; Tariboon, J.; Ntouyas, S.K. Boundary value problems for fractional difference equations with three-point fractional sum boundary conditions. *Adv. Differ. Equ.* **2013**, *2013*, 296. [CrossRef]

40. Sitthiwirattham, T. Existence and uniqueness of solutions of sequential nonlinear fractional difference equations with three-point fractional sum boundary conditions. *Math. Methods Appl. Sci.* **2015**, *38*, 2809–2815. [CrossRef]
41. Sitthiwirattham, T. Boundary value problem for p-Laplacian Caputo fractional difference equations with fractional sum boundary conditions. *Math. Methods Appl. Sci.* **2016**, *39*, 1522–1534. [CrossRef]
42. Chasreechai, S.; Kiataramkul, C.; Sitthiwirattham, T. On nonlinear fractional sum-difference equations via fractional sum boundary conditions involving different orders. *Math. Probl. Eng.* **2015**, *2015*, 519072. [CrossRef]
43. Reunsumrit, J.; Sitthiwirattham, T. Positive solutions of three-point fractional sum boundary value problem for Caputo fractional difference equations via an argument with a shift. *Positivity* **2016**, *20*, 861–876. [CrossRef]
44. Reunsumrit, J.; Sitthiwirattham, T. On positive solutions to fractional sum boundary value problems for nonlinear fractional difference equations. *Math. Methods Appl. Sci.* **2016**, *39*, 2737–2751. [CrossRef]
45. Soontharanon, J.; Jasthitikulchai, N.; Sitthiwirattham, T. Nonlocal Fractional Sum Boundary Value Problems for Mixed Types of Riemann-Liouville and Caputo Fractional Difference Equations. *Dyn. Syst. Appl.* **2016**, *25*, 409–414.
46. Laoprasittichok, S.; Sitthiwirattham, T. On a Fractional Difference-Sum Boundary Value Problems for Fractional Difference Equations Involving Sequential Fractional Differences via Different Orders. *J. Comput. Anal. Appl.* **2017**, *23*, 1097–1111.
47. Kaewwisetkul, B.; Sitthiwirattham, T. On Nonlocal Fractional Sum-Difference Boundary Value Problems for Caputo Fractional Functional Difference Equations with Delay. *Adv. Differ. Equ.* **2017**, *2017*, 219. [CrossRef]
48. Ahmad, B.; Ntouyas, S.K.; Alsaedi, A. On a coupled system of fractional differential equations with coupled nonlocal and integral boundary conditions. *Chaos Soliton Fractals* **2016**, *83*, 234–241. [CrossRef]
49. Ahmad, B.; Ntouyas, S.K. Existence results for a coupled system of Caputo type sequential fractional differential equations with nonlocal integral boundary conditions. *Appl. Math. Comput.* **2015**, *266*, 615–622. [CrossRef]
50. Henderson, J.; Luca, R. Nonexistence of positive solutions for a system of coupled fractional boundary value problems. *Bound. Value Probl.* **2015**, *2015*, 138. [CrossRef]
51. Henderson, J.; Luca, R.; Tudorache, A. On a system of fractional differential equations with coupled integral boundary conditions. *Fract. Calc. Appl. Anal.* **2015**, *18*, 361–386. [CrossRef]
52. Wang, J.R.; Zhang, Y. Analysis of fractional order differential coupled systems. *Math. Methods Appl. Sci.* **2015**, *38*, 3322–3338. [CrossRef]
53. Su, X. Boundary value problem for a coupled system of nonlinear fractional differential equations. *Appl. Math. Lett.* **2009**, *22*, 64–69. [CrossRef]
54. Ascione, G.; Leonenko, N.; Pirozzi, E. Fractional Queues with Catastrophes and Their Transient Behaviour. *Mathematics* **2018**, *6*, 159. [CrossRef]
55. Pan, Y.; Han, Z.; Sun, S.; Zhao, Y. The Existence of Solutions to a System of Discrete Fractional Boundary Value Problems. *Abstr. Appl. Anal.* **2012**, *2012*, 707631. [CrossRef]
56. Goodrich, C.S. Existence of a positive solution to a system of discrete fractional boundary value problems. *Appl. Math. Comput.* **2011**, *217*, 4740–4753. [CrossRef]
57. Dahal, R.; Duncan, D.; Goodrich, C.S. Systems of semipositone discrete fractional boundary value problems. *J. Differ. Equ. Appl.* **2014**, *20*, 473–491. [CrossRef]
58. Goodrich, C.S. Systems of discrete fractional boundary value problems with nonlinearities satisfying no growth conditions. *J. Differ. Equ. Appl.* **2015**, *21*, 437–453. [CrossRef]
59. Goodrich, C.S. Coupled systems of boundary value problems with nonlocal boundary conditions. *Appl. Math. Lett.* **2015**, *41*,17–22. [CrossRef]
60. Kunnawuttipreechachan, E.; Promsakon, C.; Sitthiwirattham, T. Nonlocal fractional sum boundary value problems for a coupled system of fractional sum-difference equations. *Dyn. Syst. Appl.* **2019**, *28*, 73–92. [CrossRef]
61. Griffel, D.H. *Applied Functional Analysis*; Ellis Horwood Publishers: Chichester, UK, 1981.

2

Hyers–Ulam Stability and Existence of Solutions for Differential Equations with Caputo–Fabrizio Fractional Derivative

Kui Liu [1], Michal Fečkan [2,3], D. O'Regan [4] and JinRong Wang [1,5,*]

1 Department of Mathematics, Guizhou University, Guiyang 550025, China; liuk180916@163.com
2 Department of Mathematical Analysis and Numerical Mathematics, Faculty of Mathematics, Physics and Informatics, Comenius University in Bratislava, Mlynská dolina, 842 48 Bratislava, Slovakia; Michal.Feckan@fmph.uniba.sk
3 Mathematical Institute, Slovak Academy of Sciences, Štefánikova 49, 814 73 Bratislava, Slovakia
4 School of Mathematics, Statistics and Applied Mathematics, National University of Ireland, H91 TK33 Galway, Ireland; donal.oregan@nuigalway.ie
5 School of Mathematical Sciences, Qufu Normal University, Qufu 273165, China
* Correspondence: jrwang@gzu.edu.cn

Abstract: In this paper, the Hyers–Ulam stability of linear Caputo–Fabrizio fractional differential equation is established using the Laplace transform method. We also derive a generalized Hyers–Ulam stability result via the Gronwall inequality. In addition, we establish existence and uniqueness of solutions for nonlinear Caputo–Fabrizio fractional differential equations using the generalized Banach fixed point theorem and Schaefer's fixed point theorem. Finally, two examples are given to illustrate our main results.

Keywords: Caputo–Fabrizio fractional differential equations; Hyers–Ulam stability

---

1. Introduction

Fractional differential operators describe mechanical and physical processes with historical memory and spatial global correlation and for the basic theory—see [1–3]. Results on existence, stability and controllability for differential equations with Caputo, Riemann–Liouville and Hilfer type fractional derivatives can be found, for example, in [4–19]. Caputo and Fabrizio [20] introduced a new nonlocal derivative without a singular kernel and Atangana and Nieto [21] studied the numerical approximation of this new fractional derivative and established a modified resistance loop capacitance (RLC) circuit model. Losada and Nieto [22] presented a fractional integral corresponding to the Caputo–Fabrizio fractional derivative and introduced Caputo–Fabrizio fractional differential equations and established existence and uniqueness results. Baleanu et al. [23] extended the study to Caputo–Fabrizio fractional integro-differential equations and obtained the approximate solution. Franc and Goufo [24] established a new Korteweg–de Vries–Burgers equation involving the Caputo–Fabrizio fractional derivative with no singular kernel and presented existence and uniqueness results and also gave numerical approximations.

Hyers–Ulam stability is a concept that provides an approximate solution for the exact solution in a simple form for differential equations. A Laplace transform method is applied to show the Hyers–Ulam stability for integer order differential equations in [25,26] and Wang and Li [27] adopted the idea and applied a Laplace transform method to show the Hyers–Ulam stability for fractional order differential equations involving Caputo derivatives. There are many papers on differential equations involving fractional derivatives–see, for example, [28–36]. However, there are only a few papers on the Hyers–Ulam stability for differential equations with the Caputo–Fabrizio fractional derivative. In [37], Wang et al. offered the Ulam stability for the fractional differential equations with the Caputo derivative.

First, we recall the well-known Caputo fractional derivative [2] of order β, given by

$$(\mathbb{D}^{\beta}y)(x) = \frac{1}{\Gamma(1-\beta)}\int_a^x \frac{\dot{f}(s)}{(x-s)^{\beta}}ds,\ 0<\beta<1,$$

where $f \in C^1(a,b), b > a$. By changing the kernel $(x-s)^{-\beta}$ with the function $\exp(-\frac{\beta}{1-\beta}(x-s))$ and $\frac{1}{\Gamma(1-\beta)}$ by $\frac{1}{\sqrt{2\pi(1-\alpha^2)}}$, we obtain the new definition of fractional derivative without a singular kernel $({}^{CF}\mathbb{D}^{\alpha}y)(x)$—see Definition 1 for details.

In this paper, we study Hyers–Ulam stability and existence and uniqueness of solutions for the following Caputo–Fabrizio fractional derivative equations:

$$({}^{CF}\mathbb{D}^{\alpha}y)(x) - \lambda({}^{CF}\mathbb{D}^{\beta}y)(x) = u(x),\ x\in[0,T],\ 0<\alpha,\beta<1, \tag{1}$$

and

$$({}^{CF}\mathbb{D}^{\alpha}y)(x) = f(x,y(x)),\ x\in[0,T],\ 0<\alpha<1, \tag{2}$$

where $({}^{CF}\mathbb{D}^{\gamma}y)(\cdot)$ denotes the Caputo–Fabrizio derivative for y with the order $0<\gamma<1$ (see Definition 1), $\lambda\in\mathbb{R}$, $u:[0,T]\to\mathbb{R}$ and $f:[0,T]\times\mathbb{R}\to\mathbb{R}$ will be specified later.

The main contributions are as follows: we obtain a simple result to check whether the approximate solution is near the exact solution for linear Equation (1), which implies Hyers–Ulam stability and generalized Hyers–Ulam stability on the finite time interval. In addition, we present a condition to derive existence and uniqueness of solutions for nonlinear Equation (2) using the generalized Banach fixed point theorem (this improves the result in (Theorem 1, [22])). In addition, we establish sufficient conditions to guarantee the existence of solutions for nonlinear Equation (2) using Schaefer's fixed point theorem. Based on the existence and uniqueness result, we prove the Hyers–Ulam stability of (2) via the Gronwall inequality.

2. Preliminaries

Let $C(I,\mathbb{R})$ be the Banach space of all continuous functions from I into $\mathbb{R}$ with the norm $\|y\|_C := \sup\{|y(x)| : x\in I\}$.

Definition 1 (see [22]). *Let $0<\alpha<1$, $h\in C^1[0,b)$ and $b>0$. The Caputo–Fabrizio fractional derivative for a function h of order α is defined by*

$${}^{CF}\mathbb{D}^{\alpha}h(\tau) = \frac{(2-\alpha)M(\alpha)}{2(1-\alpha)}\int_0^{\tau}\exp(-\frac{\alpha}{1-\alpha}(\tau-x))h'(x)dx,\ \tau\geq 0,$$

where $M(\alpha)$ is a normalization constant depending on α. Note that $({}^{CF}\mathbb{D}^{\alpha})(h) = 0$ if and only if h is a constant function.

Definition 2 (see Definition 1, [22]). *Let $0<\alpha<1$. The Caputo–Fabrizio fractional integral for a function h of order α is defined by*

$${}^{CF}I^{\alpha}h(\tau) = \frac{2(1-\alpha)}{(2-\alpha)M(\alpha)}h(\tau) + \frac{2\alpha}{(2-\alpha)M(\alpha)}\int_0^{\tau}h(x)dx,\ \tau\geq 0.$$

Theorem 1 (see [20,22]). *Let* $\alpha \in (0,1)$. *Then,*

$$\mathcal{L}[{}^{CF}\mathbb{D}^{\alpha}h(\tau)](s) = \frac{(2-\alpha)M(\alpha)}{2(s+\alpha(1-s))}(s\mathcal{L}[h(\tau)](s) - h(0)),\ s > 0.$$

Motivated by (Definition 2.3, [37]), we introduce the following definition.

Definition 3. *Let* $0 < \alpha, \beta < 1$ *and* $u : [0,T] \to \mathbb{R}$ *be a continuous function. Then,* (1) *is Hyers–Ulam stable if there exists* $K > 0$ *and* $\epsilon > 0$ *such that, for each solution* $y \in C([0,T],\mathbb{R})$ *of* (1),

$$|{}^{CF}\mathbb{D}^{\alpha}y(x) - \lambda{}^{CF}\mathbb{D}^{\beta}y(x) - u(x)| \leq \epsilon,\ \forall x \in [0,T], \tag{3}$$

and there exists a solution $z \in C([0,T),\mathbb{R})$ *of* (2) *with*

$$|y(x) - z(x)| \leq K\epsilon,\ \forall x \in [0,T].$$

Definition 4. *Let* $0 < \alpha, \beta < 1$, $u : [0,T] \to \mathbb{R}$ *be a continuous function and* $G : [0,T] \to \mathbb{R}_+$ *be continuous functions. Then,* (1) *is generalized Hyers–Ulam–Rassias stable with respect to* G *if there exists a constant* $c_{f,G} > 0$ *such that for each solution* $y \in C([0,T],\mathbb{R})$ *of* (1),

$$|{}^{CF}\mathbb{D}^{\alpha}y(x) - \lambda{}^{CF}\mathbb{D}^{\beta}y(x) - u(x))| \leq G(x),\ \forall x \in [0,T], \tag{4}$$

and there exists a solution $z \in C([0,T],\mathbb{R})$ *of* (2) *with*

$$|y(x) - z(x)| \leq c_{f,G}G(x),\ \forall x \in [0,T].$$

Definition 5. *Let* $f : [0,T] \times \mathbb{R} \to \mathbb{R}$ *be a continuous function. Then,* (2) *is Hyers–Ulam stable if there exists* $K > 0$ *and* $\epsilon > 0$ *such that for each solution* $y \in C([0,T],\mathbb{R})$ *of* (2),

$$|{}^{CF}\mathbb{D}^{\alpha}y(x) - f(x,y(x))| \leq \epsilon,\ \forall x \in [0,T], \tag{5}$$

and there exists a solution $z \in C([0,T),\mathbb{R})$ *of* (2) *with*

$$|y(x) - z(x)| \leq K\epsilon,\ \forall x \in [0,T].$$

Definition 6. *Let* $f : [0,T] \times \mathbb{R} \to \mathbb{R}$ *and* $G : [0,T] \to \mathbb{R}_+$ *be continuous functions. Then,* (2) *is generalized Hyers–Ulam–Rassias stable with respect to* G *if there exists a constant* $c_{f,G} > 0$ *such that, for each solution* $y \in C([0,T],\mathbb{R})$ *of* (2),

$$|{}^{CF}\mathbb{D}^{\alpha}y(x) - f(x,y(x))| \leq G(x),\ \forall x \in [0,T], \tag{6}$$

and there exists a solution $z \in C([0,T],\mathbb{R})$ *of* (2) *with*

$$|y(x) - z(x)| \leq c_{f,G}G(x),\ \forall x \in [0,T].$$

3. Stability Results for the Linear Equation

In this section, we study Hyers–Ulam and generalized Hyers–Ulam-Rassias stability of (1).

Theorem 2. *Let* $0 < \beta, \alpha < 1$, $\lambda \in \mathbb{R}$, *and* $u(x)$ *be a given real function on* $[0,T]$. *If a function* $y : [0,T] \to \mathbb{R}$ *satisfies the inequality*

$$|({}^{CF}\mathbb{D}^{\alpha}y)(x) - \lambda({}^{CF}\mathbb{D}^{\beta}y)(x) - u(x)| \leq \varepsilon \tag{7}$$

for each $x \in [0,T]$ and $\varepsilon > 0$, then there exists a solution $y_a : [0,T] \to \mathbb{R}$ of (1) such that

$$|y(x) - y_a(x)| \leq 2|\frac{C}{A}|\varepsilon + 2\left|\frac{AD-BC}{A^2} - \frac{\alpha\beta}{B}\right| \max\{1, \exp(-\frac{B}{A}T)\}x\varepsilon + 2|\frac{\alpha\beta}{B}|x\varepsilon, \tag{8}$$

where

$$\begin{cases} A = (1-\beta)(2-\alpha)M(\alpha) - \lambda(2-\beta)M(\beta)(1-\alpha), \\ B = (2-\alpha)M(\alpha)\beta - \lambda(2-\beta)M(\beta)\alpha, \\ C = (1-\beta)(1-\alpha), \\ D = \alpha+\beta-2\alpha\beta. \end{cases} \tag{9}$$

Proof. Let

$$F(x) = ({}^{CF}\mathbb{D}^{\alpha}y)(x) - \lambda({}^{CF}\mathbb{D}^{\beta}y)(x) - u(x),\ x \in [0,T]. \tag{10}$$

Taking the Laplace transform of (10) via Theorem 1, and we have

$$\begin{aligned} \mathcal{L}\{F(x)\}(s) &= \mathcal{L}\{({}^{CF}\mathbb{D}^{\alpha}y)(x) - \lambda({}^{CF}\mathbb{D}^{\beta}y)(x) - u(x)\}(s) \\ &= \mathcal{L}\{({}^{CF}\mathbb{D}^{\alpha}y)(x)\}(s) - \lambda\mathcal{L}\{({}^{CF}\mathbb{D}^{\beta}y)(x)\}(s) - \mathcal{L}\{u(x)\}(s) \\ &= [\frac{(2-\alpha)M(\alpha)}{2(s+\alpha(1-s))} - \lambda\frac{(2-\beta)M(\beta)}{2(s+\beta(1-s))}]s\mathcal{L}\{y(x)\}(s) \\ &\quad +[-\frac{(2-\alpha)M(\alpha)}{2(s+\alpha(1-s))} + \lambda\frac{(2-\beta)M(\beta)}{2(s+\beta(1-s))}]y(0) - \mathcal{L}\{u(x)\}(s), \end{aligned} \tag{11}$$

where $\mathcal{L}\{F\}$ denotes the Laplace transform of the function F. From (11), one has

$$\begin{aligned} &\mathcal{L}\{y(x)\}(s) \\ = &\ \frac{1}{s}y(0) + \frac{1}{s}\frac{2(s+\alpha(1-s))(s+\beta(1-s))}{(2-\alpha)M(\alpha)(s+\beta(1-s)) - \lambda(2-\beta)M(\beta)(s+\alpha(1-s))} \\ &\times\Big(\mathcal{L}\{u(x)\}(s) + \mathcal{L}\{F(x)\}(s)\Big) \\ = &\ \frac{1}{s}y(0) + 2\Big(\frac{C}{A} + \frac{AD-BC}{A^2}\frac{1}{s+\frac{B}{A}} + \frac{\alpha\beta}{B}\frac{1}{s} - \frac{\alpha\beta}{B}\frac{1}{s+\frac{B}{A}}\Big)(\mathcal{L}\{u(x)\}(s) + \mathcal{L}\{F(x)\}(s)), \end{aligned} \tag{12}$$

where A, B, C, D are defined as in (9). Set

$$y_a(x) = y(0) + 2\frac{C}{A}u(x) + 2\Big(\frac{AD-BC}{A^2} - \frac{\alpha\beta}{B}\Big)\int_0^x \exp(-\frac{B}{A}t)u(x-t)dt + 2\frac{\alpha\beta}{B}\int_0^x u(x-t)dt. \tag{13}$$

Taking the Laplace transform of (13), one has

$$\begin{aligned} &\mathcal{L}\{y_a(x)\}(s) \\ = &\ \frac{1}{s}y(0) + 2\frac{C}{A}\mathcal{L}\{u(x)\}(s) + 2\Big(\frac{AD-BC}{A^2} - \frac{\alpha\beta}{B}\Big)\frac{1}{s+\frac{B}{A}}\mathcal{L}\{u(x)\}(s) + 2\frac{\alpha\beta}{B}\frac{1}{s}\mathcal{L}\{u(x)\}(s) \\ = &\ \frac{1}{s}y(0) + 2\Big(\frac{C}{A} + \frac{AD-BC}{A^2}\frac{1}{s+\frac{B}{A}} + \frac{\alpha\beta}{B}\frac{1}{s} - \frac{\alpha\beta}{B}\frac{1}{s+\frac{B}{A}}\Big)\mathcal{L}\{u(x)\}(s). \end{aligned} \tag{14}$$

Note that

$$\begin{aligned} &\mathcal{L}\{({}^{CF}\mathbb{D}^{\alpha}y_a)(x) - \lambda({}^{CF}\mathbb{D}^{\beta}y_a)(x)\}(s) \\ = &\ \frac{(2-\alpha)M(\alpha)(s+\beta(1-s)) - \lambda(2-\beta)M(\beta)(s+\alpha(1-s))}{2(s+\alpha(1-s))(s+\beta(1-s))}(s\mathcal{L}\{y_a(x)\}(s) - y(0)). \end{aligned} \tag{15}$$

Substituting (14) into (15), we obtain

$$\mathcal{L}\{({}^{CF}\mathbb{D}^{\alpha}y_a)(x) - \lambda({}^{CF}\mathbb{D}^{\beta}y_a)(x)\}(s) = \mathcal{L}\{u(x)\},$$

which yields that $y_a(x)$ is a solution of Equation (1) since $\mathcal{L}$ is one-to-one. From (12) and (14), we have

$$\mathcal{L}\{y(x) - y_a(x)\}(s) = 2\Big(\frac{C}{A} + \frac{AD - BC}{A^2}\frac{1}{s+\frac{B}{A}} + \frac{\alpha\beta}{B}\frac{1}{s} - \frac{\alpha\beta}{B}\frac{1}{s+\frac{B}{A}}\Big)\mathcal{L}\{F(x)\}.$$

This implies that

$$y(x) - y_a(x) = 2\frac{C}{A}F(x) + 2(\frac{AD-BC}{A^2} - \frac{\alpha\beta}{B})(\exp(-\frac{B}{A}x) * F(x) + 2\frac{\alpha\beta}{B}(1 * F(x)),$$

so

$$\begin{aligned}
&|y(x) - y_a(x)| \\
= \;& \Big|2\frac{C}{A}F(x) + 2(\frac{AD-BC}{A^2} - \frac{\alpha\beta}{B})(\exp(-\frac{B}{A}x) * F(x) + 2\frac{\alpha\beta}{B}(1 * F(x))\Big| \\
\leq \;& 2|\frac{C}{A}F(x)| + 2\Big|\frac{AD-BC}{A^2} - \frac{\alpha\beta}{B}\Big||\exp(-\frac{B}{A}x) * F(x)| + 2|\frac{\alpha\beta}{B}||1 * F(x)| \\
\leq \;& 2|\frac{C}{A}||F(x)| + 2\Big|\frac{AD-BC}{A^2} - \frac{\alpha\beta}{B}\Big|\int_0^x |\exp(-\frac{B}{A}t)||F(x-t)|dt + 2|\frac{\alpha\beta}{B}|\int_0^x |F(x-t)|dt) \\
\leq \;& 2|\frac{C}{A}||F(x)| + 2\Big|\frac{AD-BC}{A^2} - \frac{\alpha\beta}{B}\Big|\varepsilon\int_0^x \max\{1, \exp(-\frac{B}{A}(T))\}dt + 2|\frac{\alpha\beta}{B}|\varepsilon\int_0^x 1dt) \\
\leq \;& 2|\frac{C}{A}|\varepsilon + 2\Big|\frac{AD-BC}{A^2} - \frac{\alpha\beta}{B}\Big|x\max\{1, \exp(-\frac{B}{A}T)\}\varepsilon + 2|\frac{\alpha\beta}{B}|x\varepsilon.
\end{aligned}$$

The proof is complete. □

Remark 1. *If $T < \infty$, then (1) is Hyers–Ulam stable with the constant*

$$K = 2|\frac{C}{A}| + 2|\frac{AD-BC}{A^2} - \frac{\alpha\beta}{B}|\max\{1, \exp(-\frac{B}{A}T)\}T + 2|\frac{\alpha\beta}{B}|T.$$

Remark 2. *Let $0 < \beta, \alpha < 1$, $\lambda \in \mathbb{R}$, and $u(x)$ be a given real function on $[0, T]$. If a function $y : [0, T] \to \mathbb{R}$ satisfies the inequality*

$$|({}^{CF}\mathbb{D}^{\alpha}y)(x) - \lambda({}^{CF}\mathbb{D}^{\beta}y)(x) - u(x)| \leq G(x), \tag{16}$$

this implies that

$$|F(x)| \leq G(x)$$

for each $x \in [0, T]$ and some function $G(x) > 0$, where F is defined in (10).

From Theorem 2, then there exists a solution $y_a : [0, T] \to \mathbb{R}$ of (1) such that

$$y(x) - y_a(x) = 2\frac{C}{A}F(x) + 2(\frac{AD-BC}{A^2} - \frac{\alpha\beta}{B})(\exp(-\frac{B}{A}x) * F(x) + 2\frac{\alpha\beta}{B}(1 * F(x)),$$

and

$$
\begin{aligned}
& |y(x) - y_a(x)| \\
\leq\ & 2|\frac{C}{A}F(x)| + 2\left|\frac{AD-BC}{A^2} - \frac{\alpha\beta}{B}\right|| \exp(-\frac{B}{A}x) * F(x)| + 2|\frac{\alpha\beta}{B}||1 * F(x)| \\
\leq\ & 2|\frac{C}{A}||F(x)| + 2\left|\frac{AD-BC}{A^2} - \frac{\alpha\beta}{B}\right| \max\{1, \exp(-\frac{B}{A}T)\}| \int_0^x F(x-t)dt| + 2|\frac{\alpha\beta}{B}|| \int_0^x F(x-t)dt| \\
\leq\ & 2|\frac{C}{A}||F(x)| + 2\left|\frac{AD-BC}{A^2} - \frac{\alpha\beta}{B}\right| \max\{1, \exp(-\frac{B}{A}T)\}|F(x)| + 2|\frac{\alpha\beta}{B}||F(x)| \\
\leq\ & 2\left[|\frac{C}{A}| + \left|\frac{AD-BC}{A^2} - \frac{\alpha\beta}{B}\right| \max\{1, \exp(-\frac{B}{A}(T))\} + |\frac{\alpha\beta}{B}|\right] G(x)
\end{aligned}
$$

provided that

$$\int_0^x F(t)dt \leq F(x)$$

for any $x \in [0, T]$, where F is defined in (10) *and A, B, C, D are defined as in* (9)*. Thus,* (2) *is generalized Hyers–Ulam stable with respect to G on $[0, T]$.*

4. Existence and Stability Results for the Nonlinear Equation

We introduce the following conditions:

$[A1] : f : [0, T] \times \mathbb{R} \to \mathbb{R}$ is continuous.

$[A2]$: There exists a $k_f > 0$ such that

$$|f(x, y) - f(x, g)| \leq k_f|y - g|, \ \ \forall y, g \in \mathbb{R}, x \in [0, T].$$

$[A3]$: There exists a constant $L > 0$ such that

$$|f(x, y)| \leq L(1 + |y|)$$

for each $x \in [0, T]$ and all $y \in \mathbb{R}$.

Let $a_\alpha = \frac{2(1-\alpha)}{2-\alpha}M(\alpha)$, $b_\alpha = \frac{2\alpha}{2-\alpha}M(\alpha)$, $y(0) = y_0$ and $C_0 = -a_\alpha f(0, y_0) + y_0$.

Theorem 3. *Let $0 < \alpha < 1$. Assume that $[A1]$ and $[A2]$ hold. If $a_\alpha k_f < 1$, then* (2) *with $y(0) = y_0$ has a unique solution.*

Proof. Consider $P : C([0, T], \mathbb{R}) \to C([0, T], \mathbb{R})$ as follows:

$$(Py)(x) = C_0 + a_\alpha f(x, y(x)) + b_\alpha \int_0^x f(s, y(s))ds. \tag{17}$$

Note P is well defined because of $[A1]$. For all $y_1, y_2 \in C([0, T], \mathbb{R})$ and all $x \in [0, T]$, using $[A2]$, we have

$$
\begin{aligned}
& |(Py_1)(x) - (Py_2)(x)| \\
\leq\ & a_\alpha|f(x, y_1(x)) - f(x, y_2(x))| + b_\alpha \int_0^x |f(s, y_1(s)) - f(x, y_2(x))|ds \\
\leq\ & a_\alpha k_f|y_1(x) - y_2(x)| + b_\alpha \int_0^x k_f|y_1(s) - y_2(s)|ds \\
=\ & a_\alpha k_f\|y_1 - y_2\|_C + b_\alpha k_f x\|y_1 - y_2\|_C.
\end{aligned}
$$

Denote $C_n^i = \frac{n!}{(n-i)!i!}$. Next,

$$\begin{aligned}
& |(P^2y_1)(x) - (P^2y_2)(x)| \\
\leq \; & a_\alpha|f(x,(Py_1)(x)) - f(x,(Py_2)(x))| + b_\alpha \int_0^x |f(s,(Py_1)(s)) - f(x,(Py_2)(x))|ds \\
\leq \; & a_\alpha k_f|Py_1(x) - Py_2(x)| + b_\alpha \int_0^x k_f|Py_1(s) - Py_2(s)|ds \\
\leq \; & a_\alpha k_f(a_\alpha k_f\|y_1 - y_2\|_C + b_\alpha k_f x\|y_1 - y_2\|_C) \\
& + b_\alpha k_f \int_0^x (a_\alpha k_f\|y_1 - y_2\|_C + b_\alpha k_f x\|y_1 - y_2\|_C)ds \\
\leq \; & \left((k_f a_\alpha)^2 + 2k_f a_\alpha(k_f b_\alpha x) + \frac{(k_f b_\alpha x)^2}{2!}\right)\|y_1 - y_2\|_C \\
= \; & \sum_{i=0}^{2} \frac{C_2^i (k_f a_\alpha)^{2-i}(k_f b_\alpha x)^i}{i!}\|y_1 - y_2\|_C.
\end{aligned}$$

For any $m \in \mathbb{N}^+$, suppose the following inequality hold

$$|(P^m y_1)(x) - (P^m y_2)(x)| \leq \sum_{i=0}^{m} \frac{C_m^i (k_f a_\alpha)^{m-i}(k_f b_\alpha x)^i}{i!}\|y_1 - y_2\|_C.$$

Then,

$$\begin{aligned}
& |(P^{m+1}y_1)(x) - (P^{m+1}y_2)(x)| \\
\leq \; & a_\alpha|f(x,(P^m y_1)(x)) - f(x,(P^m y_2)(x))| + b_\alpha \int_0^x |f(x,(P^m y_1)(s)) - f(x,(P^m y_2)(s))|ds \\
\leq \; & \left(k_f a_\alpha \sum_{i=0}^{m} \frac{C_m^i (k_f a_\alpha)^{m-i}(k_f b_\alpha x)^i}{i!} + k_f b_\alpha \int_0^x \sum_{i=0}^{m} \frac{C_m^i (k_f a_\alpha)^{m-i}(k_f b_\alpha s)^i}{i!} ds\right)\|y_1 - y_2\|_C \\
= \; & \sum_{i=0}^{m+1} \frac{C_{m+1}^i (k_f a_\alpha)^{m+1-i}(k_f b_\alpha x)^i}{i!}\|y_1 - y_2\|_C \\
\leq \; & S(m)\|y_1 - y_2\|_C,
\end{aligned}$$

where $S(m) := \sum_{i=0}^{m+1} \frac{C_{m+1}^i (k_f a_\alpha)^{m+1-i}(k_f b_\alpha T)^i}{i!}$. Thus, for any $m \in \mathbb{N}^+$,

$$\|P^{m+1}y_1 - P^{m+1}y_2\|_C \leq S(m)\|y_1 - y_2\|_C.$$

From the condition $k_f a_\alpha < 1$ via (Theorem 2.9, [38]), one has $S(m) \to 0$ as $m \to \infty$. This implies that for any large enough $m \in \mathbb{N}^+$, $S(m) < 1$. Thus, P^m is a contraction mapping. As a result, P has a fixed point. Thus, (2) with $y(0) = y_0$ has a unique solution. This proof is complete. □

Remark 3. *In (Theorem 1, [22]), an existence and uniqueness result for* (2) *with* $y(0) = y_0$ *is established by imposing a uniformly Lipschitz condition and applying Banach's fixed point theorem with the condition* $a_\alpha k_f + b_\alpha T k_f < 1$*, where* k_f *denotes the Lipschitz constant. Here, we use the generalized Banach fixed point theorem and we weaken the condition* $a_\alpha k_f + b_\alpha T k_f < 1$ *in (Theorem 1, [22]) to* $a_\alpha k_f < 1$.

Next, we show that the existence of solutions for (2) via Schaefer's fixed point theorem.

Theorem 4. *Assume that* $[A1]$ *and* $[A3]$ *hold. If* $a_\alpha L < 1$*, then* (2) *with* $y(0) = y_0$ *has at least one solution.*

Proof. Consider P as in (17). We divide our proof into several steps.

Step 1. P is continuous.
Let y_n be a sequence such that $y_n \to y$ in $C([0,T],\mathbb{R})$. For all $x \in [0,T]$, we get

$$\begin{aligned}
|Py_n(x) - Py(x)| &= |a_\alpha f(x,y_n(x)) + b_\alpha \int_0^x f(s,y_n(s))ds - a_\alpha f(x,y(x)) - b_\alpha \int_0^x f(s,y(s))ds| \\
&\leq a_\alpha |f(x,y_n(x)) - f(x,y(x))| + b_\alpha | \int_0^x f(s,y_n(s))ds - \int_0^x f(s,y(s))ds| \\
&\leq a_\alpha |f(x,y_n(x)) - f(x,y(x))| + b_\alpha \int_0^x |f(s,y_n(s)) - f(s,y(s))|ds. \\
&\leq (a_\alpha + b_\alpha T)\|f(\cdot,y_n) - f(\cdot,y)\|_C.
\end{aligned}$$

This shows that P is continuous since $\|fy_n - fy\|_C \to 0$ when $n \to \infty$.

Step 2. P maps bounded sets into bounded sets of $C([0,T],\mathbb{R})$.
Indeed, we prove that for all $r > 0$, there exists a $k > 0$ such that for every $y \in B_r = \{y \in C([0,T],\mathbb{R}) : \|y\|_C \leq r\}$, we have $\|Py\|_C \leq k$. In fact, for any $x \in [0,T]$, from $[A3]$, we have

$$\begin{aligned}
|Py(x)| &\leq |C_0| + a_\alpha |f(x,y(x))| + b_\alpha \int_0^x |f(s,y(s))|ds \\
&\leq |C_0| + a_\alpha L(1+|y|) + b_\alpha L \int_0^x (1+|y(s)|)ds \\
&\leq |C_0| + a_\alpha L(1+\|y\|_C) + b_\alpha TL|(1+\|y\|_C) \\
&\leq |C_0| + a_\alpha L(1+r) + b_\alpha TL(1+r) \\
&= |C_0| + (a_\alpha + b_\alpha T)L(1+r),
\end{aligned}$$

which implies that

$$\|Py\| \leq |C_0| + (a_\alpha + b_\alpha T)L(1+r) := k.$$

Step 3. P maps bounded sets into equicontinuous sets in $C([0,T],\mathbb{R})$.
Let $x_1, x_2 \in [0,T]$, with $0 \leq x_1 < x_2 \leq T, y \in B_r$. From $[A3]$, we have

$$\begin{aligned}
&|Py(x_1) - Py(x_2)| \\
= \; & |a_\alpha f(x_1,y(x_1)) + b_\alpha \int_0^{x_1} f(s,y(s))ds - a_\alpha f(x_2,y(x_2)) - b_\alpha \int_0^{x_2} f(s,y(s))ds| \\
\leq \; & a_\alpha |f(x_1,y(x_1)) - f(x_2,y(x_2))| + b_\alpha | \int_0^{x_1} f(s,y(s))ds - \int_0^{x_2} f(s,y(s))ds| \\
\leq \; & a_\alpha |f(x_1,y(x_1)) - f(x_1,y(x_2))| + a_\alpha |f(x_1,y(x_2)) - f(x_2,y(x_2))| + b_\alpha | \int_{x_1}^{x_2} f(s,y(s))ds| \\
\leq \; & a_\alpha |f(x_1,y(x_1)) - f(x_1,y(x_2))| + a_\alpha |f(x_1,y(x_2)) - f(x_2,y(x_2))| + b_\alpha L(1+r)(x_2 - x_1).
\end{aligned}$$

Then, as x_1 approaches x_2, the right-hand side of the above inequality tends to zero (because of $[A1]$) as $x_1 \to x_2$. Thus, P is equicontinuous.

We can conclude that P is completely continuous from Step 1–Step 3 with the Arzela–Ascoli theorem.

Step 4. A priori bounds.
Now, we show that the set $E(P) = \{y \in C([0,T],\mathbb{R}) : y = \lambda Py \text{ for some } \lambda \in (0,1)\}$ is bounded.
Let $y \in E(P)$. Then, $y = \lambda Py$ for some $\lambda \in (0,1)$. For each $x \in [0,T]$, we have

$$
\begin{aligned}
|y(x)| &\leq |C_0| + a_\alpha |f(x,y(x))| + b_\alpha \int_0^x |f(s,y(s))| ds \\
&\leq |C_0| + a_\alpha L(1+|y(x)|) + b_\alpha L \int_0^x (1+|y(s)|) ds \\
&\leq K + a_\alpha L|y(x)| + b_\alpha L \int_0^x |y(s)| ds \quad (K = |C_0| + a_\alpha L + b_\alpha LT).
\end{aligned}
$$

Using the condition $1 - a_\alpha L > 0$, one has

$$
|y(x)| \leq \frac{K}{1-a_\alpha L} + \frac{b_\alpha L}{1-a_\alpha L} \int_0^x |y(s)| ds,
$$

and Gronwall's inequality yields

$$
|y(x)| \leq \frac{K}{1-a_\alpha L} \exp\left(\frac{b_\alpha LT}{1-a_\alpha L}\right) < \infty.
$$

Then, the set $E(P)$ is bounded.

Schaefer's fixed point theorem guarantees that P has a fixed point, which is a solution of (2). The proof is finished. □

In the following, we consider (2) and (6) to discuss the generalized Ulam–Hyers–Rassias stability. We need the following condition.

$[A4]$: Let $G \in C([0,T], \mathbb{R}_+)$ be an increasing function and there exists $\lambda_G > 0$ such that

$$
\int_0^x G(s) ds \leq \lambda_G G(x), \ \forall x \in [0,T].
$$

Theorem 5. *Assumptions $[A1]$, $[A2]$ and $[A4]$ hold. If $a_\alpha k_f < 1$, then (2) is generalized Ulam–Hyers–Rassias stable with respect to G on $[0,T] (T < \infty)$.*

Proof. Let $g \in C([0,T], \mathbb{R})$ be a solution of (6). From Theorem 3,

$$
\begin{cases} {}^{CF}\mathbb{D}^\alpha y(x) = f(x,y(x)), \ 0 < \alpha < 1, \ t \in [0,T), \\ y(0) = C_0, \end{cases} \tag{18}
$$

has the unique solution

$$
y(x) = C_0 + a_\alpha f(x,y(x)) + b_\alpha \int_0^x f(s,y(s)) ds, \ x \in [0,T].
$$

From (6), we have

$$
\begin{aligned}
\left|g(x) - C_0 - a_\alpha f(x,g(x)) - b_\alpha \int_0^x f(s,g(s)) ds\right| &\leq a_\alpha G(x) + b_\alpha \int_0^x G(s) ds \\
&\leq (a_\alpha + b_\alpha \lambda_G) G(x), \ x \in [0,T].
\end{aligned}
$$

Thus,

$$\begin{aligned}
&|g(x)-y(x)| \\
\leq\; & \left|g(x)-C_0-a_\alpha f(x,y(x))-b_\alpha\int_0^x f(s,y(s))ds\right| \\
\leq\; & \Big|g(x)-C_0-a_\alpha f(x,g(x))-b_\alpha\int_0^x f(s,g(s))ds \\
& +a_\alpha f(x,y(x))+b_\alpha\int_a^x f(s,y(s))ds-a_\alpha f(x,y(x))-b_\alpha\int_0^x f(s,y(s))ds\Big| \\
\leq\; & \left|g(x)-C_0-a_\alpha f(x,g(x))-b_\alpha\int_0^x f(s,g(s))ds\right| \\
& +a_\alpha|f(x,y(x))-f(x,g(x))|+b_\alpha\int_0^x |f(s,y(s))-f(s,g(s))|ds \\
\leq\; & (a_\alpha+b_\alpha\lambda_G)G(x)+a_\alpha k_f|y(x)-g(x)|+b_\alpha k_f\int_0^x |y(s)-g(s)|ds.
\end{aligned}$$

Note that $a_\alpha k_f<1$, and so,

$$|y(x)-g(x)|\leq\frac{(a_\alpha+b_\alpha\lambda_G)G(x)}{1-a_\alpha k_f}+\frac{b_\alpha k_f}{1-a_\alpha k_f}\int_0^x |y(s)-g(s)|ds.$$

From Gronwall's inequality, we have

$$|y(x)-g(x)|\leq\left[\frac{(a_\alpha+b_\alpha\lambda_G)}{1-a_\alpha k_f}\exp(x)\right]G(x),\ x\in[0,T]. \tag{19}$$

Set $K^*=\frac{a_\alpha+b_\alpha\lambda_G}{1-a_\alpha k_f}\exp(T)$. Note that one has

$$|y(x)-g(x)|\leq K^*G(x),\ x\in[0,T].$$

From Definition 6, (2) is generalized Ulam–Hyers–Rassias stable with respect to G on $[0,T]$. The proof is complete. □

5. Examples

In this section, two examples are given to illustrate our main results.

For convenience in calculating, we suppose that $M(\cdot)$ in Definition 2 is the roots of the following equation:

$$\frac{2(1-\cdot)}{(2-\cdot)M(\cdot)}+\frac{2\cdot}{(2-\cdot)M(\cdot)}=1.$$

Then, one can derive an explicit formula $M(\alpha)=\frac{2}{2-\alpha}$ and $M(\beta)=\frac{2}{2-\beta}$ (see (p. 89, [22])).

Example 1. *Consider*

$$({}^{CF}\mathbb{D}^{\frac{1}{2}}y)(x)-\frac{1}{3}({}^{CF}\mathbb{D}^{\frac{2}{3}}y)(x)=\frac{2}{3}e^x+\frac{1}{3}e^{-2x}-\frac{2}{3},\ x\in[0,T]. \tag{20}$$

Set $\alpha=\frac{1}{2}$, $\beta=\frac{2}{3}$, $u(x)=\frac{2}{3}e^x+\frac{1}{3}e^{-2x}-\frac{2}{3}$ and $\lambda=\frac{1}{3}$. From (Definition 1, [22]), $M(\frac{1}{2})=\frac{4}{3}$ and $M(\frac{2}{3})=\frac{3}{2}$.

Let $y_1(x) = e^x$, and we have

$$({}^{CF}\mathbb{D}^{\frac{1}{2}}y_1)(x) = 2\int_0^x e^{t-x}e^t dt = e^x - e^{-x},$$

$$({}^{CF}\mathbb{D}^{\frac{2}{3}}y_1)(x) = 3\int_0^x e^{-2(x-t)}e^t dt = e^x - e^{-2x}.$$

Choose $\varepsilon = \frac{2}{3}$. Note $y_1(x) = e^x$ satisfies

$$\begin{aligned}
&\left|({}^{CF}\mathbb{D}^{\frac{1}{2}}y_1)(x) - \frac{1}{3}({}^{CF}\mathbb{D}^{\frac{2}{3}}y_1)(x) - \frac{2}{3}e^x - \frac{1}{3}e^{-2x} + \frac{2}{3}\right| \\
= &\left|e^x - e^{-x} - \frac{1}{3}e^x + \frac{1}{3}e^{-2x} - \frac{2}{3}e^x - \frac{1}{3}e^{-2x} + \frac{2}{3}\right| \\
= &|\frac{2}{3} - e^{-x}| \leq \frac{2}{3}.
\end{aligned}$$

Note $y_1(0) = 1$ and with the formulas of A, B, C, D in (9) *and* (13)*, we obtain an exact solution of Equation* (1) *as*

$$\begin{aligned}
y_a(x) &= y(0) + 2\frac{C}{A}u(x) + 2\left(\frac{AD-BC}{A^2} - \frac{\alpha\beta}{B}\right)\int_0^x \exp(-\frac{B}{A}t)u(x-t)dt \\
&\quad + 2\frac{\alpha\beta}{B}\int_0^x u(x-t)dt \\
&= 1 + \frac{2}{3}e^x - \frac{1}{3}e^{-2x} - \frac{2}{3} - \frac{4}{9}\int_0^x e^{-3t}(e^{x-t} + \frac{e^{-2(x-t)}}{2} - 1)dt \\
&\quad + \frac{4}{9}\int_0^x (e^{x-t} + \frac{e^{-2(x-t)}}{2} - 1)dt \\
&= e^x + \frac{4}{27} + \frac{5}{27}e^{-3x} - \frac{2}{3}e^{-2x} - \frac{4}{9}x.
\end{aligned}$$

Clearly,

$$\begin{aligned}
|y_1(x) - y_a(x)| &= |e^x + \frac{4}{27} + \frac{5}{27}e^{-3x} - \frac{2}{3}e^{-2x} - \frac{4}{9}x - e^x| \\
&= |\frac{4}{27} + \frac{5}{27}e^{-3x} - \frac{2}{3}e^{-2x} - \frac{4}{9}x| \\
&\leq |\frac{4}{27} - \frac{4}{9}x| \\
&\leq \frac{2}{3} + \frac{8}{9}x = (1 + \frac{4}{3}x)\frac{2}{3}.
\end{aligned}$$

Note in Theorem 2 (see Remark 1) that we have $K = 2|\frac{C}{A}| + 2|\frac{AD-BC}{A^2} - \frac{\alpha\beta}{B}|\max\{1, \exp(-\frac{B}{A}T)\}T + 2|\frac{\alpha\beta}{B}|T = 1 + \frac{4}{3}T$ and $\varepsilon = \frac{2}{3}$. Thus, Equation (20) is Hyers–Ulam stable when $T < \infty$.

Example 2. *We consider the following fractional problem:*

$$({}^{CF}\mathbb{D}^{\frac{1}{3}}y)(x) = \frac{e^{-2x}}{1+e^x}\frac{|y|}{1+|y|}, \ x \in [0,2], \tag{21}$$

and the inequality

$$\left|({}^{CF}\mathbb{D}^{\frac{1}{3}}y)(x) - \frac{e^{-2x}}{1+e^x}\frac{|y|}{1+|y|}\right| \leq G(x), \ x \in [0,2]. \tag{22}$$

Set $\alpha = \frac{1}{3}$, $T = 2$ *and* $f(x,y) = \frac{e^{-2x}}{1+e^x}\frac{|y|}{1+|y|}$, $(x,y) \in [0,2] \times \mathbb{R}$. *Clearly,* $[A1]$ *holds. Then,* $M(\frac{1}{3}) = \frac{6}{5}$, $a_{\frac{1}{3}} = \frac{24}{25}$, $b_{\frac{1}{3}} = \frac{12}{25}$. *Let* $G(x) = e^x \in C([0,2],\mathbb{R})$ *and* $\int_0^x G(s)ds = \int_0^x e^s ds = e^x - 1 \leq e^x$. *Here,* $\lambda_G = 1 > 0$.

For any $x \in [0,2]$ *and* $y_1, y_2 \in \mathbb{R}$,

$$\begin{aligned} |f(x,y_1) - f(x,y_2)| &= \frac{e^{-2x}}{1+e^x}\left|\frac{|y_1|}{1+|y_1|} - \frac{|y_2|}{1+|y_2|}\right| \leq \frac{e^{-2x}|y_1 - y_2|}{(1+e^x)(1+|y_1|)(1+|y_2|)} \\ &\leq \frac{e^{-2x}}{(1+e^x)}|y_1 - y_2| \leq \frac{e^{-2x}}{2}|y_1 - y_2| \leq \frac{1}{2}|y_1 - y_2|. \end{aligned}$$

For all $x \in [0,2]$ *and* $y \in \mathbb{R}$,

$$|f(x,y)| = \frac{e^{-2x}}{1+e^x}\frac{|y|}{1+|y|} \leq \frac{e^{-2x}}{1+e^x}|y| \leq \frac{e^{-2x}}{2}|y| \leq \frac{1}{2}|y| \leq \frac{1}{2}(1+|y|).$$

Thus, $[A2]$ *and* $[A3]$ *hold.*

Set $L = \frac{1}{2} = k_f$. *Then* $a_\alpha k_f = \frac{24}{25} \times \frac{1}{2} = \frac{12}{25} < 1$. *From Theorem 3,* (21) *has an unique solution. Thus, all the assumptions in Theorem 4 are satisfied, so our results can be applied to (21).*

Let $g \in C([0,2],\mathbb{R})$ *be a solution of* (22). *We have*

$$\left|({}^{CF}\mathbb{D}^{\frac{1}{3}}g)(x) - f(x,g(x))\right| = \left|({}^{CF}\mathbb{D}^{\frac{1}{3}}g)(x) - \frac{e^{-2x}}{1+e^x}\frac{|g|}{1+|g|}\right| \leq G(x),\ x \in [0,2]. \tag{23}$$

From Theorem 3, we see (21) *with* $y(0) = C_0$ *has the unique solution*

$$\begin{aligned} y(x) &= C_0 + a_{\frac{1}{3}}f(x,y(x)) + b_{\frac{1}{3}}\int_0^x f(s,y(s))ds \\ &= C_0 + \frac{24}{25}\frac{e^{-2x}}{1+e^x}\frac{|y|}{1+|y|} + \frac{12}{25}\int_0^x \frac{e^{-2s}}{1+e^s}\frac{|y|}{1+|y|}ds. \end{aligned}$$

Applying the fractional integrating operator ${}^{CF}I^\alpha(\cdot)$ *on both sides of* (23), *we have*

$$\begin{aligned} &\left|g(x) - C_0 - a_{\frac{1}{3}}f(x,g(x)) - b_{\frac{1}{3}}\int_0^x f(s,g(s))ds\right| \\ \leq\ & a_{\frac{1}{3}}G(x) + b_{\frac{1}{3}}\int_0^x G(s)ds \\ \leq\ & (a_{\frac{1}{3}} + b_{\frac{1}{3}}\lambda_G)G(x),\ x \in [0,2]. \end{aligned}$$

In addition,

$$|y(x) - g(x)| \leq \left[\frac{(a_{\frac{1}{3}} + b_{\frac{1}{3}}\lambda_G)}{1 - a_{\frac{1}{3}}k_f}\exp(x)\right]G(x),\ x \in [0,2].$$

Set $K^* = \frac{a_{\frac{1}{3}} + b_{\frac{1}{3}}\lambda_G}{1 - a_{\frac{1}{3}}k_f}\exp(2) = \frac{\frac{24}{25} + \frac{12}{25}\times 1}{1 - \frac{24}{25}\times\frac{1}{2}}e^2 = \frac{36e^2}{13}$. *Note that one has*

$$|y(x) - g(x)| \leq K^* G(x),\ x \in [0,2].$$

6. Conclusions

By applying the well-known Gronwall inequality and fixed point theorems, we obtain the Hyers–Ulam stability of linear and semilinear Caputo–Fabrizio fractional differential equations. Existence and uniqueness theorems of solution are established. In a forthcoming work, we shall consider the impulsive Cauchy problem with Caputo–Fabrizio fractional derivative.

Author Contributions: The contributions of all authors (K.L., M.F., D.O. and J.W.) are equal. All the main results and examples were developed together.

Acknowledgments: The authors thank the referees for their careful reading of the article and insightful comments.

References

1. Podlubny, I. *Fractional Differential Equations, Mathematics in Science and Engineering*; Academic Press: San Diego, CA, USA, 1999; Volume 198.
2. Kilbas, A.A.; Srivastava, H.M.; Trujillo, J.J. *Theory and Applications of Fractional Differential Equations*; Elsevier: Amsterdam, The Netherlands, 2006.
3. Tarasov, V.E. *Fractional Dynamics: Application of Fractional Calculuts to Dynamics of Particles, Fields and Media*; Springer: Berlin/Heidelberg, Germany, 2011.
4. Abbas, S.; Benchohra, M.; Darwish, M.A. New stability results for partial fractional differential inclusions with not instantaneous impulses. *Fract. Calc. Appl. Anal.* **2015**, *18*, 172–191. [CrossRef]
5. Li, M.; Wang, J. Exploring delayed Mittag-Leffler type matrix functions to study finite time stability of fractional delay differential equations. *Appl. Math. Comput.* **2018**, *324*, 254–265. [CrossRef]
6. Liu, S.; Wang, J.; Zhou, Y.; Fečkan, M. Iterative learning control with pulse compensation for fractional differential equations. *Math. Slov.* **2018**, *68*, 563–574. [CrossRef]
7. Wang, J.; Ibrahim, A.G.; O'Regan, D. Topological structure of the solution set for fractional non-instantaneous impulsive evolution inclusions. *J. Fixed Point Theory Appl.* **2018**, *20*, 59. [CrossRef]
8. Luo, D.; Wang, J.; Shen, D. Learning formation control for fractional-order multi-agent systems. *Math. Meth. Appl. Sci.* **2018**, *41*, 5003–5014. [CrossRef]
9. Peng, S.; Wang, J.; Yu, X. Stable manifolds for some fractional differential equations. *Nonlinear Anal. Model. Control* **2018**, *23*, 642–663. [CrossRef]
10. Chen, Y.; Wang, J. Continuous dependence of solutions of integer and fractional order non-instantaneous impulsive equations with random impulsive and junction points. *Mathematics* **2019**, *7*, 331. [CrossRef]
11. Zhang, J.; Wang, J. Numerical analysis for a class of Navier–Stokes equations with time fractional derivatives. *Appl. Math. Comput.* **2018**, *336*, 481–489
12. Zhu, B.; Liu, L.; Wu, Y. Local and global existence of mild solutions for a class of nonlinear fractional reaction-diffusion equation with delay. *Appl. Math. Lett.* **2016**, *61*, 73–79. [CrossRef]
13. Wang, Y.; Liu, L.; Wu, Y. Positive solutions for a nonlocal fractional differential equation. *Nonlinear Anal.* **2011**, *74*, 3599–3605. [CrossRef]
14. Zhang, X.; Liu, L.; Wu, Y. Existence results for multiple positive solutions of nonlinear higher order perturbed fractional differential equations with derivatives. *Appl. Math. Comput.* **2012**, *219*, 1420–1433. [CrossRef]
15. Wang, Y.; Liu, L.; Zhang, X.; Wu, Y. Positive solutions of a fractional semipositone differential system arising from the study of HIV infection models. *Appl. Math. Comput.* **2015**, *258*, 312–324.
16. Zhang, X.; Liu, L.; Wu, Y. Variational structure and multiple solutions for a fractional advection-dispersion equation. *Comput. Math. Appl.* **2014**, *68*, 1794–1805. [CrossRef]
17. Zhang, X.; Mao, C.; Liu, L.; Wu, Y. Exact iterative solution for an abstract fractional dynamic system model for bioprocess. *Qual. Theory Dyn. Syst.* **2017**, *16*, 205–222. [CrossRef]
18. Zhang, X.; Liu, L.; Wu, Y.; Wiwatanapataphee, B. Nontrivial solutions for a fractional advection dispersion equation in anomalous diffusion. *Appl. Math. Lett.* **2017**, *66*, 1–8. [CrossRef]
19. Jiang, J.; Liu, L.; Wu, Y. Multiple positive solutions of singular fractional differential system involving Stieltjes integral conditions. *Electron. J. Qual. Theory Differ. Equ.* **2012**, *43*, 1–18. [CrossRef]
20. Caputo, M.; Fabrizio, M. A new definition of fractional derivative without singular kernel. *Prog. Fract. Differ. Appl.* **2015**, *1*, 73–85.
21. Atangana, A.; Nieto, J.J. Numerical solution for the model of RLC circuit via the fractional derivative without singular kernel. *Adv. Mech. Eng.* **2015**, *7*, 1–7. [CrossRef]
22. Losada, J.; Nieto, J.J. Properties of a new fractional derivative without singular kernel. *Prog. Fract. Differ. Appl.* **2015**, *1*, 87–92.
23. Baleanu, D.; Mousalou, A.; Rezapour, S. On the existence of solutions for some infinite coefficient-symmetric

Caputo–Fabrizio fractional integro-differential equations. *Bound. Value Prob.* **2017**, *2017*, 1–9. [CrossRef]

24. Franc, E.; Goufo, D. Application of the Caputo–Fabrizio fractional derivative without singular kernel to Korteweg–de Vries–Burgers equations. *Math. Model. Anal.* **2016**, *21*, 188–198.
25. Rezaei, H.; Jung, S.M.; Rassias, T.M. Laplace transform and Hyers–Ulam stability of linear differential equations. *J. Math. Anal. Appl.* **2013**, *403*, 244–251. [CrossRef]
26. Alqifiary, Q.H.; Jung, S.M. Laplace transform and generalized Hyers–Ulam stability of linear differential equations. *Electron. J. Diff. Equ.* **2014**, *2014*, 1–11.
27. Wang, J.; Li, X.Z. A uniform method to Ulam-Hyers stability for some Linear fractional equations. *Mediterr. J. Math.* **2016**, *13*, 625–635. [CrossRef]
28. Wang, J.; Zhang, Y. Ulam-Hyers-Mittag-Leffler stability of fractional-order delay differential equations. *Optimization* **2014**, *63*, 1181–1190. [CrossRef]
29. Capelas de Oliveira, E.; da C. Sousa, J.V. Ulam-Hyers-Rassias stability for a class of fractional integro-differential equations. *Result Math.* **2018**, *73*, 111. [CrossRef]
30. da C. Sousa, J.V.; Capelas de Oliveira, E. Ulam-Hyers stability of a nonlinear fractional Volterra integro-differential equation. *Appl. Math. Lett.* **2018**, *81*, 50–56.
31. da C. Sousa, J.V.; Kucche, K.D.; Capelas de Oliveira, E. Stability of ψ-Hilfer impulsive fractional differential equations. *Appl. Math. Lett.* **2018**, *88*, 73–80.
32. Wang, J.; Zhou, Y.; Fečkan, M. Nonlinear impulsive problems for fractional differential equations and Ulam stability. *Comput. Math. Appl.* **2012**, *64*, 3389–3405. [CrossRef]
33. da C. Sousa, J.V.; Capelas de Oliveira, E. On the Ulam-Hyers-Rassias stability for nonlinear fractional differential equations using the ψ-Hilfer operator. *J. Fixed Point Theory Appl.* **2018**, *20*, 5–21.
34. Shah, K.; Ali, A.; Bushnaq, S. Hyers–Ulam stability analysis to implicit Cauchy problem of fractional differential equations with impulsive conditions. *Math. Meth. Appl. Sci.* **2018**, *41*, 8329–8343. [CrossRef]
35. Ali, Z.; Zada, A.; Shah, K. Ulam stability to a toppled systems of nonlinear implicit fractional order boundary value problem. *Bound. Value Prob.* **2018**, *2018*, 175. [CrossRef]
36. Liu, K.; Wang, J.; O'Regan, D. Ulam-Hyers-Mittag-Leffler stability for ψ-Hilfer fractional-order delay differential equations. *Adv. Differ. Equ.* **2019**, *2019*, 50. [CrossRef]
37. Wang, J.; Lv, L.; Zhou, Y. Ulam stability and data depenaence for fractional differential equations with Caputo derivative. *Electron. J. Qual. Theory Differ. Equ.* **2011**, *63*, 1–10.
38. Wang, J.; Zhou, Y.; Fečkan, M. Abstract Cauchy problem for fractional differential equations. *Nonlinear Dyn.* **2013**, *71*, 685–700. [CrossRef]

3

On Separate Fractional Sum-Difference Equations with n-Point Fractional Sum-Difference Boundary Conditions via Arbitrary Different Fractional Orders

Saowaluck Chasreechai [1] and Thanin Sitthiwirattham [2,*]

1 Department of Mathematics, Faculty of Applied Science, King Mongkut's University of Technology North Bangkok, Bangkok 10800, Thailand; saowaluck.c@sci.kmutnb.ac.th
2 Mathematics Department, Faculty of Science and Technology, Suan Dusit University, Bangkok 10700, Thailand
* Correspondence: thanin_sit@dusit.ac.th

Abstract: In this article, we study the existence and uniqueness results for a separate nonlinear Caputo fractional sum-difference equation with fractional difference boundary conditions by using the Banach contraction principle and the Schauder's fixed point theorem. Our problem contains two nonlinear functions involving fractional difference and fractional sum. Moreover, our problem contains different orders in $n+1$ fractional differences and $m+1$ fractional sums. Finally, we present an illustrative example.

Keywords: fractional sum-difference equations; boundary value problem; existence; uniqueness

---

1. Introduction

Fractional calculus has recently been an attractive field to researchers because it is a powerful tool for explaining many engineering and scientific disciplines as the mathematical modeling of systems and processes which appear in nature, for example, ecology, biology, chemistry, physics, mechanics, networks, flow in porous media, electrical, control systems, viscoelasticity, mathematical biology, fitting of experimental data, and so forth. For example, Zhang et al. [1] proposed both analytical and numerical results from studying the propagation of optical beams in the fractional Schrödinger equation with a harmonic potential. In 2015, Zingales and Failla [2] solved the fractional-order heat conduction equation by using a pertinent finite element method. For Lazopoulos's [3] work, they defined the fractional curvature of plane curves, the fractional beam small deflection, the fractional curvature is approximate. In 2017, Sumelka and Voyiadjis [4] proposed a concept of short memory connected with the definition of damage parameter evolution in terms of fractional calculus for hyperelastic materials.

Basic definitions and properties of fractional difference calculus, appear in the book [5]. In particular, fractional calculus is a powerful tool for the processes which appear in nature, e.g., ecology, biology and other areas, one may see the papers [6–8] and the references therein. The interesting papers related to discrete fractional boundary value problems can be found in [9–29] and references cited therein. For previous works, Goodrich [10] considered the discrete fractional boundary value problem

$$\begin{cases} -\Delta^{\mu_1}\Delta^{\mu_2}\Delta^{\mu_3}y(t) = f(t+\mu_1+\mu_2+\mu_3-1, y(t+\mu_1+\mu_2+\mu_3-1)), \\ y(0) = 0 = y(b+2), \end{cases} \tag{1}$$

where $t \in \mathbb{N}_{2-\mu_1-\mu_2-\mu_3, b+2-\mu_1-\mu_2-\mu_3}$, $0 < \mu_1, \mu_2, \mu_3 < 1$, $1 < \mu_2 + \mu_3 < 2$, $1 < \mu_1 + \mu_2 + \mu_3 < 2$, $f : \mathbb{N}_0 \times \mathbb{R} \to [0, +\infty)$ is a continuous function, and Δ^{μ} is the Riemann-Liouville fractional difference operator of order μ. Existence of positive solutions are obtained by the use of the Krasnosel'skii fixed point theorem.

Weidong [12] examined the sequential fractional boundary value problem with a p-Laplacian

$$\begin{cases} \Delta_C^{\beta}[\phi_p(\Delta_C^{\alpha} x)](t) = f(t+\alpha+\beta-1, x(t+\alpha+\beta-1)), & t \in \mathbb{N}_{0,b}, \\ \Delta_C^{\beta} x(\beta-1) + \Delta_C^{\beta} x(\beta+b) = 0, \\ x(\alpha+\beta-2) + x(\alpha+\beta+b) = 0, \end{cases} \tag{2}$$

where $0 < \alpha, \beta \leq 1$, $1 < \alpha + \beta \leq 2$, $f : \mathbb{N}_{\alpha+\beta-1, \alpha+\beta+T-1} \times \mathbb{R} \to \mathbb{R}$ is a continuous function, ϕ_p is the p-Laplacian operator, and Δ_C^{β} is the Caputo fractional difference operator of order β. Existence and uniqueness of solutions are obtained by using the Schaefer's fixed point theorem.

Recently, Sitthiwirattham [19,20] investigated three-point fractional sum boundary value problems for sequential fractional difference equations of the forms

$$\begin{cases} \Delta_C^{\alpha}[\phi_p(\Delta_C^{\beta} x)](t) = f(t+\alpha+\beta-1, x(t+\alpha+\beta-1)), \\ \Delta_C^{\beta} x(\alpha-1) = 0, \quad x(\alpha+\beta+T) = \rho\Delta^{-\gamma} x(\eta+\gamma), \end{cases} \tag{3}$$

and

$$\begin{cases} \Delta_{\alpha}^{\alpha}(\Delta_{\alpha+\beta-1}^{\beta} + \lambda E_{\beta}) x(t) = f(t+\alpha+\beta-1, x(t+\alpha+\beta-1)), \\ x(\alpha+\beta-2) = 0, \quad x(\alpha+\beta+T) = \rho\Delta_{\alpha+\beta-1}^{-\gamma} x(\eta+\gamma), \end{cases} \tag{4}$$

where $t \in \mathbb{N}_{0,T}$, $0 < \alpha, \beta \leq 1$, $1 < \alpha+\beta \leq 2$, $0 < \gamma \leq 1$, $\eta \in \mathbb{N}_{\alpha+\beta-1, \alpha+\beta+T-1}$, ρ is a constant, $f : \mathbb{N}_{\alpha+\beta-2, \alpha+\beta+T} \times \mathbb{R} \to \mathbb{R}$ is a continuous function, $E_{\beta} x(t) = x(t+\beta-1)$ and ϕ_p is the p-Laplacian operator. Existence and uniqueness of solutions are obtained by using the Banach fixed point theorem and the Schaefer's fixed point theorem.

The results mentioned above are the motivation for this research. In this paper, we consider a separate nonlinear Caputo fractional sum-difference equation of the form

$$\begin{aligned} \left[\Delta_C^{\alpha} + (e+1)\,\Delta_C^{\alpha-1}\right] u(t) &= \lambda F\Big(t+\alpha-1, u(t+\alpha-1), (\Upsilon^{\vartheta} u)(t+\alpha-\vartheta)\Big) \\ &\quad + \mu H\Big(t+\alpha-1, u(t+\alpha-1), (\Psi^{\gamma} u)(t+\alpha+\gamma-1)\Big), \end{aligned} \tag{5}$$

with the fractional sum-difference boundary value conditions

$$\begin{aligned} u(\alpha-n) &= \Delta_C^{\beta_1} u\,(\alpha-n-\beta_1+2) = \Delta_C^{\beta_1+\beta_2} u\,(\alpha-n-\beta_1-\beta_2+4) = \ldots \\ &= \Delta_C^{\sum_{i=1}^{n-2}\beta_i} u\left(\alpha+n-4-\sum_{i=1}^{n-2}\beta_i\right) = 0, \\ u(T+\alpha) &= \tau\,\Delta^{-\sum_{i=1}^{m}\theta_i} g\left(\eta+\sum_{i=1}^{m}\theta_i\right) u\left(\eta+\sum_{i=1}^{m}\theta_i\right), \end{aligned} \tag{6}$$

where $t \in \mathbb{N}_{0,T} := \{0, 1, \ldots, T\}$, $\tau < \frac{\Gamma\left(\sum_{i=1}^{m}\theta_i\right)\sum_{s=\alpha-n}^{T+\alpha-n+1} e^{-s}}{\sum_{r=\alpha-n}^{\eta}\sum_{s=\alpha-n}^{r-n+1}\left(\eta+\sum_{i=1}^{m}\theta_i-\sigma(r)\right)^{\underline{\sum_{i=1}^{m}\theta_i-1}} e^{-s} g\left(\eta+\sum_{i=1}^{m}\theta_i\right)}$, $\alpha \in (n-1, n]$, $\beta_i, \theta_i, \gamma \in (0, 1]$, $m, n \in \mathbb{N}_4$, $m < n$, $T > n-3$, $\sum_{i=1}^{n-2}\beta_i \in (n-3, n-2]$, $\sum_{i=1}^{m}\theta_i \in (m-1, m]$ and $\lambda, \mu \in \mathbb{R}$ are given constants; $F \in C(\mathbb{N}_{\alpha-n, T+\alpha} \times \mathbb{R} \times \mathbb{R}, \mathbb{R})$, $H \in C(\mathbb{N}_{\alpha-n, T+\alpha} \times \mathbb{R} \times \mathbb{R}, \mathbb{R})$, $g \in C(\mathbb{N}_{\alpha-n, T+\alpha}, \mathbb{R}^{+})$, and for $\varphi, \phi \in C(\mathbb{N}_{\alpha-n, T+\alpha} \times \mathbb{N}_{\alpha-n, T+\alpha}, [0, \infty))$, we defined the operators

$$
\begin{aligned}
(\Upsilon^{\vartheta}u)(t-\vartheta+1) &:= [\Delta_C^{\vartheta}\phi\, u](t-\vartheta+1)\\
&= \frac{1}{\Gamma(1-\vartheta)}\sum_{s=\alpha-n+\vartheta-1}^{t+\vartheta-1}(t-\sigma(s))^{\underline{-\vartheta}}\phi(t,s-\vartheta+1)\Delta u(s-\vartheta+1),
\end{aligned}
$$

$$
\text{and } (\Psi^{\gamma}u)(t+\gamma) := [\Delta^{-\gamma}\varphi\, u](t+\gamma) = \frac{1}{\Gamma(\gamma)}\sum_{s=\alpha-n-\gamma}^{t-\gamma}(t-\sigma(s))^{\underline{\gamma-1}}\varphi(t,s+\gamma)u(s+\gamma).
$$

The plan of this paper is as follows. In Section 2 we recall some definitions and basic lemmas. We derive a representation for the solution of (5) by converting the problem to an equivalent summation equation. In Section 3, we prove existence results of the problem (5) by using the Banach contraction principle and the Schauder's theorem. Finally, an illustrative example is presented in Section 4.

2. Preliminaries

The notations, definitions, and lemmas which are used in the main results are as follows.

Definition 1. *We define the generalized falling function by* $t^{\underline{\alpha}} := \frac{\Gamma(t+1)}{\Gamma(t+1-\alpha)}$, *for any t and α for which the right-hand side is defined. If $t+1-\alpha$ is a pole of the Gamma function and $t+1$ is not a pole, then $t^{\underline{\alpha}} = 0$.*

Lemma 1 ([16]). *Assume the factorial functions are well defined. If $t \le r$, then $t^{\underline{\alpha}} \le r^{\underline{\alpha}}$ for any $\alpha > 0$.*

Definition 2. *For $\alpha > 0$ and f defined on $\mathbb{N}_a := \{a, a+1, \ldots\}$, the α-order fractional sum of f is defined by*

$$
\Delta^{-\alpha}f(t) := \frac{1}{\Gamma(\alpha)}\sum_{s=a}^{t-\alpha}(t-\sigma(s))^{\underline{\alpha-1}}f(s),
$$

where $t \in \mathbb{N}_{a+\alpha}$ and $\sigma(s) = s+1$.

Definition 3. *For $\alpha > 0$ and f defined on $\mathbb{N}_a$, the α-order Caputo fractional difference of f is defined by*

$$
\Delta_C^{\alpha}f(t) := \Delta^{-(N-\alpha)}\Delta^N f(t) = \frac{1}{\Gamma(N-\alpha)}\sum_{s=a}^{t-(N-\alpha)}(t-\sigma(s))^{\underline{N-\alpha-1}}\Delta^N f(s),
$$

where $t \in \mathbb{N}_{a+N-\alpha}$ and $N \in \mathbb{N}$ is chosen so that $0 \le N-1 < \alpha < N$.

Lemma 2 ([14]). *Assume that $\alpha > 0$ and $0 \le N-1 < \alpha \le N$. Then*

$$
\Delta^{-\alpha}\Delta_C^{\alpha}y(t) = y(t) + C_0 + C_1 t^{\underline{1}} + C_2 t^{\underline{2}} + \ldots + C_{N-1}t^{\underline{N-1}},
$$

for some $C_i \in \mathbb{R}$, $0 \le i \le N-1$.

To investigate the solution of the boundary value problem (5) we need the following lemma involving a linear variant of the boundary value problem (5).

Lemma 3. *Let* $\tau < \frac{\Gamma(\sum_{i=1}^{m}\theta_i)\sum_{s=\alpha-n}^{T+\alpha-n+1}e^{-s}}{\sum_{r=\alpha-n}^{\eta}\sum_{s=\alpha-n}^{r-n+1}(\eta+\sum_{i=1}^{m}\theta_i-\sigma(r))^{\underline{\sum_{i=1}^{m}\theta_i-1}}e^{-s}g(\eta+\sum_{i=1}^{m}\theta_i)}$, $\alpha \in (n-1,n]$, $\beta_i, \theta_i, \gamma \in (0,1]$, *$m, n \in \mathbb{N}_4$, $m < n$, $T > n-3$, $\sum_{i=1}^{n-2}\beta_i \in (n-3, n-2]$, $\sum_{i=1}^{m}\theta_i \in (m-1, m]$ and $h \in C(\mathbb{N}_{\alpha-n,T+\alpha} \times \mathbb{R}, \mathbb{R})$, $g \in C(\mathbb{N}_{\alpha-n,T+\alpha}, \mathbb{R}^+)$ be given. Then the problem*

$$\left[\Delta_C^{\alpha} + (e-1)\,\Delta_C^{\alpha-1}\right] u(t) = h(t+\alpha-1), \quad t \in \mathbb{N}_{0,T} \tag{7}$$

$$\begin{aligned} u(\alpha-n) &= \Delta_C^{\beta_1} u\,(\alpha-n-\beta_1+2) \;=\; \Delta_C^{\beta_1+\beta_2} u\,(\alpha-n-\beta_1-\beta_2+4) \\ &= \dots = \Delta_C^{\sum_{i=1}^{n-2}\beta_i} u\left(\alpha+n-4-\sum_{i=1}^{n-2}\beta_i\right) = 0, \end{aligned} \tag{8}$$

$$u(T+\alpha) = \tau\,\Delta^{-\sum_{i=1}^{m}\theta_i} g\left(\eta+\sum_{i=1}^{m}\theta_i\right) u\left(\eta+\sum_{i=1}^{m}\theta_i\right), \tag{9}$$

has the unique solution

$$\begin{aligned} u(t) &= \frac{\mathcal{O}[h]}{\Lambda}\sum_{s=\alpha-n}^{t-n+1} e^{-s} \\ &+\frac{1}{\Gamma(\alpha-n)}\sum_{s=\alpha-n}^{t-n+1}\sum_{v=\alpha-n}^{s-1}\sum_{\xi=0}^{v-\alpha+n} e^{v-s}(v-\sigma(\xi))^{\underline{\alpha-n-1}}h(\xi+\alpha-1), \end{aligned} \tag{10}$$

where the functional $\mathcal{O}[h]$ and the constant Λ are defined by

$$\begin{aligned} \mathcal{O}[h] &= \sum_{r=\alpha-n}^{\eta}\sum_{s=\alpha-n}^{r-n+1}\sum_{v=\alpha-n}^{s-1}\sum_{\xi=0}^{v-\alpha+n}\frac{\tau\left(\eta+\sum_{i=1}^{m}\theta_i-\sigma(r)\right)^{\underline{\sum_{i=1}^{m}\theta_i-1}}}{\Gamma\left(\sum_{i=1}^{m}\theta_i\right)\Gamma(\alpha-n)}\times \\ &\quad e^{v-s}(v-\sigma(\xi))^{\underline{\alpha-n-1}} g\left(\eta+\sum_{i=1}^{m}\theta_i\right) h(\xi+\alpha-1) \\ &\quad -\frac{1}{\Gamma(\alpha-n)}\sum_{s=\alpha-n}^{T+\alpha-n+1}\sum_{v=\alpha-n}^{s-1}\sum_{\xi=0}^{v-\alpha+n} e^{v-s}(v-\sigma(x))^{\underline{\alpha-n-1}}h(x+\alpha-1), \end{aligned} \tag{11}$$

$$\begin{aligned} \Lambda &= \sum_{s=\alpha-n}^{T+\alpha-n+1} e^{-s} - \sum_{r=\alpha-n}^{\eta}\sum_{s=\alpha-n}^{r-n+1}\frac{\tau\left(\eta+\sum_{i=1}^{m}\theta_i-\sigma(r)\right)^{\underline{\sum_{i=1}^{m}\theta_i-1}}}{\Gamma\left(\sum_{i=1}^{m}\theta_i\right)}\times \\ &\quad e^{-s} g\left(\eta+\sum_{i=1}^{m}\theta_i\right), \textit{ respectively.} \end{aligned} \tag{12}$$

Proof. Using the fractional sum of order $\alpha : \Delta^{-\alpha}$ for (7), we obtain

$$\begin{aligned} u(t)+(e-1)\,\Delta^{-1}u(t) &= C_1 + C_2 t^{\underline{1}} + C_3 t^{\underline{2}} + \dots + C_n t^{\underline{n-1}} \\ &\quad +\frac{1}{\Gamma(\alpha)}\sum_{s=0}^{t-\alpha}(t-\sigma(s))^{\underline{\alpha-1}}h(s+\alpha-1), \quad t\in\mathbb{N}_{\alpha-n,T+\alpha}. \end{aligned} \tag{13}$$

For the forward difference of order $n : \Delta^n$ for (13), we have

$$\Delta^n u(t)+(e-1)\,\Delta^{n-1}u(t) = \frac{1}{\Gamma(\alpha-n)}\sum_{s=0}^{t-\alpha+n}(t-\sigma(s))^{\underline{\alpha-n-1}}h(s+\alpha-1).$$

Therefore,

$$\Delta\left[e^t\,\Delta^{n-1}u(t)\right] = \frac{e^t}{\Gamma(\alpha-n)}\sum_{s=0}^{t-\alpha+n}(t-\sigma(s))^{\underline{\alpha-n-1}}h(s+\alpha-1). \tag{14}$$

Taking the sum: Δ^{-1} to (14), we get

$$e^t\,\Delta^{n-1}u(t) = C_n + \frac{1}{\Gamma(\alpha-n)}\sum_{s=\alpha-n}^{t-1}\sum_{v=0}^{s-\alpha+n} e^s(s-\sigma(v))^{\underline{\alpha-n-1}}h(v+\alpha-1). \tag{15}$$

Next, taking the sum of order $n-1 : \Delta^{-(n-1)}$ to (15), we obtain

$$\begin{aligned} u(t) &= C_1 + C_2t^{\underline{1}} + C_3t^{\underline{2}} + \ldots + C_{n-1}t^{\underline{n-2}} + C_n\sum_{s=\alpha-n}^{t-n+1} e^{-s} \\ &+ \frac{1}{\Gamma(\alpha-n)}\sum_{s=\alpha-n}^{t-n+1}\sum_{v=\alpha-n}^{s-1}\sum_{x=0}^{v-\alpha+n} e^{v-s}(v-\sigma(x))^{\underline{\alpha-n-1}}h(x+\alpha-1),\ \ t\in\mathbb{N}_{\alpha-n,T+\alpha}. \end{aligned} \tag{16}$$

Using the Caputo fractional differences of order β_i for (16) where $i=1$ to $i=n-2$, we obtain

$$\begin{aligned} &\Delta_C^{\beta_1}u(t) \\ &= C_2\Delta_C^{\beta_1}t^{\underline{1}} + C_3\Delta_C^{\beta_1}t^{\underline{2}} + \ldots + C_{n-1}\Delta_C^{\beta_1}t^{\underline{n-2}} \\ &+C_n\sum_{r=\alpha-n}^{t+\beta_1-1}\frac{(t-\sigma(r))^{\underline{-\beta_1}}}{\Gamma(1-\beta_1)}\Delta_r\left\{\sum_{s=\alpha-n}^{r-n+1}e^{-s}\right\} \\ &+\sum_{r=\alpha-n}^{t+\beta_1-1}\frac{(t-\sigma(r))^{\underline{-\beta_1}}}{\Gamma(1-\beta_1)}\Delta_r\left\{\sum_{s=\alpha-n}^{r-n+1}\sum_{v=\alpha-n}^{s-1}\sum_{x=0}^{v-\alpha+n}e^{v-s}\frac{(v-\sigma(x))^{\underline{\alpha-n-1}}}{\Gamma(\alpha-n)}h(x+\alpha-1)\right\}, \end{aligned} \tag{17}$$

for $t\in\mathbb{N}_{\alpha-n+1-\beta_1,T+\alpha+1-\beta_1}$.

$$\begin{aligned} &\Delta_C^{\beta_1+\beta_2}u(t) \\ &= C_3\Delta_C^{\beta_1+\beta_2}t^{\underline{2}} + \ldots + C_{n-1}\Delta_C^{\beta_1+\beta_2}t^{\underline{n-2}} \\ &+C_n\sum_{r=\alpha-n}^{t+\beta_1+\beta_2-2}\frac{(t-\sigma(r))^{\underline{1-\beta_1-\beta_2}}}{\Gamma(2-\beta_1-\beta_2)}\Delta_r^2\left\{\sum_{s=\alpha-n}^{r-n+1}e^{-s}\right\} \\ &+\sum_{r=\alpha-n}^{t+\beta_1+\beta_2-2}\frac{(t-\sigma(r))^{\underline{1-\beta_1-\beta_2}}}{\Gamma(2-\beta_1-\beta_2)}\Delta_r^2\left\{\sum_{s=\alpha-n}^{r-n+1}\sum_{v=\alpha-n}^{s-1}\sum_{x=0}^{v-\alpha+n}e^{v-s}\frac{(v-\sigma(x))^{\underline{\alpha-n-1}}}{\Gamma(\alpha-n)}h(x+\alpha-1)\right\}, \end{aligned} \tag{18}$$

for $t\in\mathbb{N}_{\alpha-n+2-\beta_1-\beta_2,T+\alpha+2-\beta_1-\beta_2}$.

$$\vdots$$

$$\begin{aligned} &\Delta_C^{\sum_{i=1}^{n-2}\beta_i}u(t) \\ &= C_{n-1}\Delta_C^{\sum_{i=1}^{n-2}\beta_i}t^{\underline{n-2}} + C_n\sum_{r=\alpha-n}^{t-n+2+\sum_{i=1}^{n-2}\beta_i}\frac{(t-\sigma(r))^{\underline{n-3-\sum_{i=1}^{n-2}\beta_i}}}{\Gamma\left(n-2-\sum_{i=1}^{n-2}\beta_i\right)}\Delta_r^{n-2}\left\{\sum_{s=\alpha-n}^{r-n+1}e^{-s}\right\} \\ &+\sum_{r=\alpha-n}^{t-n+2+\sum_{i=1}^{n-2}\beta_i}\frac{(t-\sigma(r))^{\underline{n-3-\sum_{i=1}^{n-2}\beta_i}}}{\Gamma\left(n-2-\sum_{i=1}^{n-2}\beta_i\right)}\times \\ &\Delta_r^{n-2}\left\{\sum_{s=\alpha-n}^{r-n+1}\sum_{v=\alpha-n}^{s-1}\sum_{x=0}^{v-\alpha+n}e^{v-s}\frac{(v-\sigma(x))^{\underline{\alpha-n-1}}}{\Gamma(\alpha-n)}h(x+\alpha-1)\right\}, \end{aligned} \tag{19}$$

for $t \in \mathbb{N}_{\alpha-2-\sum_{i=1}^{n-2}\beta_i, T+\alpha+n-2-\sum_{i=1}^{n-2}\beta_i}$.

Employing the conditions of (8), we have the system of $n-1$ equations

$$
\begin{aligned}
&(E_1) \quad C_1 + C_2(\alpha-n) + C_3(\alpha-n)^{\underline{2}} + \ldots + C_{n-1}(\alpha-n)^{\underline{n-2}} = 0,\\
&(E_2) \quad C_2\Delta_C^{\beta_1}(\alpha-n+2-\beta_1) + \ldots + C_{n-1}\Delta_C^{\beta_1}(\alpha-n+2-\beta_1)^{\underline{n-2}} = 0,\\
&(E_3) \quad C_3\Delta_C^{\beta_1+\beta_2}(\alpha-n+4-\beta_1-\beta_2)^{\underline{2}} + \ldots + C_{n-1}\Delta_C^{\beta_1+\beta_2}(\alpha-n+4-\beta_1-\beta_2)^{\underline{n-2}} = 0,\\
&\cdots\\
&(E_{n-1}) \quad C_{n-1}\Delta_C^{\sum_{i=1}^{n-2}\beta_i}\left(\alpha+n-4-\sum_{i=1}^{n-2}\beta_i\right)^{\underline{n-2}} = 0.
\end{aligned}
\tag{20}
$$

Using the fractional sum of order $\sum_{i=1}^{m}\theta_i$ for (16), we have

$$
\begin{aligned}
&\Delta_C^{-\sum_{i=1}^{m}\theta_i}u(t)\\
&= \sum_{r=\alpha-n}^{t-\sum_{i=1}^{m}\theta_i} \frac{(t-\sigma(r))^{\underline{\sum_{i=1}^{m}\theta_i-1}}}{\Gamma(\sum_{i=1}^{m}\theta_i)}\left[C_1 + C_2\Delta_C^{\beta_1}s^{\underline{1}} + C_3\Delta_C^{\beta_1}s^{\underline{2}} + \ldots + C_{n-1}\Delta_C^{\beta_1}s^{\underline{n-2}}\right]\\
&+C_n\sum_{r=\alpha-n}^{t-\sum_{i=1}^{m}\theta_i}\sum_{s=\alpha-n}^{r-n+1} \frac{(t-\sigma(r))^{\underline{\sum_{i=1}^{m}\theta_i-1}}}{\Gamma(\sum_{i=1}^{m}\theta_i)}e^{-s}\\
&+\sum_{r=\alpha-n}^{t-\sum_{i=1}^{m}\theta_i}\sum_{s=\alpha-n}^{r-n+1}\sum_{v=\alpha-n}^{s-1}\sum_{x=0}^{v-\alpha+n} \frac{(t-\sigma(r))^{\underline{\sum_{i=1}^{m}\theta_i-1}}}{\Gamma(\sum_{i=1}^{m}\theta_i)}\frac{(v-\sigma(x))^{\underline{\alpha-n-1}}}{\Gamma(\alpha-n)}e^{v-s}h(x+\alpha-1),
\end{aligned}
\tag{21}
$$

for $t \in \mathbb{N}_{\alpha-n+\sum_{i=1}^{m}\theta_i, T+\alpha+\sum_{i=1}^{m}\theta_i}$.

By substituting $t = \eta + \sum_{i=1}^{m}\theta_i$ into (21) and using the second condition of (9), we finally get

$$
\begin{aligned}
(E_n) \quad & C_1\left\{1 - \sum_{r=\alpha-n}^{\eta}\frac{\tau(\eta+\sum_{i=1}^{m}\theta_i-\sigma(r))^{\underline{\sum_{i=1}^{m}\theta_i-1}}}{\Gamma(\sum_{i=1}^{m}\theta_i)}g\left(\eta+\sum_{i=1}^{m}\theta_i\right)\right\}\\
&+C_2\left\{(T+\alpha)^{\underline{1}} - \sum_{r=\alpha-n}^{\eta}\frac{\tau(\eta+\sum_{i=1}^{m}\theta_i-\sigma(r))^{\underline{\sum_{i=1}^{m}\theta_i-1}}s}{\Gamma(\sum_{i=1}^{m}\theta_i)}g\left(\eta+\sum_{i=1}^{m}\theta_i\right)\right\}\\
&+C_3\left\{(T+\alpha)^{\underline{2}} - \sum_{r=\alpha-n}^{\eta}\frac{\tau(\eta+\sum_{i=1}^{m}\theta_i-\sigma(r))^{\underline{\sum_{i=1}^{m}\theta_i-1}}s^{\underline{2}}}{\Gamma(\sum_{i=1}^{m}\theta_i)}g\left(\eta+\sum_{i=1}^{m}\theta_i\right)\right\}\\
&+\cdots\\
&+C_{n-1}\left\{(T+\alpha)^{\underline{n-2}} - \sum_{r=\alpha-n}^{\eta}\frac{\tau(\eta+\sum_{i=1}^{m}\theta_i-\sigma(r))^{\underline{\sum_{i=1}^{m}\theta_i-1}}s^{\underline{n-2}}}{\Gamma(\sum_{i=1}^{m}\theta_i)}g\left(\eta+\sum_{i=1}^{m}\theta_i\right)\right\}\\
&+C_n\left\{\sum_{s=\alpha-n}^{T+\alpha-n+1}e^{-s} - \sum_{r=\alpha-n}^{\eta}\sum_{s=\alpha-n}^{r-n+1}\frac{\tau(\eta+\sum_{i=1}^{m}\theta_i-\sigma(r))^{\underline{\sum_{i=1}^{m}\theta_i-1}}}{\Gamma(\sum_{i=1}^{m}\theta_i)}e^{-s}g\left(\eta+\sum_{i=1}^{m}\theta_i\right)\right\}\\
&= \sum_{r=\alpha-n}^{\eta}\sum_{s=\alpha-n}^{r-n+1}\sum_{v=\alpha-n}^{s-1}\sum_{x=0}^{v-\alpha+n}\frac{\tau(\eta+\sum_{i=1}^{m}\theta_i-\sigma(r))^{\underline{\sum_{i=1}^{m}\theta_i-1}}}{\Gamma(\sum_{i=1}^{m}\theta_i)}e^{v-s}\times\\
&\quad\frac{(v-\sigma(x))^{\underline{\alpha-n-1}}}{\Gamma(\alpha-n)}g\left(\eta+\sum_{i=1}^{m}\theta_i\right)h(x+\alpha-1)\\
&\quad-\sum_{s=\alpha-n}^{T+\alpha-n+1}\sum_{v=\alpha-n}^{s-1}\sum_{x=0}^{v-\alpha+n}e^{v-s}\frac{(v-\sigma(x))^{\underline{\alpha-n-1}}}{\Gamma(\alpha-n)}h(x+\alpha-1).
\end{aligned}
\tag{22}
$$

Solving the system of Equations $(E_1) - (E_n)$, we obtain

$$C_1 = C_2 = \ldots = C_{n-1} = 0 \quad \text{and} \quad C_n = \frac{\mathcal{O}[h]}{\Lambda},$$

where $\mathcal{O}[h], \Lambda$ are defined by (11), (12), respectively. Substituting the constants $C_1 - C_n$ into (17), we obtain (10). This completes the proof. □

3. Main Results

The goal of this section is to show the existence results for the problem (5). To accomplish this, we denote $\mathcal{C} = C(\mathbb{N}_{\alpha-n,T+\alpha}, \mathbb{R})$, the Banach space of all functions u with the norm is defined by

$$\|u\|_{\mathcal{C}} = \|u\| + \|\Delta_C^{\vartheta} u\| + \|\Delta^{-\gamma} u\|,$$

where $\|u\| = \max_{t\in\mathbb{N}_{\alpha-n,T+\alpha}} |u(t)|$, $\|\Delta_C^{\vartheta} u\| = \max_{t\in\mathbb{N}_{t\in\mathbb{N}_{\alpha-n,T+\alpha}}} |\Delta_C^{\vartheta} x(t-\vartheta+1)|$ and $\|\Delta^{-\gamma} u\| = \max_{t\in\mathbb{N}_{t\in\mathbb{N}_{\alpha-n,T+\alpha}}} |\Delta^{-\gamma} x(t+\gamma)|$. In addition, we define the operator $\mathcal{F} : \mathcal{C} \to \mathcal{C}$ by

$$\begin{aligned}
(\mathcal{F}u)(t) &= \frac{\mathcal{O}[F(u)+H(u)]}{\Lambda} \sum_{s=\alpha-n}^{t-n+1} e^{-s} + \frac{1}{\Gamma(\alpha-n)} \sum_{s=\alpha-n}^{t-n+1} \sum_{v=\alpha-n}^{s-1} \sum_{\xi=0}^{v-\alpha+n} e^{v-s} \times \\
&\quad (v-\sigma(\xi))^{\underline{\alpha-n-1}} \Big[\lambda F\Big(\xi+\alpha-1, u(\xi+\alpha-1), (\Upsilon^{\vartheta} u)(\xi+\alpha-\vartheta)\Big) \\
&\quad + \mu H\Big(\xi+\alpha-1, u(\xi+\alpha-1), (\Psi^{\gamma} u)(\xi+\alpha+\gamma-1)\Big)\Big],
\end{aligned} \tag{23}$$

where Λ is defined by (12) and the functional $\mathcal{O}[F(u)+H(u)]$ is defined by

$$\begin{aligned}
&\mathcal{O}[F(u)+H(u)] \\
&= \sum_{r=\alpha-n}^{\eta} \sum_{s=\alpha-n}^{r-n+1} \sum_{v=\alpha-n}^{s-1} \sum_{\xi=0}^{v-\alpha+n} \frac{\tau\left(\eta + \sum_{i=1}^{m} \theta_i - \sigma(r)\right)^{\underline{\sum_{i=1}^{m} \theta_i - 1}}}{\Gamma\left(\sum_{i=1}^{m} \theta_i\right) \Gamma(\alpha-n)} e^{v-s} g\left(\eta + \sum_{i=1}^{m} \theta_i\right) \times \\
&\quad (v-\sigma(\xi))^{\underline{\alpha-n-1}} \Big[\lambda F\Big(\xi+\alpha-1, u(\xi+\alpha-1), (\Upsilon^{\vartheta} u)(\xi+\alpha-\vartheta)\Big) \\
&\quad + \mu H\Big(\xi+\alpha-1, u(\xi+\alpha-1), (\Psi^{\gamma} u)(\xi+\alpha+\gamma-1)\Big)\Big] \\
&\quad - \frac{1}{\Gamma(\alpha-n)} \sum_{s=\alpha-n}^{T+\alpha-n+1} \sum_{v=\alpha-n}^{s-1} \sum_{\xi=0}^{v-\alpha+n} e^{v-s} (v-\sigma(x))^{\underline{\alpha-n-1}} \times \\
&\quad \Big[\lambda F\Big(\xi+\alpha-1, u(\xi+\alpha-1), (\Upsilon^{\vartheta} u)(\xi+\alpha-\vartheta)\Big) \\
&\quad + \mu H\Big(\xi+\alpha-1, u(\xi+\alpha-1), (\Psi^{\gamma} u)(\xi+\alpha+\gamma-1)\Big)\Big].
\end{aligned} \tag{24}$$

Clearly, the problem (5) has solutions if and only if the operator $\mathcal{F}$ has fixed points. The first show the existence and uniqueness of a solution to the problem (5) by using the Banach contraction principle.

Theorem 1. *Assume that $F, H : \mathbb{N}_{\alpha-n,T+\alpha} \times \mathbb{R} \times \mathbb{R} \to \mathbb{R}$ are continuous, $\varphi, \phi : \mathbb{N}_{\alpha-n,T+\alpha} \times \mathbb{N}_{\alpha-n,T+\alpha} \to [0,\infty)$ are continuous with $\varphi_0 = \max\{\varphi(t-1,s) : (t,s) \in \mathbb{N}_{\alpha-2,T+\alpha} \times \mathbb{N}_{\alpha-2,T+\alpha}\}$ and $\phi_0 = \max\{\phi(t-1,s) : (t,s) \in \mathbb{N}_{\alpha-2,T+\alpha} \times \mathbb{N}_{\alpha-2,T+\alpha}\}$. In addition, suppose that:*

(H_1) *there exist constants* $L_1, L_2 > 0$ *such that for each* $t \in \mathbb{N}_{\alpha-n,T+\alpha}$ *and* $u, v \in C$

$$|F(t,u(t),(\Upsilon^{\vartheta}u)(t-\vartheta+1)) - F(t,v(t),(\Upsilon^{\vartheta}v)(t-\vartheta+1))| \leq L_1|u-v| + L_2|(\Upsilon^{\vartheta}u) - (\Upsilon^{\vartheta}v)|,$$

(H_2) *there exist constants* $\ell_1, \ell_2 > 0$ *such that for each* $t \in \mathbb{N}_{\alpha-n,T+\alpha}$ *and* $u, v \in C$

$$|H(t,u(t),(\Psi^{\gamma}u)(t+\gamma)) - f(t,v(t),(\Psi^{\gamma}v)(t+\gamma))| \leq \ell_1|u-v| + \ell_2|(\Psi^{\gamma}u) - (\Psi^{\gamma}v)|,$$

(H_3) $0 < g(t) < K \neq \frac{(e^{T-n+3}-1)e^{\eta-2n+2}\Gamma(m+1)}{(\eta-\alpha+n+m)^{\underline{m}}\tau\, e^{T+\alpha-2n+2}(\eta-\alpha+n+m)^{\underline{m}}}$ *for each* $t \in \mathbb{N}_{\alpha-n,T+\alpha}$.

$$\text{If } \chi := \Big[\lambda\Big(L_1 + L_2\frac{\phi_0(T+n-\vartheta+1)^{\underline{1-\vartheta}}}{\Gamma(2-\vartheta)}\Big) + \mu\Big(\ell_1 + \ell_2\frac{\varphi_0(T+n+\gamma)^{\underline{\gamma}}}{\Gamma(\gamma+1)}\Big)\Big] \times (\Omega_1 + \Omega_2 + \Omega_3) < 1, \tag{25}$$

then the problem (5) has a unique solution on $\mathbb{N}_{\alpha-n,T+\alpha}$*, where*

$$\Omega_1 = \frac{\Theta e^{n-\alpha}(T+2)}{|\Lambda|} + \frac{e^{T-1}(T+\alpha-n+2)^{\underline{\alpha-n+2}}}{\Gamma(\alpha-n+3)}, \tag{26}$$

$$\Omega_2 = \left[\frac{\Theta e^{2n-\alpha-2}}{|\Lambda|} + \frac{e^{T-1}(T+\alpha-n+3)^{\underline{\alpha-n+2}}}{\Gamma(\alpha-n+3)}\right]\frac{(T+n-\vartheta+1)^{\underline{1-\vartheta}}}{\Gamma(2-\vartheta)}, \tag{27}$$

$$\Omega_3 = \left[\frac{\Theta e^{n-\alpha}(T+2)}{|\Lambda|} + \frac{e^{T-1}(T+\alpha-n+2)^{\underline{\alpha-n+2}}}{\Gamma(\alpha-n+3)}\right]\frac{(T+n+\gamma)^{\underline{\gamma}}}{\Gamma(\gamma+1)}, \tag{28}$$

$$\Theta = \frac{\tau K e^{\eta-\alpha-1}(\eta-\alpha+2)^{\underline{\alpha-n+2}}(\eta-\alpha+n+m)^{\underline{m}}}{\Gamma(m+1)\Gamma(\alpha-n+3)} - \frac{e^{T-1}(T+\alpha-n+2)^{\underline{\alpha-n+2}}}{\Gamma(\alpha-n+3)}. \tag{29}$$

Proof. We shall show that $\mathcal{F}$ is a contraction. For any $u, v \in C$ and for each $t \in \mathbb{N}_{\alpha-n,T+\alpha}$, we have

$$\begin{aligned}
&\Big|\mathcal{O}[F(u)+H(u)] - \mathcal{O}[F(v)+H(v)]\Big| \\
\leq\; & \sum_{r=\alpha-n}^{\eta}\sum_{s=\alpha-n}^{r-n+1}\sum_{v=\alpha-n}^{s-1}\sum_{\xi=0}^{v-\alpha+n}\frac{\tau\left(\eta+\sum_{i=1}^{m}\theta_i-\sigma(r)\right)^{\underline{\sum_{i=1}^{m}\theta_i-1}}}{\Gamma\left(\sum_{i=1}^{m}\theta_i\right)\Gamma(\alpha-n)}e^{v-s}(v-\sigma(\xi))^{\underline{\alpha-n-1}}\times \\
&\Big[\lambda\Big(L_1|u-v| + L_2|(\Upsilon^{\vartheta}u)-(\Upsilon^{\vartheta}v)|\Big) + \mu\Big(\ell_1|u-v| + \ell_2|(\Psi^{\gamma}u)-(\Psi^{\gamma}v)|\Big)\Big]\times \\
&g\left(\eta+\sum_{i=1}^{m}\theta_i\right) - \frac{1}{\Gamma(\alpha-n)}\sum_{s=\alpha-n}^{T+\alpha-n+1}\sum_{v=\alpha-n}^{s-1}\sum_{\xi=0}^{v-\alpha+n}e^{v-s}(v-\sigma(x))^{\underline{\alpha-n-1}}\times \\
&\Big[\lambda\Big(L_1|u-v| + L_2|(\Upsilon^{\vartheta}u)-(\Upsilon^{\vartheta}v)|\Big) + \mu\Big(\ell_1|u-v| + \ell_2|(\Psi^{\gamma}u)-(\Psi^{\gamma}v)|\Big)\Big] \\
\leq\; & \Big[\lambda\Big(L_1 + L_2\frac{\phi_0(T+n-\vartheta+1)^{\underline{1-\vartheta}}}{\Gamma(2-\vartheta)}\Big) + \mu\Big(\ell_1 + \ell_2\frac{\varphi_0(T+n+\gamma)^{\underline{\gamma}}}{\Gamma(\gamma+1)}\Big)\Big]\Bigg|\,\|u-v\|_C\times \\
&\Bigg|\frac{\tau K e^{\eta-\alpha-1}(\eta-\alpha+2)^{\underline{\alpha-n+2}}(\eta-\alpha+n+m)^{\underline{m}}}{\Gamma(m+1)\Gamma(\alpha-n+3)} - \frac{e^{T-1}(T+\alpha-n+2)^{\underline{\alpha-n+2}}}{\Gamma(\alpha-n+3)} \\
=\; & \Big[\lambda\Big(L_1 + L_2\frac{\phi_0(T+n-\vartheta+1)^{\underline{1-\vartheta}}}{\Gamma(2-\vartheta)}\Big) + \mu\Big(\ell_1 + \ell_2\frac{\varphi_0(T+n+\gamma)^{\underline{\gamma}}}{\Gamma(\gamma+1)}\Big)\Big]\|u-v\|_C\Theta,
\end{aligned} \tag{30}$$

and

$$\begin{aligned}
&\left|(\mathcal{F}u)(t)-(\mathcal{F}v)(t)\right| \\
&\leq \frac{1}{\Lambda}\Big|\mathcal{O}[F(u)+H(u)]-\mathcal{O}[F(v)+H(v)]\Big|\sum_{s=\alpha-n}^{t-n+1}e^{-s} \\
&+\frac{1}{\Gamma(\alpha-n)}\sum_{s=\alpha-n}^{t-n+1}\sum_{v=\alpha-n}^{s-1}\sum_{\xi=0}^{v-\alpha+n}e^{v-s}(v-\sigma(\xi))^{\underline{\alpha-n-1}}\times \\
&\Big[\lambda\Big(L_1|u-v|+L_2|(\Upsilon^{\vartheta}u)-(\Upsilon^{\vartheta}v)|\Big)+\mu\Big(\ell_1|u-v|+\ell_2|(\Psi^{\gamma}u)-(\Psi^{\gamma}v)|\Big)\Big] \\
&\leq \Big[\lambda\Big(L_1+L_2\frac{\phi_0(T+n-\vartheta+1)^{\underline{1-\vartheta}}}{\Gamma(2-\vartheta)}\Big)+\mu\Big(\ell_1+\ell_2\frac{\varphi_0(T+n+\gamma)^{\underline{\gamma}}}{\Gamma(\gamma+1)}\Big)\Big]\|u-v\|_C\times \\
&\Big\{\frac{\Theta}{\Lambda}\sum_{s=\alpha-n}^{t-n+1}e^{-s}+\frac{1}{\Gamma(\alpha-n)}\sum_{s=\alpha-n}^{t-n+1}\sum_{v=\alpha-n}^{s-1}\sum_{\xi=0}^{v-\alpha+n}e^{v-s}(v-\sigma(\xi))^{\underline{\alpha-n-1}}\Big\} \\
&\leq \Big[\lambda\Big(L_1+L_2\frac{\phi_0(T+n-\vartheta+1)^{\underline{1-\vartheta}}}{\Gamma(2-\vartheta)}\Big)+\mu\Big(\ell_1+\ell_2\frac{\varphi_0(T+n+\gamma)^{\underline{\gamma}}}{\Gamma(\gamma+1)}\Big)\Big]\|u-v\|_C\times \\
&\Big(\frac{\Theta e^{n-\alpha}(T+2)}{|\Lambda|}+\frac{e^{T-1}(T+\alpha-n+2)^{\underline{\alpha-n+2}}}{\Gamma(\alpha-n+3)}\Big) \\
&\leq \Big[\lambda\Big(L_1+L_2\frac{\phi_0(T+n-\vartheta+1)^{\underline{1-\vartheta}}}{\Gamma(2-\vartheta)}\Big)+\mu\Big(\ell_1+\ell_2\frac{\varphi_0(T+n+\gamma)^{\underline{\gamma}}}{\Gamma(\gamma+1)}\Big)\Big]\|u-v\|_C\Omega_1.
\end{aligned} \tag{31}$$

Next, we consider the following $(\Delta_C^{\vartheta}\mathcal{F}u)$ and $(\Delta^{\gamma}\mathcal{F}u)$ as

$$\begin{aligned}
&(\Delta_C^{\vartheta}\mathcal{F}u)(t-\vartheta+1) \\
&\quad=\frac{\mathcal{O}[F(u)+H(u)]}{\Lambda\Gamma(1-\vartheta)}\sum_{s=\alpha-n}^{t}(t-\vartheta+1-\sigma(s))^{\underline{-\vartheta}}\Delta_s\sum_{v=\alpha-n}^{s-n+1}e^{-v} \\
&\quad+\frac{1}{\Gamma(\alpha-n)\Gamma(1-\vartheta)}\sum_{r=\alpha-n}^{t}(t-\vartheta+1-\sigma(r))^{\underline{-\vartheta}}\Delta_r\Big\{\sum_{s=\alpha-n}^{r-n+1}\sum_{v=\alpha-n}^{s-1}\sum_{\xi=0}^{v-\alpha+n}e^{v-s}\times \\
&\quad(v-\sigma(\xi))^{\underline{\alpha-n-1}}\Big[\lambda F\Big(\xi+\alpha-1,u(\xi+\alpha-1),(\Upsilon^{\vartheta}u)(\xi+\alpha-\vartheta)\Big) \\
&\quad+\mu H\Big(\xi+\alpha-1,u(\xi+\alpha-1),(\Psi^{\gamma}u)(\xi+\alpha+\gamma-1)\Big)\Big]\Big\} \\
&\quad=\frac{\mathcal{O}[F(u)+H(u)]}{\Lambda\Gamma(1-\vartheta)}\sum_{s=\alpha-n}^{t}(t-\vartheta+1-\sigma(s))^{\underline{-\vartheta}}e^{n-s-2} \\
&\quad+\frac{1}{\Gamma(\alpha-n)\Gamma(1-\vartheta)}\sum_{r=\alpha-n}^{t}\sum_{s=\alpha-n}^{r-n+1}\sum_{v=\alpha-n}^{s-1}\sum_{\xi=0}^{v-\alpha+n}(t-\vartheta+1-\sigma(r))^{\underline{-\vartheta}}e^{v-s}\times \\
&\quad(v-\sigma(\xi))^{\underline{\alpha-n-1}}\Big[\lambda F\Big(\xi+\alpha-1,u(\xi+\alpha-1),(\Upsilon^{\vartheta}u)(\xi+\alpha-\vartheta)\Big) \\
&\quad+\mu H\Big(\xi+\alpha-1,u(\xi+\alpha-1),(\Psi^{\gamma}u)(\xi+\alpha+\gamma-1)\Big)\Big],
\end{aligned} \tag{32}$$

$$\begin{aligned}(\Delta^{-\gamma}\mathcal{F}u)(t+\gamma) &= \frac{\mathcal{O}[F(u)+H(u)]}{\Lambda\Gamma(\gamma)}\sum_{s=\alpha-n}^{t}\sum_{v=\alpha-n}^{s-n+1}(t+\gamma-\sigma(s))^{\underline{\gamma-1}}e^{-v}\\ &+\frac{1}{\Gamma(\alpha-n)\Gamma(\gamma)}\sum_{r=\alpha-n}^{t}\sum_{s=\alpha-n}^{r-n+1}\sum_{v=\alpha-n}^{s-1}\sum_{\xi=0}^{v-\alpha+n}(t+\gamma-\sigma(r))^{\underline{\gamma-1}}e^{v-s}\times\\ &(v-\sigma(\xi))^{\underline{\alpha-n-1}}\Big[\lambda F\Big(\xi+\alpha-1,u(\xi+\alpha-1),(\Upsilon^{\vartheta}u)(\xi+\alpha-\vartheta)\Big)\\ &+\mu H\Big(\xi+\alpha-1,u(\xi+\alpha-1),(\Psi^{\gamma}u)(\xi+\alpha+\gamma-1)\Big)\Big]. \end{aligned} \tag{33}$$

Similarly, we have

$$\begin{aligned}&\left|(\Delta_C{}^{\vartheta}\mathcal{F}u)(t-\vartheta+1)-(\Delta_C^{\vartheta}\mathcal{F}v)(t-\vartheta+1)\right|\\ &\le\Big[\lambda\Big(L_1+L_2\frac{\phi_0(T+n-\vartheta+1)^{\underline{1-\vartheta}}}{\Gamma(2-\vartheta)}\Big)+\mu\Big(\ell_1+\ell_2\frac{\varphi_0(T+n+\gamma)^{\underline{\gamma}}}{\Gamma(\gamma+1)}\Big)\Big]\|u-v\|_C\times\\ &\left[\frac{\Theta e^{2n-\alpha-2}}{|\Lambda|}+\frac{e^{T-1}(T+\alpha-n+3)^{\underline{\alpha-n+2}}}{\Gamma(\alpha-n+3)}\right]\frac{(T+n-\vartheta+1)^{\underline{1-\vartheta}}}{\Gamma(2-\vartheta)}\\ &\le\Big[\lambda\Big(L_1+L_2\frac{\phi_0(T+n-\vartheta+1)^{\underline{1-\vartheta}}}{\Gamma(2-\vartheta)}\Big)+\mu\Big(\ell_1+\ell_2\frac{\varphi_0(T+n+\gamma)^{\underline{\gamma}}}{\Gamma(\gamma+1)}\Big)\Big]\|u-v\|_C\Omega_2,\end{aligned} \tag{34}$$

and

$$\begin{aligned}&\left|(\Delta^{-\gamma}\mathcal{F}u)(t+\gamma)-(\Delta^{-\gamma}\mathcal{F}v)(t+\gamma)\right|\\ &\le\Big[\lambda\Big(L_1+L_2\frac{\phi_0(T+n-\vartheta+1)^{\underline{1-\vartheta}}}{\Gamma(2-\vartheta)}\Big)+\mu\Big(\ell_1+\ell_2\frac{\varphi_0(T+n+\gamma)^{\underline{\gamma}}}{\Gamma(\gamma+1)}\Big)\Big]\|u-v\|_C\times\\ &\left[\frac{\Theta e^{n-\alpha}(T+2)}{|\Lambda|}+\frac{e^{T-1}(T+\alpha-n+2)^{\underline{\alpha-n+2}}}{\Gamma(\alpha-n+3)}\right]\frac{(T+n+\gamma)^{\underline{\gamma}}}{\Gamma(\gamma+1)}\\ &\le\Big[\lambda\Big(L_1+L_2\frac{\phi_0(T+n-\vartheta+1)^{\underline{1-\vartheta}}}{\Gamma(2-\vartheta)}\Big)+\mu\Big(\ell_1+\ell_2\frac{\varphi_0(T+n+\gamma)^{\underline{\gamma}}}{\Gamma(\gamma+1)}\Big)\Big]\|u-v\|_C\Omega_3.\end{aligned} \tag{35}$$

Hence (31), (34) and (35) imply that

$$\begin{aligned}\left\|(\mathcal{F}u)(t)-(\mathcal{F}v)(t)\right\|_C &\le\Big[\lambda\Big(L_1+L_2\frac{\phi_0(T+n-\vartheta+1)^{\underline{1-\vartheta}}}{\Gamma(2-\vartheta)}\Big)\\ &+\mu\Big(\ell_1+\ell_2\frac{\varphi_0(T+n+\gamma)^{\underline{\gamma}}}{\Gamma(\gamma+1)}\Big)\Big](\Omega_1+\Omega_2+\Omega_3)\|u-v\|_C\\ &=\chi\|u-v\|_C.\end{aligned} \tag{36}$$

By (H_4), we have $\left\|(\mathcal{F}u)(t)-(\mathcal{F}v)(t)\right\|_C<\|u-v\|_C$.

Consequently, $\mathcal{F}$ is a contraction. Therefore, by the Banach fixed point theorem, we get that $\mathcal{F}$ has a fixed point which is a unique solution of the problem (5) on $t\in\mathbb{N}_{\alpha-n,T+\alpha}$. □

In the second result, we deduce the existence of at least one solution of (5) by the following, the Schauder's fixed point theorem.

Lemma 4 ([30]). *(Arzelá-Ascoli theorem) A set of function in $C[a,b]$ with the sup norm is relatively compact if and only it is uniformly bounded and equicontinuous on $[a,b]$.*

Lemma 5 ([30]). *If a set is closed and relatively compact then it is compact.*

Lemma 6 ([31]). *(Schauder fixed point theorem) Let (D, d) be a complete metric space, U be a closed convex subset of D, and $T : D \to D$ be the map such that the set $Tu : u \in U$ is relatively compact in D. Then the operator T has at least one fixed point $u^* \in U$: $Tu^* = u^*$.*

Theorem 2. *Assuming that $(H_1) - (H_3)$ hold, problem (5) has at least one solution on $\mathbb{N}_{\alpha-n,T+\alpha}$.*

Proof. We divide the proof into three steps as follows.
Step I. Verify $\mathcal{F}$ map bounded sets into bounded sets in $B_R = \{u \in \mathcal{C} : \|u\|_{\mathcal{C}} \leq R\}$. We consider $B_R = \{u \in C(\mathbb{N}_{\alpha-n,T+\alpha}) : \|u\|_{\mathcal{C}} \leq R\}$.

Let $\max_{t\in\mathbb{N}_{\alpha-n,T+\alpha}} |F(t,0,0)| = M$, $\max_{t\in\mathbb{N}_{\alpha-n,T+\alpha}} |H(t,0,0)| = N$ and choose a constant

$$R \geq \frac{(M+N)(\Omega_1+\Omega_2+\Omega_3)}{1-(\Omega_1+\Omega_2+\Omega_3)\left\{\lambda\left(L_1+L_2\frac{\varphi_0(T+n-\vartheta+1)^{\underline{1-\vartheta}}}{\Gamma(2-\vartheta)}\right)+\mu\left(\ell_1+\ell_2\frac{\varphi_0(T+n+\gamma)^{\underline{\gamma}}}{\Gamma(\gamma+1)}\right)\right\}}. \tag{37}$$

Noting that

$$\begin{aligned}|\mathcal{S}(t,u,0)| &= \Big|F(t+\alpha-1,u(t+\alpha-1),\Delta_C^{\vartheta}u(t+\alpha-\vartheta))-F(t+\alpha-1,0,0)| \\ &\quad +|F(t+\alpha-1,0,0)|\Big|, \\ |\mathcal{T}(t,u,0)| &= \Big|H(t+\alpha-1,u(t+\alpha-1),\Delta^{-\gamma}u(t+\alpha+\gamma-1))-H(t+\alpha-1,0,0)| \\ &\quad +|H(t+\alpha-1,0,0)|\Big|,\end{aligned}$$

for each $u \in B_R$, we obtain

$$\begin{aligned}&\Big|\mathcal{O}[F(u)+H(u)]\Big| \\ &\leq \sum_{r=\alpha-n}^{\eta}\sum_{s=\alpha-n}^{r-n+1}\sum_{v=\alpha-n}^{s-1}\sum_{\xi=0}^{v-\alpha+n}\frac{\tau\left(\eta+\sum_{i=1}^m\theta_i-\sigma(r)\right)^{\underline{\sum_{i=1}^m\theta_i-1}}}{\Gamma\left(\sum_{i=1}^m\theta_i\right)\Gamma(\alpha-n)}e^{v-s}(v-\sigma(\xi))^{\underline{\alpha-n-1}}\times \\ &\quad \Big[|\mathcal{S}(\xi,u,0)|+\mu|\mathcal{T}(\xi,u,0)|\Big] \\ &\leq \sum_{r=\alpha-n}^{\eta}\sum_{s=\alpha-n}^{r-n+1}\sum_{v=\alpha-n}^{s-1}\sum_{\xi=0}^{v-\alpha+n}\frac{\tau\left(\eta+\sum_{i=1}^m\theta_i-\sigma(r)\right)^{\underline{\sum_{i=1}^m\theta_i-1}}}{\Gamma\left(\sum_{i=1}^m\theta_i\right)\Gamma(\alpha-n)}e^{v-s}(v-\sigma(\xi))^{\underline{\alpha-n-1}}\times \\ &\quad \Big[\lambda\Big(L_1\|u\|+L_2\|\Upsilon^{\vartheta}u\|+M\Big)+\mu\Big(\ell_1\|u\|+\ell_2\|\Psi^{\gamma}u\|+N\Big)\Big] \\ &\quad -\frac{1}{\Gamma(\alpha-n)}\sum_{s=\alpha-n}^{T+\alpha-n+1}\sum_{v=\alpha-n}^{s-1}\sum_{\xi=0}^{v-\alpha+n}e^{v-s}(v-\sigma(x))^{\underline{\alpha-n-1}}\times \\ &\quad \Big[\lambda\Big(L_1|u-v|+L_2|(\Upsilon^{\vartheta}u)-(\Upsilon^{\vartheta}v)|+M\Big)+\mu\Big(\ell_1|u-v|+\ell_2|(\Psi^{\gamma}u)-(\Psi^{\gamma}v)|+N\Big)\Big] \\ &\leq \Big\{\lambda\Big[\Big(L_1+L_2\frac{\varphi_0(T+n-\vartheta+1)^{\underline{1-\vartheta}}}{\Gamma(2-\vartheta)}\Big)\|u\|_{\mathcal{C}}+M\Big] \\ &\quad +\mu\Big[\Big(\ell_1+\ell_2\frac{\varphi_0(T+n+\gamma)^{\underline{\gamma}}}{\Gamma(\gamma+1)}\Big)\|u\|_{\mathcal{C}}+N\Big]\Big\}\times \\ &\quad \left|\frac{\tau K e^{\eta-\alpha-1}(\eta-\alpha+2)^{\underline{\alpha-n+2}}(\eta-\alpha+n+m)^{\underline{m}}}{\Gamma(m+1)\Gamma(\alpha-n+3)}-\frac{e^{T-1}(T+\alpha-n+2)^{\underline{\alpha-n+2}}}{\Gamma(\alpha-n+3)}\right| \\ &\leq \left\{\lambda\Big[\Big(L_1+L_2\frac{\varphi_0(T+n-\vartheta+1)^{\underline{1-\vartheta}}}{\Gamma(2-\vartheta)}\Big)\|u\|_{\mathcal{C}}+M\Big]\right. \\ &\quad \left.+\mu\Big[\Big(\ell_1+\ell_2\frac{\varphi_0(T+n+\gamma)^{\underline{\gamma}}}{\Gamma(\gamma+1)}\Big)\|u\|_{\mathcal{C}}+N\Big]\right\}\Theta,\end{aligned} \tag{38}$$

and

$$
\begin{aligned}
&\big|(\mathcal{F}u)(t)\big| \\
&\le \frac{1}{\Lambda}\Big|\mathcal{O}[F(u)+H(u)]\Big|\sum_{s=\alpha-n}^{t-n+1}e^{-s} \\
&\quad+\frac{1}{\Gamma(\alpha-n)}\sum_{s=\alpha-n}^{t-n+1}\sum_{v=\alpha-n}^{s-1}\sum_{\xi=0}^{v-\alpha+n}e^{v-s}(v-\sigma(\xi))^{\underline{\alpha-n-1}}\Big[\lambda\big|\mathcal{S}(\xi,u,0)\big|+\mu\big|\mathcal{T}(\xi,u,0)\big|\Big] \\
&\le \Big[\lambda\Big(L_1\|u\|+L_2\|\Upsilon^{\vartheta}u\|+M\Big)+\mu\Big(\ell_1\|u\|+\ell_2\|\Psi^{\gamma}u\|+N\Big)\Big]\times \\
&\quad\left\{\frac{\Theta}{|\Lambda|}\sum_{s=\alpha-n}^{t-n+1}e^{-s}+\frac{1}{\Gamma(\alpha-n)}\sum_{s=\alpha-n}^{t-n+1}\sum_{v=\alpha-n}^{s-1}\sum_{\xi=0}^{v-\alpha+n}e^{v-s}(v-\sigma(\xi))^{\underline{\alpha-n-1}}\right\} \\
&\le \left\{\lambda\Big[\Big(L_1+L_2\frac{\varphi_0(T+n-\vartheta+1)^{\underline{1-\vartheta}}}{\Gamma(2-\vartheta)}\Big)\|u\|_C+M\Big]\right. \\
&\quad\left.+\mu\Big[\Big(\ell_1+\ell_2\frac{\varphi_0(T+n+\gamma)^{\underline{\gamma}}}{\Gamma(\gamma+1)}\Big)\|u\|_C+N\Big]\right\}\left(\frac{\Theta e^{n-\alpha}(T+2)}{|\Lambda|}+\frac{e^{T-1}(T+\alpha-n+2)^{\underline{\alpha-n+2}}}{\Gamma(\alpha-n+3)}\right) \\
&\le \left\{\lambda\Big[\Big(L_1+L_2\frac{\varphi_0(T+n-\vartheta+1)^{\underline{1-\vartheta}}}{\Gamma(2-\vartheta)}\Big)\|u\|_C+M\Big]\right. \\
&\quad\left.+\mu\Big[\Big(\ell_1+\ell_2\frac{\varphi_0(T+n+\gamma)^{\underline{\gamma}}}{\Gamma(\gamma+1)}\Big)\|u\|_C+N\Big]\right\}\Omega_1.
\end{aligned}
\tag{39}
$$

Furthermore, we have

$$
\begin{aligned}
\big|(\Delta_C^{\vartheta}\mathcal{F}u)(t-\vartheta+1)\big| \le &\left\{\lambda\Big[\Big(L_1+L_2\frac{\varphi_0(T+n-\vartheta+1)^{\underline{1-\vartheta}}}{\Gamma(2-\vartheta)}\Big)\|u\|_C+M\Big]\right. \\
&\left.+\mu\Big[\Big(\ell_1+\ell_2\frac{\varphi_0(T+n+\gamma)^{\underline{\gamma}}}{\Gamma(\gamma+1)}\Big)\|u\|_C+N\Big]\right\}\Omega_2,
\end{aligned}
\tag{40}
$$

and

$$
\begin{aligned}
\big|(\Delta^{-\gamma}\mathcal{F}u)(t+\gamma)\big| \le &\left\{\lambda\Big[\Big(L_1+L_2\frac{\varphi_0(T+n-\vartheta+1)^{\underline{1-\vartheta}}}{\Gamma(2-\vartheta)}\Big)\|u\|_C+M\Big]\right. \\
&\left.+\mu\Big[\Big(\ell_1+\ell_2\frac{\varphi_0(T+n+\gamma)^{\underline{\gamma}}}{\Gamma(\gamma+1)}\Big)\|u\|_C+N\Big]\right\}\Omega_3.
\end{aligned}
\tag{41}
$$

Hence (39)–(41) imply that

$$
\begin{aligned}
\big\|(\mathcal{F}u)(t)\big\|_C \le &\left\{\lambda\Big[\Big(L_1+L_2\frac{\varphi_0(T+n-\vartheta+1)^{\underline{1-\vartheta}}}{\Gamma(2-\vartheta)}\Big)\|u\|_C+M\Big]\right. \\
&\left.+\mu\Big[\Big(\ell_1+\ell_2\frac{\varphi_0(T+n+\gamma)^{\underline{\gamma}}}{\Gamma(\gamma+1)}\Big)\|u\|_C+N\Big]\right\}(\Omega_1+\Omega_2+\Omega_3) \\
\le &\ R.
\end{aligned}
\tag{42}
$$

So, $\|\mathcal{F}u\|_C \le R$. This implies that $\mathcal{F}$ is uniformly bounded.

Step II. Since F and H are continuous, the operator $\mathcal{F}$ is continuous on B_R.

Step III. Examine $\mathcal{F}$ is equicontinuous on B_R. For any $\epsilon > 0$, there exists a positive constant $\rho^* = \max\{\delta_1, \delta_2, \delta_3, \delta_4\}$ such that for $t_1, t_2 \in \mathbb{N}_{\alpha-n,T+\alpha}$

$$\begin{aligned}
&\left|t_2 - t_1\right| < \frac{\epsilon\Lambda}{6e^{T-1}\Theta\left[\lambda\|F\|+\mu\|H\|\right]}, \qquad \text{whenever } |t_2-t_1| < \delta_1,\\
&\left|(t_2-n+1)^{\underline{\alpha-n+2}} - (t_1-n+1)^{\underline{\alpha-n+2}}\right| < \frac{\epsilon\Gamma(\alpha-n+3)}{6e^{n-\alpha}\Theta\left[\lambda\|F\|+\mu\|H\|\right]}, \qquad \text{whenever } |t_2-t_1| < \delta_2,\\
&\left|(t_2-\alpha+n-\vartheta+1)^{\underline{1-\vartheta}} - (t_1-\alpha+n-\vartheta+1)^{\underline{1-\vartheta}}\right|\\
&\quad < \frac{\epsilon}{3[\lambda\|F\|+\mu\|H\|]\left(\frac{\Theta e^{2n-\alpha-2}}{\Lambda\Gamma(2-\vartheta)} + \frac{e^{T-1}(T+\alpha-n+3)^{\underline{\alpha-n+2}}}{\Gamma(\alpha-n+3)\Gamma(2-\vartheta)}\right)}\\
&\quad < \frac{\epsilon\,\Gamma(1-\nu)\Gamma(\beta)\Gamma(\alpha)}{4\widetilde{\Theta}_R}, \qquad \text{whenever } |t_2-t_1| < \delta_3,\\
&\left|(t_2-\alpha+n+\gamma)^{\underline{\gamma}} - (t_1-\alpha+n+\gamma)^{\underline{\gamma}}\right| < \frac{\epsilon}{3\left[\lambda\|F\|+\mu\|H\|\right]\left(\frac{\Theta e^{n-\alpha}(T+2)}{\Lambda\Gamma(\gamma+1)} + \frac{e^{T-1}(T+\alpha-n+2)^{\underline{\alpha-n+2}}}{\Gamma(\alpha-n+3)\Gamma(\gamma+1)}\right)},\\
&\qquad \text{whenever } |t_2-t_1| < \delta_4.
\end{aligned}$$

Then we have

$$\begin{aligned}
&\left|(\mathcal{F}x)(t_2) - (\mathcal{F}x)(t_1)\right|\\
&\le \frac{1}{|\Lambda|}\mathcal{O}[F(u)+H(u)]\left|\sum_{s=\alpha-n}^{t_2-n+1} e^{-s} - \sum_{s=\alpha-n}^{t_1-n+1} e^{-s}\right| + \frac{1}{\Gamma(\alpha-n)}\left|\sum_{s=\alpha-n}^{t_2-n+1}\sum_{v=\alpha-n}^{s-1}\sum_{\xi=0}^{v-\alpha+n} e^{v-s}\times\right.\\
&\quad (v-\sigma(\xi))^{\underline{\alpha-n-1}}\Big[\lambda F\Big(\xi+\alpha-1, u(\xi+\alpha-1), (\Upsilon^{\vartheta}u)(\xi+\alpha-\vartheta)\Big)\\
&\quad + \mu H\Big(\xi+\alpha-1, u(\xi+\alpha-1), (\Psi^{\gamma}u)(\xi+\alpha+\gamma-1)\Big)\Big] - \sum_{s=\alpha-n}^{t_1-n+1}\sum_{v=\alpha-n}^{s-1}\sum_{\xi=0}^{v-\alpha+n} e^{v-s}\times\\
&\quad (v-\sigma(\xi))^{\underline{\alpha-n-1}}\Big[\lambda F\Big(\xi+\alpha-1, u(\xi+\alpha-1), (\Upsilon^{\vartheta}u)(\xi+\alpha-\vartheta)\Big)\\
&\quad \left. + \mu H\Big(\xi+\alpha-1, u(\xi+\alpha-1), (\Psi^{\gamma}u)(\xi+\alpha+\gamma-1)\Big)\Big]\right|\\
&\le [\lambda\|F\|+\mu\|H\|]\left\{\frac{\Theta e^{n-\alpha}}{|\Lambda|}\left|t_2-t_1\right| + \frac{e^{T-1}}{\Gamma(\alpha-n+3)}\left|(t_2-n+1)^{\underline{\alpha-n+2}} - (t_1-n+1)^{\underline{\alpha-n+2}}\right|\right\}\\
&< \frac{\epsilon}{6} + \frac{\epsilon}{6} = \frac{\epsilon}{3}.
\end{aligned} \tag{43}$$

Furthermore, we have

$$\begin{aligned}
&\left|(\Delta_C^{\vartheta}\mathcal{F}u)(t_2-\vartheta+1) - (\Delta_C^{\vartheta}\mathcal{F}u)(t_1-\vartheta+1)\right|\\
&\le [\lambda\|F\|+\mu\|H\|]\left\{\left[\frac{\Theta e^{2n-\alpha-2}}{|\Lambda|\Gamma(2-\vartheta)} + \frac{e^{T-1}(T+\alpha-n+3)^{\underline{\alpha-n+2}}}{\Gamma(\alpha-n+3)\Gamma(2-\vartheta)}\right]\times\right.\\
&\qquad \left.\left|(t_2-\alpha+n-\vartheta+1)^{\underline{1-\vartheta}} - (t_1-\alpha+n-\vartheta+1)^{\underline{1-\vartheta}}\right|\right\} \le \frac{\epsilon}{3},
\end{aligned} \tag{44}$$

and

$$\begin{aligned}
&\left|(\Delta^{-\gamma}\mathcal{F}u)(t_2+\gamma)-(\Delta^{-\gamma}\mathcal{F}u)(t_1+\gamma)\right| \\
&\leq [\lambda\|F\|+\mu\|H\|]\left\{\left[\frac{\Theta e^{n-\alpha}(T+2)}{|\Lambda|\Gamma(\gamma+1)}+\frac{e^{T-1}(T+\alpha-n+2)^{\underline{\alpha-n+2}}}{\Gamma(\alpha-n+3)\Gamma(\gamma+1)}\right]\times\right. \\
&\left.\left|(t_2-\alpha+n+\gamma)^{\underline{\gamma}}-(t_1-\alpha+n+\gamma)^{\underline{\gamma}}\right|\right\} \leq \frac{\epsilon}{3}.
\end{aligned} \tag{45}$$

Hence

$$\left\|(\mathcal{F}x)(t_2)-(\mathcal{F}x)(t_1)\right\|_{\mathcal{C}} < \frac{\epsilon}{3}+\frac{\epsilon}{3}+\frac{\epsilon}{3} = \epsilon. \tag{46}$$

This implies that the set $\mathcal{F}(B_R)$ is an equicontinuous set. As a consequence of Steps I to III together with the Arzelá-Ascoli theorem, we find that $\mathcal{F}:\mathcal{C}\to\mathcal{C}$ is completely continuous. By Schauder fixed point theorem, we can conclude that problem (5) has at least one solution. The proof is completed. □

4. An Example

In order to study the existence of a solution to our problem, we obtain the conditions provided in Section 3. Since our designated problem is a theoretical problem, it is rare to find the application related to our results. However, for thorough explanation, we provide the following example to illustrate our results. Consider the following fractional difference boundary value problem

$$\begin{aligned}
\left[\Delta_C^{\frac{9}{2}}+(e+1)\Delta_C^{\frac{7}{2}}\right]u(t) =& \frac{e^{-\cos^2(2\pi(t+\frac{7}{2})+10)}}{100+e^{\sin^2(2\pi(t+\frac{7}{2}))}}\cdot\frac{|u(t+\frac{7}{2})|+\left|\Upsilon^{\frac{1}{3}}u(t+\frac{26}{6})\right|}{[1+|u(t+\frac{7}{2})|]} \\
&+\frac{(t+\frac{227}{2})^{-2}|u(t+\frac{7}{2})|+\left|\Psi^{\frac{4}{5}}u\left(t+\left(t+\frac{43}{10}\right)\right)\right|}{(t+\frac{27}{2})^3[1+|u(t+\frac{7}{2})|]}, \\
u\left(-\frac{1}{2}\right) =& D_C^{\frac{1}{4}}u\left(\frac{5}{4}\right)=D_C^{\frac{1}{2}}u\left(\frac{11}{4}\right)=D_C^{\frac{3}{4}}u(4)=0 \\
u\left(\frac{27}{2}\right) =& \frac{1}{e^5}\Delta^{-\frac{77}{60}}e^{-sin(\frac{587\pi}{60})}u\left(\frac{587}{60}\right),
\end{aligned} \tag{47}$$

where $$(\Psi^{\frac{1}{3}}u)\left(t+\frac{26}{6}\right)=\sum_{s=-\frac{7}{6}}^{t-\frac{2}{3}}\frac{(t-\sigma(s))^{\underline{-\frac{1}{3}}}}{\Gamma(\frac{2}{3})}\frac{e^{-(s+\frac{2}{3})}}{(t+50)^3}\Delta u\left(s+\frac{2}{3}\right),$$

$$(\Psi^{\frac{4}{5}}u)\left(t+\frac{4}{5}\right)=\sum_{s=-\frac{13}{10}}^{t-\frac{4}{5}}\frac{(t-\sigma(s))^{\underline{-\frac{1}{5}}}}{\Gamma\left(\frac{4}{5}\right)}\frac{e^{-(s+\frac{4}{5})}}{(t+100)^2}u\left(s+\frac{4}{5}\right).$$

Set $\alpha=\frac{9}{2}$, $n=5$, $\vartheta=\frac{1}{3}$, $\gamma=\frac{4}{5}$, $\beta_1=\frac{1}{4}$, $\beta_2=\frac{1}{2}$, $\beta_3=\frac{3}{4}$, $\lambda=e^{-10}$, $\mu=1$, $T=6$, $\eta=\frac{17}{2}$, $T=6$, $\tau=e^{-5}$, $m=4$, $\theta_1=\frac{1}{2}$, $\theta_2=\frac{1}{3}$, $\theta_3=\frac{1}{4}$, $\theta_4=\frac{1}{5}$, $g(t)=e^{\sin(\pi t)}$, $\phi(t,s-\vartheta+1)=\frac{e^{-(s+\frac{2}{3})}}{(t+50)^3}$ and $\varphi(t,s+\gamma)=\frac{e^{-s+\frac{4}{5}}}{(t+100)^2}$.

We can show that

$$\Theta=545.5721,\ |\Lambda|=202.553,\ \Omega_1=2189.264,\ \Omega_2=15715.32$$
$$\Omega_3=17049.09 \text{ and } \phi_0=\left(\tfrac{2}{99}\right)^3e^{-\frac{1}{6}},\ \varphi_0\leq\left(\tfrac{2}{199}\right)^2e^{-\frac{3}{10}}.$$

Nothing that $(H_1)-(H_3)$ hold, for each $t\in\frac{1}{2}\mathbb{N}_{-\frac{1}{2},\frac{27}{2}}$, we obtain

$$\left|F[t,u,\Upsilon^{\frac{1}{3}}u]-F[t,v,\Upsilon^{\frac{1}{3}}v]\right|\leq\tfrac{1}{101}|u-v|+\tfrac{1}{101}|\Upsilon^{\frac{1}{3}}u-\Upsilon^{\frac{1}{3}}v|,$$

$$\left|F[t,u,\Psi^{\frac{4}{5}}u]-F[t,v,\Psi^{\frac{4}{5}}v]\right| \leq \left(\tfrac{2}{199}\right)^2\left(\tfrac{2}{19}\right)^3|u-v|+\left(\tfrac{2}{19}\right)^3|\Psi^{\frac{4}{5}}u-\Psi^{\frac{4}{5}}v|,$$
$$\text{and } \tfrac{1}{e}<g(t)<e,$$

so, $L_1 = L_2 = 0.0099$, $\ell_1 = 1.178 \times 10^{-7}, \ell_2 = 0.0012$, $K = e$.

Finally, we find that

$$\chi = 0.0443 < 1.$$

Hence, by Theorem 1, the problem (47) has a unique solution on $\frac{1}{2}\mathbb{N}_{-\frac{1}{2},\frac{27}{2}}$.

5. Conclusions

We study the existence and unique results of the solution for a separate nonlinear Caputo fractional sum-difference equation with fractional sum-difference boundary conditions. Some conditions are obtained when Banach contraction principle is used as a tool. In addition, the conditions for the case of at least one solution are obtained by using the Schauder fixed point theorem.

Author Contributions: These authors contributed equally to this work.

Acknowledgments: The last author of this research was supported by Suan Dusit University.

References

1. Zhang, Y.; Liu, X.; Belić, M.R.; Zhong, W.; Zhang, Y.; Xiao, M. Propagation Dynamics of a Light Beam in a Fractional Schrödinger Equation. *Phys. Rev. Lett.* **2015**, *115*, 180403. [CrossRef] [PubMed]
2. Zingales, M.; Failla, G. The finite element method for fractional non-local thermal energy transfer in non-homogeneous rigid conductors. *Commun. Nonlinear Sci. Numer. Simul.* **2015**, *29*, 116–127. [CrossRef]
3. Lazopoulos, K.A.; Lazopoulos, A.K. On fractional bending of beams. *Arch. Appl. Mech.* **2016**, *86*, 1133–1145. [CrossRef]
4. Sumelka, W.; Voyiadjis, G.Z. A hyperelastic fractional damage material model with memory. *Int. J. Solids Struct.* **2017**, *124*, 151–160. [CrossRef]
5. Goodrich, C.S.; Peterson, A.C. *Discrete Fractional Calculus*; Springer: New York, NY, USA, 2015.
6. Wu, G.C.; Baleanu, D. Discrete fractional logistic map and its chaos. *Nonlinear Dyn.* **2014**, *75*, 283–287. [CrossRef]
7. Wu, G.C.; Baleanu, D. Chaos synchronization of the discrete fractional logistic map. *Signal Process.* **2014**, *102*, 96–99. [CrossRef]
8. Wu, G.C.; Baleanu, D.; Xie, H.P.; Chen, F.L. Chaos synchronization of fractional chaotic maps based on stability results. *Phys. A Stat. Mech. Its Appl.* **2016**, *460*, 374–383. [CrossRef]
9. Agarwal, R.P.; leanu, D.; Rezapour, S.; Salehi, S. The existence of solutions for some fractional finite difference equations via sum boundary conditions. *Adv. Differ. Equ.* **2014**, *2014*, 282. [CrossRef]
10. Goodrich, C.S. On discrete sequential fractional boundary value problems. *J. Math. Anal. Appl.* **2012**, *385*, 111–124. [CrossRef]
11. Goodrich, C.S. On a discrete fractional three-point boundary value problem. *J. Differ. Equ. Appl.* **2012**, *18*, 397–415. [CrossRef]
12. Weidong, L. Existence of solutions for discrete fractional boundary value problems witha p-Laplacian operator. *Adv. Differ. Equ.* **2012**, *2012*, 163.
13. Ferreira, R. Existence and uniqueness of solution to some discrete fractional boundary value problems of order less than one. *J. Differ. Equ. Appl.* **2013**, *19*, 712–718. [CrossRef]
14. Abdeljawad, T. On Riemann and Caputo fractional differences. *Comput. Math. Appl.* **2011**, *62*, 1602–1611. [CrossRef]

15. Atici, F.M.; Eloe, P.W. Two-point boundary value problems for finite fractional difference equations. *J. Differ. Equ. Appl.* **2011**, *17*, 445–456. [CrossRef]
16. Atici, F.M.; Eloe, P.W. A transform method in discrete fractional calculus. *Int. J. Differ. Equ.* **2007**, *2*, 165–176.
17. Sitthiwirattham, T.; Tariboon, J.; Ntouyas, S.K. Existence Results for fractional difference equations with three-point fractional sum boundary conditions. *Discret. Dyn. Nat. Soc.* **2013**, *2013*, 104276. [CrossRef]
18. Sitthiwirattham, T.; Tariboon, J.; Ntouyas, S.K. Boundary value problems for fractional difference equations with three-point fractional sum boundary conditions. *Adv. Differ. Equ.* **2013**, *2013*, *296*. [CrossRef]
19. Sitthiwirattham, T. Existence and uniqueness of solutions of sequential nonlinear fractional difference equations with three-point fractional sum boundary conditions. *Math. Method. Appl. Sci.* **2015**, *38*, 2809–2815. [CrossRef]
20. Sitthiwirattham, T. Boundary value problem for p-Laplacian Caputo fractional difference equations with fractional sum boundary conditions. *Math. Method Appl. Sci.* **2016**, *39*, 1522–1534. [CrossRef]
21. Chasreechai, S.; Kiataramkul, C.; Sitthiwirattham, T. On nonlinear fractional sum-difference equations via fractional sum boundary conditions involving different orders. *Math. Probl. Eng.* **2015**, *2015*, 519072. [CrossRef]
22. Reunsumrit, J.; Sitthiwirattham, T. Positive solutions of three-point fractional sum boundary value problem for Caputo fractional difference equations via an argument with a shift. *Positivity* **2016**, *20*, 761–1014. [CrossRef]
23. Reunsumrit, J.; Sitthiwirattham, T. On positive solutions to fractional sum boundary value problems for nonlinear fractional difference equations. *Math. Method. Appl. Sci.* **2016**, *39*, 2737–2751. [CrossRef]
24. Soontharanon, J.; Jasthitikulchai, N.; Sitthiwirattham, T. Nonlocal Fractional Sum Boundary Value Problems for Mixed Types of Riemann-Liouville and Caputo Fractional Difference Equations. *Dyn. Syst. Appl.* **2016**, *25*, 409–414.
25. Laoprasittichok, S.; Sitthiwirattham, T. On a Fractional Difference-Sum Boundary Value Problems for Fractional Difference Equations Involving Sequential Fractional Differences via Different Orders. *J. Comput. Anal. Appl.* **2017**, *23*, 1097–1111.
26. Kaewwisetkul, B.; Sitthiwirattham, T. On Nonlocal Fractional Sum-Difference Boundary Value Problems for Caputo Fractional Functional Difference Equations with Delay. *Adv. Differ. Equ.* **2017**, *2017*, 219. [CrossRef]
27. Reunsumrit, J.; Sitthiwirattham, T. A New Class of Four-Point Fractional Sum Boundary Value Problems for Nonlinear Sequential Fractional Difference Equations Involving Shift Operators. *Kragujevac J. Math.* **2018**, *42*, 371–387. [CrossRef]
28. Chasreechai, S.; Sitthiwirattham, T. Existence Results of Initial Value Problems for Hybrid Fractional Sum-Difference Equations. *Discret. Dyn. Nat. Soc.* **2018**, *2018*, 5268528. [CrossRef]
29. Chasreechai, S.; Sitthiwirattham, T. On Nonlocal Boundary Value Problems for Hybrid Fractional Sum-Difference Equations Involving Different Orders. *J. Nonlinear Funct. Anal.* **2018**. [CrossRef]
30. Griffel, D.H. *Applied Functional Analysis*; Ellis Horwood Publishers: Chichester, UK, 1981.
31. Guo, D.; Lakshmikantham, V. *Nonlinear Problems in Abstract Cone*; Academic Press: Orlando, FL, USA, 1988.

4

Application of Fixed-Point Theory for a Nonlinear Fractional Three-Point Boundary-Value Problem

Ehsan Pourhadi [1,2], Reza Saadati [2] and Sotiris K. Ntouyas [3,4,*]

1 International Center for Mathematical Modelling in Physics and Cognitive Sciences, Department of Mathematics, Linnaeus University, SE-351 95 Växjö, Sweden; epourhadi@alumni.iust.ir
2 Department of Mathematics, Iran University of Science and Technology, Narmak, Tehran 16846-13114, Iran; rsaadati@eml.cc
3 Department of Mathematics, University of Ioannina, 451 10 Ioannina, Greece
4 Nonlinear Analysis and Applied Mathematics (NAAM)-Research Group, Department of Mathematics, Faculty of Science, King Abdulaziz University, P.O. Box 80203, Jeddah 21589, Saudi Arabia
* Correspondence: sntouyas@uoi.gr

Abstract: Throughout this paper, via the Schauder fixed-point theorem, a generalization of Krasnoselskii's fixed-point theorem in a cone, as well as some inequalities relevant to Green's function, we study the existence of positive solutions of a nonlinear, fractional three-point boundary-value problem with a term of the first order derivative

$$({}_a^C D^{\alpha} x)(t) = f(t, x(t), x'(t)), \quad a < t < b, \quad 1 < \alpha < 2,$$

$$x(a) = 0, x(b) = \mu x(\eta), \quad a < \eta < b, \quad \mu > \lambda,$$

where $\lambda = \dfrac{b-a}{\eta-a}$ and ${}_a^C D^{\alpha}$ denotes the Caputo's fractional derivative, and $f : [a,b] \times \mathbb{R} \times \mathbb{R} \to \mathbb{R}$ is a continuous function satisfying the certain conditions.

Keywords: three-point boundary-value problem; Caputo's fractional derivative; Riemann-Liouville fractional integral; fixed-point theorems

---

1. Introduction

In the last decade, questions on positive solutions to two-point, three-point, and multi-point boundary value problems (BVPs) and integral boundary-value problems for nonlinear ordinary and fractional differential equations have attracted much interest. The investigation of three-point BVPs for nonlinear integer-order ordinary differential equations was initially begun by Gupta [1]. Since then, several authors have put their focus on the existence and multiplicity of solutions (or positive solutions) of three-point BVPs for nonlinear integer-order ordinary differential equations. Several papers are available in regard to the setting of integer orders of differential equations in the literature. In 2000, applying the fixed-point index theorems, the Leray-Schauder degree, and upper and lower solutions, Ma [2] studied a class of second-order three-point boundary value problems with a nonlinear term $f(x)$. In 2002, He and Ge [3], with the help of the Leggett-Williams fixed-point theorem [4], investigated the multiplicity of positive solutions of a problem with the nonlinear term $f(t,x)$ (see [5–15] and the references therein).

In recent years, multi-point boundary value problems have also been considered for fractional-order differential equations. For instance, employing the superlinearity and sublinearity, together with the well-known Guo-Lakshmikantham fixed-point theorem in cones, Ntouyas and Pourhadi [16] studied the existence of positive solutions to the boundary-value problem with a fractional order, $1 < \alpha < 2$. Furthermore, they investigated the convexity and concavity of the solutions with respect to the behavior of a given function as a coefficient of the subjected problem (see also [17–21]).

There were only a few papers available which focused on the existence of solutions for nonlinear fractional differential equations associated with three-point boundary conditions, which served as motivation for this work. The key idea of the current paper is that a term of the first-order derivative is involved in the subjected nonlinear problem, while most works (either fractional or ordinary differential equations) are done under the assumption that the first-order derivative is not involved explicitly in the nonlinear term.

In this paper, an analogy with a boundary-value problem for differential equations of integer orders via the Schauder fixed-point theorem, a generalized version of Krasnoselskii's fixed-point theorem in a cone [22], and also using the associated Green's function for the relevant problem, the existence of positive solutions for a fractional three-point boundary-value problem is investigated.

$$\begin{cases} ({}_a^C D^\alpha x)(t) = f(t, x(t), x'(t)), & a < t < b, \quad 1 < \alpha < 2, \\ x(a) = 0, \quad x(b) = \mu x(\eta), & a < \eta < b, \quad \mu > \lambda, \end{cases} \tag{1}$$

where $\lambda = \dfrac{b-a}{\eta - a}$ and ${}_a^C D^\alpha$ stands for the Caputo's fractional derivative, and $f : [a,b] \times \mathbb{R} \times \mathbb{R} \to \mathbb{R}$ is a continuous function which will be specified later on.

The organization of this paper is as follows. In Section 2, we recall some auxiliary facts and preliminaries. In Section 3, we first find the Green's function associated with (1), and then, using the inequalities related with this function and two well-known fixed-point theorems, we present our main results. An illustrative example is also given.

2. Preliminaries

This section is devoted to recall and gathering of some essential definitions and auxiliary facts in fractional calculus, as well as the results needed further on, which can be found in [23–25].

Definition 1. *Let $\alpha \geq 0$ and f be a real function defined in $[a,b]$. The Riemann-Liouville fractional integral of order α for a continuous function $f : (a, \infty) \to \mathbb{R}$ is defined by $({}_aI^0 f)(t) = f(t)$ and*

$$({}_aI^\alpha f)(t) = \frac{1}{\Gamma(\alpha)} \int_a^t (t-s)^{\alpha-1} f(s) ds, \quad \alpha > 0, \quad t \in [a,b],$$

where $\Gamma(\cdot)$ is the Gamma function.

Definition 2. *For a continuous function $f : (a, \infty) \to \mathbb{R}$, the Riemann-Liouville fractional derivative of fractional order $\alpha > 0$ is defined by*

$${}^{RL}D_{a+}^\alpha f(t) = \frac{1}{\Gamma(n-\alpha)} \left(\frac{d}{dt}\right)^n \int_a^t (t-s)^{n-\alpha-1} f(s) ds, \ \ n = [\alpha] + 1,$$

where $[\alpha]$ denotes the integer part of the real number α.

For $\alpha < 0$ and the convenience of the reader, we use the denotation $D^\alpha y = I^{-\alpha} y$. Moreover, for $\beta \in [0, \alpha)$, it is valid that $D^\beta I^\alpha y = I^{\alpha-\beta} y$.

Definition 3. *Caputo's fractional derivative of order $\alpha \geq 0$ is given by $({}_a^C D^0 f)(t) = f(t)$, and $({}_a^C D^\alpha f)(t) = ({}_a I^{m-\alpha} D^m f)(t)$ for $\alpha > 0$, where m is the smallest integer greater or equal to α. Besides, it can be formulated by*

$$ {}^C D_{a+}^{\alpha} f(t) = \frac{1}{\Gamma(n-\alpha)} \int_a^t (t-s)^{n-\alpha-1} f^{(n)}(s) ds, \quad n = [\alpha]+1, \quad \text{for } f \in AC^n([a,b]), $$

where $\alpha \notin \mathbb{N}_0$ and $AC^n([a,b])$ represents the space of all absolutely continuous functions having an absolutely continuous derivative up to $(n-1)$ (see also [23]).

In the sequel, the associated Green's function for the three-point BVP (1) is formulated by utilizing a crucial lemma derived by Zhang [26] as follows:

Lemma 1. *Let $\alpha > 0$; then, in $C(0,T) \cap L(0,T)$, the differential equation*

$$ {}^C D_{0+}^{\alpha} u(t) = 0 $$

has solutions $u(t) = c_0 + c_1 t + c_2 t^2 + \cdots + c_n t^{n-1}$, $c_i \in \mathbb{R}$, $i = 0, 1, \cdots, n$, $n = [\alpha]+1$.

Furthermore, it has been proved that $I_{0+}^{\alpha} D_{0+}^{\alpha} u(t) = u(t) + c_0 + c_1 t + c_2 t^2 + \cdots + c_n t^{n-1}$ for some $c_i \in \mathbb{R}$, $i = 0, 1, \cdots, n$, $n = [\alpha]+1$ (see Lemma 2.3 in [26]).

3. Main Results

In the following, we present a pivotal lemma which will play a crucial role in our next analysis and direct our attention to a variant of Problem (1).

Lemma 2. *Let $\Delta := \mu(\eta - a) - (b-a) > 0$. Then, $x \in C^1(I, \mathbb{R})$ is the solution of fractional three-point BVP (1) if, and only if x satisfies the integral equation*

$$ x(t) = \int_a^b G(t,s) f(s, x(s), x'(s)) ds, \qquad t \in I := [a,b] \tag{2} $$

where the Green's function $G(t,s) := G_1(t,s) + G_2(t,s)$ is given by

$$ G_1(t,s) = \begin{cases} \dfrac{(t-s)^{\alpha-1}}{\Gamma(\alpha)}, & a \leq s \leq t \leq b \\ 0, & a \leq t \leq s \leq b \end{cases} \tag{3} $$

$$ G_2(t,s) = \begin{cases} \dfrac{t-a}{\Delta\Gamma(\alpha)} \left((b-s)^{\alpha-1} - \mu(\eta-s)^{\alpha-1} \right), & a \leq s \leq \eta, \quad t \in I, \\ \dfrac{(b-s)^{\alpha-1}(t-a)}{\Delta\Gamma(\alpha)}, & \eta \leq s \leq b, \quad t \in I. \end{cases} \tag{4} $$

Moreover,

$$ \max_{t,s \in I} G(t,s) \leq \frac{\mu(b-a)^\alpha}{\lambda\Delta\Gamma(\alpha)}. \tag{5} $$

Proof. By employing the Riemann-Liouville fractional integral ${}_a I^\alpha$ for Equation (1), the imposed boundary conditions, and the knowledge received from the fractional calculus theory, we observe that $x \in C^1[a,b]$ is a solution of (1) if, and only if

$$ x(t) = c_0 + c_1(t-a) + \frac{1}{\Gamma(\alpha)} \int_a^t (t-s)^{\alpha-1} f(s, x(s), x'(s)) ds \tag{6} $$

for some real constants c_0 and c_1 (see Lemma 1). Since $x(a) = 0$, we immediately derive $c_0 = 0$. Now,

$$\begin{aligned} x(b) = \mu x(\eta) \quad \Leftrightarrow \quad & c_1(b-a) + \frac{1}{\Gamma(\alpha)} \int_a^b (b-s)^{\alpha-1} f(s, x(s), x'(s)) ds \\ & = c_1 \mu(\eta - a) + \frac{\mu}{\Gamma(\alpha)} \int_a^\eta (\eta - s)^{\alpha-1} f(s, x(s), x'(s)) ds \\ \Leftrightarrow \quad & c_1 = \frac{1}{\Delta\Gamma(\alpha)} \bigg(\int_a^b (b-s)^{\alpha-1} f(s, x(s), x'(s)) ds \\ & - \mu \int_a^\eta (\eta - s)^{\alpha-1} f(s, x(s), x'(s)) ds \bigg) \end{aligned}$$

which, together with (6), implies that

$$\begin{aligned} x(t) = & \frac{1}{\Gamma(\alpha)} \int_a^t (t-s)^{\alpha-1} f(s, x(s), x'(s)) ds \\ & + \frac{t-a}{\Delta\Gamma(\alpha)} \bigg(\int_a^b (b-s)^{\alpha-1} f(s, x(s), x'(s)) ds - \mu \int_a^\eta (\eta - s)^{\alpha-1} f(s, x(s), x'(s)) ds \bigg). \end{aligned} \tag{7}$$

This is also equivalent to

$$\begin{aligned} x(t) = & \frac{1}{\Gamma(\alpha)} \int_a^t (t-s)^{\alpha-1} f(s, x(s), x'(s)) ds \\ & + \frac{t-a}{\Delta\Gamma(\alpha)} \bigg(\int_a^\eta \Big((b-s)^{\alpha-1} - \mu(\eta - s)^{\alpha-1} \Big) f(s, x(s), x'(s)) ds \\ & + \int_\eta^b (b-s)^{\alpha-1} f(s, x(s), x'(s)) ds \bigg). \end{aligned} \tag{8}$$

Now, (8) can be rewritten as follows:

$$x(t) = \int_a^b G(t,s) f(s, x(s), x'(s)) ds, \qquad t \in I = [a, b],$$

where the associated Green's function $G(t,s) = G_1(t,s) + G_2(t,s)$ is defined by (3) and (4). Furthermore, for any $s \in I$,

$$\begin{aligned} \max_{t \in I} G(t,s) & = \frac{(b-s)^{\alpha-1}}{\Gamma(\alpha)} + \max_{t \in I} \left(\frac{(b-s)^{\alpha-1}(t-a)}{\Delta \cdot \Gamma(\alpha)} \right) \\ & = \frac{(b-s)^{\alpha-1}}{\Gamma(\alpha)} \left(1 + \frac{b-a}{\Delta} \right) \\ & = \frac{(b-s)^{\alpha-1}}{\Delta\Gamma(\alpha)} \Big(\mu(\eta - a) \Big) \\ & \leq \frac{\mu(b-a)^\alpha}{\lambda\Delta\Gamma(\alpha)}. \end{aligned} \tag{9}$$

Therefore, the inequality (5) is proved. □

Throughout the remainder of this paper, we employ two well-known fixed-point results to study Equation (1).

3.1. Existence of Positive Solution with the Schauder Fixed-Point Principle

In the following, we investigate Equation (1) via the Schauder fixed-point theorem.

Theorem 1 (Schauder fixed-point Theorem, [27]). *Let U be a nonempty and convex subset of a normed space X. Let T be a continuous mapping of U into a compact set $K \subset U$. Then, T has a fixed point.*

In the sequel, we suppose the following condition:

(C_0) f satisfies Carathéodory-type conditions. That is, $f(\cdot, u, v)$ is measurable for the fixed u, v, and $f(t, \cdot, \cdot)$ is continuous for a.e. $t \in I$. Moreover, if $u \geq 0$, then $f(t, u, v) \geq 0$.

Under this condition, the equivalent representation for Equation (2) is given by

$$x(t) = [\mathcal{F}x](t), \qquad t \in I = [a, b],$$

where $\mathcal{F}$ is an operator defined by

$$[\mathcal{F}x](t) = \int_a^b G(t,s) f(s, x(s), x'(s)) ds, \qquad t \in I.$$

It is obvious to see that $x(t)$ is a solution to the problem (1) if it is a fixed point of the operator $\mathcal{F}$.

Theorem 2. *Suppose that f satisfies the condition (C_0) and the followings:*

(C_1) *There exists an L^1-function $\varphi : I \to \mathbb{R}^+$, such that*

$$|f(t, x(t), x'(t))| \leq \varphi(t)\Omega(\|x\|), \qquad x \in C^1(I, \mathbb{R}), \quad t \in I,$$

where $\Omega : I \to [0, \infty)$ is a non-decreasing continuous function and $\| \cdot \|$ denotes the supremum norm on I.

(C_2) *The point $\eta \in (a, b)$ is taken sufficiently close to a, such that*

$$\mu \int_a^\eta (\eta - s)^{\alpha-1} f(s, x(s), x'(s)) ds \leq \int_a^b (b - s)^{\alpha-1} f(s, x(s), x'(s)) ds, \quad \text{for all } x \in C^1(I, \mathbb{R}).$$

Moreover, suppose that there exists a continuous function p defined on I satisfying the following inequality:

$$\frac{\mu(b-a)^\alpha}{\lambda \Delta \Gamma(\alpha)} \|\varphi\|_1 \Omega(\|p\|) \leq \|p\|. \tag{10}$$

Then, Equation (1) has at least one positive solution in $C^1(I, \mathbb{R})$, bounded above by $\|p\|$.

Proof. Let us define

$$S = \left\{ x \in C^1(I, \mathbb{R}) \;\middle|\; 0 \leq x(t) \leq \|p\| \text{ for } t \in I \right\},$$

where $\| \cdot \|_1$ denotes the L^1-norm on I, and p is a function satisfying the condition (C_2). Clearly, the set S is a non-empty, closed, bounded, and convex subset of $C^1(I, \mathbb{R})$. To establish that Equation (1) has a positive solution, it only suffices to show that the operator $\mathcal{F}$ has a fixed point in S. We first show that S is $\mathcal{F}$-invariant. Let $x(t)$ be a non-negative function; then, following condition (C_0), one finds that $f(t, x(t), x'(t))$ is non-negative too, and the right-hand side of (7), together with conditions (C_0), (C_2) and the fact that $\Delta > 0$ imply that $[\mathcal{F}x](t) \geq 0$.

On the other hand, using (2), (9), (10) and condition (C_1), one can see that

$$|[\mathcal{F}x](t)| \leq \|p\|, \qquad \text{for } t \in I.$$

Hence, $\mathcal{F}S \subset S$. Furthermore, to show the continuity of the operator $\mathcal{F} : S \to S$, we have

$$|[\mathcal{F}x_n](t) - [\mathcal{F}x](t)| \to 0 \quad \Longleftrightarrow \quad \int_a^b |G(t,s)| \cdot |f(s, x(s), x'(s)) - f(s, x_n(s), x_n'(s))| ds \to 0$$

for $x, x_n \in S \subset C^1(I, \mathbb{R})$.

Next, we show that $\mathcal{F}S$ is equicontinuous. Assume that $a \leq t_1 < t_2 \leq b$. Following the definition of $\mathcal{F}$ and the condition (C_1), we have

$$|[\mathcal{F}x](t_2) - [\mathcal{F}x](t_1)| \leq \|\varphi\|_1 \Omega(\|x\|) \cdot \max_{s \in I} |G(t_2, s) - G(t_1, s)|,$$

which tends to zero, as $t_1 \to t_2$. Consequently, we conclude that $\mathcal{F}S$ is equicontinuous. Furthermore, the equicontinuity of the set of functions $[\mathcal{F}S]' = \{y' : y = \mathcal{F}x, x \in S\}$ can also be shown. Indeed, suppose that $a \leq t_1 < t_2 \leq b$; then,

$$|[\mathcal{F}x]'(t_2) - [\mathcal{F}x]'(t_1)| \leq \|\varphi\|_1 \Omega(\|x\|) \cdot \max_{s \in I} |\frac{\partial}{\partial t} G(t_2, s) - \frac{\partial}{\partial t} G(t_1, s)| \to 0$$

whenever $t_1 \to t_2$. Therefore, we conclude that $[\mathcal{F}S]'$ is equicontinuous.

Besides, S is totally bounded (since every sequence in S has a Cauchy subsequence), so S is compact and $\mathcal{F}S$ is compact. Now, all the conditions of the Schauder fixed point are fulfilled; thus, the operator $\mathcal{F}$ as a self-map on S possesses a fixed point in this set, which yields that Equation (1) has a positive solution bounded above by $\|p\|$. □

3.2. Existence of Positive Solution via the Krasnoselskii Type Fixed-Point Theorem

In what follows, we recall a generalization of Krasnoselskii's fixed-point theorem of cone expansion and compression of a norm type. To do this, let us suppose $(X, \|\cdot\|)$ is a Banach space, and P is the cone in X. Assume that $\overline{\alpha}, \overline{\beta} : X \to \mathbb{R}^+$ are two continuous non-negative functionals that satisfy

$$\overline{\alpha}(rx) \leq |r|\overline{\alpha}(x), \quad \overline{\beta}(rx) \leq |r|\overline{\beta}(x), \quad \text{for } x \in X,\ r \in [0,1], \tag{11}$$

and

$$M_1 \max\{\overline{\alpha}(x), \overline{\beta}(x)\} \leq \|x\| \leq M_2 \max\{\overline{\alpha}(x), \overline{\beta}(x)\}, \quad \text{for } x \in X, \tag{12}$$

where M_1, M_2 are two positive constants.

The following lemma is understood as a special case of a result derived by Bai and Ge (see [22] Theorem 2.1).

Lemma 3. *Let $r_2 > r_1 > 0$, $L_2 > L_1 > 0$ be constants and*

$$\Omega_i = \{x \in X \mid \overline{\alpha}(x) < r_i,\ \overline{\beta}(x) < L_i\},\ i = 1,2$$

be two open subsets in X, such that $\theta \in \Omega_1 \subset \overline{\Omega}_1 \subset \Omega_2$. In addition, let

$$C_i = \{x \in X \mid \overline{\alpha}(x) = r_i,\ \overline{\beta}(x) \leq L_i\},\ i = 1,2;$$
$$D_i = \{x \in X \mid \overline{\alpha}(x) \leq r_i,\ \overline{\beta}(x) = L_i\},\ i = 1,2.$$

Assume $T : P \to P$ is a completely continuous operator satisfying

(S_1) $\overline{\alpha}(Tx) \leq r_1, x \in C_1 \cap P$; $\overline{\beta}(Tx) \leq L_1, x \in D_1 \cap P$;
$\overline{\alpha}(Tx) \geq r_2, x \in C_2 \cap P$; $\overline{\beta}(Tx) \geq L_2, x \in D_2 \cap P$;

or

(S_2) $\overline{\alpha}(Tx) \geq r_1, x \in C_1 \cap P$; $\overline{\beta}(Tx) \geq L_1, x \in D_1 \cap P$;
$\overline{\alpha}(Tx) \leq r_2, x \in C_2 \cap P$; $\overline{\beta}(Tx) \leq L_2, x \in D_2 \cap P$;

then, T has at least one fixed point in $(\overline{\Omega_2} \setminus \Omega_1) \cap P$.

To apply the recent fixed-point theorem, let us consider the following settings.

Let X be a Banach space in $C^1(I, \mathbb{R})$, with

$$\|x\| = \max\{\max_{t\in I} |x(t)|, \max_{t\in I} |x'(t)|\}, \quad x \in X.$$

Define a cone P by

$$P = \left\{ x \in X \;\middle|\; x(t) \geq 0, \quad \text{for all } t \in I \right\},$$

and functionals

$$\overline{\alpha}(x) = \max_{t\in I} |x(t)|, \qquad \overline{\beta}(x) = \max_{t\in I} |x'(t)|, \quad \forall x \in X.$$

With the help of (11) and (12), $\overline{\alpha}$ and $\overline{\beta}$ are two continuous non-negative functionals, such that $\|x\| = \max\{\overline{\alpha}(x), \overline{\beta}(x)\}$. Let us consider the following notations:

$$L := \max_{t\in I} \left| \int_a^b G(t,s)ds \right|,$$

$$N := \left(\frac{(b-a)^{\alpha-1}}{\Gamma(\alpha)} + \frac{1}{\Delta\Gamma(\alpha+1)} \left[2(b-\xi)^\alpha - 2\mu(\eta-\xi)^\alpha + \mu(\eta-a)^\alpha - (b-a)^\alpha \right] \right),$$

where $\xi = \eta - \dfrac{b-\eta}{\mu^{\frac{1}{\alpha-1}} - 1} \in [a, \eta]$.

Accounting on condition (C_2), we get that the operator $\mathcal{F}$ (as defined before) transforms P into itself; moreover, a standard argument shows that it is completely continuous. In fact, $\mathcal{F}$ is continuous and maps any bounded subset of P into a relatively compact subset of P.

In the following result, we suppose that η is sufficiently close to a such that the Green function G is non-negative. For the possibility, we refer to Example (1).

Theorem 3. *Suppose there are four constants,* $k_2 > k_1 > 0$, $l_2 > l_1 > 0$, *such that* $\max\left\{\frac{k_1}{L}, \frac{l_1}{M}\right\} \leq \min\left\{\frac{k_2}{L}, \frac{l_2}{N}\right\}$, *and the following assumptions hold:*

(C_3) *There is an* L^1*-function* $\psi : I \to \mathbb{R}^+$ *which satisfies the following condition:*

$$\int_a^b (b-s)^{\alpha-1} f(s, x(s), x'(s))ds - \mu \int_a^\eta (\eta-s)^{\alpha-1} f(s, x(s), x'(s))ds \geq \int_a^b \psi(s) f(s, x(s), x'(s))ds,$$

for all $x \in C^1(I, \mathbb{R}^+)$.

(C_4) $f(t,u,v) \geq \max\left\{\frac{k_1}{L}, \frac{l_1}{M}\right\}$, *for* $(t,u,v) \in I \times [0, k_1] \times [-l_1, l_1]$;

(C_5) $f(t,u,v) \leq \min\left\{\frac{k_2}{L}, \frac{l_2}{N}\right\}$, *for* $(t,u,v) \in I \times [0, k_2] \times [-l_2, l_2]$, *where* $M = \dfrac{\|\psi\|_1}{\Delta\Gamma(\alpha)}$. *Then, problem* (1) *has at least one positive solution* $x(t)$*, such that*

$$k_1 \le \max_{t\in I} x(t) \le k_2 \qquad or \qquad l_1 \le \max_{t\in I} |x'(t)| \le l_2.$$

Proof. Let us take the following subsets of $X = C^1(I, \mathbb{R})$

$$\begin{aligned}
\Omega_i &= \{x \in X \mid \overline{\alpha}(x) < k_i, \quad \overline{\beta}(x) < l_i\}, \quad i = 1,2;\\
\mathcal{P}_i &= \{x \in X \mid \overline{\alpha}(x) = k_i, \quad \overline{\beta}(x) \le l_i\}, \quad i = 1,2;\\
\mathcal{Q}_i &= \{x \in X \mid \overline{\alpha}(x) \le k_i, \quad \overline{\beta}(x) = l_i\}, \quad i = 1,2.
\end{aligned}$$

For $x \in \mathcal{P}_1 \cap P$, by (C_4), there exists

$$\overline{\alpha}(\mathcal{F}x) = \max_{t\in I}\left|\int_a^b G(t,s)f(s,x(s),x'(s))ds\right| \ge \frac{k_1}{L}\max_{t\in I}\left|\int_a^b G(t,s)ds\right| = k_1. \tag{13}$$

Since η is taken sufficiently close to a such that the Green function G is non-negative, the inequality (13) holds. Moreover, taking into account the continuity and properties of $\mathcal{F}$, we derive

$$\begin{aligned}
(\mathcal{F}x)'(t) &= \frac{\alpha-1}{\Gamma(\alpha)}\int_a^t (t-s)^{\alpha-2} f(s,x(s),x'(s))ds\\
&\quad + \frac{1}{\Delta\Gamma(\alpha)}\left(\int_a^b (b-s)^{\alpha-1} f(s,x(s),x'(s))ds - \mu\int_a^\eta (\eta-s)^{\alpha-1} f(s,x(s),x'(s))ds\right),\\
(\mathcal{F}x)''(t) &= -\frac{(\alpha-1)(2-\alpha)}{\Gamma(\alpha)}\int_a^t (t-s)^{\alpha-3} f(s,x(s),x'(s))ds \le 0, \qquad t \in I.
\end{aligned}$$

Therefore, $(\mathcal{F}x)(t)$ is concave on I, and so the absolute value of $(\mathcal{F}x)'$ takes its maximum only at the endpoints of I. That is,

$$\max_{t\in I} |(\mathcal{F}x)'(t)| = \max_{t\in I}\{|(\mathcal{F}x)'(a)|, |(\mathcal{F}x)'(b)|\} = |(\mathcal{F}x)'(b)|.$$

Therefore, for $x \in \mathcal{Q}_1 \cap P$, followed by (C_3) and (C_4), one can see that

$$\begin{aligned}
\beta(\mathcal{F}x) &= \max_{t\in I}\{|(\mathcal{F}x)'(a)|, |(\mathcal{F}x)'(b)|\}\\
&\ge |(\mathcal{F}x)'(a)|\\
&= \frac{1}{\Delta\Gamma(\alpha)}\left(\int_a^b (b-s)^{\alpha-1} f(s,x(s),x'(s))ds - \mu\int_a^\eta (\eta-s)^{\alpha-1} f(s,x(s),x'(s))ds\right)\\
&\ge \frac{1}{\Delta\Gamma(\alpha)}\int_a^b \psi(s) f(s,x(s),x'(s))ds\\
&\ge \frac{l_1}{M\Delta\Gamma(\alpha)}\|\psi\|_1 = l_1.
\end{aligned}$$

Now, assuming $x \in \mathcal{P}_2 \cap P$, by (C_5), there is

$$\begin{aligned}
\overline{\alpha}(\mathcal{F}x) &= \max_{t\in I}\left|\int_a^b G(t,s)f(s,x(s),x'(s))ds\right|\\
&\le \frac{k_2}{L}\max_{t\in I}\left|\int_a^b G(t,s)ds\right| = k_2.
\end{aligned}$$

Finally, for $x \in \mathcal{Q}_2 \cap P$, by ($C_5$), one can find

$$\begin{aligned}
\beta(\mathcal{F}x) &= \max_{t\in I}\{|(\mathcal{F}x)'(a)|, |(\mathcal{F}x)'(b)|\} \\
&= |(\mathcal{F}x)'(b)| \\
&= \frac{\alpha-1}{\Gamma(\alpha)}\int_a^b (b-s)^{\alpha-2} f(s,x(s),x'(s))ds \\
&\quad + \frac{1}{\Delta\Gamma(\alpha)}\left(\int_a^\eta \left((b-s)^{\alpha-1} - \mu(\eta-s)^{\alpha-1}\right) f(s,x(s),x'(s))ds\right. \\
&\quad \left. + \int_\eta^b (b-s)^{\alpha-1} f(s,x(s),x'(s))ds\right) \\
&\leq \frac{l_2}{N}\left(\frac{(b-a)^{\alpha-1}}{\Gamma(\alpha)} + \frac{1}{\Delta\Gamma(\alpha)}\int_a^\eta \left|(b-s)^{\alpha-1} - \mu(\eta-s)^{\alpha-1}\right| ds + \frac{(b-\eta)^\alpha}{\Delta\Gamma(\alpha+1)}\right) \\
&= \frac{l_2}{N}\left(\frac{(b-a)^{\alpha-1}}{\Gamma(\alpha)} + \frac{1}{\Delta\Gamma(\alpha)}\left(\int_a^\xi \left[\mu(\eta-s)^{\alpha-1} - (b-s)^{\alpha-1}\right] ds\right.\right. \\
&\quad \left.\left. + \int_\xi^\eta \left[(b-s)^{\alpha-1} - \mu(\eta-s)^{\alpha-1}\right] ds\right) + \frac{(b-\eta)^\alpha}{\Delta\Gamma(\alpha+1)}\right) \\
&= \frac{l_2}{N}\left(\frac{(b-a)^{\alpha-1}}{\Gamma(\alpha)} + \frac{1}{\Delta\Gamma(\alpha+1)}\left[2(b-\xi)^\alpha - 2\mu(\eta-\xi)^\alpha + \mu(\eta-a)^\alpha - (b-a)^\alpha\right]\right) \\
&= l_2.
\end{aligned}$$

Now, all conditions of Lemma 3 are satisfied, and it implies that there exists $x \in (\overline{\Omega_2 \setminus \Omega_1}) \cap P$, such that $x = \mathcal{F}x$. That is, the problem (1) has at least one positive solution $x(t)$, such that

$$k_1 \leq \overline{\alpha}(x) \leq k_2 \qquad \text{or} \qquad l_1 \leq \overline{\beta}(x) \leq l_2.$$

In other words,

$$k_1 \leq \max_{t\in I} x(t) \leq k_2 \qquad \text{or} \qquad l_1 \leq \max_{t\in I} |x'(t)| \leq l_2,$$

which completes the proof. □

In the following, we illustrate the said result with an example.

Example 1. *Consider the boundary value problem:*

$$\begin{cases} ({}_a^C D^{\frac{3}{2}} x)(t) = f(t,x(t),x'(t)), & 0 < t < 1, \\ x(0) = 0, \quad x(1) = \mu x(\eta), & 0 < \eta < 1, \quad \mu\eta > 1, \end{cases} \tag{14}$$

where $\mu = a\eta^{-r}$ for some $a > 1$, $0 < r < 1.5$, and $f : [0,1] \times \mathbb{R} \times \mathbb{R} \to \mathbb{R}$ is given by

$$f(t,u,v) = \lambda_1 \sin^2 u + \lambda_2 \cos^2 v + \lambda_3 t + \lambda_4, \qquad t \in I = [0,1],\ u,v \in \mathbb{R}$$

such that $\lambda_i \geq 0$, $\lambda_4 > 0$, $i = 1,2,3$, are constant. Since f takes both supremum and infimum over domain D, let us set

$$\inf_D f(t,u,v) = M_1, \qquad \sup_D f(t,u,v) = M_2.$$

A direct computation shows that

$$L = \frac{4\mu\lambda}{3(\mu\lambda - 1)\sqrt{\pi}}.$$

On the other hand, by considering η *as being sufficiently close to* 0 *and* $\psi(t) < \frac{2M_1}{3M_2}$, *we see that*

$$\int_0^1 \sqrt{1-s} f(s,x(s),x'(s))ds > \int_0^1 \psi(s) f(s,x(s),x'(s))ds.$$

Thus, condition (C_3) *is satisfied. To give more detail, if* $\eta \to 0$, *then using the fact that* $\mu = a\eta^{-r}$, *together with Leibniz's rule, we see that*

$$\begin{aligned}
\lim_{\eta\to 0}\left|\mu \int_0^\eta \sqrt{\eta-s} f(s,x(s),x'(s))ds\right| &= \frac{a}{2r}\lim_{\eta\to 0}\frac{\left|\int_0^\eta \frac{1}{\sqrt{\eta-s}} f(s,x(s),x'(s))ds\right|}{\eta^{r-1}} \\
&\le \frac{a|M_2|}{2r}\lim_{\eta\to 0}\frac{\int_0^\eta \frac{1}{\sqrt{\eta-s}}ds}{\eta^{r-1}} \\
&= \frac{a|M_2|}{r}\lim_{\eta\to 0}\eta^{1.5-r} = 0,
\end{aligned}$$

which shows that the second integral term in the left-hand side of the inequality in condition (C_3) *vanishes for* η *sufficiently close to* 0.

Furthermore, $M = \frac{2\|\psi\|_1}{(\mu\eta-1)\sqrt{\pi}} < L$ *and*

$$N = \frac{2}{\sqrt{\pi}} + \frac{L}{\mu\eta}\left(2(1-\xi)^{1.5} - 2\mu(\eta-\xi)^{1.5} + \mu\eta^{1.5} - 1\right) > L,$$

where $\xi = \frac{\eta\mu^2-1}{\mu^2-1} \in (0,1)$. *Next, to check the conditions* (C_4) *and* (C_5), *choose* $k_2 > l_2 > l_1 > k_1 > 0$, *such that* $l_1 = M \cdot M_1$ *and* $l_2 = N \cdot M_2$. *Then, one can derive the followings:*

$$f(t,u,v) \ge \max\left\{\frac{k_1}{L}, \frac{l_1}{M}\right\} = \frac{l_1}{M}, \quad \text{for } (t,u,v) \in I \times [0,k_1] \times [-l_1,l_1];$$

$$f(t,u,v) \le \min\left\{\frac{k_2}{L}, \frac{l_2}{N}\right\} = \frac{l_2}{N}, \quad \text{for } (t,u,v) \in I \times [0,k_2] \times [-l_2,l_2].$$

That is to say, all the assumptions of Theorem 3 are fulfilled, then problem (14) *has at least one positive solution* x, *such that*

$$k_1 \le \max_{t\in I} x(t) \le k_2 \quad \text{or} \quad M \cdot M_1 \le \max_{t\in I}|x'(t)| \le N \cdot M_2.$$

Author Contributions: Formal Analysis, E.P., R.S. and S.K.N.

Acknowledgments: The authors thank the reviewers for their useful remarks on our work.

References

1. Gupta, C.P. Solvability of a three-point nonlinear boundary value problem for a second order ordinary differential equation. *J. Math. Anal. Appl.* **1992**, *168*, 540–551. [CrossRef]
2. Ma, R. Multiplicity of positive solutions for second-order three-point boundary value problems. *Comput. Math. Appl.* **2000**, *40*, 193–204. [CrossRef]
3. He, X.; Ge, W. Triple solutions for second-order three-point boundary value problems. *J. Math. Anal. Appl.* **2002**, *268*, 256–265. [CrossRef]

4. Leggett, R.W.; Williams, L.R. Multiple positive fixed-points of nonlinear operators on ordered Banach spaces. *Indiana Univ. Math. J.* **1979**, *28*, 673–688. [CrossRef]
5. Bai, Z. Solvability for a class of fractional m-point boundary value problem at resonance. *Comput. Math. Appl.* **2011**, *62*, 1292–1302. [CrossRef]
6. Bai, Z.; Zhang, Y. Solvability of fractional three-point boundary value problems with nonlinear growth. *Appl. Math. Comput.* **2011**, *218*, 1719–1725. [CrossRef]
7. Cui, Y.; Sun, J. Positive solutions for second-order three-point boundary value problems in Banach spaces. *Acta Math. Sin.* **2011**, *4*, 743–751.
8. Guo, Y.; Ge, W. Positive solutions for three-point boundary value problems with dependence on the first order derivative. *J. Math. Anal. Appl.* **2004**, *290*, 291–301. [CrossRef]
9. Il'in, V.A.; Moiseev, E.I. Nonlocal boundary value problem of the first kind for a Sturm-Liouville operator in its differential and finite difference aspects. *Differ. Equ.* **1987**, *23*, 803–810.
10. Il'in, V.A.; Moiseev, E.I. Nonlocal boundary value problem of the second kind for a Sturm-Liouville operator. *Differ. Equ.* **1987**, *23*, 979–987.
11. Ji, D.; Bai, Z.; Ge, W. The existence of countably many positive solutions for singular multipoint boundary value problems. *Nonlinear Anal. Theory Methods Appl.* **2010**, *72*, 955–964. [CrossRef]
12. Li, H. Existence of nontrivial solutions for superlinear three-point boundary value problems. *Acta Math. Appl. Sin.* **2017**, *33*, 1043–1052. [CrossRef]
13. Ma, R. Positive solutions of nonlinear three-point boundary value problem. *Electron. J. Differ. Equations* **1999**, *34*, 1–8. [CrossRef]
14. Marano, S. A remark on a second order three-point boundary value problem. *J. Math. Anal. Appl.* **1994**, *183*, 518–522. [CrossRef]
15. Webb, J.R.L. Positive solutions of some three point boundary value problems via fixed-point index theory. *Nonlinear Anal. Theory Methods Appl.* **2001**, *47*, 4319–4332. [CrossRef]
16. Ntouyas, S.K.; Pourhadi, E. Positive solutions of nonlinear fractional three-point boundary-value problem. *Le Mat.* **2018**, *73*, 139–154.
17. Sudsutad, W.; Tariboon, J.; Ntouyas, S.K. Positive solutions for fractional differential equations with three-point multi-term fractional integral boundary conditions. *Adv. Differ. Equ.* **2014**, *2014*, 28.
18. Wang, G.; Ntouyas, S.K.; Zhang, L. Positive solutions of the three-point boundary value problem for fractional-order differential equations with an advanced argument. *Adv. Differ. Equ.* **2011**, *2011*, 2. [CrossRef]
19. Wang, G.; Zhang, L.; Ntouyas, S.K. Multiplicity of positive solutions for fractional order three-point boundary value problems. *Commun. Appl. Nonlinear Anal.* **2013**, *20*, 41–53.
20. Zhang, Y. Existence results for a coupled system of nonlinear fractional multi-point boundary value problems at resonance. *J. Inequal. Appl.* **2018**, *2018*, 198. [CrossRef]
21. Zou, Y.; Liu, L.; Cui, Y. The existence of solutions for four-point coupled boundary value problems of fractional differential equations at resonance. *Abstr. Appl. Anal.* **2014**, *2014*, 1–8. [CrossRef]
22. Bai, Z.; Ge, W. Existence of positive solutions to fourth-order quasilinear boundary value problems. *Acta Math. Sin.* **2006**, *22*, 1825–1830. [CrossRef]
23. Kilbas, A.A.; Srivastava, H.M.; Trujillo, J.J. *Theory and Applications of Fractional Differential Equations*; North-Holland Math. Stud.; Elsevier: Amsterdam, The Netherlands, 2006; Volume 204.
24. Podlubny, I. *Fractional Differential Equations*; Mathematics in Science and Engineering; Academic Press: New York, NY, USA; London, UK; Toronto, ON, Canada, 1999; Volume 198.
25. Samko, S.G.; Kilbas, A.A.; Marichev, O.I. *Fractional Integral and Derivatives (Theory and Applications)*; Gordon and Breach: Yverdon, Switzerland, 1993.
26. Zhang, S. Positive solutions for boundary-value problems of nonlinear fractional differential equations. *Electron. J. Differ. Equ.* **2006**, *2006*, 1–12. [CrossRef]
27. Smart, D.R. *Fixed-Point Theorems*; Cambridge University Press: London, UK; New York, NY, USA, 1974.

5

Existence of Solutions for Anti-Periodic Fractional Differential Inclusions Involving ψ-Riesz-Caputo Fractional Derivative

Dandan Yang * and Chuanzhi Bai

School of Mathematical Science, Huaiyin Normal University, Huaian 223300, Jiangsu, China
* Correspondence: ydd@hytc.edu.cn

Abstract: In this paper, we investigate the existence of solutions for a class of anti-periodic fractional differential inclusions with ψ-Riesz-Caputo fractional derivative. A new definition of ψ-Riesz-Caputo fractional derivative of order α is proposed. By means of Contractive map theorem and nonlinear alternative for Kakutani maps, sufficient conditions for the existence of solutions to the fractional differential inclusions are given. We present two examples to illustrate our main results.

Keywords: fractional differential inclusions; ψ-Riesz-Caputo derivative; existence of solutions; fixed point theorem; anti-periodic boundary value problems

---

1. Introduction

Fractional order models, providing excellent description of memory and hereditary processes, are more adequate than integer order ones. Some recent contributions to fractional differential equations and inclusions have been carried out, see the monographs [1–8], and the references cited therein. The study of fractional differential equations or inclusions with anti-periodic boundary problems, which are applied in different fields, such as physics, chemical engineering, economics, populations dynamics and so on, have recently received considerable attention, see the references ([9,10]) and papers cited therein.There are several definitions of fractional differential derivatives and integrals, such like Caputo type, Rimann-Liouville type, Hadamard type and Erdelyi-Kober type and so on. In order to develop the fractional calculus, some different and special form of differential operators are chosen, for example, see [11–15] and the references therein. The α order ψ-Caputo fractional derivative was first introduced by Almeida in [3]. Some properties, like semigroup law, Taylor's Theorem, Fermat's Thorem, etc., were presented. This newly defined fractional derivative could model more accurately the process using differential kernels for the fractional operator. In 2018, Samet and Aydi in [16] considered the following fractional differential equation with anti-periodic boundary conditions:

$$\begin{cases} {}^cD^{\alpha,\psi}u(x) + f(x,u(x)) = 0, \quad a < x < b, \\ u(a) + u(b) = 0, u'(a) + u'(b) = 0 \end{cases} \tag{1}$$

where $(a,b) \in R^2, a < b, 1 < \alpha < 2, \psi \in C^2([a,b]), \psi'(x) > 0, x \in [a,b]\ {}^cD^{\alpha,\psi}$ is the ψ-Caputo fractional derivative of order α, and $f : [a,b] \times R \to R$ is a given function. A Lyapunov-type inequality is established for problem (1). The authors also give some examples to illustrate the applications of their main results.

Very recently, Chen et al. in [10] studied the following anti-periodic boundary problem involving the Riesz-Caputo derivative

$$\begin{cases} {}_0^{RC}D_T^{\gamma} y(\tau) = g(\tau, y(\tau)), \quad \tau \in [0,T], 1 < \gamma \leq 2, \\ y(0) = -y(T), \quad y'(0) = -y'(T), \end{cases} \tag{2}$$

where ${}_0^{RC}D_T^{\gamma}$ is a Riesz-Caputo derivative, which can reflect both the past and the future nonlocal memory effects and $g : [0,T] \times R \to R$ is a continuous function with respect to τ and y. Some existence results of solutions are given based on the Lipschitz condition, the growth condition and the comparison condition. Most of the present work are concerned with fractional differential equations or inclusions involving Riemann-Liouville or Caputo fractional derivative, merely reflecting the past or future memory effect. Riesz derivative is a two-sided fractional operator, whose advantage is that it could reflect both the past and the future memory effects. We take anomalous diffusion problem for example. The fractional differential equation with the Riesz derivative is adopted to describe the anomalous diffusion problem, in which the Riesz derivative stands for the nonlocality and the dependence on path of the diffusion concentration. Some applications of Riesz derivative about anomalous diffusion, we refer the reader to [17,18]. Another typical example is stocks. According to the price trend of the past and future time, investors would buy or sell a stock at an agreed-on price within a period of time. This process depends on both past state and its development in the future, which is the characteristic of Riesz derivative. There are some other applications of this derivative, and we refer the reader to [10,19,20]. In 2009, Ahamad and Otero-Espinar [1] investigated the following fractional inclusions with anti-periodic boundary conditions

$$\begin{cases} {}^cD^q x(t) \in F(t, x(t)), \quad t \in [0,T], 1 < q \leq 2, \\ x(0) = -x(T), \quad x'(0) = -x'(T), \end{cases} \tag{3}$$

where ${}^cD^q x(t)$ is the standard Caputo derivative of order q, $F : [0,T] \times R \to \mathcal{P}(R)$ is a multivalued map, $\mathcal{P}(R)$ is the family of all subsets of R. Some sufficient conditions for the existence of solutions are given by means of Bohnenblust-Karlin fixed point theorem.

Inspired by the above-mentioned works, in this paper, we are concerned with the following anti-periodic fractional inclusions with ψ-Riesz-Caputo derivative:

$$\begin{cases} {}_a^{RC}D_b^{\alpha,\psi} u(x) \in F(x, u(x)), \quad a < x < b, \\ u(a) + u(b) = 0, u'(a) + u'(b) = 0, \end{cases} \tag{4}$$

where $(a,b) \in R^2, a < b, 1 < \alpha < 2$, $\psi \in C^2([a,b])$, $\psi'(x) > 0, x \in [a,b]$. ${}_a^{RC}D_b^{\alpha,\psi}$ is the ψ-Riesz-Caputo fractional derivative of order α, and $F : [a,b] \times R \to \mathcal{P}(R)$ is a multivalued map. Sufficient conditions for the existence of solutions are given in view of the fixed point theorems for multi-valued mapping. The aim of this paper is to develop the calculus of fractional derivatives. We shall combine the two definitions of Riesz-Caputo derivative and ψ-Caputo fractional derivative. Then we investigate the existence of solutions of anti-periodic inclusions (4). The rest of this paper is organized as follows. We first present some basic definitions of fractional calculus, ψ-Caputo derivative, Riesz-Caputo derivative and multi-valued maps, and then a new definition of ψ-Riesz-Caputo fractional derivative of order α is given. In Section 3, the main results on the existence of solutions for anti-periodic boundary value problem (4) are provided. We present two examples in order to illustrate our main results in last section. Our results generalize some published known results. There is no literature to research the fractional differential inclusions with ψ-Riesz-Caputo fractional derivative. If we take $F(x,u) = \{f(x,u)\}$, where $f : [a,b] \times R \to R$ is a given continuous function, then the problem (4) corresponds to the single-valued problem (1). If we take $a = 0, b = T, \psi(x) = x, F(x,u) = \{g(x,u)\}$, where $g : [0,T] \times R \to R$ is a given continuous function, then the problem (4) corresponds to the single-valued problem (2).

2. Preliminaries

In this section, we recall some notation, definitions and preliminaries about fractional calculus [6,7,21], ψ-Caputo fractional calculus [3–5,22,23], and Riesz or Riesz-Caputo fractional derivative [17–19].

Definition 1 ([6]). *The left Caputo fractional derivative order α ($1 < \alpha \leq 2$) of a function $f \in C^2([a,b])$ is given by*

$$ {}_a^C D_x^{\alpha} f(x) = ({}_a I_x^{2-\alpha} f'')(x), a < x < b, $$

that is,

$$ {}_a^C D_x^{\alpha} f(x) = \frac{1}{\Gamma(2-\alpha)} \int_a^x (x-t)^{1-\alpha} f''(t) dt, a < x < b. $$

Similarly, the right Caputo fractional integral order α ($1 < \alpha \leq 2$) of a function $f \in C^2([a,b])$ is given by

$$ {}_x^C D_b^{\alpha} f(x) = \frac{1}{\Gamma(2-\alpha)} \int_x^b (t-x)^{1-\alpha} f''(t) dt, a < x < b. $$

Definition 2 ([6]). *The fractional left, right and Riemann-Liouville integrals of order $\beta > 0$ are defined as*

$$ {}_a I_\tau^\beta g(\tau) = \frac{1}{\Gamma(\beta)} \int_a^\tau (\tau - s)^{\beta-1} g(s) ds, $$

$$ {}_\tau I_b^\beta g(\tau) = \frac{1}{\Gamma(\beta)} \int_\tau^b (s - \tau)^{\beta-1} g(s) ds, $$

$$ {}_a I_b^\beta g(\tau) = \frac{1}{\Gamma(\beta)} \int_a^b |s - \tau|^{\beta-1} g(s) ds. $$

Let $\psi \in C^2([a,b])$ be a given function such that

$$ \psi'(x) > 0, a \leq x \leq b. $$

Definition 3 ([3]). *The fractional left, right integral of order $\alpha > 0$ of a function $f \in C([a,b])$ with respect to ψ are defined by*

$$ ({}_a I_x^{\alpha,\psi}) f(x) = \frac{1}{\Gamma(\alpha)} \int_a^x \psi'(t)(\psi(x) - \psi(t))^{\alpha-1} f(t) dt, a \leq x \leq b, \tag{5} $$

$$ ({}_x I_b^{\alpha,\psi}) f(x) = \frac{1}{\Gamma(\alpha)} \int_x^b \psi'(t)(\psi(t) - \psi(x))^{\alpha-1} f(t) dt, a \leq x \leq b. \tag{6} $$

Definition 4 ([3]). *The left, right ψ-Caputo fractional derivative of order α ($1 < \alpha \leq 2$) of a function $f \in C^2([a,b])$ are defined as*

$$ {}_a^C D_x^{\alpha,\psi} f(x) = \frac{1}{\Gamma(2-\alpha)} \int_a^x \psi'(t)(\psi(x) - \psi(t))^{1-\alpha} \left(\frac{1}{\psi'(t)} \frac{d}{dt}\right)^2 f(t) dt, a < x < b, \tag{7} $$

$$ {}_x^C D_b^{\alpha,\psi} f(x) = \frac{1}{\Gamma(2-\alpha)} \int_x^b \psi'(t)(\psi(t) - \psi(x))^{1-\alpha} \left(\frac{1}{\psi'(t)} \frac{d}{dt}\right)^2 f(t) dt, a < x < b. \tag{8} $$

Remark 1. *Consider $\psi(x) = x$, $\psi(x) = \ln x$, the Riemann-Liouville and Hadamard fractional operators are obtained.*

Inspired by the above definitions, we shall present a new definition of ψ-Riesz-Caputo fractional derivative of order α, which is a combination of ψ-Caputo fractional derivative and Riesz-Caputo fractional derivative.

Definition 5. *Let $f \in C^2([a,b])$. For $x \in [a,b]$, the ψ-Riesz-Caputo fractional derivative ${}^{RC}_{a}D^{\alpha,\psi}_{b}f(x)$ of order α $(1 < \alpha \leq 2)$ could be defined by*

$$ {}^{RC}_{a}D^{\alpha,\psi}_{b}f(x) = \frac{1}{2}\left({}^{C}_{a}D^{\alpha,\psi}_{x} + {}^{C}_{x}D^{\alpha,\psi}_{b}\right)f(x). \tag{9} $$

If we take $\psi(x) = x$, it follows from (7)–(9) that the classic Riesz-Caputo derivative fractional order α $(1 < \alpha \leq 2)$ of a function $f \in C^2([a,b])$ is given by

$$ {}^{RC}_{a}D^{\alpha}_{b}f(x) = \frac{1}{2}\left({}^{C}_{a}D^{\alpha}_{x} + {}^{C}_{x}D^{\alpha}_{b}\right)f(x), \tag{10} $$

which is defined as in [19]. For convenience, denote

$$ P_{cl}(X) = \{Y \in \mathcal{P}(X) : Y \quad \text{is closed}\}, $$

$$ P_b(X) = \{Y \in \mathcal{P}(X) : Y \quad \text{is bounded}\}, $$

$$ P_{cp}(X) = \{Y \in \mathcal{P}(X) : Y \quad \text{is compact}\}, $$

$$ P_{cp,c}(X) = \{Y \in \mathcal{P}(X) : Y \quad \text{is convex and compact}\}. $$

The following are definitions and properties concerning multi-valued maps [24–28] which will be used in the remainder of this paper.

Definition 6 ([28])**.** *A multivalued map $G : X \to \mathcal{P}(X)$:*

(a) *denote the set $Gr(G) = \{(x,y) \in X \times Y, y \in G(x)\}$ as the graph of G,*

$$ t \mapsto d(y, G(t)) = \inf\{|y - z| : z \in G(t)\} $$

is measurable.

(b) *if $G : X \to \mathcal{P}_{cl}(X)$ is called $\gamma-$Lipschitz if and only if there exists $\gamma > 0$ such that*

$$ H_d(N(x), N(y)) \leq \gamma d(x,y), \text{for each } x, y \in X. $$

(c) *if $G : X \to \mathcal{P}_{cl}(X)$ is called contraction if and only if it is $\gamma-$Lipschitz with $\gamma < 1$.*

(d) *G is said to be measurable if for every $y \in R$, the function*

Definition 7 ([26])**.** *Assume that $F : J \times R \to \mathcal{P}(R)$ is a multivalued map with nonempty compact values. Denote a multivalued operator $\mathcal{F} : C(J \times R) \to \mathcal{P}(L^1(J,R)$ associated with F as*

$$ \mathcal{F}(x) = \{w \in L^1(J,R) : w(t) \in F(t, x(t))\} $$

for a.e. $t \in J := [a,b]$ is a closed interval from a to b.

Definition 8 ([26])**.** *Assume that Y is a separable metric space and $N : Y \to \mathcal{P}(L^1(J,R))$ is a multivalued operator. If N is lower semi-continuous(l.s.c.) and has nonempty closed and decomposable values, we say N has a property (BC) .*

Definition 9 ([28])**.** *For each $u \in C(J,R)$, $t \in J = [a,b]$, denote the selection set of F as*

$$ S_{F,y} := \{f \in L^1(J,R) : f(t) \in F(t, u(t)) \quad a.e. \quad t \in J\}. $$

Definition 10 ([28])**.** *Let $A, B \in \mathcal{P}_{cl}(X.)$ The Pompeiu-Hausdorff distance of A, B is defined by*

$$ H_d(A,B) = \max\{\sup_{a \in A} d(a,B), \sup_{b \in B} d(A,b)\}, $$

where $d(A,b) = \inf_{a\in A} d(a,b)$, $d(a,B) = \inf_{b\in B} d(a,b)$.

Property 1 ([24]). *Let G be a completely continuous multi-valued map with nonempty compact values, then T is u.s.c.* $\Longleftrightarrow$ *G has a closed graph.*

The following lemmas play important roles in the proof of our main results.

Lemma 1 ([28]). *(Nonlinear alternative for Kakutani maps). Assume that E is a Banach space, C is a closed convex subset of E, and U is an open subset of C with* $0 \in U$. *Let* $F : \bar{U} \to \mathcal{P}_{c,cv}(C)$ *be a upper semicontinuous compact map. Then either*

(i) *F has a fixed point in* $\bar{U}$, *or*

(ii) *there exist a* $u \in \partial U$ *and* $\lambda \in (0,1)$ *satisfying* $u \in \lambda F(u)$.

Lemma 2 ([29]). *Let* (X,d) *be a complete metric space. If* $N : X \to \mathcal{P}_{cl}(X)$ *is a contraction, then* $FixN \neq \emptyset$.

Lemma 3 ([30]). *Let X be a Banach space, and* $F : J \times X \to (P)(X)$ *be a* L^1-*Carathédory set-valued map with* $S_F \neq \emptyset$ *and let* $\Theta : L^1(J,X) \to C(J,X)$ *be a linear continuous mapping. Then the set-valued map* $\Gamma \circ S_F : C(J,X) \to \mathcal{P}(C(J,X))$ *defined by*

$$(\Theta \circ S_F)(u) : C(J \times X) \to \mathcal{P}_{cp,c}(C(J,X)),\, x \mapsto (\Theta \circ S_F)(u) = \Theta(S_{F,u})$$

is a closed graph operator in $C(J,X) \times C(J,X)$.

Lemma 4 ([20]). *Assume that Y is a separable metric space and* $N : Y \to \mathcal{P}(L^1(J,R))$ *is a multivalued operator with the property (BC). Then there exists a continuous single-valued function* $g : Y \to L^1(J,R)$ *satisfying* $g(x) \in N(x)$ *for every* $x \in Y$, *i.e., N has a continuous selection.*

From [6], we have

Lemma 5. *If* $1 < \beta \leq 2$ *and* $g \in C^2[a,b]$, *then*

$$ {}_aI_{\tau}^{\beta}{}_a^CD_{\tau}^{\beta} g(\tau) = g(\tau) - g(a) - g'(a)(\tau - a),$$
$$ {}_{\tau}I_{b}^{\beta}{}_{\tau}^CD_{b}^{\beta} g(\tau) = g(\tau) - g(b) + g'(b)(b - \tau).$$

From (10) and Lemma 2.1 in [15], for $u \in C^2[a,b]$, *and* $1 < \alpha \leq 2$, *we have that*

$$\begin{aligned} {}_aI_{b}^{\alpha}{}_a^CD_{b}^{\alpha} u(\tau) &= \frac{1}{2}\left({}_aI_{\tau}^{\alpha}{}_a^CD_{\tau}^{\alpha} + {}_{\tau}I_{b}^{\alpha}{}_{\tau}^CD_{b}^{\alpha}\right) u(\tau) \\ &= u(\tau) - \frac{1}{2}(u(a) + u(b)) - \frac{1}{2}u'(a)(\tau - a) + \frac{1}{2}u'(b)(b - \tau). \end{aligned} \tag{11}$$

By (11), similar to the proof of Lemma 2.2 in [10], we have the following lemma.

Lemma 6. *Assume that* $h \in C[a,b]$. *A function* $u \in C^2[a,b]$ *given by*

$$\begin{aligned} u(t) = &-\frac{b-a}{2\Gamma(\alpha-1)}\int_a^b (b-s)^{\alpha-2}h(s)ds \\ &+\frac{1}{\Gamma(\alpha)}\int_a^t (t-s)^{\alpha-1}h(s)ds + \frac{1}{\Gamma(\alpha)}\int_t^b (s-t)^{\alpha-1}h(s)ds, \end{aligned} \tag{12}$$

is a unique solution of the following anti-periodic boundary value problem

$$\begin{cases} ({}_a^{RC}D_b^{\alpha}u)(t) = h(t), \quad t \in (a,b), 1 < \alpha \leq 2, \\ u(a) + u(b) = 0, \quad u'(a) + u'(b) = 0. \end{cases} \tag{13}$$

As the same argument of Lemma 2.1 in [16], we can easily obtain the following result, which plays a very important role in proving the main results.

Lemma 7. *If $f, \psi \in C^2([a,b])$, and $\psi'(x) > 0$ for each $x \in [a,b]$, then*

$$({}_a^{RC}D_x^{\alpha,\psi} f)(\psi^{-1}(y)) = ({}_{\psi(a)}^{RC}D_{\psi(x)}^{\alpha}(f \circ \psi^{-1}))(y), \psi(a) < y < \psi(b), \tag{14}$$

and

$$({}_x^{RC}D_b^{\alpha,\psi} f)(\psi^{-1}(y)) = ({}_{\psi(x)}^{RC}D_{\psi(b)}^{\alpha}(f \circ \psi^{-1}))(y), \psi(a) < y < \psi(b). \tag{15}$$

Moreover, we have

$$({}_a^{RC}D_b^{\alpha,\psi} f)(\psi^{-1}(y)) = ({}_{\psi(a)}^{RC}D_{\psi(b)}^{\alpha}(f \circ \psi^{-1}))(y), \psi(a) < y < \psi(b). \tag{16}$$

Lemma 8. *If $f : [a,b] \times R \to R$, $\psi \in C^2[a,b]$ with $\psi'(x) > 0$, and $\psi'(a) = \psi'(b)$, then the problem*

$$\begin{cases} {}_a^{RC}D_b^{\alpha,\psi} u(x) = f(x, u(x)), \quad a < x < b, \\ u(a) + u(b) = 0, u'(a) + u'(b) = 0, \end{cases} \tag{17}$$

could be transformed into the following problem

$$\begin{cases} {}_{\psi(a)}^{RC}D_{\psi(b)}^{\alpha} v(y) = f(\psi^{-1}(y), v(y)), \quad \psi(a) < y < \psi(b), \\ v(\psi(a)) + v(\psi(b)) = 0, \quad v'(\psi(a)) + v'(\psi(b)) = 0. \end{cases} \tag{18}$$

A nontrivial solution to (18) is given by $\quad v(y) = -\dfrac{B-A}{2\Gamma(\alpha-1)} \displaystyle\int_A^B (B-s)^{\alpha-2} f(\psi^{-1}(s), v(s))ds$

$$+\frac{1}{\Gamma(\alpha)} \int_A^y (y-s)^{\alpha-1} f(\psi^{-1}(s), v(s))ds + \frac{1}{\Gamma(\alpha)} \int_y^B (s-y)^{\alpha-1} f(\psi^{-1}(s), v(s))ds, \tag{19}$$

where $A = \psi(a)$ and $B = \psi(b)$.

Proof. We introduce the function $v : [\psi(a), \psi(b)] \to R$, defined by

$$v(y) = u(\psi^{-1}(y)), \psi(a) \le y \le \psi(b).$$

In virtue of (16), one has

$${}_{\psi(a)}^{RC}D_{\psi(b)}^{\alpha} v(\psi^{-1}(y)) = {}_a^{RC}D_b^{\alpha,\psi} u(x), \quad \psi(a) < y < \psi(b). \tag{20}$$

By a chain rule, we have

$$v'(y) = \frac{1}{\psi'(\psi^{-1}(y))} u'(\psi^{-1}(y)), \psi(a) \le y \le \psi(b).$$

Thus, we have

$$v'(\psi(a)) = \frac{1}{\psi'(a)} u'(a), \text{and } v'(\psi(b)) = \frac{1}{\psi'(b)} u'(b).$$

From boundary condition (17) and condition $\psi'(a) = \psi'(b)$, we have that

$$v(\psi(a)) + v(\psi(b)) = 0, v'(\psi(a)) + v'(\psi(b)) = 0. \tag{21}$$

Therefore, the problem (17) could be transformed into problem (18). By virtue of Lemma 6, we obtain $v \in C^2[A, B]$ is a nontrivial solution to (18).

From Lemma 8, we can easily know that

$$u(x) = -\frac{\psi(b)-\psi(a)}{2\Gamma(\alpha-1)}\int_a^b (\psi(b)-\psi(t))^{\alpha-2}\psi'(t)f(t,u(t))dt$$

$$+\frac{1}{\Gamma(\alpha)}\int_a^x (\psi(x)-\psi(t))^{\alpha-1}\psi'(t)f(t,u(t))dt + \frac{1}{\Gamma(\alpha)}\int_x^b (\psi(t)-\psi(x))^{\alpha-1}\psi'(t)f(t,u(t))dt \quad (22)$$

is a unique solution of problem (17). □

3. Main Results

We pose the following hypotheses:

(H_1) $F : [a,b] \times R \to \mathcal{P}(R)$ is Carathéodory and it has nonempty compact and convex values;

(H_2) there exist a continuous nondecreasing function $q : [0,\infty) \to [0,\infty)$ and a function $p \in C([a,b], R^+)$ satisfying

$$\|F(t,u)\| := \sup\{|f| : f \in F(t,u)\} \leq p(t)q(\|u\|), \text{ for each } (t,x) \in [a,b] \times R.$$

(H_3) $1 < \alpha < 2$, $\psi \in C^2([a,b])$, $\psi'(x) > 0$, $x \in [a,b]$.

(H_4) $\psi'(a) = \psi'(b)$.

(H_5) $F : [a,b] \times R \to \mathcal{P}_{cp}(R)$ is such that, for every $u \in R$, $F(\cdot,u)$ is measurable.

(H_6) There exists $m \in L^1([a,b)], R^+)$ for almost all $t \in [a,b]$, such that

$$d_H(F(t,u), F(t,\bar{u})) \leq m(t)|u-\bar{u}|, \quad \forall u, \bar{u} \in R$$

with $d(0,F(t,0)) \leq m(t)$ for almost all $t \in [a,b]$.

(H_7) $F : [a,b] \times R \to \mathcal{P}(R)$ is a nonempty compact-valued multivalued map such that

(a) $(x,u) \mapsto F(x,u)$ is $\mathcal{L} \bigotimes \mathcal{B}$ is measurable.

(b) $u \mapsto F(x,u)$ is lower semicontinuous for each $x \in [a,b]$,

Now we are in the position to state our main results. The first theorem is dealing with the Carathéodory case.

Theorem 1. *Assume that (H_1)–(H_4) hold. Moreover, if there exists a constant $M > 0$, such that*

$$M\left[q(M)\left(\frac{\psi(b)-\psi(a)}{2\Gamma(\alpha-1)}\int_a^b (\psi(b)-\psi(s))^{\alpha-2}p(s)\psi'(s)ds\right.\right.$$
$$\left.\left.+\frac{2}{\Gamma(\alpha)}(\psi(b)-\psi(a))^{\alpha-1}\int_a^b p(s)\psi'(s)ds\right)\right]^{-1} > 1. \quad (23)$$

Then (4) has at least one solution on $[a,b]$.

Proof. The operator $T : C([a,b],R) \to \mathcal{P}(C[A,B],R)$ is defined as follows:

$$T(u) = \{h \in C([a,b],R) : h(t) = -\frac{\psi(b)-\psi(a)}{2\Gamma(\alpha-1)}\int_a^b (\psi(b)-\psi(s))^{\alpha-2}\psi'(s)f(s)ds$$

$$+\frac{1}{\Gamma(\alpha)}\int_a^t (\psi(t)-\psi(s))^{\alpha-1}\psi'(s)f(s)ds + \frac{1}{\Gamma(\alpha)}\int_t^b (\psi(s)-\psi(t))^{\alpha-1}\psi'(s)f(s)ds, \quad f \in S_{F,u}\}.(24)$$

We divide the proof into 5 parts, which shows that T satisfies all the conditions of Lemma 1.

Part (i). T maps the bounded sets into bounded sets of $C([a,b],R)$. Set $B_r = \{v \in C([a,b],R) : \|v\| \leq r, \quad r > 0\}$, which is a bounded ball in $C([a,b],R)$, then for $h \in T(u)$, $u \in B_r$, there exists $f \in S_{F,u}$ such that

$$h(t) = -\frac{\psi(b)-\psi(a)}{2\Gamma(\alpha-1)}\int_a^b (\psi(b)-\psi(s))^{\alpha-2}\psi'(s)f(s)ds$$

$$+\frac{1}{\Gamma(\alpha)}\int_a^t(\psi(t)-\psi(s))^{\alpha-1}\psi'(s)f(s)ds+\frac{1}{\Gamma(\alpha)}\int_t^b(\psi(s)-\psi(t))^{\alpha-1}\psi'(s)f(s)ds. \quad (25)$$

Then

$$\begin{aligned}|h(t)| \leq & \frac{\psi(b)-\psi(a)}{2\Gamma(\alpha-1)}\int_a^b(\psi(b)-\psi(s))^{\alpha-2}\psi'(s)|f(s)|ds\\ &+\frac{1}{\Gamma(\alpha)}\int_a^t(\psi(t)-\psi(s))^{\alpha-1}\psi'(s)|f(s)|ds+\frac{1}{\Gamma(\alpha)}\int_t^b(\psi(s)-\psi(t))^{\alpha-1}\psi'(s)|f(s)|ds.\\ \leq & \, q(\|u\|)\left[\frac{\psi(b)-\psi(a)}{2\Gamma(\alpha-1)}\int_a^b(\psi(b)-\psi(s))^{\alpha-2}\psi'(s)p(s)ds\right.\\ &\left.+\frac{1}{\Gamma(\alpha)}\int_a^t(\psi(t)-\psi(s))^{\alpha-1}\psi'(s)p(s)ds+\frac{1}{\Gamma(\alpha)}\int_t^b(\psi(s)-\psi(t))^{\alpha-1}\psi'(s)p(s)ds\right]\\ \leq & \, q(r)\left(\frac{\psi(b)-\psi(a)}{2\Gamma(\alpha-1)}\int_a^b(\psi(b)-\psi(s))^{\alpha-2}\psi'(s)p(s)ds\right.\\ &\left.+\frac{2}{\Gamma(\alpha)}(\psi(b)-\psi(a))^{\alpha-1}\int_a^b\psi'(s)p(s)ds\right). \end{aligned} \quad (26)$$

Part (ii). T maps bounded set into equicontinuous sets. Let $u \in B_r$, $t_1, t_2 \in [a,b]$, $t_1 < t_2$, where B_r is a bounded set in $C([a,b],R)$, for $u \in T(u)$, we have

$$\begin{aligned}|h(t_2)-h(t_1)| \leq & \frac{1}{\Gamma(\alpha)}\int_a^{t_1}[(\psi(t_2)-\psi(s))^{\alpha-1}-(\psi(t_1)-\psi(s))^{\alpha-1}]\psi'(s)|f(s)|ds\\ &+\frac{1}{\Gamma(\alpha)}\int_{t_1}^{t_2}[(\psi(t_2)-\psi(s))^{\alpha-1}-(\psi(t_1)-\psi(s))^{\alpha-1}]\psi'(s)|f(s)|ds\\ &+\frac{1}{\Gamma(\alpha)}\int_{t_2}^{b}[(\psi(t_2)-\psi(s))^{\alpha-1}-(\psi(t_1)-\psi(s))^{\alpha-1}]\psi'(s)|f(s)|ds\\ \leq & \frac{q(r)}{\Gamma(\alpha)}\int_a^{t_1}[(\psi(t_2)-\psi(s))^{\alpha-1}-(\psi(t_1)-\psi(s))^{\alpha-1}]\psi'(s)p(s)ds\\ &+\frac{q(r)}{\Gamma(\alpha)}\int_{t_1}^{t_2}[(\psi(t_2)-\psi(s))^{\alpha-1}-(\psi(t_1)-\psi(s))^{\alpha-1}]\psi'(s)p(s)ds\\ &+\frac{q(r)}{\Gamma(\alpha)}\int_{t_2}^{b}[(\psi(t_2)-\psi(s))^{\alpha-1}-(\psi(t_1)-\psi(s))^{\alpha-1}]\psi'(s)p(s)ds, \end{aligned} \quad (27)$$

independent of $u \in B_r$ as $t_1 \to t_2$, the right side hand of above inequality tends to 0. According to the Ascoli-Arzelá Theorem, T is completely continuous.

Part (iii). T has a closed graph. Set $u_n \to u_*$, $h_n \in T(u_n)$ and $h_n \to h_*$. Then, we shall show that $h_* \in T(u_*)$. For $h_n \in T(u_n)$, there exist $f_n \in S_{F,u_n}$ such that

$$h_n(t) = -\frac{\psi(b)-\psi(a)}{2\Gamma(\alpha-1)}\int_a^b(\psi(b)-\psi(s))^{\alpha-2}\psi'(s)f_n(s)ds \quad (28)$$

$$+\frac{1}{\Gamma(\alpha)}\int_a^t(\psi(t)-\psi(s))^{\alpha-1}\psi'(s)f_n(s)ds+\frac{1}{\Gamma(\alpha)}\int_t^b(\psi(s)-\psi(t))^{\alpha-1}\psi'(s)f_n(s)ds.$$

Hence, it suffices to show that there exists $f_* \in S_{F,u_*}$, such that for each $t \in [a,b]$,

$$h_*(t) = -\frac{\psi(b)-\psi(a)}{2\Gamma(\alpha-1)}\int_a^b(\psi(b)-\psi(s))^{\alpha-2}\psi'(s)f_*(s)ds$$

$$+\frac{1}{\Gamma(\alpha)}\int_a^t(\psi(t)-\psi(s))^{\alpha-1}\psi'(s)f_*(s)ds+\frac{1}{\Gamma(\alpha)}\int_t^b(\psi(s)-\psi(t))^{\alpha-1}\psi'(s)f_*(s)ds. \quad (29)$$

Define the continuous linear the operator $\Phi : L^1([a,b],R) \to C([a,b],R)$:

$$f \mapsto \Phi(f)(t) = -\frac{\psi(b)-\psi(a)}{2\Gamma(\alpha-1)}\int_a^b(\psi(b)-\psi(s))^{\alpha-2}\psi'(s)f(s)ds$$

$$+\frac{1}{\Gamma(\alpha)}\int_a^t(\psi(t)-\psi(s))^{\alpha-1}\psi'(s)f(s)ds+\frac{1}{\Gamma(\alpha)}\int_t^b(\psi(s)-\psi(t))^{\alpha-1}\psi'(s)f(s)ds. \quad (30)$$

We have $\|h_n - h\| \to 0$, as $n \to \infty$. Thus, in light of Lemma 3, $\Phi \circ S_F$ is a closed graph operator. Furthermore, we have $h_n(t) \in \Phi(S_{F,u_n})$. By $u_n \to u_*$, we obtain

$$h_*(t) = -\frac{\psi(b)-\psi(a)}{2\Gamma(\alpha-1)}\int_a^b(\psi(b)-\psi(s))^{\alpha-2}\psi'(s)f_*(s)ds$$

$$+\frac{1}{\Gamma(\alpha)}\int_a^t(\psi(t)-\psi(s))^{\alpha-1}\psi'(s)f_*(s)ds+\frac{1}{\Gamma(\alpha)}\int_t^b(\psi(s)-\psi(t))^{\alpha-1}\psi'(s)f_*(s)ds, \quad (31)$$

for some $f_* \in S_{F,u_*}$.

Part (iv). T is convex for each $x \in C([a,b],R)$. Since $S_{F,u}$ is convex, it is obviously true.

Part (v). We show that there exists a open set $U \subset C([a,b],R)$, with $u \notin T(u)$ for any $\eta \in (0,1)$ and all $u \in \partial U$. Let $\eta \in (0,1)$, $u \in \eta T(u)$. Then for $t \in [a,b]$, there exists $f \in S_{F,u}$ such that

$$h(t) = -\frac{\psi(b)-\psi(a)}{2\Gamma(\alpha-1)}\int_a^b(\psi(b)-\psi(s))^{\alpha-2}\psi'(s)f(s)ds$$

$$+\frac{1}{\Gamma(\alpha)}\int_a^t(\psi(t)-\psi(s))^{\alpha-1}\psi'(s)f(s)ds+\frac{1}{\Gamma(\alpha)}\int_t^b(\psi(s)-\psi(t))^{\alpha-1}\psi'(s)f(s)ds. \quad (32)$$

A similar discussion as in part (i), we have

$$\|h\| \le q(\|u\|)\left(\frac{\psi(b)-\psi(a)}{2\Gamma(\alpha-1)}\int_a^b(\psi(b)-\psi(s))^{\alpha-2}\psi'(s)p(s)ds\right.$$

$$\left.+\frac{2}{\Gamma(\alpha)}(\psi(b)-\psi(a))^{\alpha-1}\int_a^b\psi'(s)p(s)ds\right). \quad (33)$$

Consequently, we have

$$\frac{\|u\|}{q(\|u\|)\left(\frac{\psi(b)-\psi(a)}{2\Gamma(\alpha-1)}\int_a^b(\psi(b)-\psi(s))^{\alpha-2}\psi'(s)p(s)ds+\frac{2}{\Gamma(\alpha)}(\psi(b)-\psi(a))^{\alpha-1}\int_a^b\psi'(s)p(s)ds\right)} \le 1. \quad (34)$$

By (23), there exists M such that $\|u\| \ne M$. Let

$$U = \{x \in C([a,b],R) : \|u\| < M\}.$$

It is clear that the operator $T : \bar{U} \to \mathcal{P}(C([a,b],R))$ is upper semicontinuous and completely continuous. If we choose U properly, for some $\eta \in (0,1)$, there is no $u \in \partial U$ such that $u \in \eta T(u)$. Thus, by means of Lemma 1, we can get the conclusion that thereexists a fixed point $u \in \bar{U}$, that is, it is a solution of problem (4). We complete the proof. □

We shall give the second theorem which is concerned with the Lipschitz case.

Theorem 2. *Suppose that the conditions* (H_3)–(H_6) *are satisfied. Moreover, if*

$$\gamma := \frac{\psi(b)-\psi(a)}{2\Gamma(\alpha-1)}\int_a^b(\psi(b)-\psi(s))^{\alpha-2}\psi'(s)m(s)ds+\frac{2}{\Gamma(\alpha)}(\psi(b)-\psi(a))^{\alpha-1}\int_a^b\psi'(s)m(s)ds < 1 \quad (35)$$

then problem (4) has at least a solution on $[a,b]$.

Proof. By (22), we define the operator $T: C([a,b],R) \to \mathcal{P}(C[a,b],R)$ as follows:

$$T(u) = \{h \in C([a,b],R) : h(t) = -\frac{\psi(b)-\psi(a)}{2\Gamma(\alpha-1)}\int_a^b (\psi(b)-\psi(s))^{\alpha-2}\psi'(s)g(s)ds$$

$$+\frac{1}{\Gamma(\alpha)}\int_a^t (\psi(t)-\psi(s))^{\alpha-1}\psi'(s)g(s)ds$$

$$+\frac{1}{\Gamma(\alpha)}\int_t^b (\psi(s)-\psi(t))^{\alpha-1}\psi'(s)g(s)ds, \quad g \in S_{F,u}\}. \tag{36}$$

Obviously, the fixed point of T is the solution of (4). Our aim is to prove that the operator T satisfies all the conditions in Lemma 2. The proof will be given in two claims.

Claim 1. For each $h \in C([a,b],\mathbb{R})$ the operator T is closed. Let $\{h_n\}_{n\geq 0} \in T(u)$ be such that $h_n \to h (n \to \infty)$ in $C([a,b],R)$. Then $h \in C([a,b],\mathbb{R})$, and there exists $v_n \in S_{F,u}$ such that for each $t \in [a,b]$,

$$h_n(t) = -\frac{\psi(b)-\psi(a)}{2\Gamma(\alpha-1)}\int_a^b (\psi(b)-\psi(s))^{\alpha-2}\psi'(s)v_n(s)ds$$

$$+\frac{1}{\Gamma(\alpha)}\int_a^t (\psi(t)-\psi(s))^{\alpha-1}\psi'(s)v_n(s)ds + \frac{1}{\Gamma(\alpha)}\int_t^b (\psi(s)-\psi(t))^{\alpha-1}\psi'(s)v_n(s)ds. \tag{37}$$

For F has compact values, we get a subsequence v_n which converges to $v \in L^1([a,b],R)$. Thus, $v \in S_{F,u}$, and for each $t \in [a,b]$, one has

$$h_n(t) \to h(t) = -\frac{\psi(b)-\psi(a)}{2\Gamma(\alpha-1)}\int_a^b (\psi(b)-\psi(s))^{\alpha-2}\psi'(s)v(s)ds$$

$$+\frac{1}{\Gamma(\alpha)}\int_a^t (\psi(t)-\psi(s))^{\alpha-1}\psi'(s)v(s)ds + \frac{1}{\Gamma(\alpha)}\int_t^b (\psi(s)-\psi(t))^{\alpha-1}\psi'(s)v(s)ds. \tag{38}$$

Therefore, $h \in T(u)$.

Claim 2. We shall show that there exists $\gamma < 1$ such that

$$H_d(F(t,u),F(t,\bar{u})) \leq \gamma\|u-\bar{u}\|.$$

Let $u,\bar{u} \in C([a,b],R)$ and $h_1 \in T(u)$. There exists $v_1(t) \in F(t,u(t))$ such that for each $t \in [a,b]$,

$$h_1(t) = -\frac{\psi(b)-\psi(a)}{2\Gamma(\alpha-1)}\int_a^b (\psi(b)-\psi(s))^{\alpha-2}\psi'(s)v_1(s)ds$$

$$+\frac{1}{\Gamma(\alpha)}\int_a^t (\psi(t)-\psi(s))^{\alpha-1}\psi'(s)v_1(s)ds + \frac{1}{\Gamma(\alpha)}\int_t^b (\psi(s)-\psi(t))^{\alpha-1}\psi'(s)v_1(s)ds. \tag{39}$$

By (H_6), there exists $w \in F(t,\bar{u}(t))$ such that

$$|v_1(t)-w(t)| \leq m(t)|u(t)-\bar{u}(t)|, t \in [a,b].$$

$U: [a,b] \to \mathcal{P}(R)$ is defined as

$$U(t) := \{w \in R : |v_1(t)-w(t)| \leq m(t)|u(t)-\bar{u}(t)|\}.$$

The multivalued operator $U(t) \cap F(t,\bar{u}(t))$ is measurable, so there exits a measurable selection for $U(t) \cap F(t,\bar{u}(t))$. We denote this function as $v_2(t)$. For each $t \in [a,b]$, one has

$$|v_1(t)-v_2(t)| \leq m(t)|u(t)-\bar{u}(t)|.$$

Then, we define for each $t \in [a,b]$,

$$h_2(t) = -\frac{\psi(b)-\psi(a)}{2\Gamma(\alpha-1)} \int_a^b (\psi(b)-\psi(t))^{\alpha-2}\psi'(s)v_2(s)ds$$
$$+\frac{1}{\Gamma(\alpha)} \int_a^t (\psi(t)-\psi(s))^{\alpha-1}\psi'(s)v_2(s)ds + \frac{1}{\Gamma(\alpha)} \int_t^b (\psi(s)-\psi(t))^{\alpha-1}\psi'(s)v_2(s)ds, \quad (40)$$

it follows that

$$\begin{aligned}
|h_1(t)-h_2(t)| \le & \frac{\psi(b)-\psi(a)}{2\Gamma(\alpha-1)} \int_a^b (\psi(b)-\psi(t))^{\alpha-2}\psi'(s)|v_1(s)-v_2(s)|ds \\
&+\frac{1}{\Gamma(\alpha)} \int_a^t (\psi(t)-\psi(s))^{\alpha-1}\psi'(s)|v_1(s)-v_2(s)|ds \\
&+\frac{1}{\Gamma(\alpha)} \int_t^b (\psi(s)-\psi(t))^{\alpha-1}\psi'(s)|v_1(s)-v_2(s)|ds \\
\le & \frac{\psi(b)-\psi(a)}{2\Gamma(\alpha-1)} \|u-\bar{u}\| \int_a^b (\psi(b)-\psi(t))^{\alpha-2}\psi'(s)m(s)ds \\
&+\frac{1}{\Gamma(\alpha)} \|u-\bar{u}\| \int_a^t (\psi(t)-\psi(s))^{\alpha-1}\psi'(s)m(s)ds \\
&+\frac{1}{\Gamma(\alpha)} \|u-\bar{u}\| \int_t^b (\psi(s)-\psi(t))^{\alpha-1}\psi'(s)m(s)ds \\
\le & \left[\frac{\psi(b)-\psi(a)}{2\Gamma(\alpha-1)} \int_a^b (\psi(b)-\psi(t))^{\alpha-2}\psi'(s)m(s)ds \right. \\
&\left. + \frac{2}{\Gamma(\alpha)}(\psi(b)-\psi(a))^{\alpha-1} \int_a^b \psi'(s)m(s)ds \right] \|u-\bar{u}\|. \\
= & \gamma\|u-\bar{u}\|. \qquad (41)
\end{aligned}$$

Therefore,

$$\|h_1-h_2\| \le \gamma\|u-\bar{u}\|.$$

Interchanging u and $\bar{u}$ yields

$$H_d(F(t,\bar{u}),F(t,u)) \le \gamma\|u-\bar{u}\|.$$

Thus, T is a contraction by $\gamma < 1$. Since Lemma 2, we conclude that T admits a fixed point which is a solution to problem (4). □

The third theorem is about the lower semicontinuous case.

Theorem 3. *Assume that (H_1)–(H_4) hold, if (H_7) is also satisfied, then the anti-periodic boundary problem (4) has at least one solution on $[a,b]$.*

Proof. It is clear that F is of l.s.c. type as condition (H_7) is satisfied. By means of Lemma 4, there exists a continuous function $f : C(J,R) \to L^1(J,R)$ such that $f(u) \in \mathcal{F}(u)$ for all $u \in C(J,R)$.

Next, we shall consider the following problem

$$\begin{cases} ({}^{RC}_{\psi(a)}D^{\alpha}_{\psi(b)}u)(x) = f(u(x)), \quad \psi(a) < x < \psi(b), \\ u(\psi(a)) + u(\psi(b)) = 0, \quad u'(\psi(a)) + u'(\psi(b)) = 0, \end{cases} \qquad (42)$$

Note that if $u \in C^2([a,b],R)$ is a solution to (42), then u is a solution to the problem (4). we define the operator $\bar{\mathcal{T}}$ as

$$\bar{\mathcal{T}}u(x) = -\frac{\psi(b)-\psi(a)}{2\Gamma(\alpha-1)}\int_a^b (\psi(b)-\psi(s))^{\alpha-2}\psi'(s)f(u(s))ds$$
$$+\frac{1}{\Gamma(\alpha)}\int_a^t (\psi(t)-\psi(s))^{\alpha-1}\psi'(s)f(u(s))ds + \frac{1}{\Gamma(\alpha)}\int_t^b (\psi(s)-\psi(t))^{\alpha-1}\psi'(s)f(u(s))ds. \quad (43)$$

We transform the problem (42) into a fixed point problem. Obviously, the operator $\bar{\mathcal{T}}$ is continuous and completely continuous. As the remainder of the proof is similar to that of Theorem 1, we omit it here. □

Remark 2. *If we take $F(x,u) = \{f(x,u)\}$, where $f : [a,b] \times R \to R$ is a given continuous function, then the problem (4) corresponds to the single-valued problem (1).*

Remark 3. *If we take $a = 0$, $b = T$, $\psi(x) = x$, $F(x,u) = \{g(x,u)\}$, where $g : [0,T] \times R \to R$ is a given continuous function, then the problem (4) corresponds to the single-valued problem (2).*

4. Applications

Example 1. *Consider the fractional differential inclusion involving ψ-Riesz-Caputo derivative with anti-periodic boundary value conditions*

$$\begin{cases} {}^{RC}_{-1}D_1^{\frac{3}{2},\psi}u(x) \in F(x,u(x)), \\ u(-1)+u(1) = 0, u'(-1)+u'(1) = 0, \end{cases} \quad (44)$$

where $\psi(x) = \sinh(x)$, $-1 \le x \le 1$. $\alpha = \frac{3}{2}$. Observe that $\psi \in C^2([-1,1])$, $\psi'(x) = \cosh(x) > 0$, $-1 \le x \le 1$. Moreover, we have

$$\psi'(-1) = \cosh(-1) = \cosh(1) = \psi'(1),$$

which implies condition (H_3)–(H_4) hold.

$$x \to F(x,u(x)) := \left[\frac{|u|^5}{|u|^5+3}+x^2+1, \frac{|u|}{|u|+1}+x^3+1\right], u \in R,$$

and

$$\|F(x,u)\| := \sup |v| : v \in F(x,u) \le 3 := p(x)q(\|u\|)\ u \in R.$$

Obviously, condition (H_1) is satisfied. And $p(x) = 1$, $q(\|u\|) = 3$, we can find a positive constant M such that

$$M\left[3\left(\frac{|\sinh(1)-\sinh(-1)|}{2\Gamma(1/2)}\int_{-1}^1 (\sinh(1)-\sinh(s))^{-1/2}\cosh(s))ds\right.\right.$$
$$\left.\left.+\frac{2(\sinh(1)-\sinh(-1))^{1/2}}{\Gamma(3/2)}\int_{-1}^1 \cosh(s)ds\right)\right]^{-1} > 1,$$

that is, $M > 30.486$. All the conditions in Theorem 1 are satisfied. Therefore, the fractional differential inclusion with anti-periodic boundary value conditions (44) has at least one solution.

Example 2. *Consider the fractional differential inclusion involving ψ-Riesz-Caputo derivative with anti-periodic boundary value conditions*

$$\begin{cases} {}^{RC}_{-1}D_1^{\frac{6}{5},\psi}u(x) \in F(x,u(x)), \\ u(-1)+u(1) = 0, u'(-1)+u'(1) = 0, \end{cases} \quad (45)$$

where $\psi(x) = \sin(x)$, $-1 \le x \le 1$. $\alpha = \frac{6}{5}$. *Observe that* $\psi \in C^2([-1,1])$, $\psi'(x) = \cos(x) > 0$, $-1 \le x \le 1$. *Moreover, we have*

$$\psi'(-1) = \cos(-1) = \cos(1) = \psi'(1),$$

which implies condition (H_3)–(H_4) *hold.*

$$x \to F(x, u(x)) := \left[0, \frac{|x|}{3}\frac{|u|}{|u|+1}\right], u \in R,$$

and

$$d_H(F(t,u), F(t,\bar{u})) \le \frac{|t|}{3}|u - \bar{u}|, u \in R.$$

we can find out that

$$\begin{aligned}
\gamma &= \frac{\psi(1)-\psi(-1)}{2\Gamma(\frac{1}{5})}\int_{-1}^{1}(\psi(1)-\psi(s))^{-\frac{4}{5}}\psi'(s)\frac{|s|}{3}ds + \frac{2}{\Gamma(\frac{1}{5})}(\psi(1)-\psi(-1))^{\frac{1}{5}}\int_{-1}^{1}\psi'(s)\frac{|s|}{3}ds \\
&= \frac{\sin(1)-\sin(-1)}{2\Gamma(\frac{1}{5})}\int_{-1}^{1}(\sin(1)-\sin(s))^{-\frac{4}{5}}\cos(s)\frac{|s|}{3}ds \\
&+ \frac{2}{\Gamma(\frac{6}{5})}(\sin(1)-\sin(-1))^{\frac{1}{5}}\int_{-1}^{1}\cos(s)\frac{|s|}{3}ds \\
&< 0.967,
\end{aligned}$$

that is, $\gamma < 1$. *All the conditions in Theorem 2 are satisfied. Therefore, the fractional differential inclusion with anti-periodic boundary value conditions (45) has at least one solution.*

5. Conclusions

Riesz derivative, which is different from one-sided fractional derivative, as the Caputo or Riemann-Liouville derivative, is a two-sided fractional operator. It is of great use due to its reflecting both the past and the future memory effects. We study the existence of solutions for a class of anti-periodic fractional differential inclusions with ψ-Riesz-Caputo fractional derivative in this paper. Firstly, combining ψ-Caputo derivative with Riesz-Caputo derivative, we give a new definition of ψ-Riesz-Caputo fractional derivative of order α. Then, in virtue of fixed-point theorems for multi-valued maps, some sufficient conditions for the existence of solutions to the fractional differential inclusions are presented. Last but not least, we present two examples to illustrate our main results.

Author Contributions: Both authors contributed equally and significantly in writing this paper. Both authors read and approved the final manuscript.

Acknowledgments: The authors thanks anonymous referees for their remarkable comments, suggestion, and ideas that help to improve this paper.

References

1. Ahmad, B.; Otero-Espinar, V. Existence of solutions for fractional differential inclusions with antiperiodic boundary conditions. *Bound. Value Probl.* **2009**, *2009*, 625347. [CrossRef]
2. Ahmad, B.; Ntouyas, S.K.; Zhou, Y.; Alsaedi, A. A Study of Fractional Differential Equations and Inclusions with Nonlocal Erdelyi-Kober Type Integral Boundary Conditions. *Bull. Iran. Math. Soc.* **2018**, *44*, 1315–1328. [CrossRef]

3. Almeida, R. A Caputo fractional derivative of a function with respect to another function. *Commun. Nonlinear Sci. Numer. Simula* **2017**, *44*, 460–481. [CrossRef]
4. Almeida, R.; Malinowska, A.B.; Monteiro, M.T.T. Fractional differential equations with a Caputo derivative with respect to a kernel function and their applications. *Math. Meth. Appl. Sci.* **2018**, *41*, 336–352. [CrossRef]
5. Almeida, R.; Malinowska, A.B.; Odzijewicz, T. On systems of fractional differential equations with the ψ-Caputo derivative and their applications. *Math. Meth. Appl. Sci.* **2019**. [CrossRef]
6. Kilbas, A.A.; Srivastava, H.M.; Trujillo, J.J. *Theory and Applications of Fractional Differential Equations*; North-Holl and Mathematics Studies: Elsevier: Amsterdam, The Netherlands, 2006.
7. Podlubny, I. *Fractional Differential Equations*; Academic Press: San Diego, CA, USA, 1999.
8. Wang, J.R.; Ibrahim, A.G.; O'Regan, D. Topological structure of the solution set for fractional non-instantaneous impulsive evolution inclusions. *J. Fixed Point Theory Appl.* **2018**, *20*, 59. [CrossRef]
9. Ahmad, B.; Ntouyas, S.K.; Alsaedi, A. On fractional differential inclusions with anti-periodic type integral boundary conditions. *Bound. Value Probl.* **2013**, *2013*, 82. [CrossRef]
10. Chen, F.L.; Chen, A.P.; Wu, X. Anti-periodic boundary value problems with Riesz-Caputo derivative. *Adv. Differ. Equ.* **2019**, *2019*, 119. [CrossRef]
11. Ahmad, B.; Ntouyas, S.K. On Hadmard fractional integro-differential boundary value problems. *J. Appl. Math. Camput.* **2015**, *47*, 119–131. [CrossRef]
12. Baghani, H. Existence and uniqueness of solutions to fractional Langevin equations involving two fractional orders. *J. Fixed Point Theory Appl.* **2013**, *20*, 63. [CrossRef]
13. Baleanu, D.; Mousalou, A.; Rezapour, S. A new method for investigating approximate solutions of some fractional integro-differential equations involving the Caputo-Fabrizio derivative. *Adv. Differ. Equ.* **2017**, *2017*, 51. [CrossRef]
14. da C. Sousa, J.V.; Rodrigues, F.G.; de Oliveira, E.C. Stability of the fractional Volterra integro-differential equation by means of ψ-Hilfer operator. *Math. Meth. Appl. Sci.* **2019**, *42*, 3033–3043.
15. Vivek, D.; Kanagarajan, K.; Elsayed, E.M. Some existence and stability results for Hilfer-fractional implicit differential equations with nonlocal conditions. *Mediterr. J. Math.* **2018**, *15*, 15. [CrossRef]
16. Samet, B.; Aydi, H. Lyapunov-type inequalities for an anti-periodic fractional boundary value problem involving ψ-Caputo fractional derivative. *J. Inequal. Appl.* **2018**, *2018*, 286. [CrossRef] [PubMed]
17. Wu, G.; Baleanu, D.; Deng, Z.; Zeng, S. Lattice fractional diffusion equation in terms of a Riesz-Caputo difference. *Physica A* **2015**, *438*, 335–339. [CrossRef]
18. Yang, Q.; Liu, F.; Turner, I. Numerical methods for fractional partial differential equations with Riesz space fractional derivatives. *Appl. Math. Model.* **2013**, *34*, 200–218. [CrossRef]
19. Agrawal, O.P. Fractional variational calculus in terms of Riesz fractional derivatives. *J. Phys.* **2007**, *40*, 6287–6303. [CrossRef]
20. Chen, F.L.; Baleanu, D.; Wu, G.C. Existence results of fractional differential equations with Riesz-Caputo derivative. *Eur. Phys. J. Spec. Top.* **2017**, *226*, 3411–3425. [CrossRef]
21. Lakshmikantham, V.; Vatsala, A.S. Basic theory of fractional differential equations. *Nonlinear Anal.* **2008**, *69*, 2677–2682. [CrossRef]
22. Awadalla, M.; Yameni, Y.Y. Modeling exponential growth and exponential decay real phenomena by ψ-Caputo fractional derivative. *J. Adv. Math. Comput. Sci.* **2018**, *28*, 1–13. [CrossRef]
23. Luo, D.F.; Luo, Z.G. Existence and finite-time stability of solutions for a class of nonlinear fractional differential equations with time-varying delays and non-instantaneous impulses. *Adv. Dffer. Equ.* **2019**, *2019*, 155. [CrossRef]
24. Bohnenblust, H.F.; Karlin, S. On a theorem of Ville. In *Contributions to the Theory of Games, Vol. I*; Annals of Mathematics Studies; Princeton University Press: Princeton, NJ, USA, 1950; Volume 24, pp. 155–160.
25. Bressan, A.; Colombo, G. Extensions and selections of maps with decomposable values. *Studia Math.* **1988**, *90*, 69–86. [CrossRef]
26. Deimling, K. *Multivalued Differential Equations*; De Gruyter: Berlin, Germany, 1992.
27. Granas, A.; Dugundji, J. *Fixed Point Theory*; Springer: New York, NY, USA, 2005.
28. Papageorgiou; Shouchuan, H.; Nikolaos, S. *Handbook of Multivalued Analysis Theory I*; Kluwer: Dordrecht, The Netherlands, 1997.

29. Covitz, H.; Nadler, S.B., Jr. Multivalued contraction mappings in generalized metric spaces. *Israel J. Math.* **1970**, *8*, 5–11. [CrossRef]
30. Lasota, A.; Opial, Z. An application of the Kakutani-Ky Fan theorem in the theory of ordinary differential equations. *Bull. Acad. Pol. Sci. Ser. Sci. Math. Astronom. Phys.* **1965**, *13*, 781–786.

6

Ostrowski Type Inequalities Involving ψ-Hilfer Fractional Integrals

Yasemin Basci [1,*] and Dumitru Baleanu [2,3]

1 Department of Mathematics, Faculty of Arts and Sciences, Bolu Abant Izzet Baysal University, 14280 Golkoy, Bolu, Turkey
2 Department of Mathematics, Faculty of Arts and Sciences, Çankaya University, 06530 Ankara, Turkey
3 Institute of Space Sciences, 077125 Bucharest-Magurele, Romania
* Correspondence: basci_y@ibu.edu.tr

Abstract: In this study we introduce several new Ostrowski-type inequalities for both left and right sided fractional integrals of a function g with respect to another function ψ. Our results generalized the ones presented previously by Farid. Furthermore, two illustrative examples are presented to support our results.

Keywords: fractional calculus; fractional integrals; Ostrowski type inequality

---

1. Introduction and Preliminaries

Since 1695 [1–3], fractional calculus has been studied by many researchers from both theoretical and applied viewpoints [4,5]. Particularly, fractional calculus is used to generalize classical inequalities. Studies involving integral inequalities are important in several areas such as mathematics, physics, chemistry, biology, engineering and others [6–15]. We recall that there are many definitions of fractional operators, including Riemann–Liouville (RL), Hadamard, Liouville, Weyl (see [16–19]). From such fractional integrals, one can obtain generalizations of the inequalities: Hadamard, Hermite–Hadamard, Hardy, Opial, Gruss, and Montgomery, among others [20–32].

We mention that the following inequality was developed by Ostrowski [33]:

Theorem 1. *Let $g : I \to \mathbb{R}$ be a mapping differentiable in I° such that I is an interval in $\mathbb{R}$, I° is the interior of I and $a_1, b_1 \in I^\circ$, $a_1 < b_1$. If $|g'(\xi)| \leq M$ for all $\xi \in [a_1, b_1]$, then the integral inequality holds*

$$\left|g(x) - \frac{1}{b_1 - a_1}\int_{a_1}^{b_1} g(\xi)\, d\xi\right| \leq \left[\frac{1}{4} + \frac{\left(x - \frac{a_1+b_1}{2}\right)^2}{(b_1 - a_1)^2}\right](b_1 - a_1)\, M, \tag{1}$$

for all $x \in [a_1, b_1]$.

In the literature, the inequality (1) is called the Ostrowski inequality, see [34]. This inequality has a great importance while studying the error bounds of different numerical quadrature rules. In recent years, such inequalities have been generalized and developed by many researchers. Various authors obtained new Ostrowski-type inequalities for different fractional operators, see [16–19,35–47] and the references therein.

In 2009, Anastassiou et al. [20] obtained Montgomery identities for fractional integrals and a generalization for double fractional integrals. For fractional integrals they discussed both Ostrowski and Grüss inequalities. In 2010, Alomari and Darus [36] presented some Ostrowski-type inequalities for the class of convex (or concave) functions. In 2012, Set [41] obtained some new fractional Ostrowski-type inequalities. In the same year, Liu [40] established some Ostrowski-type inequalities involving RL fractional integrals for the h-convex function. His results are generalizations of [41,42]. He also provided new estimates on Ostrowski-type inequalities for fractional integrals.

In 2013, Yue [38] obtained Ostrowski inequalities for both fractional integrals and associated fractional integrals. In 2014, Aljinević [16] studied Montgomery identities for fractional integrals of a function f with respect to another function g. Also, he gave the Ostrowski inequality for fractional integrals for functions whose first derivatives belong to L_p spaces. In the same year, Yıldırım and Kırtay [46] established new generalizations for Ostrowski inequalities by using the generalized RL fractional integral.

Yıldız et al. [47] used the RL fractional integrals to obtain several new generalizations of Ostrowski type inequalities. Farid [35] found a new version of Ostrowski type inequalities in a very simple way for RL fractional integrals. He also derived some related results. Recently, Dragomir studied several generalizations of the Ostrowski type integral inequality involving RL fractional integrals of bounded variation: Hölder and Lipschitzian functions, see [17–19]. In 2018, Yaldız and Set [45] obtained some new Ostrowski type inequalities for generalized fractional integral operators.

Recently, Sousa and Oliveira [43] introduced the left and right sided fractional integrals and the so-called ψ-Hilfer fractional derivative with respect to another function. They studied Gronwall inequalities and the Cauchy-type problem by means of the ψ- Hilfer operator in [44]. Consequently, they opened a window for new applications.

The following definitions are special approaches for when the kernel is unknown, involving a function ψ. Let $\alpha_1 > 0$ and $I = [a_1, b_1]$ be a finite or infinite interval. Also, let the function g be integrable defined on I, and the function ψ be increasing and positive monotone on $(a_1, b_1]$, having a continuous derivative $\psi'(x)$ on (a_1, b_1).

The expressions of the left sided and right sided fractional integrals of a function g with respect to another function ψ can be seen [4,5], respectively:

$$I_{a_1^+}^{\alpha_1;\psi} g(x) := \frac{1}{\Gamma(\alpha_1)} \int_{a_1}^{x} \psi'(\xi) (\psi(x) - \psi(\xi))^{\alpha_1 - 1} g(\xi) d\xi \tag{2}$$

and

$$I_{b_1^-}^{\alpha_1;\psi} g(x) := \frac{1}{\Gamma(\alpha_1)} \int_{x}^{b_1} \psi'(\xi) (\psi(\xi) - \psi(x))^{\alpha_1 - 1} g(\xi) d\xi. \tag{3}$$

If we take $\psi(x) = x$ and $\psi(x) = \ln x$, then we obtain RL and Hadamard fractional integrals, respectively.

The organization of this manuscript is as follows. In Section 1, we give the introduction and preliminaries. Motivated by [35,43], several Ostrowski-type inequalities for the left sided and right sided fractional integrals of a function g with respect to another function ψ are established in Section 2. Illustrative examples are presented in Section 3 to support our results. Section 4 deals with our conclusions.

2. Main Results

Below, we will show several new Ostrowski-type inequalities for both left and right sided fractional integrals of a function g with respect to another function ψ.

Theorem 2. *Assume that the conditions of the Theorem 1 are satisfied. Also, suppose that the function $\psi \in C^1(I)$ is increasing and positive monotone, and $\psi'(x) \geq 1$ $(\forall x \in I)$. Let $I_{a_1^+}^{\alpha_1;\psi}$ and $I_{b_1^-}^{\beta_1;\psi}$ be defined as (2) and (3), respectively. Then the following inequality holds:*

$$\Big| \Big((\psi(b_1) - \psi(x))^{\beta_1} + (\psi(x) - \psi(a_1))^{\alpha_1} \Big) g(x)$$

$$- \left(\Gamma(\beta_1 + 1) I_{b_1^-}^{\beta_1;\psi} g(x) + \Gamma(\alpha_1 + 1) I_{a_1^+}^{\alpha_1;\psi} g(x) \right) \Bigg|$$

$$\leq M \left(\frac{\beta_1}{\beta_1 + 1} (\psi(b_1) - \psi(x))^{\beta_1 + 1} + \frac{\alpha_1}{\alpha_1 + 1} (\psi(x) - \psi(a_1))^{\alpha_1 + 1} \right), \tag{4}$$

where $\alpha_1, \beta_1 > 0$ *and* $x \in [a_1, b_1]$.

Proof of Theorem 2. Taking into account that ψ is an increasing and positive monotone function, for $\alpha_1 > 0$ and $\xi \in [a_1, x]$ we get

$$(\psi(x) - \psi(\xi))^{\alpha_1} \leq (\psi(x) - \psi(a_1))^{\alpha_1}. \tag{5}$$

Utilizing (5) and the given condition on g', we obtain

$$\int_{a_1}^{x} (M\psi'(\xi) - g'(\xi)) (\psi(x) - \psi(\xi))^{\alpha_1} d\xi \leq (\psi(x) - \psi(a_1))^{\alpha_1} \int_{a_1}^{x} (M\psi'(\xi) - g'(\xi)) d\xi$$

and

$$\int_{a_1}^{x} (M\psi'(\xi) + g'(\xi)) (\psi(x) - \psi(\xi))^{\alpha_1} d\xi \leq (\psi(x) - \psi(a_1))^{\alpha_1} \int_{a_1}^{x} (M\psi'(\xi) + g'(\xi)) d\xi.$$

If the above integrals are calculated, we obtain the following inequalities, respectively:

$$(\psi(x) - \psi(a_1))^{\alpha_1} g(x) - \Gamma(\alpha_1 + 1) I_{a_1^+}^{\alpha_1;\psi} g(x) \leq \frac{M\alpha_1}{\alpha_1 + 1} (\psi(x) - \psi(a_1))^{\alpha_1 + 1} \tag{6}$$

and

$$\Gamma(\alpha_1 + 1) I_{a_1^+}^{\alpha_1;\psi} g(x) - (\psi(x) - \psi(a_1))^{\alpha_1} g(x) \leq \frac{M\alpha_1}{\alpha_1 + 1} (\psi(x) - \psi(a_1))^{\alpha_1 + 1}. \tag{7}$$

By using (6) and (7), we report the following inequality:

$$\left| (\psi(x) - \psi(a_1))^{\alpha_1} g(x) - \Gamma(\alpha_1 + 1) I_{a_1^+}^{\alpha_1;\psi} g(x) \right| \leq \frac{M\alpha_1}{\alpha_1 + 1} (\psi(x) - \psi(a_1))^{\alpha_1 + 1}. \tag{8}$$

On the other hand, if ψ is an increasing and positive function, for $\xi \in [x, b_1]$ and $\beta_1 > 0$ we get

$$(\psi(\xi) - \psi(x))^{\beta_1} \leq (\psi(b_1) - \psi(x))^{\beta_1}. \tag{9}$$

By using (9) and the given condition on g', we conclude

$$\int_{x}^{b_1} (M\psi'(\xi) - g'(\xi)) (\psi(\xi) - \psi(x))^{\beta_1} d\xi \leq (\psi(b_1) - \psi(x))^{\beta_1} \int_{x}^{b_1} (M\psi'(\xi) - g'(\xi)) d\xi$$

and

$$\int_{x}^{b_1} (M\psi'(\xi) + g'(\xi)) (\psi(\xi) - \psi(x))^{\beta_1} d\xi \leq (\psi(b_1) - \psi(x))^{\beta_1} \int_{x}^{b_1} (M\psi'(\xi) + g'(\xi)) d\xi.$$

If the above integrals are calculated, we obtain the following inequalities, respectively:

$$\Gamma(\beta_1 + 1) I_{b_1^-}^{\beta_1;\psi} g(x) - (\psi(b_1) - \psi(x))^{\beta_1} g(x) \leq \frac{M\beta_1}{\beta_1 + 1} (\psi(b_1) - \psi(x))^{\beta_1 + 1} \tag{10}$$

and

$$(\psi(b_1) - \psi(x))^{\beta_1} g(x) - \Gamma(\beta_1 + 1) I_{b_1^-}^{\beta_1;\psi} g(x) \leq \frac{M\beta_1}{\beta_1 + 1} (\psi(b_1) - \psi(x))^{\beta_1 + 1}. \tag{11}$$

By using (10) and (11), the following inequality will appear:

$$\left|(\psi(b_1)-\psi(x))^{\beta_1}g(x)-\Gamma(\beta_1+1)\,I_{b_1^-}^{\beta_1;\psi}g(x)\right|\le\frac{M\beta_1}{\beta_1+1}(\psi(b_1)-\psi(x))^{\beta_1+1}. \tag{12}$$

So, by utilizing (8) and (12), we obtain (4). □

Theorem 3. *Let $g:I\to\mathbb{R}$ be a mapping differentiable in I° (the interior of I) such that I is an interval in $\mathbb{R}$ and $a_1,b_1\in I^\circ$, $a_1<b_1$. Assume that the function $\psi\in C^1(I)$ is increasing and positive monotone, and $\psi'(x)\ge 1$ ($\forall x\in I$). Also, let $I_{a_1^+}^{\alpha_1;\psi}$ and $I_{b_1^-}^{\beta_1;\psi}$ be defined as (2) and (3), respectively. If $m\le g'(t)\le M$ for $M\ge 0$, $m\le 0$ and all $\xi\in[a_1,b_1]$, then the following inequalities hold:*

$$\begin{aligned}&\left((\psi(x)-\psi(a_1))^{\alpha_1}-(\psi(b_1)-\psi(x))^{\beta_1}\right)g(x)\\&\qquad-\left(\Gamma(\alpha_1+1)\,I_{a_1^+}^{\alpha_1;\psi}g(x)-\Gamma(\beta_1+1)\,I_{b_1^-}^{\beta_1;\psi}g(x)\right)\\&\le M\left(\frac{\alpha_1}{\alpha_1+1}(\psi(x)-\psi(a_1))^{\alpha_1+1}+\frac{\beta_1}{\beta_1+1}(\psi(b_1)-\psi(x))^{\beta_1+1}\right)\end{aligned} \tag{13}$$

and

$$\begin{aligned}&\left((\psi(b_1)-\psi(x))^{\beta_1}-(\psi(x)-\psi(a_1))^{\alpha_1}\right)g(x)\\&\qquad+\left(\Gamma(\alpha_1+1)\,I_{a_1^+}^{\alpha_1;\psi}g(x)-\Gamma(\beta_1+1)\,I_{b_1^-}^{\beta_1;\psi}g(x)\right)\\&\le -m\left(\frac{\beta_1}{\beta_1+1}(\psi(b_1)-\psi(x))^{\beta_1+1}+\frac{\alpha_1}{\alpha_1+1}(\psi(x)-\psi(a_1))^{\alpha_1+1}\right),\end{aligned} \tag{14}$$

where $\alpha_1,\beta_1>0$ and $x\in[a_1,b_1]$.

Proof of Theorem 3. Using the given comparing conditions on g', the proof is similar to one of Theorem 2. That is, from (5) and by using the given condition on g', we conclude

$$\int_{a_1}^{x}(M\psi'(\xi)-g'(\xi))(\psi(x)-\psi(\xi))^{\alpha_1}\,d\xi\le(\psi(x)-\psi(a_1))^{\alpha_1}\int_{a_1}^{x}(M\psi'(\xi)-g'(\xi))\,d\xi$$

and

$$\int_{a_1}^{x}(g'(\xi)-m\psi'(\xi))(\psi(x)-\psi(\xi))^{\alpha_1}\,d\xi\le(\psi(x)-\psi(a_1))^{\alpha_1}\int_{a_1}^{x}(g'(\xi)-m\psi'(\xi))\,d\xi.$$

If the above integrals are calculated, we obtain the following inequalities, namely:

$$(\psi(x)-\psi(a_1))^{\alpha_1}g(x)-\Gamma(\alpha_1+1)\,I_{a_1^+}^{\alpha_1;\psi}g(x)\le\frac{M\alpha_1}{\alpha_1+1}(\psi(x)-\psi(a_1))^{\alpha_1+1} \tag{15}$$

and

$$\Gamma(\alpha_1+1)I_{a_1^+}^{\alpha_1;\psi}g(x)-(\psi(x)-\psi(a_1))^{\alpha_1}g(x)\le-\frac{m\alpha_1}{\alpha_1+1}(\psi(x)-\psi(a_1))^{\alpha_1+1}. \tag{16}$$

On the other hand, by using (9) and the given condition on g', we have

$$\int_{x}^{b_1}(M\psi'(\xi)-g'(\xi))(\psi(\xi)-\psi(x))^{\beta_1}\,d\xi\le(\psi(b_1)-\psi(x))^{\beta_1}\int_{x}^{b_1}(M\psi'(\xi)-g'(\xi))\,d\xi$$

and

$$\int_x^{b_1} (g'(\xi) - m\psi'(\xi)) (\psi(\xi) - \psi(x))^{\beta_1} d\xi \le (\psi(b_1) - \psi(x))^{\beta_1} \int_x^{b_1} (g'(\xi) - m\psi'(\xi)) d\xi.$$

If the above integrals are calculated, we obtain the following inequalities, namely:

$$\Gamma(\beta_1+1) I_{b_1^-}^{\beta_1;\psi} g(x) - (\psi(b_1) - \psi(x))^{\beta_1} g(x) \le \frac{M\beta_1}{\beta_1+1} (\psi(b_1) - \psi(x))^{\beta_1+1} \tag{17}$$

and

$$(\psi(b_1) - \psi(x))^{\beta_1} g(x) - \Gamma(\beta_1+1) I_{b_1^-}^{\beta_1;\psi} g(x) \le -\frac{m\beta_1}{\beta_1+1} (\psi(b_1) - \psi(x))^{\beta_1+1}, \tag{18}$$

respectively. By using (15) and (17), we obtain (13). In addition, by using (16) and (18), we provide (14). □

Theorem 4. *Let $g : I \to \mathbb{R}$ be a mapping differentiable in I° (the interior of I) such that I is an interval in $\mathbb{R}$ and $a_1, b_1 \in I^\circ$, $a_1 < b_1$. Assume that the function $\psi \in C^1(I)$ is increasing and positive monotone, and $\psi'(x) \ge 1$ $(\forall x \in I)$. Also, let $I_{a_1^+}^{\alpha_1;\psi}$ and $I_{b_1^-}^{\beta_1;\psi}$ be defined as (2) and (3), respectively. If $m \le g'(t) \le M$ for $M \ge 0$, $m \le 0$ and all $\xi \in [a_1, b_1]$, then the following inequalities hold:*

$$\begin{aligned} &\left((\psi(x) - \psi(a_1))^{\alpha_1} + (\psi(b_1) - \psi(x))^{\beta_1}\right) g(x) \\ &\quad - \left(\Gamma(\alpha_1+1) I_{a_1^+}^{\alpha_1;\psi} g(x) + \Gamma(\beta_1+1) I_{b_1^-}^{\beta_1;\psi} g(x)\right) \\ &\le \frac{M\alpha_1}{\alpha_1+1} (\psi(x) - \psi(a_1))^{\alpha_1+1} - \frac{m\beta_1}{\beta_1+1} (\psi(b_1) - \psi(x))^{\beta_1+1} \end{aligned} \tag{19}$$

and

$$\begin{aligned} &-\left((\psi(b_1) - \psi(x))^{\beta_1} + (\psi(x) - \psi(a_1))^{\alpha_1}\right) g(x) \\ &\quad + \left(\Gamma(\alpha_1+1) I_{a_1^+}^{\alpha_1;\psi} g(x) + \Gamma(\beta_1+1) I_{b_1^-}^{\beta_1;\psi} g(x)\right) \\ &\le \frac{M\beta_1}{\beta_1+1} (\psi(b_1) - \psi(x))^{\beta_1+1} - \frac{m\alpha_1}{\alpha_1+1} (\psi(x) - \psi(a_1))^{\alpha_1+1}, \end{aligned} \tag{20}$$

where $\alpha_1, \beta_1 > 0$ and $x \in [a_1, b_1]$.

Proof of Theorem 4. Proof is constructed in the same line as the proof of Theorem 3. By using (15) and (18), we obtain (19). In addition, from (16) and (17), we get (20). □

Theorem 5. *Suppose that the conditions of the Theorem 2 are satisfied. Also, assume that the function $\psi \in C^1(I)$ is increasing and positive monotone, and $\psi'(x) \ge 1$ $(\forall x \in I)$. Let $I_{a_1^+}^{\alpha_1;\psi}$ and $I_{b_1^-}^{\beta_1;\psi}$ be defined as (2) and (3), respectively. Then the following inequality holds:*

$$\begin{aligned} &\Big|\left((\psi(b_1) - \psi(x))^{\beta_1} g(b_1) + (\psi(x) - \psi(a_1))^{\alpha_1} g(a_1)\right) \\ &\quad - \left(\Gamma(\beta_1+1) I_{x^+}^{\beta_1;\psi} g(b_1) + \Gamma(\alpha_1+1) I_{x^-}^{\alpha_1;\psi} g(a_1)\right)\Big| \\ &\le M\left(\frac{\beta_1}{\beta_1+1} (\psi(b_1) - \psi(x))^{\beta_1+1} + \frac{\alpha_1}{\alpha_1+1} (\psi(x) - \psi(a_1))^{\alpha_1+1}\right), \end{aligned} \tag{21}$$

where $\alpha_1, \beta_1 > 0$ and $x \in [a_1, b_1]$.

Proof of Theorem 5. Recalling ψ is an increasing and positive monotone function, for $\alpha_1 > 0$ and $\xi \in [a_1, x]$ we obtain

$$(\psi(\xi) - \psi(a_1))^{\alpha_1} \leq (\psi(x) - \psi(a_1))^{\alpha_1}. \tag{22}$$

By using (22) and the given condition on g', we have

$$\int_{a_1}^{x} (M\psi'(\xi) - g'(\xi))(\psi(\xi) - \psi(a_1))^{\alpha_1} d\xi \leq (\psi(x) - \psi(a_1))^{\alpha_1} \int_{a_1}^{x} (M\psi'(\xi) - g'(\xi))\, d\xi$$

and

$$\int_{a_1}^{x} (M\psi'(\xi) + g'(\xi))(\psi(\xi) - \psi(a_1))^{\alpha_1} d\xi \leq (\psi(x) - \psi(a_1))^{\alpha_1} \int_{a_1}^{x} (M\psi'(\xi) + g'(\xi))\, d\xi.$$

If the above integrals are calculated, we obtain the following inequalities, respectively:

$$\Gamma(\alpha_1 + 1) I_{x^-}^{\alpha_1;\psi} g(a_1) - (\psi(x) - \psi(a_1))^{\alpha_1} g(a_1) \leq \frac{M\alpha_1}{\alpha_1 + 1} (\psi(x) - \psi(a_1))^{\alpha_1 + 1} \tag{23}$$

and

$$(\psi(x) - \psi(a_1))^{\alpha_1} g(a_1) - \Gamma(\alpha_1 + 1) I_{x^-}^{\alpha_1;\psi} g(a_1) \leq \frac{M\alpha_1}{\alpha_1 + 1} (\psi(x) - \psi(a_1))^{\alpha_1 + 1}. \tag{24}$$

By utilizing (23) and (24), the following inequality holds:

$$\left| (\psi(x) - \psi(a_1))^{\alpha_1} g(a_1) - \Gamma(\alpha_1 + 1) I_{x^-}^{\alpha_1;\psi} g(a_1) \right| \leq \frac{M\alpha_1}{\alpha_1 + 1} (\psi(x) - \psi(a_1))^{\alpha_1 + 1}. \tag{25}$$

Using the fact that ψ is an increasing and positive monotone function, for $\xi \in [x, b_1]$ and $\beta_1 > 0$ we get

$$(\psi(b_1) - \psi(\xi))^{\beta_1} \leq (\psi(b_1) - \psi(x))^{\beta_1}. \tag{26}$$

By using (26) and the given condition on g', we have

$$\int_{x}^{b_1} (M\psi'(\xi) - g'(\xi))(\psi(b_1) - \psi(\xi))^{\beta_1} d\xi \leq (\psi(b_1) - \psi(x))^{\beta_1} \int_{x}^{b_1} (M\psi'(\xi) - g'(\xi))\, d\xi$$

and

$$\int_{x}^{b_1} (M\psi'(\xi) + g'(\xi))(\psi(b_1) - \psi(\xi))^{\beta_1} d\xi \leq (\psi(b_1) - \psi(x))^{\beta_1} \int_{x}^{b_1} (M\psi'(\xi) + g'(\xi))\, d\xi.$$

If the above integrals are calculated, we obtain the following inequalities, respectively:

$$(\psi(b_1) - \psi(x))^{\beta_1} g(b_1) - \Gamma(\beta_1 + 1) I_{x^+}^{\beta_1;\psi} g(b_1) \leq \frac{M\beta_1}{\beta_1 + 1} (\psi(b_1) - \psi(x))^{\beta_1 + 1} \tag{27}$$

and

$$\Gamma(\beta_1 + 1) I_{x_1^+}^{\beta_1;\psi} g(b_1) - (\psi(b_1) - \psi(x))^{\beta_1} g(b_1) \leq \frac{M\beta_1}{\beta_1 + 1} (\psi(b_1) - \psi(x))^{\beta_1 + 1}. \tag{28}$$

By making use of (27) and (28), the following inequality holds:

$$|(\psi(b_1) - \psi(x))^{\beta_1} g(b_1) - \Gamma(\beta_1 + 1) I_{x^+}^{\beta_1;\psi} g(b_1)| \leq \frac{M\beta_1}{\beta_1 + 1} (\psi(b_1) - \psi(x))^{\beta_1 + 1}. \tag{29}$$

So, from (25) and (29), we obtain (21). □

Corollary 1. *If $\beta_1 = \alpha_1$ in Theorem 2, then the following fractional Ostrowski inequality holds:*

$$\begin{aligned}&\left|\left((\psi(b_1)-\psi(x))^{\alpha_1}+(\psi(x)-\psi(a_1))^{\alpha_1}\right)g(x)\right.\\&\qquad\left.-\Gamma(\alpha_1+1)\left(I_{b_1^-}^{\alpha_1;\psi}g(x)+I_{a_1^+}^{\alpha_1;\psi}g(x)\right)\right|\\&\qquad\le\frac{M\alpha_1}{\alpha_1+1}\left((\psi(b_1)-\psi(x))^{\alpha_1+1}+(\psi(x)-\psi(a_1))^{\alpha_1+1}\right),\end{aligned}$$

where $\alpha_1 > 0$ and $x \in [a_1, b_1]$.

Corollary 2. *If $\alpha_1 = \beta_1 = 1$ and $\psi(x) = x$, then we lead to the Ostrowski inequality (1).*

Corollary 3. *If $\alpha_1 = \beta_1$ in Theorem 5, then we obtain the following fractional Ostrowski inequality:*

$$\begin{aligned}&\left|\left((\psi(b_1)-\psi(x))^{\alpha_1}g(b_1)+(\psi(x)-\psi(a_1))^{\alpha_1}g(a_1)\right)\right.\\&\qquad\left.-\Gamma(\alpha_1+1)\left(I_{x^+}^{\alpha_1;\psi}g(b_1)+I_{x^-}^{\alpha_1;\psi}g(a_1)\right)\right|\\&\qquad\le\frac{M\alpha_1}{\alpha_1+1}\left((\psi(b_1)-\psi(x))^{\alpha_1+1}+(\psi(x)-\psi(a_1))^{\alpha_1+1}\right),\end{aligned}$$

where $\alpha_1 > 0$ and $x \in [a_1, b_1]$.

Remark 1. *If we take $\psi(x) = x$, then Theorem 2, Theorem 3, and Theorem 5 reduce to Theorem 1.2–Theorem 1.4 in Farid [35], respectively. But, in [35], $-m$ should be M in the first inequality in Theorem 1.3. Also, M should be $-m$ in the second inequality.*

Remark 2. *After following the steps of the proof of Theorem 2 with $\psi(x) = x$ and $\alpha_1 = \beta_1 = 1$, an alternative proof of the Ostrowski inequality is obtained (see [37]).*

3. Examples

In this section, we support our main results by presenting two examples.

Example 1. *Let $\alpha_1 = 0.5$, $\beta_1 = 2.2$, $\psi(x) = e^x$, $g(x) = \sin x$ and $[a_1, b_1] = [0, \pi]$. Then, we obtain $|g'(x)| = |\cos x| \le 1$, that is, $M = 1$. Also, $\psi(x) = e^x$ is an increasing continuous derivative and positive monotone function with $\psi'(x) = e^x \ge 1$ for all $x \in [0, \pi]$. Then, using Theorem 2, for $[0, \pi]$ we obtain the following Ostrowski type inequality:*

$$\begin{aligned}&\left|\left((e^\pi-e^x)^{2.2}+(e^x-1)^{0.5}\right)\sin x-\left(\Gamma(3.2)\,I_{\pi^-}^{2.2;\psi}\sin x+\Gamma(1.5)\,I_{0^+}^{0.5;\psi}\sin x\right)\right|\\&\qquad\le\frac{11}{16}(e^\pi-e^x)^{3.2}+\frac{1}{3}(e^x-1)^{1.5}.\end{aligned}$$

Example 2. *Let $\alpha_1 = 0.5$, $\beta_1 = 2.2$, $\psi(x) = 6\sqrt{x+2}$, $g(x) = (x-1)^2$ and $[a_1, b_1] = [0, 2]$. Then, we get $g'(x) = 2(x-1)$. Let $m = -2$ and $M = 2$. Also, $\psi(x) = 6\sqrt{x+2}$ is an increasing continuous derivative and positive monotone function with $\psi'(x) = \frac{3}{\sqrt{x+2}} \ge 1$ for all $x \in I = [0, 2]$.Then, using Theorem 3, for $x \in [0, 2]$ we obtain the following Ostrowski type inequality:*

$$\begin{aligned}&\left(\left(6\sqrt{x+2}-6\sqrt{2}\right)^{0.5}-\left(12-6\sqrt{x+2}\right)^{2.2}\right)(x-1)^2-\left(\Gamma(1.5)\,I_{0^+}^{0.5;\psi}(x-1)^2-\Gamma(3.2)\,I_{2^-}^{2.2;\psi}(x-1)^2\right)\\&\qquad\le 2\left(\frac{1}{3}\left(6\sqrt{x+2}-6\sqrt{2}\right)^{1.5}+\frac{11}{16}\left(12-6\sqrt{x+2}\right)^{3.2}\right)\end{aligned}$$

and

$$\left(\left(12-6\sqrt{x+2}\right)^{2.2}-\left(6\sqrt{x+2}-6\sqrt{2}\right)^{1.5}\right)(x-1)^2+\left(\Gamma(1.5)\, I_{0^+}^{0.5;\psi}(x-1)^2-\Gamma(3.2)\, I_{2^-}^{2.2;\psi}(x-1)^2\right)$$
$$\leq 2\left(\frac{11}{16}\left(6\sqrt{2}-6\sqrt{x+2}\right)^{3.2}+\frac{1}{3}\left(6\sqrt{x+2}-6\sqrt{2}\right)^{1.5}\right).$$

4. Conclusions

Studies involving integral inequalities play an important role in several areas of science and engineering. During recent years, such inequalities have been generalized and developed by many researchers. Ostrowski inequalities have a great importance while studying the error bounds of different numerical quadrature rules, for example the midpoint rule, Simpson's rule, the trapezoidal rule and other generalized Riemann types. In this paper, by generalizing the inequalities in [35], we proposed, within four theorems and their related corollaries, several new Ostrowski-type integral inequalities for the left and right sided fractional integrals of a function g with respect to another function ψ. Finally, we investigated in detail two examples to show the reported results.

Author Contributions: All authors contributed to each part of this work equally, and they all read and approved the final manuscript.

Acknowledgments: The authors would like to thank to the referees for their useful comments and remarks.

References

1. Leibniz, G.W. Letter from Hanover, Germany to G.F.A L'Hospital, September 30, 1695. In *Leibniz Mathematische Schriften*; Olms-Verlag: Hildesheim, Germany, 1962; pp. 301–302.
2. Leibniz, G.W. Letter from Hanover, Germany, to Johann Bernoulli, December 28, 1695. In *Leibniz Mathematische Schriften*; Olms-Verlag: Hildesheim, Germany, 1962; p. 226.
3. Leibniz, G.W. Letter from Hanover, Germany, to John Wallis, May 30, 1697. In *Leibniz Mathematische Schriften*; Olms-Verlag: Hildesheim, Germany, 1962; p. 25.
4. Kilbas, A.A.; Srivastava, H.M.; Trujillo, J.J. Theory and applications of fractional differential equations. In *North-Holland Mathematics Studies*; Elsevier: Amsterdam, The Netherlands, 2006; p. 204.
5. Samko, S.G.; Kilbas, A.A.; Marichev, O.I. *Fractional Integrals and Derivatives*; Gordon and Breach: Yverdon, Switzerland, 1993.
6. Atanackovic, T.M.; Pilipovic, S.; Stankovic, B.; Zorica, D. *Fractional Calculus with Applications in Mechanics: Vibrations and Diffusion Processes*; Wiley: London, UK; Hoboken, NJ, USA, 2014.
7. Bandle, C.; Gilányi, A.; Losonczi, L.; Páles, Z.; Plum, M. *Inequalities and Applications: Conference on Inequalities and Applications, Noszvaj (Hungary), September 2007*; Springer: Berlin/Heidelberg, Germany, 2008; p. 157.
8. Bainov, D.D.; Simeonov, P.S. *Integral Inequalities and Applications*; Springer: Berlin/Heidelberg, Germany, 2013; p. 57.
9. Corlay, S.; Lebovits, J.; Véhel, J.L. Multifractional stochastic volatility models. *Math. Finance* **2014**, *24*, 364–402. [CrossRef]
10. Herrmann, R. *Fractional Calculus: An Introduction for Physicists*; World Scientific Publishing Company: Singapore, 2011.
11. Magin, R.L.; Ingo, C; Colon-Perez, L.; Triplett, W.; Mareci, T.H. Characterization of anomalous diffusion in porous biological tissues using fractional order derivatives an, entropy. *Microporous Mesoporous Mater.* **2013**, *178*, 39–43. [CrossRef] [PubMed]
12. Magin, R.L. Fractional calculus models of complex dynamics in biological tissues. *Comput. Math. Appl.* **2010**, *59*, 1586–1593. [CrossRef]
13. Meral, F.; Royston, T.; Magin, R. Fractional calculus in viscoelasticity: An experimental study. *Commun. Nonlinear Sci. Numer. Simul.* **2010**, *15*, 939–945. [CrossRef]

14. Silva Costa, F.; Soares, J.C.S.; Gomez Plata, A.R.; Capelas de Oliveira, E. On the fractional Harry Dym equation. *Comput. Appl. Math.* **2018**, *37*, 2862–2876. [CrossRef]
15. Silva Costa, F.; Contharteze Grigoletto, E.; Vaz Jr, J.; Capelas de Oliveira, E. Slowing-down of neutrons: A fractional model. *Commun. Appl. Ind. Math.* **2015**, *6*. [CrossRef]
16. Aljinević, A.A. Montgomery identity and Ostrowski type inequalities for Riemann-Liouville fractional integral. *J. Math. Vol.* **2014**, 503195. [CrossRef]
17. Dragomir, S.S. Ostrowski type inequalities for Riemann-Liouville fractional integrals of bounded variation, Hölder and Lipschitzian functions. *RGMIA Res. Rep. Coll.* **2017**, *20*, 48.
18. Dragomir, S.S. Ostrowski type inequalities for generalized Riemann-Liouville fractional integrals of functions with bounded variation. *RGMIA Res. Rep. Coll.* **2017**, *20*, 58.
19. Dragomir, S.S. Further Ostrowski and Trapezoid type inequalities for Riemann-Liouville fractional integrals of functions with bounded variation. *RGMIA Res. Rep. Coll.* **2017**, *20*, 84.
20. Anastassiou, G.M.; Hooshmandasl, R.; Ghasemi, A.; Moftakharzahed, F. Montgomery identities for fractional integrals and related fractional inequalities. *J. Inequal. Pure Appl. Math.* **2009**, *10*, 97.
21. Chen, F. Extensions of the Hermite-Hadamard inequality for convex functions via fractional integrals. *J. Math. Inequal.* **2016**, *10*, 75–81. [CrossRef]
22. Chen, H.; Katugampola, U.N. Hermite-Hadamard and Hermite-Hadamard-Fejér type inequalities for generalized fractional integrals. *J. Math. Anal. Appl.* **2013**, *446*, 1274–1291. [CrossRef]
23. Iqbal, S.; Krulić, K.; Pečarić, J. Weighted Hardy-type inequalities for monotone convex functions with some applications. *Fract. Differ. Calc.* **2013**, *3*, 31–53. [CrossRef]
24. Iqbal, S.; Krulić, K.; Pečarić, J. On refined Hardy-type inequalities with fractional integrals and fractional derivatives. *Math. Slovaca* **2014**, *64*, 879–892. [CrossRef]
25. Iqbal S, Krulić K, Pečarić J. On a new class of Hardy-type inequalities with fractional integrals and fractional derivatives. *Rad Hazu. Matematičke Znanosti* **2014**, *18=519*, 91–106.
26. Iqbal, S.; Pečarić, J.; Samraiz, M.; Tomovski, Z. Hardy-type inequalities for generalized fractional integral operators. *Tbilisi Math. J.* **2017**, *10*, 75–90. [CrossRef]
27. Iqbal, S.; Pečarić, J.; Samraiz, M.; Tomovski, Z. On some Hardy-type inequalities for fractional calculus operators. *Banach J. Math. Anal.* **2017**, *11*, 438–457. [CrossRef]
28. Jain, R.; Pathan, M.A. On Weyl fractional integral operators. *Tamkang J. Math.* **2004**, *35*, 169–173. [CrossRef]
29. Nasibullin, R. Hardy-type inequalities for fractional integrals and derivatives of Riemann-Liouville. *Lobachevskii J. Math.* **2017**, *38*, 709–718. [CrossRef]
30. Sarıkaya, M.Z.; Budak, H. New inequalities of Opial type for conformable fractional integrals. *Turk. J. Math.* **2017**, *41*, 1164–1173. [CrossRef]
31. Set, E.; İscan, I.; Sarıkaya, M.Z.S.; Özdemir, M.E. On new inequalities of Hermite-Hadamard Fejér type for convex functions via fractional integrals. *Appl. Math. Comput.* **2015**, *259*, 875–881. [CrossRef]
32. Wang, J.; Li, X.; Fekan, M.; Zhou, Y. Hermite-Hadamard-type inequalities for Riemann-Liouville fractional integrals via two kinds of convexity. *Appl. Anal.* **2013**, *92*, 2241–2253. [CrossRef]
33. Ostrowski, A. Über die Absolutabweichung einer differentiierbaren Funktion von ihrem Integralmittelwert. *Comment. Math. Helv.* **1937**, *10*, 226–227. [CrossRef]
34. Mitrinovic, D.S.; Pecaric, J.; Fink, A.M. *Inequalities Involving Functions and Their Integrals and Derivatives*; Mathematics and its Applications (East European Series), *53*, Kluwer Academic Publishers: Dortrecht, The Netherlands, 1991.
35. Farid, G. Some new Ostrowski type inequalities via fractional integrals. *Int. J. Anal. Appl.* **2017**, *14*, 64–68.
36. Alomari, M.; Darus, M. Some Ostrowski type inequalities for convex functions with applications. *RGMIA* **2010**, *13*, 3.
37. Farid, G. Straightforward proofs of Ostrowski inequality and some related results. *Int. J. Anal.* **2016**, *2016*, 3918483. [CrossRef]
38. Yue, Hu. Ostrowski inequality for fractional integrals and related fractional inequalities. *Transylv. J. Math. Mech.* **2013**, *5*, 85–89.
39. Khan, M.A.; Begum, S.; Khurshid, Y.; Chu, Y.-M. Ostrowski type inequalities involving conformable fractional integral. *J. Inequal. Appl.* **2018**, *2018*, 70. [CrossRef]
40. Liu, W. Some Ostrowski type inequalities via Riemann-Liouville fractional integrals for *h*-convex functions. *J. Comput. Anal. Appl.* **2012**, *16*, 998–1004.

41. Set, E. New inequalities for Ostrowski type for mappings whose derivatives are s-convex in the second sense via fractional integrals. *Comput. Math. Appl.* **2012**, *63*, 1147–1154. [CrossRef]
42. Tunç, M. Ostrowski type inequalities via h-convex functions with applications for special means. *J. Inequal. Appl.* **2013**, *2013*, 326. [CrossRef]
43. Vanterler da C. Sousa, J.; Capelas de Oliveira, E. On the ψ-Hilfer fractional derivative. *Commun. Nonlinear Sci. Numer. Simul.*, **2018**, *60*, 72–91. [CrossRef]
44. Vanterler da C. Sousa, J.; Capelas de Oliveira, E. A Gronwall inequality and the Cauchy-type problem by means of ψ- Hilfer operator. *Differ. Equ. Appl.* **2019** *11*, 87–106. [CrossRef]
45. Yaldız, H.; Set, E. Some new Ostrowski type inequalities for generalized fractional integrals. *AIP Conf. Proc.* **2018**, *1991*, 020018. [CrossRef]
46. Yıldırım, H.; Kırtay, Z. Ostrowski inequality for generalized fractional integral and related inequalities. *Malaya J. Mat.* **2014**, *2*, 322–329.
47. Yıldız, Ç.; Özdemir, M.E.; Sarıkaya, M.Z. New generalizations of Ostrowski-Like type inequalities for fractional integral. *Kyungpook Math. J.* **2016**, *56*, 161–172. [CrossRef]

7

Hermite–Hadamard-Type Inequalities for Convex Functions via the Fractional Integrals with Exponential Kernel

Xia Wu [1], **JinRong Wang** [2,3,*] **and Jialu Zhang** [1]

1 School of Mathematics and Finance, Xiangnan University, Chenzhou 411105, China; 230169714@seu.edu.cn (X.W.); zhangjl@xnu.edu.cn (J.Z.)
2 Department of Mathematics, Guizhou University, Guiyang 550025, China
3 School of Mathematical Sciences, Qufu Normal University, Qufu 273165, China
* Correspondence: jrwang@gzu.edu.cn

Abstract: In this paper, we establish three fundamental integral identities by the first- and second-order derivatives for a given function via the fractional integrals with exponential kernel. With the help of these new fractional integral identities, we introduce a few interesting Hermite–Hadamard-type inequalities involving left-sided and right-sided fractional integrals with exponential kernels for convex functions. Finally, some applications to special means of real number are presented.

Keywords: convex functions, Hermite–Hadamard-type inequalities, fractional integrals, exponential kernel

---

1. Introduction

Let $h : [a,b] \subset \Re \rightarrow \Re$ be a convex function. Then, h meets the following classic Hermite–Hadamard inequality (see [1])

$$h\left(\frac{a+b}{2}\right) \leq \frac{1}{b-a}\int_a^b h(s)ds \leq \frac{h(a)+h(b)}{2}. \tag{1}$$

If h is a concave function, the inequalities in (1) are presented in the negative direction. The Hermite–Hadamard inequality provides us the estimates for the integral average of a continuous convex function on a compact interval.

For the latest results on generalizing, improving, and extending this classical Hermite–Hadamard inequality, one can see [2–9] and the references therein.

In [10], Dragomir and Agarwal proved the following result connected with the right part of (1). In [11], Alomart also elicited the similar result for functions whose second derivatives absolute values are convex.

Lemma 1 (see [10], Theorem 2.2). *Assuming $h : [a,b] \subseteq \Re \rightarrow \Re$ is a differentiable function, $h' \in L[a,b]$ and $|h''|$ is convex on $[a,b]$. Then, the bellow inequality holds:*

$$\left|\frac{h(a)+h(b)}{2} - \frac{1}{b-a}\int_a^b h(s)ds\right| \leq \frac{(b-a)}{8}(|h'(a)|+|h'(b)|). \tag{2}$$

Lemma 2 (see [11], Theorem 3). *Assuming $h : [a,b] \subseteq \Re \to \Re$ is a twice differentiable function, $h'' \in L[a,b]$ and $|h''|$ is convex on $[a,b]$. Then, the following inequality holds*

$$\left|\frac{h(a)+h(b)}{2} - \frac{1}{b-a}\int_a^b h(s)ds\right| \le \frac{(b-a)^2}{24}(|h''(a)| + |h''(b)|). \tag{3}$$

Now, fractional calculus has turned into an enchanting field of mathematics. Many extensive investigations have been carried out in this area. Due to the wide applications of Hermite–Hadamard inequalities and fractional integrals, many researchers have extended their research to Hermite–Hadamard inequalities involving fractional integrals rather than integer integrals, see [12–22]. Sarikaya et al. [12] have deduced an amusing inequality of Hermite–Hadamard-type involving fractional integrals in the place of ordinary integrals. This research fascinates many researchers to consider this respect. As a result, some new integral inequalities by the approach of fractional calculus have been obtained in the literature until now. In addition, Ahmad et al. [16] gave the new fractional integral operators with an exponential kernel and proved similar inequalities.

Definition 1 (see [16], Definition 2). *Let $h \in L[a,b]$. The fractional left-side integral $\mathcal{J}^{\alpha}_{a+}h$ and right-side integral $\mathcal{J}^{\alpha}_{b-}h$ of order $\alpha \in (0,1)$ are, respectively, defined by*

$$\mathcal{J}^{\alpha}_{a+}h(x) = \frac{1}{\alpha}\int_a^x e^{-\frac{1-\alpha}{\alpha}(x-s)}h(s)ds, \;\; x > a, \tag{4}$$

and

$$\mathcal{J}^{\alpha}_{b-}h(x) = \frac{1}{\alpha}\int_x^b e^{-\frac{1-\alpha}{\alpha}(x-s)}h(s)ds, \;\; x < b. \tag{5}$$

Lemma 3 (see [16], Theorem 1). *Let $h : [a,b] \to \Re$ be a positive convex function with $0 \le a < b$ and $h \in L[a,b]$. The following inequality for fractional integrals* (4) *and* (5) *holds:*

$$h(\frac{a+b}{2}) \le \frac{1-\alpha}{2\,(1-e^{-\rho})}[\mathcal{J}^{\alpha}_{a+}h(b) + \mathcal{J}^{\alpha}_{b-}h(a)] \le \frac{h(a)+h(b)}{2}. \tag{6}$$

Remark 1. *In* (6)*, note that*

$$\rho = \frac{1-\alpha}{\alpha}(b-a).$$

In addition, Ahmad et al. [16] derived the bound estimate of the difference between the mean value of the endpoints and the average of the fractional integrals.

Lemma 4 (see [16], Theorem 3). *Assuming $h : [a,b] \subseteq \Re \to \Re$ is differentiable, $h' \in L[a,b]$ and $|h'|$ are convex on $[a,b]$. Then, the following inequality holds:*

$$|Q_{mr}| \le \frac{(b-a)}{2\rho}\tanh(\frac{\rho}{4})(|h'(a)| + |h'(b)|),$$

where

$$Q_{mr} := \frac{h(a)+h(b)}{2} - \frac{1-\alpha}{2\,(1-e^{-\rho})}[\mathcal{J}^{\alpha}_{a+}h(b) + \mathcal{J}^{\alpha}_{b-}h(a)]$$

denotes the bound estimate of the difference between the mean value of the endpoints and the average of the fractional integrals.

However, the bound for the left of the Hermite–Hadamard inequality (6) has not been studied. It will be interesting to find

$$|Q_{ml}| \leq \textbf{what?}$$

Here,

$$Q_{ml} := \frac{1-\alpha}{2\,(1-e^{-\rho})}[\mathcal{J}^{\alpha}_{a^+}h(b) + \mathcal{J}^{\alpha}_{b^-}h(a)] - h(\frac{a+b}{2})$$

denotes the bound estimate of the difference between the value of the midpoint and the average of the fractional integrals.

Furthermore, if $|h''|$ is convex, it is natural to study the right- and left-type Hermite–Hadamard inequality via the fractional integral with an exponential kernel similar to Lemma 2, i.e., we want to find the constants ρ_1 and ρ_2 satisfying the following inequities:

$$|Q_{mr}| \leq \rho_1 \cdot (|h''(a)| + |h''(b)|),$$

and

$$|Q_{ml}| \leq \rho_2 \cdot (|h''(a)| + |h''(b)|).$$

Motivated by [12,15,16], we will demonstrate three new fractional-type integral identities and set up their corresponding Hermite–Hadamard-type inequalities involving left-sided and right-sided fractional integrals for convex functions, respectively.

2. New Fractional Integral Identity and Hermite–Hadamard-Type Inequality for First Order Derivative

We firstly prove the following lemma in order to attest the following result.

Lemma 5. *Assuming $h : [a,b] \subseteq \Re \to \Re$ is a differentiable mapping and $h' \in L[a,b]$. Then, the following equality for the fractional integrals* (4) *and* (5) *holds:*

$$\begin{aligned} Q_{ml} &= \frac{b-a}{2}\int_0^1 kh'(sa+(1-s)b)ds \\ &\quad -\frac{b-a}{2(1-e^{-\rho})}\left[\int_0^1 e^{-\rho s}h'(sa+(1-s)b)ds - \int_0^1 e^{-\rho(1-s)}h'(sa+(1-s)b)ds\right], \end{aligned} \tag{7}$$

where

$$k = \begin{cases} 1, & 0 \leq s < \frac{1}{2}, \\ -1, & \frac{1}{2} \leq s \leq 1. \end{cases}$$

Proof. Define

$$\begin{aligned} V &:= \int_0^1 e^{-\rho s}h'(sa+(1-s)b)ds - \int_0^1 e^{-\rho(1-s)}h'(sa+(1-s)b)ds \\ &= V_1 - V_2, \end{aligned} \tag{8}$$

where

$$\begin{aligned} V_1 &= \int_0^1 e^{-\rho s}h'(sa+(1-s)b)ds, \\ V_2 &= \int_0^1 e^{-\rho(1-s)}h'(sa+(1-s)b)ds. \end{aligned}$$

Integrating by parts, one has

$$\begin{aligned} V_1 &= \int_0^1 e^{-\rho s} h'(sa+(1-s)b)ds \\ &= \frac{1}{a-b}\int_0^1 e^{-\rho s} d\left(h(sa+(1-s)b)\right) \\ &= \frac{1}{a-b}\left[e^{-\rho s} f(sa+(1-s)b)\Big|_0^1 - \int_0^1 h(sa+(1-s)b)d\left(e^{-\rho s}\right)\right] \\ &= \frac{1}{a-b}\left[e^{-\rho}h(a)-h(b)+\rho\int_0^1 h(sa+(1-s)b)e^{-\rho s}ds\right] \\ &= \frac{e^{-\rho}h(a)-h(b)}{a-b} + \frac{\rho}{(a-b)^2}\int_b^a h(x)e^{-\rho\frac{x-b}{a-b}}dx \\ &= \frac{e^{-\rho}h(a)-h(b)}{a-b} - \frac{\rho}{(a-b)^2}\int_a^b e^{-\frac{1-\alpha}{\alpha}(b-x)}h(x)dx \\ &= \frac{e^{-\rho}h(a)-h(b)}{a-b} - \frac{\rho}{(a-b)^2}\alpha\mathcal{J}_{a+}^{\alpha}h(b) \\ &= \frac{e^{-\rho}h(a)-h(b)}{a-b} - \frac{1-\alpha}{(b-a)}\mathcal{J}_{a+}^{\alpha}h(b), \end{aligned} \tag{9}$$

and

$$\begin{aligned} V_2 &= \int_0^1 e^{-\rho(1-s)} h'(sa+(1-s)b)ds \\ &= \frac{1}{a-b}\int_0^1 e^{-\rho(1-s)} d\left(h(sa+(1-s)b)\right) \\ &= \frac{1}{a-b}\left[e^{-\rho(1-s)} h(sa+(1-s)b)\Big|_0^1 - \int_0^1 h(sa+(1-s)b)d\left(e^{-\rho(1-s)}\right)\right] \\ &= \frac{1}{a-b}\left[h(a)-e^{-\rho}h(b)-\rho\int_0^1 h(sa+(1-s)b)e^{-\rho(1-s)}ds\right] \\ &= \frac{h(a)-e^{-\rho}h(b)}{a-b} - \frac{\rho}{(a-b)^2}\int_b^a h(x)e^{-\rho\frac{a-x}{a-b}}dx \\ &= \frac{h(a)-e^{-\rho}h(b)}{a-b} + \frac{\rho}{(a-b)^2}\int_a^b e^{-\frac{1-\alpha}{\alpha}(x-a)}h(x)dx \\ &= \frac{h(a)-e^{-\rho}h(b)}{a-b} + \frac{\rho}{(a-b)^2}\alpha\mathcal{J}_{b-}^{\alpha}h(a) \\ &= \frac{h(a)-e^{-\rho}h(b)}{a-b} + \frac{1-\alpha}{(b-a)}\mathcal{J}_{b-}^{\alpha}h(a). \end{aligned} \tag{10}$$

Substituting (9) and (10) into (8), we get that

$$\begin{aligned} V &= V_1 - V_2 \\ &= \frac{\left(1-e^{-\rho}\right)\left(h(a)+h(b)\right)}{b-a} - \frac{1-\alpha}{(b-a)}\left[\mathcal{J}_{a+}^{\alpha}h(b)+\mathcal{J}_{b-}^{\alpha}h(a)\right]. \end{aligned} \tag{11}$$

Note

$$\begin{aligned}
&\frac{b-a}{2}\int_0^1 kh'(sa+(1-s)b)dt \\
=\ &\frac{b-a}{2}\int_0^{\frac{1}{2}} h'(sa+(1-s)b)ds-\frac{b-a}{2}\int_{\frac{1}{2}}^1 h'(sa+(1-s)b)ds \\
=\ &\frac{f(b)-h\left(\frac{a+b}{2}\right)}{2}+\frac{h(a)-h\left(\frac{a+b}{2}\right)}{2} \\
=\ &\frac{h(a)+h(b)}{2}-h\left(\frac{a+b}{2}\right).
\end{aligned} \tag{12}$$

Substituting (12) and (11) into the right-side of (7), we obtain the left of (7). This testifies the proof. □

Then, we can declare the first theorem including Hermite–Hadmard-type inequality.

Theorem 1. *If $h:[a,b]\subseteq\Re\to\Re$ is differentiable, $|h'|$ is convex on [a,b], and $h'\in L[a,b]$, then the following inequality about the fractional integrals* (4) *and* (5) *holds:*

$$|Q_{ml}|\le\frac{b-a}{2}\left[\frac{1}{2}-\frac{\tanh(\frac{\rho}{4})}{\rho}\right](|h'(a)|+|h'(b)|). \tag{13}$$

Proof. Using Lemma 5, convexity of $|h'|$, and $e^{-\rho s}\ge e^{-\rho}$ and $e^{-\rho(1-s)}\ge e^{-\rho}$ for any $s\in[0,1]$, we obtain

$$\begin{aligned}
|Q_{ml}| &= \left|\frac{b-a}{2}\int_0^1 kh'(sa+(1-s)b)ds\right. \\
&\quad\left.-\frac{b-a}{2(1-e^{-\rho})}\left[\int_0^1 e^{-\rho s}h'(sa+(1-s)b)ds-\int_0^1 e^{-\rho(1-s)}h'(sa+(1-s)b)ds\right]\right| \\
&= \frac{b-a}{2\left(1-e^{-\rho}\right)}\left|\int_0^{\frac{1}{2}}[(1-e^{-\rho})-e^{-\rho s}+e^{-\rho(1-s)}]h'(sa+(1-s)b)ds\right. \\
&\quad\left.-\int_{\frac{1}{2}}^1[(1-e^{-\rho})-e^{-\rho(1-s)}+e^{-\rho s}]h'(sa+(1-s)b)ds\right| \\
&\le \frac{b-a}{2\left(1-e^{-\rho}\right)}[\int_0^{\frac{1}{2}}(1-e^{-\rho}-e^{-\rho s}+e^{-\rho(1-s)})(s|h'(a)|+(1-s)|h'(b)|)ds \\
&\quad+\int_{\frac{1}{2}}^1(1-e^{-\rho}-e^{-\rho(1-s)}+e^{-\rho s})(s|h'(a)|+(1-s)|h'(b)|)ds] \\
&= \frac{b-a}{2\left(1-e^{-\rho}\right)}[\int_0^{\frac{1}{2}}(1-e^{-\rho}-e^{-\rho s}+e^{-\rho(1-s)})(s|h'(a)|+(1-s)|h'(b)|)ds \\
&\quad+\int_0^{\frac{1}{2}}(1-e^{-\rho}-e^{-\rho s}+e^{-\rho(1-s)})((1-s)|h'(a)|+s|h'(b)|)ds] \\
&= \frac{b-a}{2\left(1-e^{-\rho}\right)}\int_0^{\frac{1}{2}}(1-e^{-\rho}-e^{-\rho s}+e^{-\rho(1-s)})(|h'(a)|+|h'(b)|)ds \\
&= \frac{b-a}{2\left(1-e^{-\rho}\right)}\left[\frac{1-e^{-\rho}}{2}-\frac{1}{\rho}(1-e^{-\frac{\rho}{2}})^2\right](|h'(a)|+|h'(b)|) \\
&= \frac{b-a}{2}\left[\frac{1}{2}-\frac{\tanh(\frac{\rho}{4})}{\rho}\right](|h'(a)|+|h'(b)|).
\end{aligned}$$

The proof is completed. □

3. New Fractional Integral Identity and Hermite–Hadamard-Type Inequality for Second Order Derivative

In [16] Lemma 4, Ahmad et al. gave the equality

$$Q_{mr} = \frac{(b-a)}{2(1-e^{-\rho})}\left[\int_0^1 e^{-\rho s}h'(sa+(1-s)b)ds - \int_0^1 e^{-\rho(1-s)}h'(sa+(1-s)b)ds\right]. \tag{14}$$

By (14), we will prove the Hermite–Hadamard-type inequality of the order derivatives via the fractional integrals with an exponential kernel for convex functions. Before we prove our main results in this section, we give the following lemmas.

Lemma 6. *Assuming $h : [a,b] \to \Re$ is a twice differentiable function. If $h'' \in L[a,b]$, then the following equality for fractional integrals holds:*

$$Q_{mr} = \frac{(b-a)^2}{2\rho(1-e^{-\rho})}\int_0^1 \left(1+e^{-\rho}-e^{-\rho s}-e^{-\rho(1-s)}\right)h''(sa+(1-s)b)ds. \tag{15}$$

Proof. By using equality (14), we note

$$\begin{aligned} K_1 &= \int_0^1 e^{-\rho s}h'(sa+(1-s)b)ds = -\frac{1}{\rho}\int_0^1 h'(sa+(1-s)b)d(e^{-\rho s}) \\ &= \frac{1}{\rho}\left[h'(b)-e^{-\rho}h'(a)+(a-b)\int_0^1 e^{-\rho s}h''(sa+(1-s)b)ds\right], \end{aligned} \tag{16}$$

and

$$\begin{aligned} K_2 &= \int_0^1 e^{-\rho(1-s)}h'(sa+(1-s)b)ds = \frac{1}{\rho}\int_0^1 h'(sa+(1-s)b)d(e^{-\rho(1-s)}) \\ &= \frac{1}{\rho}\left[h'(a)-e^{-\rho}h'(b)-(a-b)\int_0^1 e^{-\rho(1-s)}h''(sa+(1-s)b)ds\right]. \end{aligned} \tag{17}$$

Inserting the values of K_1 and K_2 in (14), we obtain

$$\begin{aligned} &\frac{h(a)+h(b)}{2} - \frac{1-\alpha}{2\left(1-e^{-\rho}\right)}[\mathcal{J}^{\alpha}_{a+}h(b)+\mathcal{J}^{\alpha}_{b-}h(a)] \\ &= \frac{b-a}{2\rho(1-e^{-\rho})}\Bigg[\left(1+e^{-\rho}\right)\left(h'(b)-h'(a)\right) \\ &\quad -(b-a)\int_0^1\left(e^{-\rho s}+e^{-\rho(1-s)}\right)h''(sa+(1-s)b)ds\Bigg] \\ &= \frac{b-a}{2\rho(1-e^{-\rho})}\Bigg[(1+e^{-\rho})(b-a)\int_0^1 h''(sa+(1-s)b)ds \\ &\quad -(b-a)\int_0^1\left(e^{-\rho s}+e^{-\rho(1-s)}\right)h''(sa+(1-s)b)ds\Bigg] \\ &= \frac{(b-a)^2}{2\rho(1-e^{-\rho})}\int_0^1\left(1+e^{-\rho}-e^{-\rho s}-e^{-\rho(1-s)}\right)h''(sa+(1-s)b)ds. \end{aligned} \tag{18}$$

This completes the proof. □

Lemma 7. *Assuming $h : [a,b] \to \Re$ is a twice differentiable function. If $h'' \in L[a,b]$, then the following equality for fractional integrals holds:*

$$Q_{ml} = \frac{(b-a)^2}{2}\int_0^1 m(s)h''(sa+(1-s)b)ds, \tag{19}$$

where

$$m(s)=\begin{cases} s-\frac{1+e^{-\rho}-e^{-\rho s}-e^{-\rho(1-s)}}{\rho(1-e^{-\rho})}, & 0\le s<\frac{1}{2},\\ \\ (1-s)-\frac{1+e^{-\rho}-e^{-\rho s}-e^{-\rho(1-s)}}{\rho(1-e^{-\rho})}, & \frac{1}{2}\le s<1. \end{cases}$$

Proof. By using the proof of the Lemma 5, we can get

$$\begin{aligned} Q_{ml} &= \frac{b-a}{2}\int_0^1 kh'(sa+(1-s)b)ds \\ &-\frac{b-a}{2(1-e^{-\rho})}\left[\int_0^1 e^{-\rho s}h'(sa+(1-s)b)ds-\int_0^1 e^{-\rho(1-s)}h'(sa+(1-s)b)ds\right]. \end{aligned} \tag{20}$$

Thus,

$$\begin{aligned} &\frac{b-a}{2}\int_0^1 kh'(sa+(1-s)b)ds \\ =\ &\frac{b-a}{2}\int_0^{\frac{1}{2}} h'(sa+(1-s)b)ds-\frac{b-a}{2}\int_{\frac{1}{2}}^1 h'(sa+(1-s)b)ds \\ =\ &\frac{b-a}{2}\left[sh'(sa+(1-s)b)\Big|_0^{\frac{1}{2}}-(a-b)\int_0^{\frac{1}{2}} sh''(sa+(1-s)b)ds\right] \\ &-\frac{b-a}{2}\left[sh'(sa+(1-s)b)\Big|_{\frac{1}{2}}^1-(a-b)\int_{\frac{1}{2}}^1 sh''(sa+(1-s)b)ds\right] \\ =\ &\frac{b-a}{2}\left[\frac{1}{2}h'(\frac{a+b}{2})-(a-b)\int_0^{\frac{1}{2}} sh''(sa+(1-s)b)ds\right] \\ &-\frac{b-a}{2}\left[h'(a)-\frac{1}{2}h'(\frac{a+b}{2})-(a-b)\int_{\frac{1}{2}}^1 sh''(sa+(1-s)b)ds\right] \\ =\ &\frac{b-a}{2}\left[h'(\frac{a+b}{2})-h'(a)\right] \\ &+\frac{(b-a)^2}{2}\int_0^{\frac{1}{2}} sh''(sa+(1-s)b)ds-\frac{(b-a)^2}{2}\int_{\frac{1}{2}}^1 sh''(sa+(1-s)b)ds \\ =\ &\frac{(b-a)^2}{2}\int_{\frac{1}{2}}^1 h''(sa+(1-s)b)ds \\ &+\frac{(b-a)^2}{2}\int_0^{\frac{1}{2}} sh''(sa+(1-s)b)ds-\frac{(b-a)^2}{2}\int_{\frac{1}{2}}^1 sh''(sa+(1-s)b)ds \\ =\ &\frac{(b-a)^2}{2}\int_0^{\frac{1}{2}} sh''(sa+(1-s)b)ds+\frac{(b-a)^2}{2}\int_{\frac{1}{2}}^1 (1-s)h''(sa+(1-s)b)ds. \end{aligned} \tag{21}$$

Submitting (16), (17), and (21) to (20), we get (19). This completes the proof. □

Now, we can prove our Hermite–Hadamard-type inequalities by the second order derivatives.

Theorem 2. *Assuming $h:[a,b]\to\Re$ is a twice differentiable function. If $h''\in L[a,b]$ and $|h''|$ is convex on $[a,b]$, the following inequality for fractional integrals with exponential kernel holds:*

$$|Q_{mr}|\le\frac{(b-a)^2}{2\rho(1-e^{-\rho})}\left(\frac{1+e^{-\rho}}{2}-\frac{1-e^{-\rho}}{\rho}\right)(|h''(a)|+|h''(b)|). \tag{22}$$

Proof. Note that

$$\begin{aligned}
& \int_0^1 \left(1+e^{-\rho}-e^{-\rho s}-e^{-\rho(1-s)}\right) sds \\
= & (1+e^{-\rho}) \int_0^1 sds - \int_0^1 se^{-\rho s}ds - \int_0^1 se^{-\rho(1-s)}ds \\
= & \frac{1+e^{-\rho}}{2} + \frac{1}{\rho}\left(e^{-\rho} - \frac{1-e^{-\rho}}{\rho}\right) - \frac{1}{\rho}\left(1 - \frac{1-e^{-\rho}}{\rho}\right) \\
= & \frac{1+e^{-\rho}}{2} - \frac{1-e^{-\rho}}{\rho},
\end{aligned} \tag{23}$$

and

$$\begin{aligned}
& \int_0^1 \left(1+e^{-\rho}-e^{-\rho s}-e^{-\rho(1-s)}\right)(1-s)ds \\
= & \int_0^1 \left(1+e^{-\rho}-e^{-\rho s}-e^{-\rho(1-s)}\right) ds - \int_0^1 s\left(1+e^{-\rho}-e^{-\rho s}-e^{-\rho(1-s)}\right) ds \\
= & (1+e^{-\rho})\Big|_0^1 + \frac{e^{-\rho s}}{\rho}\Big|_0^1 - \frac{e^{-\rho(1-s)}}{\rho}\Big|_0^1 - (\frac{1+e^{-\rho}}{2} - \frac{1-e^{-\rho}}{\rho}) \\
= & \frac{1+e^{-\rho}}{2} - \frac{1-e^{-\rho}}{\rho}.
\end{aligned} \tag{24}$$

According to (15), (23), (24), and the convex of $|h''|$, we can get

$$\begin{aligned}
|Q_{mr}| = & \left|\frac{(b-a)^2}{2\rho(1-e^{-\rho})} \int_0^1 \left(1+e^{-\rho}-e^{-\rho s}-e^{-\rho(1-s)}\right) h''(sa+(1-s)b)ds\right| \\
\leq & \frac{(b-a)^2}{2\rho(1-e^{-\rho})} \int_0^1 \left|1+e^{-\rho}-e^{-\rho s}-e^{-\rho(1-s)}\right| \left|h''(sa+(1-s)b)\right| ds \\
\leq & \frac{(b-a)^2}{2\rho(1-e^{-\rho})} \int_0^1 \left(1+e^{-\rho}-e^{-\rho s}-e^{-\rho(1-s)}\right) \left(s|h''(a)|+(1-s)|h''(b)|\right) ds \\
= & \frac{(b-a)^2}{2\rho(1-e^{-\rho})} \left(\frac{1+e^{-\rho}}{2} - \frac{1-e^{-\rho}}{\rho}\right) (|h''(a)|+|h''(b)|).
\end{aligned}$$

The proof is finished. □

Remark 2. *$\alpha \to 1$ in* (22) *of Theorem 2, then $\rho = \frac{1-\alpha}{\alpha}(b-a) \to 0$, one obtains*

$$\lim_{\rho\to 0} \frac{1-\alpha}{2(1-e^{-\rho})} = \lim_{x\to 1} \frac{1-\frac{b-a}{b-a-\ln x}}{2(1-x)} = \lim_{x\to 1} \frac{-\ln x}{2(1-x)(b-a-\ln x)} = \frac{1}{2(b-a)}, \tag{25}$$

and

$$\lim_{\rho\to 0} \frac{1}{\rho(1-e^{-\rho})} \left(\frac{1+e^{-\rho}}{2} - \frac{1-e^{-\rho}}{\rho}\right) = \lim_{\rho\to 0} \frac{\rho+\rho e^{-\rho}-2+2e^{-\rho}}{2\rho^2(1-e^{-\rho})} = \frac{1}{12}. \tag{26}$$

So (22) *is transformed to*

$$\left|\frac{h(a)+h(b)}{2} - \frac{1}{b-a}\int_a^b h(s)ds\right| \leq \frac{(b-a)^2}{24}(|h''(a)|+|h''(b)|).$$

This result coincides the conclusion in [11], Theorem 3.

Theorem 3. *Assuming $h:[a,b]\to\Re$ is a twice differentiable function. If $h''\in L[a,b]$ and $|h''|$ is convex on $[a,b]$, then the following inequality for fractional integrals with exponential kernel holds:*

$$|Q_{ml}|\leq\frac{(b-a)^2}{2}\left[\frac{1}{8}+\frac{1+e^{-\rho}}{2\rho(1-e^{-\rho})}-\frac{1}{\rho^2}\right](|h''(a)|+|h''(b)|). \tag{27}$$

Proof. According to Lemma 7 and the convex of $|h''|$, we can get

$$\begin{aligned}
|Q_{ml}| &= \left|\frac{(b-a)^2}{2}\int_0^1 m(s)h''(sa+(1-s)b)ds\right|\leq\frac{(b-a)^2}{2}\int_0^1|m(s)|\left|h''(sa+(1-s)b)\right|ds\\
&\leq \frac{(b-a)^2}{2}\int_0^{\frac{1}{2}}\left(s+\frac{1+e^{-\rho}-e^{-\rho s}-e^{-\rho(1-s)}}{\rho(1-e^{-\rho})}\right)(s|h''(a)|+(1-s)|h''(b)|)\,ds\\
&\quad+\frac{(b-a)^2}{2}\int_{\frac{1}{2}}^1\left(1-s+\frac{1+e^{-\rho}-e^{-\rho s}-e^{-\rho(1-s)}}{\rho(1-e^{-\rho})}\right)(s|h''(a)|+(1-s)|h''(b)|)\,ds\\
&= \frac{(b-a)^2}{2}\left[\int_0^{\frac{1}{2}}\left(s^2|h''(a)|+t(1-t)|h''(b)|\right)ds+\int_{\frac{1}{2}}^1\left(s(1-s)|h''(a)|ds+(1-s)^2|h''(b)|\right)ds\right.\\
&\quad\left.+\frac{1}{\rho(1-e^{-\rho})}\int_0^1\left(1+e^{-\rho}-e^{-\rho s}-e^{-\rho(1-s)}\right)(s|h''(a)|+(1-s)|h''(b)|)\,ds\right]\\
&= \frac{(b-a)^2}{2}\left[\frac{1}{24}|h''(a)|+\frac{1}{12}|h''(b)|+\frac{1}{12}|h''(a)|+\frac{1}{24}|h''(b)|\right.\\
&\quad\left.+\frac{1}{\rho(1-e^{-\rho})}\left(\frac{1+e^{-\rho}}{2}-\frac{1-e^{-\rho}}{\rho}\right)(|h''(a)|+|h''(b)|)\right]\\
&= \frac{(b-a)^2}{2}\left[\frac{1}{8}+\frac{1+e^{-\rho}}{2\rho(1-e^{-\rho})}-\frac{1}{\rho^2}\right](|h''(a)|+|h''(b)|).
\end{aligned}$$

This completes the proof. □

Remark 3. *Let $\alpha\to 1$ in (27), one has*

$$\left|\frac{1}{b-a}\int_a^b h(s)ds-h(\frac{a+b}{2})\right|\leq\frac{5(b-a)^2}{48}(|h''(a)|+|h''(b)|).$$

4. Application to Special Means

Think on the following particular means [23] for $\forall p,q\in R, p\neq q$ as follows:

$$(i)\quad H(p,q)=\frac{2}{\frac{1}{p}+\frac{1}{q}},\quad p,q\in R\backslash\{0\};$$

$$(ii)\quad A(p,q)=\frac{p+q}{2},\quad p,q\in R;$$

$$(iii)\quad L(p,q)=\frac{q-p}{\ln|p|-\ln|q|},\quad |p|\neq|q|, pq\neq 0;$$

$$(iv)\quad L_m(p,q)=\left[\frac{q^{m+1}-p^{m+1}}{(m+1)(q-p)}\right]^{\frac{1}{m}},\quad m\in Z\backslash\{-1,0\}, p,q\in R, p\neq q.$$

Next, making use of the acquired results in Section 3, we give some applications to particular means of real number.

Proposition 1. *Let $p,q\in\Re,\ p<q,\ pq>0$ and $m\in\mathbb{Z},\ |m|\geq 2$. Then,*

$$|L_m^m(p,q)-A^m(p,q)|\leq\frac{5}{24}(q-p)^2|m(m-1)|A(|p|^{m-2},|q|^{m-2}). \tag{28}$$

Proof. Applying Remark 3 for $h(x) = x^m$, we can get the conclusion immediately. □

The upper bound is smaller than the result of Proposion 3.1 in [5] when $|q-p| \leq 1$ and $|p|, |q| > 1$ obviously.

Proposition 2. *Let $p, q \in \Re$, $p < q$, $pq > 0$. Then,*

$$\left| L^{-1}(p,q) - A^{-1}(p,q) \right| \leq \frac{5}{12}(q-p)^2 A(|p|^{-3}, |q|^{-3}). \tag{29}$$

Proof. The inference follows from Remark 3 used for $h(x) = \frac{1}{x}$. □

Proposition 3. *Let $p, q \in \Re$, $p < q$, $pq > 0$ and $m \in \mathbb{Z}$, $|m| \geq 2$. Then, we have*

$$\left| L_m^m(q^{-1}, p^{-1}) - H^{-m}(q,p) \right| \leq \frac{5}{24}\left(\frac{1}{p} - \frac{1}{q}\right)^2 |m(m-1)| H^{-1}(|p|^{m-2}, |q|^{m-2}), \tag{30}$$

and

$$\left| L^{-1}(p,q) - H(p,q) \right| \leq \frac{5}{12}(q-p)^2 H^{-1}(|p|^3, |q|^3). \tag{31}$$

Proof. Doing he replacement $q^{-1} \to p, p^{-1} \to q$ in the inequalities (28) and (29), we can obtain the required inequalities (30) and (31), respectively. Here, we have observed $A^{-1}(p^{-1}, q^{-1}) = H(p,q) = 2/(\frac{1}{p} + \frac{1}{q}), q^{-1} < p^{-1}$. □

At last, we will present an application to a midpoint formula. In [23], let w be a division $p = s_0 < s_1 \cdots < s_{m-1} < s_m = q$ of the interval $[p,q]$ and inspect the quadrature formula

$$\int_p^q h(s)ds = T(h,w) + E(h,w), \tag{32}$$

where

$$T(h,w) = \sum_{i=0}^{m-1} h(\frac{s_i + s_{i+1}}{2})(s_{i+1} - s_i)$$

is the midpoint version and $E(h,w)$ refers to the approximation error. Here, we deduce the error estimate for the midpoint formula.

Proposition 4. *Let $h : [p,q] \to \Re$ be a twice differentiable mapping on (p,q) with $p < q$. If $h'' \in L[a,b]$ and $|h''|$ is convex on $[p,q]$, then in (32), for every division w of $[p,q]$, the following inequality holds:*

$$|E(f,w)| \leq \frac{5}{48} \sum_{i=0}^{m-1} (s_{i+1} - s_i)^3 \left(|h''(s_i)| + |h''(s_{i+1})|\right).$$

Proof. Applying Remark 3 on subinterval $[s_i, s_{i+1}] (i = 0, 1, \cdots, m-1)$ of the division w, we derive

$$\left| \int_{s_i}^{s_{i+1}} h(s)ds - h(\frac{s_i + s_{i+1}}{2})(s_i - s_{i+1}) \right| \leq \frac{5}{48}(s_{i+1} - s_i)^3 \left(|h''(s_i)| + |h''(s_{i+1})|\right).$$

Summing over from 0 to $m-1$ and making use of the convexity of $|h''|$, we infer that

$$\begin{aligned}\left|\int_p^q h(s)ds - T(h,w)\right| &= \left|\sum_{i=0}^{m-1}\left[\int_{s_i}^{s_{i+1}} h(s)ds - f(\frac{s_i+s_{i+1}}{2})(s_i-s_{i+1})\right]\right| \\ \leq & \sum_{i-=0}^{m-1}\left|\int_{s_i}^{s_{i+1}} h(s)ds - h(\frac{s_i+s_{i+1}}{2})(s_i-s_{i+1})\right| \\ \leq & \frac{5}{48}\sum_{i=0}^{m-1}(s_{i+1}-s_i)^3\left(|h''(s_i)|+|h''(s_{i+1})|\right).\end{aligned}$$

The proof is completed. □

5. Conclusions

Based on the above interpretation, we acquire the bound estimates of the difference between the average of the fractional integrals with an exponential kernel and the mean values of the endpoints and the midpoint.

By comparing these bound estimates, we have obtained the following conclusions:

(i) With the first and second order derivatives of a given function, the Hermite–Hadamard-type inequalities involving left-sided and right-sided, the fractional integrals are different. The Hermite–Hadamard-type inequalities with the second order derivatives of a given function are more accurate.

(ii) With the same order derivatives of a given function, the Hermite–Hadamard-type inequalities involving different fractional integrals finally tend to be same when $\alpha \to 1$.

Author Contributions: Conceptualization and supervision, J.W.; Formal analysis and writing-original draft preparation, X.W., J.Z.; Writing-review and editing, X.W., J.W..

Acknowledgments: The authors thank the referees for their careful reading of the article and insightful comments.

References

1. Mitrinović, D.S.; Lackovixcx, I.B. Hermite and convexity. *Aequ. Math.* **1985**, *28*, 229–232. [CrossRef]
2. Abramovich, S.; Barić, J.; Pexcxarixcx, J. Fejer and Hermite-Hadamard type inequalities for superquadratic functions. *J. Math. Anal. Appl.* **2008**, *344*, 1048–1056. [CrossRef]
3. Cal, J.; Carcamob, J.; Escauriaza, L. A general multidimensional Hermite-Hadamard type inequality. *J. Math. Anal. Appl.* **2009**, *356*, 659–663. [CrossRef]
4. Ödemir, M.E.; Avci, M.; Set, E. On some inequalities of Hermite-Hadamard type via m-convexity. *Appl. Math. Lett.* **2010**, *23*, 1065–1070. [CrossRef]
5. Xiao, Z.; Zhang, Z.; Wu, Y. On weighted Hermite-Hadamard inequalities. *Appl. Math. Comput.* **2011**, *218*, 1147–1152. [CrossRef]
6. Niculescu, C.P. The Hermite-Hadamard inequality for log-convex functions. *Nonlinear Anal. TMA* **2012**, *75*, 662–669. [CrossRef]
7. Dragomir, S.S. Inequalities of Hermite-Hadamard type for h-convex functions on linear spaces. *Proyecc. J. Math.* **2015**, *34*, 323–341. [CrossRef]
8. Muhammad, A.K.; Chu, Y.; Tahir, U.K.; Jamroz, K. Some new inequalities of Hermite-Hadamard type for s-convex functions with applications. *Open Math.* **2017**, *15*, 1414–1430.
9. Budak, H.; Sarikaya, M.Z. Journal of Mathematical Extension, Some new generalized Hermite-Hadamard inequalities for generalized convex functions and applications. *Math. Ext.* **2018**, *12*, 51–56.

10. Dragomir, S.S.; Agarwal, R.P. Two inequalities for differentiable mappings and applications to special means of real numbers and to trapezoidal formula. *Appl. Math. Lett.* **1998**, *11*, 91–95. [CrossRef]
11. Alomari, M.; Darus, M.; Dragomir, S.S. New inequalities of Hermite-Hadamard type for functions whose second derivates absolute values are quasiconvex. *RGMIA Res. Rep. Coll.* **2010**, *12*. [CrossRef]
12. Sarikaya, M.Z.; Set, E.; Yaldiz, H.; Başak, N. Hermite-Hadamard's inequalities for fractional integrals and related fractional inequalities. *Math. Comput. Modell.* **2013**, *57*, 2403–2407. [CrossRef]
13. İŞcan, I.; Wu, S. Hermite-Hadamard type inequalities for harmonically convex functions via fractional integrals. *Appl. Math. Comput.* **2014**, *238*, 237–244. [CrossRef]
14. Jleli, M.; O'Regan, M.; Samet, B. On Hermite-Hadamard type inequalities via generalized fractional integrals. *Turk. J. Math.* **2016**, *40*, 1221–1230. [CrossRef]
15. Noor, M.A.; Noor, K.I.; Awan, M.U.; Khan, S. Fractional Hermite-Hadamard inequalities for some new classes of Godunova-Levin functions. *Appl. Math. Inf. Sci.* **2014**, *8*, 2865–2872. [CrossRef]
16. Ahmad, B.; Alsaedi, A.; Kirane, M.; Torebek, B.T. Hermite-Hadamard, Hermite-Hadamard-Fejér, Dragomir-Agarwal and Pachpatte type inequalities for convex functions via new fractional integrals. *J. Comput. Appl. Math.* **2019**, *353*, 120–129. [CrossRef]
17. Wang, J.; Zhu, C.; Zhou, Y. New generalized Hermite-Hadamard type inequalities and applications to specialmeans. *J. Inequ. Appl.* **2013**, *325*, 1–15.
18. Set, E.; Çelik, B. Generaliaed Fractional Hermite-Hadamard type inequalities for m-Convex and (α, m)-Convex funciton. *Commun. Fac. Sci. Univ. Ank. Ser. Math. Stat.* **2018**, *67*, 333–344.
19. Chen, H.; Katugampola, U.N. Hermite-Hadamard and Hermite-Hadamard-Fejér type inequalities for generalized fractional integrals. *J. Math. Anal. Appl.* **2017**, *446*, 1274–1291. [CrossRef]
20. Set, E.; Noor, M.A.; Awan, M.U.; Gözpinar, A. Generalized Hermite-Hadamard type inequalities involving fractional integral operators. *J. Inequ. Appl.* **2017**, *169*, 1–10. [CrossRef]
21. Wang, J.; Li, X.; Zhou, Y. Hermite-Hadamard inequalities involving Riemann-Liouville fractional integrals via s-convex functions and applications to special means. *Filomat* **2016**, *30*, 1143–1150,. [CrossRef]
22. Liu, W.; Wen, W.; Park, J. Hermite-Hadamard type inequalities for MT-convex functions via classical integrals and fractional integrals. *J. Nonlinear Sci. Appl.* **2016**, *9*, 766–777. [CrossRef]
23. Pearce, C.E.M.; Pećarixcx, J. Inequalities for Differentiable Mappings with Application to Special Means and Quadrature Formula. *Appl. Math. Lett.* **2000**, *13*, 51–55. [CrossRef]

8

On Ulam Stability and Multiplicity Results to a Nonlinear Coupled System with Integral Boundary Conditions

Kamal Shah [1], Poom Kumam [2,*] and Inam Ullah [1]

1 Department of Mathematics, University of Malakand, Chakdara Dir(L), Khyber Pakhtunkhwa 18800, Pakistan; kamalshah408@gmail.com (K.S.); inamullahkhan829@gmail.com (I.U.)
2 KMUTT Fixed Point Research Laboratory, KMUTT-Fixed Point Theory and Applications Research Group, Theoretical and Computational Science Center (TaCS), Science Laboratory Building, Faculty of Science, King Mongkut's University of Technology Thonburi (KMUTT), 126 Pracha-Uthit Road, Bang Mod, Thrung Khru, Bangkok 10140, Thailand
* Correspondence: poom.kum@kmutt.ac.th or poom.kumam@mail.kmutt.ac.th

Abstract: This manuscript is devoted to establishing existence theory of solutions to a nonlinear coupled system of fractional order differential equations (FODEs) under integral boundary conditions (IBCs). For uniqueness and existence we use the Perov-type fixed point theorem. Further, to investigate multiplicity results of the concerned problem, we utilize Krasnoselskii's fixed-point theorems of cone type and its various forms. Stability analysis is an important aspect of existence theory as well as required during numerical simulations and optimization of FODEs. Therefore by using techniques of functional analysis, we establish conditions for Hyers-Ulam (HU) stability results for the solution of the proposed problem. The whole analysis is justified by providing suitable examples to illustrate our established results.

Keywords: arbitrary order differential equations; multiple positive solution; Perov-type fixed point theorem; HU stability

---

1. Introduction

Fractional order differential equations (FODEs) emerge in the scientific demonstration of numerous frameworks and different fields of science such as physics, chemistry , economics, polymer rheology, aerodynamics, electrodynamics of complicated medium, blood flow phenomena, biophysics, etc. (see [1–5]). Recently, many authors have studied FODEs from different aspects, one is the numerical and scientific techniques for finding solutions and the other is the theoretical perspective of uniqueness and existence of solutions. The interest of the researchers in the investigation of FODEs lies in the incontrovertible fact that fractional-order models (FOM) are found to be highly realistic and practical, compared to the integer order models. Because there are additional degrees of opportunity in the FOM, in consequence, the subject of FODEs is gaining more attention from researchers. Another facet of research, which has been completely studied for integer order differential equations is devoted to uniqueness and existence of solutions to boundary value problems (BVPs). The mentioned aspect has been very well studied for FODEs, we refer the readers [6–10]. Uniqueness and existence results of solutions to multi-point BVPs have been studied via classical fixed point theorems such as the Schauder fixed point theorem and the Banach contraction principle, see [11–17].

FODEs under integral boundary conditions (IBCs) have been investigated very well because these type of equations are increasingly used in fluid-mechanics and dynamical problems. Jankowski [18] studied the ordinary differential equation under IBCs given by

$$\begin{cases} y'(\vartheta) = F(\vartheta, y(\vartheta)), \quad \vartheta \in [0,T], \quad T > 0, \\ y(\vartheta)|_{\vartheta=0} = \delta \int_0^T y(s)ds + d_0, \quad d_0 \in R, \end{cases}$$

where $F \in C([0,T] \times R, R)$ and $\delta = 1$ or -1. He developed a sufficient condition for iterative approximate solutions to the above problem.

Nanware and Dhaigude [19] have investigated the aforementioned BVP under the IBCs for FODE as given by

$$\begin{cases} D^{\sigma}_{+0}y(\vartheta) = F(\vartheta, y(\vartheta)), \quad \vartheta \in [0,T], \quad T > 0, \\ y(\vartheta)|_{\vartheta=0} = \delta \int_0^T y(s)ds + d_0, \quad d_0 \in R, \end{cases}$$

where $0 < \sigma \leq 1, \delta$ is 1 or -1 and $F \in C([0,T] \times R, R)$, D^{σ}_{+0} is Riemann-Liouville fractional derivative of order σ is defined in (2). The aforementioned author also studied the iterative approximate solution to the above FODEs.

In the same line Cabada and Wang [20] studied the following problem under IBCs as

$$\begin{cases} {}^{C}D^{\sigma}_{+0}y(\vartheta) + \varphi(\vartheta, y(\vartheta)) = 0; \quad \vartheta \in (0,1), \\ y(0) = y''(0) = 0, \quad y(1) = \delta \int_0^1 y(s)ds, \end{cases}$$

where $\sigma \in (2,3], \delta \in (0,2)$ and $y : [0,1] \times [0,\infty] \to [0,\infty]$ are the continuous functions. Also we remark that ${}^{C}D^{\sigma}_{+0}$ stands for Caputo's fractional derivative.

Inspired from the aforementioned work, in this article we investigate a system of nonlinear FODEs with IBCs as

$$\begin{cases} D^{\sigma}_{+0}y(\vartheta) + \varphi(\vartheta, y(\vartheta), z(\vartheta)) = 0; \quad \vartheta \in (0,1); \quad m-1 < \sigma \leq m, \\ D^{æ}_{+0}z(\vartheta) + \chi(\vartheta, y(\vartheta), z(\vartheta)) = 0; \quad \vartheta \in (0,1); \quad m-1 < æ \leq m, \\ y(0) = y'(0) = y''(0) = \cdots = y^{(m-2)}(0) = 0, \quad y(1) = \delta \int_0^1 y(s)ds, \\ z(0) = z'(0) = z''(0) = 0 \cdots = z^{(m-2)}(0) = 0, \quad z(1) = \varrho \int_0^1 z(s)ds, \end{cases} \tag{1}$$

such that $m \geq 3$, $\delta, \varrho \in (0,2)$, the functions $\varphi, \chi : [0,1] \times [0,\infty] \times [0,\infty] \to [0,\infty]$ are continuous functions and $D^{\sigma}_{+0}, D^{æ}_{+0}$ stand for Riemann-Liouville fractional derivatives is defined in (2). We claim that such a system of FODEs are very rarely considered for stability as well as multiplicity results. Our analysis is devoted to the existence theory of a solution, multiplicity results and stability analysis of the suggested problem.

During the last few decades another part of research, which has been considered for FODEs and got much attention from the researchers is stability analysis. Numerous forms of stabilities have been studied in literature which are Mittag-Leffer stability, exponential stability, Lyapunov stability etc., we refer [21–23].

The Ulam stability was first presented by Ulam in 1940 and then brilliantly explained by Hyers in 1941. For more information about HU stability, we refer [24,25]. The HU stability results were generalized and extended by many researchers for FODEs under IBCs. In 1978, Jung studied the said stability for ODEs. Oblaz, Benchohra, etc., have studied the said stability for FODEs but their investigation was limited to initial value problems, we refer to [26–28]. To the best of our information and knowledge, the HU stability has been very rarely studied for coupled system of FODEs under

IBCs. Therefore in this article we investigate HU stability to the considered problem. Here we remark that we also provide some necessary results for nonexistence of solution. Finally a series of examples are provided to support our analysis.

2. Axillary Results

In the current section, we review some fundamental definitions and useful results of functional analysis, fractional calculus and fixed point theory (see reference [1,2,8,29–32]). Here, first of all, we define the Banach space which is utilized throughout in this article.

Let us define $\mathrm{E} = \{\mathrm{y}(\vartheta) | \mathrm{y} \in C[0,1]\}$ with the norm $||\mathrm{y}|| = \max_{\vartheta\in[0,1]} |\mathrm{y}(\vartheta)|$. We define the norm for the product space as $||(\mathrm{y},z)|| = ||\mathrm{y}|| + ||z||$. Obviously $(\mathrm{E}\times\mathrm{E}, ||\cdot||)$ is a Banach space.
Let $K = [\theta, 1-\theta]$ for each $\theta \in (0,1)$, then, we define the cone $\mathrm{C} \subset \mathrm{E}\times\mathrm{E}$ by

$$\mathrm{C} = \{(\mathrm{y},z) \in \mathrm{E}\times\mathrm{E} : \min_{\vartheta\in K}[\mathrm{y}(\vartheta) + z(\vartheta)] \geq \lambda||(\mathrm{y},z)||\}.$$
$$\mathrm{C}_r = \{(\mathrm{y},z) \in \mathrm{C} : \|(\mathrm{y},z)\| \leq r\},\ \partial\mathrm{C}_r = \{(\mathrm{y},z) \in K : \|(\mathrm{y},z)\| = r\}.$$

As in [31], we define positive solution as follows.

Definition 1. *A pair of functions* $(\mathrm{y},z) \in \mathrm{E}\times\mathrm{E}$ *is called a positive solution of problem* (1) *under the given IBCs if* $\mathrm{D}^{\sigma}_{+0}\mathrm{y}, \mathrm{D}^{\rho}_{+0}z \in L^1[0,1]$ *with* $(\mathrm{y},z) > (0,0)$ *on* $(0,1]\times(0,1]$, *where the functions* y, z *satisfy the IBCs given in* (1) *respectively, for all* $\vartheta \in [0,1]$.

Definition 2. *The Riemann-Liouville fractional derivative of order* $\sigma > 0$ *of a continuous function* $\mathrm{y} : (0,\infty) \to \mathrm{R}$ *is defined as*

$$\mathrm{D}^{\sigma}_{+0}\mathrm{y}(\vartheta) = \frac{1}{\Gamma(\mathrm{m}-\sigma)}\left(\frac{\mathrm{d}}{\mathrm{d}\vartheta}\right)^{\mathrm{m}} \int_0^{\vartheta} (\vartheta - \mathrm{s})^{\mathrm{m}-\sigma-1}\mathrm{y}(\mathrm{s})\mathrm{ds}, \tag{2}$$

where $\mathrm{m} = [\sigma] + 1$ *and* $[\sigma]$ *denotes the integer part of* σ.

Definition 3. *The Riemann-Liouville fractions of integration of order* $\sigma > 0$ *of a continuous function* $\mathrm{y} : (0,\infty) \to \mathrm{R}$ *is defined by*

$$\mathrm{I}^{\sigma}_{+0}\mathrm{y}(\vartheta) = \frac{1}{\Gamma(\sigma)} \int_0^{\vartheta} (\vartheta - \mathrm{s})^{\sigma-1}\mathrm{y}(\mathrm{s})\mathrm{ds}, \tag{3}$$

where the integral is point-wise defined on $(0,\infty)$.

Lemma 1. *Let* $\sigma > o$, *then the FODE*

$$\mathrm{D}^{\sigma}_{+0}\mathrm{y}(\vartheta) = 0 \tag{4}$$

has a solution given by

$$\mathrm{y}(\vartheta) = \sum_{i=1}^{\mathrm{m}} \frac{\mathrm{y}^i(0)}{i!}\vartheta^{-i}. \tag{5}$$

Lemma 2. *Let* $\sigma > 0$. *Then we have*

$$\mathrm{I}^{\sigma}_{+0}[\mathrm{D}^{\sigma}_{+0}\mathrm{y}(\vartheta)] = \mathrm{y}(\vartheta) - \sum_{i=0}^{\mathrm{m}} \frac{\mathrm{y}^i(0)}{i!}\vartheta^{-i}. \tag{6}$$

Lemma 3. *[2] Let* $\sigma > o$ *and* $\vartheta \in \mathrm{C}(0,1) \cap \mathrm{L}(0,1)$, *then the FODE*

$$\mathrm{D}^{\sigma}_{+0}\mathrm{y}(\vartheta) = \mathrm{h}(\vartheta)$$

has a solution given by

$$y(\vartheta) = c_1\vartheta^{\sigma-1} + c_2\vartheta^{\sigma-2} + \cdots + c_m\vartheta^{\sigma-m} + I^{\sigma}_{+0}h(\vartheta),$$

where $c_i \in R$ *for* $i = 0, 1, 2, \ldots, m$ *and* $m = [\sigma] + 1$.

Definition 4. *[32,33] On the Banach space* E *defined afore, the mapping* $d : E \times E \to R^n$ *is called a generalized metric on* E *if* $\forall\, x, y,$ *and* $y, z \in E$ *with* $y \neq x$, $z \neq y$, $z \neq y$, *then the following hold*

$(A1)$ $d(y,z) = 0 \Leftrightarrow y = z,\ \forall\, y, z \in E$
$(A2)$ $d(y,z) = d(z,y),\ \forall\, y, z \in E$
$(A3)$ $d(x,y) = d(x,z) + d(z,y) + d(y,y),\ \forall\, x, y, y, z, \in E.$

Further the pair (E, d) *is called a generalized metric space.*

Definition 5. *[32,33] Let* $M = \{M_{m,m} \in R_+^{m\times m}\}$*, for any matrix* $\mathbf{B} \in M$ *the spectral radius is defined by* $æ(\mathbf{B}) = \sup\{|\widehat{\lambda}_i|, i = 1, 2, ..., m\}$*, where* $\widehat{\lambda}_i$*, for* $i = 1, 2, ..., m$ *are the eigenvalues of the matrix* $\mathbf{B}$ *and the matrix will converge to zero if* $æ(\mathbf{B}) < 1$.

Lemma 4. *[32,33] A complete generalized metric space* (M, d)*, with operator* $B : M \to M$ *such that there* $\exists$ *a matrix* $\mathbf{B} \in M$ *with*

$$d(By, Bz) \leq Bd(y,z), \textit{for all } y, z \in M,$$

if $æ(\mathbf{B}) < 1$*, then* B *has a fixed point in* M.

Lemma 5. *[32,33] Consider a Banach space* E *with cone* $C \subseteq E$ *and* $y \subset C$ *is relatively open set with* $0 \in y$ *and* $B : \overline{y} \to y$ *be a completely continuous mapping. Then one of the following hold*

$(A1)$ *The mapping* B *has a fixed point in* $\overline{y}$
$(A2)$ *There exist* $y \in \partial y$ *and* $\eta \in (0,1)$ *with* $y = \eta By$.

Lemma 6. *[33,34] Consider a cone* C *in the Banach space* E *and if* $\mathfrak{A}_1$ *and* $\mathfrak{A}_2$ *be two bounded open sets in* E*, such that* $0 \in \mathfrak{A}_1 \subset \overline{\mathfrak{A}}_1 \subset \mathfrak{A}_2$. *Let* $B : C \cap (\overline{\mathfrak{A}}_2 \setminus \mathfrak{A}_1) \to C$ *be completely continuous operator and one of the following satisfied:*

(1) $\|By\| \leq \|y\|\ \ \forall\, y \in C \cap \partial\mathfrak{A}_1; \|B\| \geqslant \|y\|, \forall\, y \in C \cap \partial\mathfrak{A}_2$
(2) $\|By\| \geqslant \|y\|\ \ \forall\, y \in C \cap \partial\mathfrak{A}_1; \|B\| \leq \|y\|, \forall\, y \in C \cap \partial\mathfrak{A}_2$

Then B has at least one fixed point in $C \cap (\overline{\mathfrak{A}}_2 \setminus \mathfrak{A}_1)$.

3. Existence of at Least One Solution

Lemma 7. *Let* $h \in C[0,1]$*, then the BVP*

$$\begin{cases} D^{\sigma}_{+0}y(\vartheta) + h(\vartheta) = 0; \quad \vartheta \in (0,1); \quad m-1 < \sigma \leq m, \\ y(0) = y'(0) = y''(0) = \cdots = y^{(m-2)}(0) = 0, \quad y(1) = \delta\displaystyle\int_0^1 y(s)ds,, \end{cases} \tag{7}$$

where $\delta \in (0,2)$, has the following unique solution

$$y(\vartheta) = \int_0^1 H_{\sigma}(\vartheta, s)h(s)ds,$$

where H_σ is the Green's function given by

$$H_\sigma(\vartheta,s)=\begin{cases}\dfrac{\vartheta^{\sigma-1}(1-s)^{\sigma-1}(\sigma-\delta+\delta s)-(\sigma-\delta)(\vartheta-s)^{\sigma-1}}{(\sigma-\delta)\Gamma(\delta)}, & 0\le s\le\vartheta\le 1,\\ \dfrac{\vartheta^{\sigma-1}(1-s)^{\sigma-1}(\sigma-\delta+\delta s)}{(\sigma-\delta)\Gamma(\delta)}, & 0\le\vartheta\le s\le 1.\end{cases}\tag{8}$$

Proof. Thanks to Lemma 3 for (7), one has

$$y(\vartheta)=-I^{\sigma}_{+0}h(\vartheta)+c_1\vartheta^{\sigma-1}+c_2\vartheta^{\sigma-2}+\cdots+c_m\vartheta^{\sigma-m}.\tag{9}$$

By using initial condition $y(0)=y'(0)=y''(0)=\cdots=y^{(m-2)}(0)=0$, we get $c_2=c_3=\cdots=c_m=0$. Therefore (9) implies that

$$y(\vartheta)=c_1\vartheta^{\sigma-1}-I^{\sigma}_{+0}h(\vartheta).\tag{10}$$

By using boundary condition $y(1)=\delta\int_0^1 y(s)ds$ in (10), we get

$$c_1=\int_0^1\frac{(\vartheta-s)^{\sigma-1}}{\Gamma(\sigma)}h(s)ds+\delta\int_0^1 y(s)ds.$$

Hence we have the following solution to (1)

$$y(\vartheta)=-\int_0^\vartheta\frac{(\vartheta-s)^{\sigma-1}}{\Gamma(\sigma)}h(s)ds+\vartheta^{\sigma-1}\int_0^1\frac{(1-s)^{\sigma-1}}{\Gamma(\sigma)}h(s)ds+\delta\vartheta^{\sigma-1}\int_0^1 y(s)ds.\tag{11}$$

Let $B=\int_0^1 y(s)ds$, then from Equation (11), we have

$$B=-\int_0^1\int_0^\vartheta\frac{(\vartheta-s)^{\sigma-1}}{\Gamma(\sigma)}h(s)ds+\int_0^1\int_0^1\frac{\vartheta^{\sigma-1}(\vartheta-s)^{\sigma-1}}{\Gamma(\sigma)}h(s)ds+\int_0^1\delta B\vartheta^{\sigma-1}ds$$

$$B=-\int_0^1\frac{(1-s)^{\sigma}}{\sigma\Gamma(\sigma)}h(s)ds+\int_0^1\frac{(1-s)^{\sigma-1}}{\sigma\Gamma(\sigma)}h(s)ds+\frac{1}{\sigma}\delta B\tag{12}$$

implies Equation (12), so we get

$$B=-\frac{1}{\sigma-\delta}\int_0^1\frac{(1-s)^{\sigma}}{\Gamma(\sigma)}h(s)ds+\frac{1}{\sigma-\delta}\int_0^1\frac{(1-s)^{\sigma-1}}{\Gamma(\sigma)}h(s)ds.$$

Replacing this valve in (11), we get

$$\begin{aligned}y(\vartheta)=&-\int_0^t\frac{(\vartheta-s)^{\sigma-1}}{\Gamma(\sigma)}h(s)ds+\vartheta^{\sigma-1}\int_0^1\frac{(1-s)^{\sigma-1}}{\Gamma(\sigma)}h(s)ds-\frac{\delta}{\sigma-\delta}\int_0^1\frac{\vartheta^{\sigma-1}(1-s)^{\sigma}}{\sigma\Gamma(\sigma)}h(s)ds\\&+\frac{\delta}{\sigma-\delta}\int_0^1\frac{\vartheta^{\sigma-1}(1-s)^{\sigma-1}}{\Gamma(\sigma)}h(s)ds.\\=&-\int_0^\vartheta\frac{(\vartheta-s)^{\sigma-1}}{\Gamma(\sigma)}h(s)ds+\int_0^1\frac{\vartheta^{\sigma-1}(1-s)^{\sigma-1}(\sigma-\delta+\delta s)}{(\sigma-\delta)\Gamma(\sigma)}h(s)ds\\=&\int_0^\vartheta\frac{\vartheta^{\sigma-1}(1-s)^{\sigma-1}(\sigma-\delta+\delta s)-(\sigma-\delta)(\vartheta-s)^{\sigma-1}}{(\sigma-\delta)\Gamma(\sigma)}h(s)ds\\&+\int_\vartheta^1\frac{\vartheta^{\sigma-1}(1-s)^{\sigma-1}(\sigma-\delta+\delta s)}{(\sigma-\delta)\Gamma(\sigma)}h(s)ds\\=&\int_0^1 H_\sigma(\vartheta,s)h(s)ds,\end{aligned}$$

where $H_\sigma(\vartheta, s)$ is the Green's function of BVP (7). Similarly we can obtain $z(\vartheta) = \int_0^1 H_{æ}(\vartheta, s)h(s)ds$, where $H_{æ}(\vartheta, s)$ is the Green's function for the second equation of the system (1) and is given by

$$H_{æ}(\vartheta, s) = \begin{cases} \dfrac{\vartheta^{æ-1}(1-s)^{æ-1}(æ - \varrho + \varrho s) - (æ - \varrho)(\vartheta - s)^{æ-1}}{(æ - \varrho)\Gamma(æ)}, & 0 \le s \le \vartheta \le 1, \\ \dfrac{\vartheta^{æ-1}(1-s)^{æ-1}(æ - \varrho + \varrho s)}{(æ - \varrho)\Gamma(æ)}, & 0 \le \vartheta \le s \le 1. \end{cases} \tag{13}$$

□

Lemma 8. *Let* $H(\vartheta, s) = (H_\sigma(\vartheta, s),\ H_{æ}(\vartheta, s))$ *be the Green's function of* (1) *defined in Equations* (8) *and* (13). *This* $H(\vartheta, s)$ *has the given properties*

(F_1) $H(\vartheta, s)$ *is continuous function on the unit square* $\forall\ (\vartheta, s) \in [0,1] \times [0,1]$
(F_2) $H(\vartheta, s) \ge 0\ \forall\ \vartheta, s \in [0,1]$ *and* $H(\vartheta, s) > 0\ \forall\ \vartheta, s \in (0,1)$
(F_3) $\max_{0 \le \vartheta \le 1} H(\vartheta, s) = H(1, s), \forall s \in [0,1]$
(F_4) $\min_{\vartheta \in [\theta, 1-\theta]} H(\vartheta, s) \ge \lambda(s) H(1, s)$ *for each* $\theta,\ s \in (0,1)$,
where $\lambda = \min\{\lambda_\sigma = \theta^{\sigma-1}, \lambda_{æ} = \theta^{æ-1}\}$.

Now according to Lemma 7, we can write system (1) as follows

$$\begin{cases} y(\vartheta) = \int_0^1 H_\sigma(\vartheta, s)\varphi(s, y(s), z(s))ds, \\ z(\vartheta) = \int_0^1 H_{æ}(\vartheta, s)\chi(s, y(s), z(s))ds. \end{cases} \tag{14}$$

Let $B : E \times E \to E \times E$ be the operator defined as

$$\begin{aligned} B(y,z)(\vartheta) &= \left(\int_0^1 H_\sigma(\vartheta, s)\varphi(s, y(s), z(s))ds,\ \int_0^1 H_{æ}(\vartheta, s)\chi(s, y(s), z(s))ds \right) \\ &= \Big(B_1(y,z)(\vartheta), B_2(y,z)(\vartheta) \Big). \end{aligned} \tag{15}$$

Then the fixed point of operator B coincides with the solution of the coupled system (1).

Theorem 1. *Consider that* $u, v : [0,1] \times [0,\infty) \times [0,\infty) \to [0,\infty)$ *are continuous. Then* $B(C) \subset C$ *and* $B : C \to C$ *is completely continuous, where* B *is defined in* (15).

Proof. To prove that $B(C) \subset C$, let $(y,z) \in C$, then by Lemma 8, we have $B(y,z) \in C$ and from (F_4) and $\forall\ \vartheta \in K$, we obtain

$$B_1(y(\vartheta), z(\vartheta)) = \int_0^1 H_\sigma(\vartheta, s)\varphi(s, y(s), z(s))ds \ge \lambda_\sigma \int_0^1 H_\sigma(1, s)\varphi(s, y(s)), z(s)ds. \tag{16}$$

Also from (F_3), we obtain

$$B_1(y(\vartheta), z(\vartheta)) = \int_0^1 H_\sigma(\vartheta, s)\varphi(s, y(s), z(s))ds \le \int_0^1 H_\sigma(1, s)\varphi(s, y(s)), z(s)ds. \tag{17}$$

Thus from (16) and (17), we have

$$B_1(y(\vartheta), z(\vartheta)) \geq \lambda \|B_1(y, z)\|, \text{ for all } \vartheta \in K.$$

Similarly, one can write that

$$B_2(y(\vartheta), z(\vartheta)) \geq \lambda \|B_2(y, z)\|, \text{ for all } \vartheta \in K.$$

Thus

$$B_1(y(\vartheta), z(\vartheta)) + B_2(y(\vartheta), z(\vartheta)) \geq \lambda \|B(y, z)\|, \text{ for all } \vartheta \in K,$$

$$\min_{\vartheta \in K}[B_1(y(\vartheta), z(\vartheta)) + B_2(y(\vartheta), z(\vartheta))] \geq \lambda \|B(y, z)\|.$$

Hence we have $B(y, z) \in C \Rightarrow B(C) \subset C$. Next, like the proof of Theorem 1 of [35], and applying the Arzelà-Ascoli's theorem, it can be easily proven that $B : C \to C$ is completely continuous □

Theorem 2. *Consider that φ and χ are continuous on $[0,1] \times [0, \infty) \times (0, \infty) \to [0, \infty)$, and there exist $f_i(\vartheta), H_i(\vartheta), (i = 1, 2) : (0, 1) \to [0, \infty)$ that satisfy*

(A_1) $|\varphi(\vartheta, y, z) - \varphi(\vartheta, \bar{y}, \bar{z})| \leq u_1(\vartheta)|y - \bar{y}| + v_1(\vartheta)|z - \bar{z}|$, *for* $\vartheta \in (0, 1)$ *and* $y, z, \bar{y}, \bar{z} \geq 0$
(A_2) $|\chi(\vartheta, y, z) - \chi(\vartheta, \bar{y}, \bar{z})| \leq u_2(\vartheta)|y - \bar{y}| + v_2(\vartheta)|z - \bar{z}|$, *for* $\vartheta \in (0, 1)$ *and* $y, z, \bar{y}, \bar{z} \geq 0$
(A_3) $æ(\mathbf{B}) < 1$, *where* $\mathbf{B} \in \{M_{2,2} \in R_+^{2\times 2}\}$ *is a matrix given by*

$$\begin{bmatrix} \int_0^1 H_\sigma(1, s)u_1(s)ds & \int_0^1 H_\sigma(1, s)v_1(s)ds \\ \int_0^1 H_æ(1, s)u_2(s)ds & \int_0^1 H_æ(1, s)v_2(s)ds \end{bmatrix}.$$

Then the system (1) *has a unique positive solution* $(y, z) \in C$.

Proof. Let us define a generalized metric $d : E^2 \times E^2 \to R^2$ by

$$d((y, z), (\bar{y}, \bar{z})) = \begin{pmatrix} ||y - \bar{y}|| \\ ||z - \bar{z}|| \end{pmatrix}, \text{ for all } (y, z), (\bar{y}, \bar{z}) \in E \times E.$$

Obviously $(E \times E, d)$ is a generalized complete metric space. Then for any $(y, z), (\bar{y}, \bar{z}) \in E \times E$ and using property (F_3) we get

$$\begin{aligned} |B_1(y, z)(\vartheta) - B_1(\bar{y}, \bar{z})(\vartheta)| &\leq \max_{\vartheta \in [0,1]} \int_0^1 |H_\sigma(\vartheta, s)|[|\varphi(s, y(s), z(s)) - \varphi(s, \bar{y}(s), \bar{z}(s))|]ds \\ &\leq \int_0^1 H_\sigma(1, s)[u_1(s)\|y - \bar{y}\| + v_1(s)\|z - \bar{z}\|]ds \\ &\leq \int_0^1 u_1(s)H_\sigma(1, s)ds\|y - \bar{y}\| + \int_0^1 v_1(s)H_\sigma(1, s)ds\|z - \bar{z}\|. \end{aligned}$$

Similarly we can show that

$$|B_2(y, z) - B_2(\bar{y}, \bar{z})| \leq \int_0^1 u_2(s)H_æ(1, s)ds\|y - \bar{y}\| + \int_0^1 v_2(s)H_æ(1, s)ds\|z - \bar{z}\|.$$

Thus we have

$$|B(y, z) - B(\bar{y}, \bar{z})| \leq \mathbf{B}d\left((y, z), (\bar{y}, \bar{z})\right), \forall (y, z), (\bar{y}, \bar{z}) \in E \times E,$$

where

$$\mathbf{B} = \begin{bmatrix} \int_0^1 \mathrm{H}_\sigma(1,s)u_1(s)ds & \int_0^1 \mathrm{H}_\sigma(1,s)v_1(s)ds \\ \int_0^1 \mathrm{H}_æ(1,s)u_2(s)ds & \int_0^1 \mathrm{H}_æ(1,s)v_2(s)ds \end{bmatrix}.$$

As $æ(\mathbf{B}) < 1$, in the light of Lemma 4, system (1) has a unique positive solution. □

Theorem 3. *Consider that φ and χ are continuous on $[0,1] \times [0,\infty) \times (0,\infty) \to [0,\infty)$ and there exist $a_i, b_i, c_i (i = 1,2) : (0,1) \to [0,\infty)$ satisfying:*

(A_4) $\varphi(\vartheta, y(\vartheta), z(\vartheta)) \le a_1(\vartheta) + b_1(\vartheta)y(\vartheta) + c_1(\vartheta)z(\vartheta),\ \vartheta \in (0,1), y, z \ge 0$

(A_5) $\chi(\vartheta, y(\vartheta), z(\vartheta)) \le a_2(\vartheta) + b_2(\vartheta)y(\vartheta) + c_2(\vartheta)z(\vartheta),\ \vartheta \in (0,1), y, z \ge 0$

(A_6) $\Lambda_1 = \int\limits_0^1 \mathrm{H}_\sigma(1,s)a_1(s)ds < \infty,\ \Delta_1 = \int\limits_0^1 \mathrm{H}_\sigma(1,s)[b_1(s) + c_1(s)]ds < \frac{1}{2}$

(A_7) $\Lambda_2 = \int\limits_0^1 \mathrm{H}_æ(1,s)a_2(s)ds < \infty,\ \Delta_2 = \int\limits_0^1 \mathrm{H}_æ(s,s)[b_2(s) + c_2(s)]ds < \frac{1}{2}.$

Then the system (1) *has at least one positive solution in*

$$\Big\{(y,z) \in C : \|(y,z)\| \le r\Big\},\ \textit{where}\ \max\left\{\frac{\Lambda_1}{\frac{1}{2} - \Delta_1}, \frac{\Lambda_2}{\frac{1}{2} - \Delta_2}\right\} < r.$$

Proof. Define $\Omega = \Big\{(y,z) \in C : \|(y,z)\| < r\Big\}$ with $\max\left\{\frac{\Lambda_1}{\frac{1}{2} - \Delta_1}, \frac{\Lambda_2}{\frac{1}{2} - \Delta_2}\right\} < r.$

According to the Theorem 1, the operator $B : \overline{\Omega} \to C$ is completely continuous. Let $(y,z) \in \Omega$, such that $\|(y,z)\| < r$. Then, we have

$$\begin{aligned} \|B_1(y,z)\| &= \max_{\vartheta\in[0,1]} \left| \int_0^1 \mathrm{H}_\sigma(\vartheta,s)\varphi(s,y(s),z(s)) \right| ds \\ &\le \left(\int_0^1 \mathrm{H}_\sigma(1,s)a_1(s)ds + \int_0^1 \mathrm{H}_\sigma(1,s)b_1(s)|y(s)|ds + \int_0^1 \mathrm{H}_\sigma(1,s)c_1(s)|z(s)|ds \right) \\ &\le \int_0^1 \mathrm{H}_\sigma(1,s)a_1(s)ds + r\left[\int_0^1 \mathrm{H}_\sigma(1,s)[b_1(s) + c_1(s)]ds \right] \\ &\le \Lambda_1 + r\Delta_1 < \frac{r}{2}, \end{aligned}$$

Similarly, $\|B_2(y,z)\| < \frac{r}{2}$, thus $\|B(y,z)\| < r$. Therefore, thanks to Lemma 5, we have $B(y,z) \in \overline{\Omega}$, thus $B : \overline{\Omega} \to \overline{\Omega}$. Let there exist $\varsigma \in (0,1)$ and $(y,z) \in \partial\Omega$ such that $(y,z) = \varsigma B(y,z)$. Then in the light of assumptions (A_4), (A_5) and by (F_4) of Lemma 8, we get $\forall\ \vartheta \in [0,1]$

$$\begin{aligned} |z(\vartheta)| &\le \varsigma \int_0^1 \mathrm{H}_\sigma(\vartheta,s)|\varphi(s,y(s),z(s))|ds \\ &\le \varsigma\left(\int_0^1 \mathrm{H}_\sigma(1,s)a_1(s)ds + \int_0^1 \mathrm{H}_\sigma(1,s)b_1(s)y(s)ds + \int_0^1 \mathrm{H}_\sigma(1,s)c_1(s)z(s)ds \right) \\ &\le \zeta\Big(\Delta_1 + r\Lambda_1\Big) \\ &< \varsigma\frac{r}{2} \end{aligned}$$

which implies that $\|y\| < \varsigma\frac{r}{2}$. Similarly, it can be proved that $\|z\| < \varsigma\frac{r}{2}$. From which, we have $\|(y,z)\| < \varsigma r$, with $\varsigma \in (0,1)$ which is a contradiction that $(y,z) \in \partial\Omega$ as $r = \|(y,z)\|$. Thus, according to Lemma 5, B has at least one fixed point $(y,z) \in \overline{\Omega}$. □

Next the following assumptions and notations will be used:

(C_1) $\varphi, \chi : [0,1] \times [0,\infty) \times [0,\infty) \to [0,\infty)$ are continuous and $\varphi(\vartheta,0,0) = \chi(\vartheta,0,0) = 0$ uniformly with respect to ϑ on $[0,1]$

(C_2) $H_\sigma(1,s), H_æ(1,s)$ defined in Lemma 8 satisfy

$$0 < \int_0^1 H_\sigma(1,s)ds < \infty,\ 0 < \int_0^1 H_æ(1,s)ds < \infty$$

(C_3) Let these limits hold

$$\begin{aligned} \varphi^\alpha &= \lim_{(y,z)\to(\alpha,\alpha)} \sup_{\vartheta\in[0,1]} \frac{\varphi(\vartheta,y,z)}{y+z},\ \chi^\alpha = \lim_{(y,z)\to(\alpha,\alpha)} \sup_{\vartheta\in[0,1]} \frac{\chi(\vartheta,y,z)}{y+z}, \\ \varphi_\alpha &= \lim_{(y,z)\to(\alpha,\alpha)} \in f_{\vartheta\in[0,1]} \frac{\varphi(\vartheta,y,z)}{y+z},\ \chi_æ = \lim_{(y,z)\to(\alpha,\alpha)} \in f_{\vartheta\in[0,1]} \frac{\chi(\vartheta,y,z)}{y+z},\ \text{where } \alpha \in \{0,\infty\}. \end{aligned} \tag{18}$$

$$\alpha_\sigma = \max_{\vartheta\in[0,1]} \int_0^1 H_\sigma(\vartheta,s)ds,\ \alpha_æ = \max_{\vartheta\in[0,1]} \int_0^1 H_æ(\vartheta,s)ds. \tag{19}$$

Theorem 4. *If the assumptions $(C_1) - (C_2)$ hold and one of the following conditions is also satisfied:*

(D_1) $\varphi_0\left(\lambda^2 \int_\theta^{1-\theta} H_\sigma(1,s)ds\right) > 1,\ \varphi^\infty\alpha_\sigma < 1$ *and* $\chi_0\left(\lambda^2 \int_\theta^{1-\theta} H_æ(1,s)ds\right) > 1,\ \chi^\infty\alpha_æ < 1.$
Moreover, $\varphi_0 = \chi_0 = \infty$ *and* $\varphi^\infty = \chi^\infty = 0$

(D_2) *There exist two constants η_1, η_2 with $0 < \eta_1 \le \eta_2$ such that $\varphi(\vartheta,\cdot,\cdot)$ and $\chi(\vartheta,\cdot,\cdot)$ are nondecreasing on $[0,\eta_2]$ $\forall\, \vartheta \in [0,1]$,*

$$\begin{aligned} &\varphi(\vartheta,\lambda_\sigma\eta_1,\lambda_æ\eta_1) \ge \frac{\eta_1}{2}\left(\lambda_\sigma \int_\theta^{1-\theta} H_\sigma(1,s)ds\right)^{-1}, \\ &\chi(\vartheta,\lambda_\sigma\eta_1,\lambda_æ\eta_1) \ge \frac{\eta_1}{2}\left(\lambda_æ \int_\theta^{1-\theta} H_æ(1,s)ds\right)^{-1} \\ &\text{and } \varphi(\vartheta,\eta_2,\eta_2) \le \frac{\eta_2}{2\alpha_\sigma},\ \chi(\vartheta,\eta_2,\eta_2) \le \frac{\eta_2}{2\alpha_æ},\ \text{for all } \vartheta \in [0,1], \end{aligned}$$

where $\lambda, H_\sigma(1,s), H_æ(1,s)$ defined in Lemma 8 and $\varphi_0, \chi_0, \varphi^\infty, \chi^\infty, \alpha_\sigma, \sigma_\alpha$ defined in Equations (18) *and* (19)*. Then the coupled system* (1) *has at least one positive solution.*

Proof. B as defined in (15) is completely continuous.

Case I. Let the condition (D_1) hold. Taking $\varphi_0\left(\lambda_\sigma^2 \int_\theta^{1-\theta} H_\sigma(1,s)ds\right) > 1$, then there exists a constant $\kappa_1 > 0$ such that

$$\varphi(\vartheta,y,z) \ge (\varphi_0 - r_1)(y(\vartheta)+z(\vartheta)),\ \chi(\vartheta,y,z) \ge (\chi_0 - r_2)(y(\vartheta)+z(\vartheta)),\ \text{for all } \vartheta \in [0,1], y,z \in [0,\kappa_1],$$

where $r_1 > 0$, and satisfies the conditions

$$(\varphi_0 - r_1)\frac{\lambda_\sigma^2}{2}\int_\theta^{1-\theta} H_\sigma(1,s)ds \geq 1, \ \ (\chi_0 - r_1)\frac{\lambda_æ^2}{2}\int_\theta^{1-\theta} H_æ(1,s)ds \geq 1.$$

So for $\vartheta \in [0,1], (y,z) \in \partial C_{\kappa_1}$, we have

$$\begin{aligned} B_1(y,z)(\vartheta) &= \int_0^1 H_\sigma(\vartheta,s)\varphi(s,y(s),z(s))ds \geq \lambda_\sigma \int_0^1 H_\sigma(1,s)\varphi(s,y(s),z(s))ds \\ &\geq (\varphi_0 - r_1)\frac{\lambda_\sigma^2}{2}\int_0^1 H_\sigma(1,s)ds\|(y,z)\| \geq \frac{\|(y,z)\|}{2}. \end{aligned}$$

Analogously

$$\begin{aligned} B_2(y,z)(\vartheta) &= \int_0^1 H_æ(\vartheta,s)\chi(s,y(s),z(s))ds \geq \lambda_æ \int_0^1 H_æ(1,s)\varphi(s,y(s),z(s))ds \\ &\geq (\chi_0 - r_2)\frac{\lambda_æ^2}{2}\int_0^1 H_æ(1,s)ds\|(y,z)\| \geq \frac{\|(y,z)\|}{2}. \end{aligned}$$

Therefore, we have

$$\|B(y,z)\| \geq \|B_1(y,z)\| + \|B_2(y,z)\| \geq \|(y,z)\|. \tag{20}$$

Also for $\varphi^\infty\alpha_\sigma < 1$ and $\chi^\infty\alpha_æ < 1$, there exists a constant say $\bar{\kappa_2} > 0$ such that $\varphi(\vartheta,y,z) \leq (\varphi^\infty + r_2)(y+z), \chi(\vartheta,y,z) \leq (\chi^\infty + r_2)(y+z)$, for $\vartheta \in [0,1], y,z \in (\bar{\kappa_2},\infty)$, where $r_2 > 0$ satisfies the conditions $\alpha_\sigma(\varphi^\infty + r_2) \leq 1$, $\alpha_æ(\chi^\infty + r_2) \leq 1$. Let $J = \max_{\vartheta\in[0,1],y,z\in[0,\bar{\kappa_2}]} \varphi(\vartheta,y,z), L = \max_{\vartheta\in[0,1],y,z\in[0,\bar{\kappa_2}]} \chi(\vartheta,y,z)$, then $\varphi(\vartheta,y,z) \leq J + (\varphi^\infty + r_2)(y,z), \chi(\vartheta,y,z) \leq L + (\chi^\infty + r_2)(y,z)$. Now setting $\max\{\kappa_1,\bar{\kappa_2}, J\alpha_\sigma(1-\alpha_\sigma(\varphi^\infty + r_2))^{-1}\} \leq \frac{\kappa_2}{2}, \max\{\kappa_1,\bar{\kappa_2}, L\alpha_æ(1-\alpha_æ(\chi^\infty + r_2))^{-1}\} \leq \frac{\kappa_2}{2}$.

So for any $\vartheta \in [0,1], (y,z) \in \partial C_{\kappa_2}$, we obtain

$$\begin{aligned} B_1(y,z)(\vartheta) &= \int_0^1 H_\sigma(\vartheta,s)\varphi(s,y(s),z(s))ds \leq \lambda_\sigma \int_0^1 H_\sigma(1,s)\varphi(s,y(s),z(s))ds \\ &\leq \int_0^1 H_\sigma(1,s)(J + (\varphi^\infty + r_2)[u(s) + z(s)]ds \\ &\leq J\int_0^1 H_\sigma(1,s)ds + (\varphi^\infty + r_2)\int_0^1 H_\sigma(1,s)ds\|(y,z)\| \\ &< \frac{\kappa_2}{2} - \alpha_\sigma(\varphi^\infty + r_2)\frac{\kappa_2}{2} + (\varphi^\infty + r_2)\alpha_\sigma\|(y,z)\| < \frac{\kappa_2}{2}. \end{aligned}$$

Similarly $B_2(y,z)(\vartheta) < \frac{\kappa_2}{2}$, as $(y,z) \in \partial C_{\kappa_2}$, thus we have

$$\|B(y,z)\| < \|(y,z)\|. \tag{21}$$

Case II. If assumptions in (D_2) hold, then in light of the definition of C for $(y,z) \in \partial C_{\eta_1}$, we have $\|(y,z)\| = \eta_1$, for $\vartheta \in K$. Then from (D_2), we have

$$B_1(y,z)(\vartheta) = \int_0^1 H_\sigma(\vartheta,s)\varphi(s,y(s),z(s))ds \geq \lambda_\sigma \int_\theta^{1-\theta} H_\sigma(1,s)\varphi(s,y(s),z(s))ds$$
$$\geq \left(\lambda_\sigma \int_\theta^{1-\theta} H_\sigma(1,s)ds\right)\frac{\eta_1}{2}\left(\lambda_\sigma \int_\theta^{1-\theta} H_\sigma(1,s)ds\right)^{-1} = \frac{\eta_1}{2}.$$

Similarly it can also be obtained that $B_2(y,z)(\vartheta) \geq \frac{\eta_1}{2}$, for $(y,z) \in \partial C_{\eta_1}$, and we get

$$\|B(y,z)\| = \|B_1(y,z)\| + \|B_2(y,z)\| \geq \|(y,z)\|. \tag{22}$$

Also for $(y,z) \in \partial C_{\eta_2}$, we get that $\|(y,z)\| = \eta_2$ for $\vartheta \in [0,1]$. Then from (D_2), one can get

$$B_1(y,z)(\vartheta) = \int_0^1 H_\sigma(\vartheta,s)\varphi(s,y(s),z(s))ds \leq \int_0^1 H_\sigma(1,s)\varphi(s,y(s),z(s))ds$$
$$\leq \frac{\eta_2}{2\alpha_\sigma}\int_0^1 H_\sigma(1,s)ds = \frac{\eta_2}{2}.$$

Similarly, it can also obtained that $B_2(y,z)(\vartheta) \leq \frac{\eta_2}{2}, (y,z) \in \partial C_{\eta_2}$. Hence, we have

$$\|B(y,z)\| = \|B_1(y,z)\| + \|B_2(y,z)\| \leq \|(y,z)\|. \tag{23}$$

Now according to the application of Lemma 6 to (20) and (21) or (22) and (23) implies that B has a fixed point $(y_1,z_1) \in \bar{C}_{K,\eta}$ or $(y_1,z_1) \in \bar{C}_{K_i,\eta_i}(i = 1,2)$ such that $y_1(\vartheta) \geq \lambda_\sigma\|y_1\| > 0$ and $z_1(\vartheta) \geq \lambda_æ\|z_1\| > 0, \vartheta \in [0,1]$. From which it follows that the coupled system (1) has at least one positive solution. □

Theorem 5. *Under the conditions* $(C_1) - (C_3)$ *and if the following assumptions hold*

(D_3) *If* $\varphi^0\alpha_\sigma < 1; \varphi^\infty\left(\lambda_\sigma^2 \int_\theta^{1-\theta} H_\sigma(1,s)ds\right) > 1$ *and* $\chi^0\alpha_æ < 1; \chi^\infty\left(\lambda_æ^2 \int_\theta^{1-\theta} H_\sigma(1,s)ds\right) > 1,$

then the coupled system (1) *has at least one positive solution. Further, if* $\varphi^0 = \chi^0 = 0$ *and* $\varphi^\infty = \chi^\infty = \infty$, *where* $\lambda, H_\sigma(1,s), H_æ(1,s)$ *defined in Lemma 8 and* $\varphi_0, \chi_0, \varphi^\infty, \chi^\infty, \alpha_\sigma, \sigma_\alpha$ *defined in Equations* (18) *and* (19), *then the the considered system* (1) *has at least one positive solution.*

Proof. Proof can be obtained as proof of Theorem 4. □

4. Existence of More Than One Solutions

Theorem 6. *Consider that* $(C_1) - (C_3)$ *hold and the following conditions are satisfied:*

(D_4) *If* $\varphi_0\left(\lambda_\sigma^2 \int_\theta^{1-\theta} H_\sigma(1,s)ds\right) > 1, \varphi_\infty\left(\lambda_\sigma^2 \int_\theta^{1-\theta} H_\sigma(1,s)ds\right) > 1$ *and*

$\chi_0\left(\lambda_æ^2 \int_\theta^{1-\theta} H_æ(1,s)ds\right) > 1, \chi_\infty\left(\lambda_æ^2 \int_\theta^{1-\theta} H_\sigma(1,s)ds\right) > 1.$

Moreover, $\varphi_0 = \chi_0 = \varphi_\infty = \chi_\infty = \infty$ *also hold:*

(D_5) *there exists* $a > 0$ *such that* $\max_{\vartheta\in[0,1],(y,z)\in\partial C_a} \varphi(\vartheta,y,z) < \frac{a}{2\alpha_\sigma}$ *and* $\max_{\vartheta\in[0,1],(y,z)\in\partial C_a} \chi(\vartheta,y,z) < \frac{a}{2\alpha_æ}$.

Then the coupled system (1) *has at least two positive solutions* $(y,z),(\bar{y},\bar{z})$ *such that*

$$0 < \|(y,z)\| < a < \|(\bar{y},\bar{z})\|. \tag{24}$$

Where $\lambda, H_\sigma(1,s), H_æ(1,s)$ are defined in Lemma 8 and $\varphi_0,\chi_0,\varphi^\infty,\chi^\infty,\alpha_\sigma,\sigma_\alpha$ defined in Equations (18) and (19)

Proof. Let (D_4) hold. Select κ,η such that $0<\kappa<\mu<\eta$. Now if $\varphi_0\left(\lambda_\sigma^2\int\limits_\theta^{1-\theta}H_\sigma(1,s)ds\right) > 1$ and $\chi_0\left(\lambda_æ^2\int\limits_\theta^{1-\theta}H_æ(1,s)ds\right)>1$, then like the proof of Theorem 4, we have

$$\|B(y,z)\| \geq |(y,z)\|,\ \text{for } (y,z)\in\partial C_\kappa. \tag{25}$$

Now, if $\varphi_\infty\left(\lambda_\sigma^2\int\limits_\theta^{1-\theta}H_\sigma(1,s)ds\right)>1$ and $\chi_\infty\left(\lambda_æ^2\int\limits_\theta^{1-\theta}H_æ(1,s)ds\right)>1,$
then like the proof of Theorem 4, we have

$$\|B(y,z)\| \geq \|(y,z)\|,\ \text{for } (y,z)\in\partial C_\eta. \tag{26}$$

Also from (D_5), $(y,z)\in\partial C_\mu$, we get

$$\begin{aligned} B_1(y,z)(\vartheta) &= \int_0^1 H_\sigma(\vartheta,s)\varphi(s,y(s),z(s))ds \\ &\leq \int_0^1 H_\sigma(1,s)\varphi(s,y(s),z(s))ds < \frac{\mu}{2\alpha_\sigma}\int_0^1 H_\sigma(1,s)ds = \frac{\mu}{2}. \end{aligned}$$

Similarly, we have $B_1(y,z)(\vartheta) < \frac{\mu}{2}$ as $(y,z)\in\partial C_\mu$. Hence, we have

$$\|B(y,z)\| < |(y,z)\|,\ \text{for } (y,z)\in\partial C_\mu. \tag{27}$$

Now according to Lemma 6 for (25) and (27), we have gives that B has a fixed point $(y,z)\in\partial\overline{C}_{\kappa,\mu}$ and a fixed point in $(\bar{y},\bar{z})\in\partial\overline{C}_{\mu,\eta}$. Therefore system (1) has at least two positive solutions $(y,z),(\bar{y},\bar{z})$ such that $\|(y,z)\|\neq\mu$ and $\|(\bar{y},\bar{z})\|\neq\mu$. Thus the relation (24) holds. □

Theorem 7. *Consider that* $(C_1)-(C_3)$ *hold together with the given conditions*

(D_6) $\alpha_\sigma\varphi_0<1$ *and* $\varphi_\infty\alpha_\sigma<1$; $\alpha_æ\chi_0<1$, *and* $\chi_\infty\alpha_æ<1$
(D_7) *there exist* $\mu>0$ *such that*

$$\max_{\vartheta\in K,(y,z)\in\partial C_\mu}\varphi(\vartheta,y,z) > \frac{\mu}{2}\left(\lambda_\sigma^2\int\limits_\theta^{1-\theta}H_\sigma(1,s)ds\right)^{-1},$$

$$\max_{\vartheta\in K,(y,z)\in\partial C_\mu}\chi(\vartheta,y,z) > \frac{\mu}{2}\left(\lambda_æ^2\int\limits_\theta^{1-\theta}H_æ(1,s)ds\right)^{-1},$$

such that

$$0<\|(y,z)\|<\mu<\|(\bar{y},\bar{z})\|,$$

where $\lambda, H_\sigma(1,s), H_æ(1,s)$ *defined in Lemma 8 and* $\varphi_0,\chi_0,\varphi^\infty,\chi^\infty,\alpha_\sigma,\sigma_\alpha$ *defined in Equations* (18) *and* (19). *Thus the system* (1) *has at least two positive solutions.*

Proof. We left the proof out, as it similar to the proof of Theorem 6. □

In same line for multiple solutions we give the following results.

Theorem 8. *Let* $(C_1) - (C_3)$ *hold. If there exist* $2m$ *positive numbers* $\mathbf{u}_L, \hat{\mathbf{u}}_L,\ L = 1,2\ldots m$ *with* $\mathbf{u}_1 < \lambda_\sigma \hat{\mathbf{u}}_1 < \hat{\mathbf{u}}_1 < \mathbf{u}_2 < \lambda_\sigma \hat{\mathbf{u}}_2 < \hat{\mathbf{u}}_2 \ldots \mathbf{u}_m < \lambda_\sigma \hat{\mathbf{u}}_m < \hat{\mathbf{u}}_m$ *and* $\mathbf{u}_1 < \lambda_æ \hat{\mathbf{u}}_1 < \hat{\mathbf{u}}_1 < \mathbf{u}_2 < \lambda_æ \hat{\mathbf{u}}_2 < \hat{\mathbf{u}}_2 \ldots \mathbf{u}_m < \lambda_æ \hat{\mathbf{u}}_m < \hat{\mathbf{u}}_m$*, such that*

(D_8) $\varphi(\vartheta, y(\vartheta), z(\vartheta)) \geq \mathbf{u}_L \left(\lambda_\sigma \int_0^1 H_\sigma(1,s)ds\right)^{-1}$, *for* $(\vartheta, y, z) \in [0,1] \times [\lambda_\sigma \mathbf{u}_L, \mathbf{u}_L] \times [\lambda_æ \mathbf{u}_L, \mathbf{u}_L]$, *and*

$$\varphi(\vartheta, y(\vartheta), z(\vartheta)) \leq \alpha_\sigma^{-1} \hat{\mathbf{u}}_L, \text{for } (\vartheta, y, z) \in [0,1] \times [\lambda_\sigma \hat{\mathbf{u}}_L, \hat{\mathbf{u}}_L] \times [\lambda_æ \mathbf{u}_L, \mathbf{u}_L], L = 1,2\ldots m,$$

(D_9) $\chi(\vartheta, y(\vartheta), z(\vartheta)) \geq \mathbf{u}_L \left(\lambda_æ \int_0^1 H_æ(1,s)ds\right)^{-1}$, *for* $(\vartheta, y, z) \in [0,1] \times [\lambda_æ \mathbf{u}_L, \mathbf{u}_L] \times [\lambda_\sigma \mathbf{u}_L, \mathbf{u}_L]$, *and*

$$\chi(\vartheta, y(\vartheta), z(\vartheta)) \leq \alpha_æ^{-1} \hat{\mathbf{u}}_L, \text{for } (\vartheta, y, z) \in [0,1] \times [\lambda_\sigma \mathbf{u}_L, \mathbf{u}_L] \times [\lambda_æ \hat{\mathbf{u}}_L, \hat{\mathbf{u}}_L],\ L = 1,2\ldots m.$$

where $\lambda, H_\sigma(1,s), H_æ(1,s)$ *defined in Lemma 8.*

Then the coupled system (1) *has at least* m*-positive solutions* (y_L, z_L)*, satisfying*

$$\mathbf{u}_L \leq \|(y_L, z_L)\| \leq \hat{\mathbf{u}}_L,\ L = 1,2\ldots m.$$

Theorem 9. *Suppose that* $(C_1) - (C_3)$ *holds. If there exist* $2m$ *positive numbers* $\mathbf{u}_L, \hat{\mathbf{u}}_L,\ L = 1,2\ldots m$*, with* $\mathbf{u}_1 < \hat{\mathbf{u}}_1 < \mathbf{u}_2 < \hat{\mathbf{u}}_2 \ldots < \mathbf{u}_m < \hat{\mathbf{u}}_m$ *such that*

(D_{10}) φ *and* χ *are non-decreasing on* $[0, \hat{\mathbf{u}}_m]$ $\forall\, \vartheta \in [0,1]$;

(D_{11}) $\varphi(\vartheta, y(\vartheta), z(\vartheta)) \geq \mathbf{u}_L \left(\lambda_\sigma \int_\theta^{1-\theta} H_\sigma(1,s)ds\right)^{-1}$, $\varphi(\vartheta, y(\vartheta), z(\vartheta)) \leq \frac{\hat{\mathbf{u}}_L}{\alpha_\sigma}$, $L = 1,2\ldots m$,

$$\chi(\vartheta, y(\vartheta), z(\vartheta)) \geq \mathbf{u}_L \left(\lambda_æ \int_\theta^{1-\theta} H_æ(1,s)ds\right)^{-1}, \ \chi(\vartheta, y(\vartheta), z(\vartheta)) \leq \frac{\hat{\mathbf{u}}_L}{\alpha_æ},\ L = 1,2\ldots m.$$

Hence we conclude that there exist at least m *positive solutions* (y_L, z_L)*, corresponding to coupled system* (1) *which satisfy*

$$\mathbf{u}_L \leq \|(y_L, z_L)\| \leq \hat{\mathbf{u}}_L,\ L = 1,2\ldots m.$$

5. Hyers-Ulam Stability

Definition 6. *[30] Let* $B_1, B_2 : E \times E \to E \times E$ *be the two operators. Then the system of operator equations*

$$\begin{cases} y(\vartheta) = B_1(y,z)(\vartheta) \\ z(\vartheta) = B_2(y,z)(\vartheta) \end{cases} \tag{28}$$

is called the HU stability if we can find $J_i (i = 1,2,3,4) > 0$*, with* $æ_i (i = 1,2) > 0$ *and for each solution* $(y^*, z^*) \in E \times E$ *of the inequalities given by*

$$\begin{cases} ||y^* - \varphi(y^*, z^*)||_{E \times E} \leq æ_1, \\ ||z^* - \chi(y^*, z^*)||_{E \times E} \leq æ_2, \end{cases} \tag{29}$$

there exists a solution $(\overline{y}, \overline{z}) \in E \times E$ *of system* (28) *such that*

$$\begin{cases} ||y^* - \overline{y}||_{E \times E} \leq k_1 æ_1 + k_2 æ_2, \\ ||z^* - \overline{z}||_{E \times E} \leq k_3 æ_1 + k_4 æ_2, \end{cases} \tag{30}$$

Theorem 10. *[30] Let* $B_1, B_2 : E \times E \to E \times E$ *be the two operators such that*

$$\begin{cases} ||B_1(y,z) - B_1(y^*,z^*)||_{E\times E} \leq k_1||y - y^*||_{E\times E}ds + k_2||z - z^*||_{E\times E}ds, \\ ||B_2(y,z) - B_2(y^*,z^*)||_{E\times E} \leq k_3||y - y^*||_{E\times E}ds + k_4||z - z^*||_{E\times E}ds, \\ \textit{for all} \quad (y,z),(y^*,z^*) \in E \times E, \end{cases} \tag{31}$$

and if the matrix

$$\mathbf{B} = \begin{bmatrix} k_1 & k_2 \\ k_3 & k_3 \end{bmatrix}$$

converges to zero, then the fixed points corresponding to operator system (28) *are HU-stable. Further, the given condition holds* (M_{11}) *under the continuity of* $\varphi_i, i = 1,2$*, there exist* $f_i, H_i \in C(0,1), i = 1,2$ *and* $(y,z),(\overline{y},\overline{z})$ *such that*

$$|\varphi_i(\vartheta,y,z) - \varphi_i(\vartheta,\overline{y},\overline{z})| \leq f_i(\vartheta)|y - \overline{y}| + H_i(\vartheta)|z - \overline{z}|, i = 1,2.$$

In this section, we study HU stability for the solutions of our proposed system.

Theorem 11. *Suppose that the assumption* (M_{11}) *along with condition that matrix*

$$\mathbf{B} = \begin{bmatrix} \int_0^1 H_\sigma(1,s)u_1(s)ds & \int_0^1 H_\sigma(1,s)v_1(s)ds \\ \int_0^1 H_æ(1,s)u_2(s)ds & \int_0^1 H_æ(1,s)v_2(s)ds \end{bmatrix}.$$

is converging to zero. Then, the solutions of (1) *are HU-stable.*

Proof. Thanks to Theorem 2, we have

$$\begin{cases} ||B_1(y,z) - B_1(y^*,z^*)||_{E\times E} \leq \int_0^1 H_\sigma(1,s)u_1(s)||y - y^*||_{E\times E}ds + \int_0^1 H_\sigma(1,s)v_1(s)||z - z^*||_{E\times E}ds, \\ ||B_2(y,z) - B_2(y^*,z^*)||_{E\times E} \leq \int_0^1 H_æ(1,s)u_2(s)||y - y^*||_{E\times E}ds + \int_0^1 H_æ(1,s)v_2(s)||z - z^*||_{E\times E}ds. \end{cases}$$

From which we get

$$\begin{cases} ||B_1(y,z) - B_1(y^*,z^*)||_{E\times E} \leq \left[\int_0^1 H_\sigma(1,s)u_1(s)ds\right] ||y - y^*||_{E\times E} + \left[\int_0^1 H_\sigma(1,s)v_1(s)ds\right] ||z - z^*||_{E\times E}, \\ ||B_2(y,z) - B_2(y^*,z^*)||_{E\times E} \leq \left[\int_0^1 H_æ(1,s)u_2(s)ds\right] ||y - y^*||_{E\times E} + \left[\int_0^1 H_æ(1,s)v_2(s)ds\right] ||z - z^*||_{E\times E}. \end{cases} \tag{32}$$

Analogously one has

$$||P(y,z) - P(y^*,z^*)||_{E\times E} \leq \mathbf{B}||(y,z) - (y^*,z^*)||_{E\times E}, \tag{33}$$

such that

$$\mathbf{B} = \begin{bmatrix} \int_0^1 H_\sigma(1,s)u_1(s)ds & \int_0^1 H_\sigma(1,s)v_1(s)ds \\ \int_0^1 H_æ(1,s)u_2(s)ds & \int_0^1 H_æ(1,s)v_2(s)ds \end{bmatrix}.$$

Hence, we get the required results. □

6. Example

To verify the aforesaid established analysis we provide some test problems here in the given sequel.

Example 1. *Take the system of given BVPs with IBCs as*

$$\begin{cases} D_{+0}^{\frac{7}{2}}y(\vartheta)+\dfrac{\vartheta+1}{4}[\Gamma(\dfrac{5}{2})|y(\vartheta)|+\cos|z(\vartheta)|]=0,\ \vartheta\in[0,1],y,z\geq 0\\ D_{+0}^{\frac{7}{2}}z(\vartheta)+\dfrac{\vartheta^2+1}{4}[\sin|y(\vartheta)|+|z(\vartheta)|]=0,\ \vartheta\in[0,1],y,z\geq 0\\ y(0)=y'(0)=y''(0)=0\quad y(1)=\dfrac{1}{2}\displaystyle\int_0^1 y(s)ds,\\ z(0)=z'(0)=z''(0)=0\quad z(1)=\dfrac{1}{3}\displaystyle\int_0^1 z(s)ds. \end{cases} \tag{34}$$

Since $\varphi(\vartheta,y(\vartheta),z(\vartheta))=\dfrac{\vartheta+1}{4}[\Gamma(\frac{5}{2})|y(\vartheta)|+\cos|z(\vartheta)|]$, $\chi(\vartheta,y(\vartheta),z(\vartheta))=\dfrac{\vartheta^2+1}{4}[\sin|y(\vartheta)|+|z(\vartheta)|]$.

Also as $m=[3.5]+1=4,\delta=\frac{1}{2}$ *and* $\varrho=\frac{1}{3}$.

Then

$$\begin{aligned}|\varphi(\vartheta,y_2,z_2)-\varphi(\vartheta,y_1,z_1)|&\leq\Gamma(\frac{5}{2})\frac{\vartheta+1}{4}|y_2-y_1|+\frac{\vartheta+1}{4}|z_2-z_1|,\\ |\chi(\vartheta,y_2,z_2)-\chi(\vartheta,y_1,z_1)|&\leq\frac{\vartheta^2+1}{4}|y_2-y_1|+\frac{\vartheta^2+1}{4}|z_2-z_1|.\end{aligned}$$

where $u_1(\vartheta)=\frac{\vartheta+1}{4}\Gamma(\frac{5}{2}),v_1(\vartheta)=\frac{\vartheta+1}{4}$, $u_2(\vartheta)=v_2(\vartheta)=\dfrac{\vartheta^2+1}{4}$, *so one can get*

$$\mathbf{B}=\begin{bmatrix}\int\limits_0^1 H_\sigma(1,s)u_1(s)ds & \int\limits_0^1 H_\sigma(1,s)v_1(s)ds\\ \int\limits_0^1 H_æ(1,s)u_2(s)ds & \int\limits_0^1 H_æ(1,s)v_2(s)ds\end{bmatrix}=\begin{bmatrix}\dfrac{8}{11} & \dfrac{8}{165\sqrt{\pi}}\\ \dfrac{496}{11583\sqrt{\pi}} & \dfrac{496}{11583\sqrt{\pi}}\end{bmatrix}.$$

$$det(\mathbf{B}-\widehat{\lambda}I)=\begin{bmatrix}\dfrac{8}{11}-\widehat{\lambda} & \dfrac{8}{165\sqrt{\pi}}\\ \dfrac{496}{11583\sqrt{\pi}} & \dfrac{496}{11583\sqrt{\pi}}-\widehat{\lambda}\end{bmatrix}.$$

We get $\widehat{\lambda}_1=0.728$ *and* $\widehat{\lambda}_2=0.024$ *since* $æ(\mathbf{B})=\sup\{|\widehat{\lambda}_i|,i=1,2\}=0.728<1$. *Therefore due to Theorem 2, BVPs* (34) *has a unique positive solution given by*

$$\begin{cases} y(\vartheta)=\displaystyle\int\limits_0^1 H_{\frac{7}{2}}(\vartheta,s)\frac{s+1}{4}[\Gamma(\frac{5}{2})|y(s)|+\cos|z(s)|]ds,\\ z(\vartheta)=\displaystyle\int\limits_0^1 H_{\frac{7}{2}}(\vartheta,s)\frac{s^2+1}{4}[\sin|y(s)|+|z(s)|]ds, \end{cases} \tag{35}$$

where $H_{\frac{7}{2}}(\vartheta,s)$ *and* $H_{\frac{7}{2}}(\vartheta,s)$ *are the Green's functions given by*

$$H_{\frac{7}{2}}(\vartheta,s)=\begin{cases}\dfrac{\vartheta^{\frac{5}{2}}(1-s)^{\frac{5}{2}}(3+\frac{1}{2}s)-3(\vartheta-s)^{\frac{5}{2}}}{3\Gamma(\frac{7}{2})}, & 0\leq s\leq\vartheta\leq 1,\\ \dfrac{\vartheta^{\frac{5}{2}}(1-s)^{\frac{5}{2}}(3+\frac{1}{2}s)}{3\Gamma(\frac{7}{2})}, & 0\leq\vartheta\leq s\leq 1.\end{cases}$$

$$H_{\frac{7}{2}}(\vartheta,s)=\begin{cases}\dfrac{\vartheta^{\frac{5}{2}}(1-s)^{\frac{5}{2}}(\frac{19}{6}+\frac{1}{3}s)-\frac{19}{6}(\vartheta-s)^{\frac{5}{2}}}{\frac{19}{6}\Gamma(\frac{7}{2})}, & 0\le s\le\vartheta\le 1,\\ \dfrac{\vartheta^{\frac{5}{2}}(1-s)^{\frac{5}{2}}(\frac{19}{6}+\frac{1}{3}s)}{\frac{19}{6}\Gamma(\frac{7}{2})}, & 0\le\vartheta\le s\le 1.\end{cases}$$

Further, by the use of Theorem 11, the solution is HU-stable.

Example 2. *Taking a system of FODEs with IBCs as*

$$\begin{cases}D^{\frac{10}{3}}_{+0}y(\vartheta)+a(\vartheta)\sqrt{y(\vartheta)+z(\vartheta)}=0,\ D^{\frac{7}{2}}_{+0}z(\vartheta)+b(\vartheta)\sqrt[3]{y(\vartheta)+z(\vartheta)}=0,\ \vartheta\in(0,1),\\ y(0)=y'(0)=y''(0)=y'''(0)=0\quad y(1)=\int_0^1 y(s)ds,\\ z(0)=z'(0)=z''(0)=z'''(0)=0\quad z(1)=\int_0^1 z(s)ds.\end{cases}\tag{36}$$

From the given system one has

$$\varphi(\vartheta,y,z)=a(\vartheta)\sqrt{y(\vartheta)+z(\vartheta)}$$

and

$$\chi(\vartheta,y,z)=b(\vartheta)\sqrt[3]{y(\vartheta)+z(\vartheta)},\ m=4,\ \delta=\varrho=1.$$

Also $a,\ b:[0,1]\to[0,\infty)$ are continuous. Now $\varphi^0=\lim_{(y,z)\to 0}\dfrac{\varphi(\vartheta,y,z)}{y+z}=\infty$, similarly $\chi^0=\infty$. Obviously we compute $\varphi^\infty=0=\chi^\infty$. Hence due to Theorem 4, system (36) has at least one positive solution.

Example 3. *Taking another test problem with IBCs as*

$$\begin{cases}D^{\frac{9}{2}}_{+0}y(\vartheta)+(1-\vartheta^2)[y(\vartheta)+z(\vartheta)]^2=0,\ D^{\frac{14}{3}}_{+0}z(\vartheta)+[y(\vartheta)+z(\vartheta)]^3=0,\ \vartheta\in(0,1),\\ y(0)=y'(0)=y''(0)=y'''(0)=y''''(0)=0,\quad y(1)=\dfrac{3}{2}\int_0^1 y(s)ds,\\ z(0)=z'(0)=z''(0)=z'''(0)=z''''(0)=0,\quad z(1)=\dfrac{3}{2}\int_0^1 z(s)ds.\end{cases}\tag{37}$$

From the considered problem (37), one has $\delta=\varrho=\frac{3}{2}$, as $m=5$. It is easy to see that $\varphi^0=\chi^0=0$ and $\varphi^\infty=\chi^\infty=\infty$. Therefore thanks to Theorem 5, the given system (37) has a positive solution.

Example 4. *Further we take another system of FODEs with IBCs as*

$$\begin{cases}D^{\frac{11}{2}}_{+0}y(\vartheta)+\dfrac{(1+\vartheta^2)[u^2(\vartheta)+z(\vartheta)]}{(4\vartheta^2+4)\alpha_\sigma}=0,\ \vartheta\in(0,1),\\ D^{\frac{16}{3}}_{+0}z(\vartheta)+\dfrac{(\vartheta^3+1)[y(\vartheta)+v^2(\vartheta)]}{(4\vartheta^3+4)\alpha_{æ}}=0,\ \vartheta\in(0,1),\\ y(0)=y'(0)=y''(0)=y'''(0)=y''''(0)=0,\quad y(1)=\dfrac{3}{2}\int_0^1 y(s)ds,\\ z(0)=z'(0)=z''(0)=z'''(0)=z''''(0)=0,\quad z(1)=\dfrac{3}{2}\int_0^1 z(s)ds.\end{cases}\tag{38}$$

where $\delta=\varrho=\frac{3}{2}$ and $m=6$. It is easy to obtain $\varphi_0=\chi_0=\infty$ and $\varphi_\infty=\chi_\infty=\infty$.

Further $\forall$ $(\vartheta, y, z) \in [0,1] \times [0,1] \times [0,1]$, *we have*

$$\varphi(\vartheta, y, z) \leq \frac{(\vartheta^2+1)2}{4(\vartheta^2+1)\alpha_\sigma} = \frac{\alpha_\sigma^{-1}}{2}, \ \chi(\vartheta, y, z) \leq \frac{(\vartheta^3+1)2}{4(\vartheta^3+1)\alpha_æ} = \frac{\alpha_æ^{-1}}{2}.$$

Hence all the conditions of Theorem 6 hold. Thanks to Theorem 6, the given system (38) *has at least two positive solutions* (y_1, z_1) *and* (y_2, z_2) *which satisfy*

$$0 < \|(y_1, z_1)\| < 1 < \|(y_2, z_2)\|.$$

7. Non-Existence of Positive Solution

Here some conditions are developed under which the coupled system (1) with given IBCs has no solution.

Theorem 12. *Consider that* $(C_1) - (C_3)$ *hold and* $\varphi(\vartheta, y, z) < \frac{\|(y,z)\|}{2\alpha_\sigma}$ *and* $\chi(\vartheta, y, z) < \frac{\|(y,z)\|}{2\alpha_æ}$ *for all* $\vartheta \in [0,1]$, $y > 0$, $z > 0$, *then there is no positive solution for BVPs* (1).

Proof. Consider (y, z) to be the positive solution of BVPs (1). Then, $(y, z) \in C$ for $0 < \vartheta < 1$ and

$$\begin{aligned}
\|(y,z)\| &= \|y\| + \|z\| \\
&= \max_{\vartheta\in[0,1]} |y(\vartheta)| + \max_{\vartheta\in[0,1]} |z(\vartheta)| \\
&\leq \max_{\vartheta\in[0,1]} \int_0^1 H_\sigma(\vartheta, s)|\varphi(s, y(s), z(s))|ds + \max_{\vartheta\in[0,1]} \int_0^1 H_æ(\vartheta, s)|\chi(s, y(s), z(s)|ds \\
&< \int_0^1 H_\sigma(1, s)\frac{\|(y,z)\|}{2\alpha_æ}ds + \int_0^1 H_æ(1, s)\frac{\|(y,z)\|}{2\alpha_æ}ds \\
\Rightarrow \|(y,z)\| &< \|(y,z)\|,
\end{aligned}$$

which is contradiction. Hence the considered system (1) has no solution. □

Theorem 13. *Let the hypothesis* $(C_1) - (C_3)$ *hold along with the conditions*

$$\begin{aligned}
\varphi(\vartheta, y(\vartheta), z(\vartheta)) &> \frac{\|(y,z)\|}{2}\left(\lambda_\alpha^2 \int_\theta^{1-\theta} H_\sigma(1, s)ds\right)^{-1}, \\
\chi(\vartheta, y(\vartheta), z(\vartheta)) &> \frac{\|(y,z)\|}{2}\left(\lambda_\beta^2 \int_\theta^{1-\theta} H_æ(1, s)ds\right)^{-1}, \ \textit{for all } \vartheta \in [0,1],\ y > 0 \textit{and } z > 0.
\end{aligned}$$

Then there does not exist positive solution to BVPs (1).

To demonstrate the results of Theorems 12 and 13 respectively, we give the following example.

Example 5. *Taking the given system of FODEs with given IBCs as*

$$\begin{cases} D^{\frac{5}{2}}_{+0}y(\vartheta) = 5 - 4\left(y + z + \frac{\pi}{3}\right)^{\frac{-5}{2}}, \ \vartheta \in [0,1], \\ D^{\frac{5}{2}}_{+0}z(\vartheta) = \left(30 + \frac{30}{\sqrt{y+z}}\right)^{\frac{-3}{2}} + \frac{1}{50}, \ \vartheta \in [0,1], \\ y(0) = y'(0) = y''(0) = 0, \ y(1) = \frac{1}{2}\int_0^1 y(\vartheta)d\vartheta, \\ z(0) = z'(0) = z''(0) = 0, \ z(1) = \frac{1}{2}\int_0^1 z(\vartheta)d\vartheta. \end{cases} \tag{39}$$

Also as $(C_1) - (C_3)$ *hold, where* $m = [2.5] + 1 = 3$ *and* $\delta = \varrho = \frac{1}{2}$. *We calculate*

$$\begin{aligned} \varphi^0 = 5 - \left(\frac{3}{\pi}\right)^{\frac{5}{2}}, \chi^0 &= \frac{1}{50}, \ \varphi^\infty = 5, \ \chi^\infty = 51, \\ \left(5 - \left(\frac{3}{\pi}\right)^{\frac{5}{2}}\right)\|(y,z)\| &< \varphi(\vartheta, y(\vartheta), z(\vartheta)) < 5\|(y,z)\|, \\ \frac{1}{50}\|(y,z)\| &< \chi(\vartheta, y(\vartheta), z(\vartheta)) < 51\|(y,z)\|. \end{aligned}$$

Therefore we have

$$5 - \left(\frac{3}{\pi}\right)^{\frac{5}{2}}\|(y,z)\| < \varphi(\vartheta, y(\vartheta), z(\vartheta)) < 5\|(y,z)\| \text{ and } \chi(\vartheta, y(\vartheta), z(\vartheta)) < 5\|(y,z)\| < \frac{\|(y,z)\|}{\alpha_\sigma},$$

where $\alpha_\sigma \approx 0.32239$ *and* $\alpha_{æ} \approx 0.32239$.

Case I: *Now*

$$\varphi(\vartheta, y(\vartheta), z(\vartheta)) < \frac{\|(y,z)\|}{\alpha_\sigma} \approx 1.1413\|(y,z)\|$$

yields that

$$\varphi(\vartheta, y(\vartheta), z(\vartheta)) < 5\|(y,z)\| \approx 3.1018\|(y,z)\|$$

and

$$\chi(\vartheta, y(\vartheta), z(\vartheta)) < 51\|(y,z)\| \approx 3.1018\|(y,z)\|.$$

Hence under the condition of Theorem 12, there is no solution corresponding to problem (39).

Case II: *Also*

$$\varphi(\vartheta, y(\vartheta), z(\vartheta)) > \left(5 - \left(\frac{3}{\pi}\right)^{\frac{5}{2}}\right)\|(y,z)\| > \|(y,z)\|\left(\lambda_\alpha^2 \int_{\frac{1}{100}}^{\frac{99}{100}} H_\alpha(1,s)ds\right)^{-1} \approx 0.615\|(y,z)\|$$

and

$$\chi(\vartheta, y(\vartheta), z(\vartheta)) > \frac{1}{50}\|(y,z)\| > \|(y,z)\|\left(\lambda_{æ}^2 \int_{\frac{1}{100}}^{\frac{99}{100}} H_{æ}(1,s)ds\right)^{-1} \approx 0.615\|(y,z)\|.$$

Hence under the condition of Theorem 13, there is no solution corresponding to coupled system (39).

8. Conclusions

In the above research work we have successfully investigated a coupled system of nonlinear FODEs with IBCs for multiplicity results. Further, the aforesaid investigation has been strengthened by developing some conditions under which the solutions of the proposed system are HU-stable. Further some results which demonstrate the conditions of nonexistence of solutions have been established. The whole results have been verified by considering some examples where needed.

Author Contributions: All authors have equal contribution in this paper.

Acknowledgments: All authors are very thankful to the referees for their useful comments and suggestions.

References

1. Miller, K.S.; Ross, B. *An Introduction to the Fractional Calculus and Fractional Differential Equations*; Wiley: New York, NY, USA, 1993.
2. Kilbas, A.A.; Srivastava, H.M.; Trujillo, J.J. Theory and Applications of Fractional Differential Equations. In *North-Holland Mathematics Studies*; Elsevier: Amsterdam, The Nederlands, 2006; Volume 204.
3. Podlubny, I. *Fractional Differential Equations, Mathematics in Science and Engineering*; Academic Press: New York, NY, USA, 1999.
4. Hilfer, R. *Applications of Fractional Calculus in Physics*; World Scientific: Singapore, 2000.
5. Lakshmikantham, V.; Vatsala, A.S. Basic theory of fractional differential equations. *Nonlinear Anal.* **2008**, *69*, 2677–2682. [CrossRef]
6. Benchohra, M.; Graef, J.R.; Hamani, S. Existence results for boundary value problems with nonlinear fractional differential equations. *Appl. Anal.* **2008**, *87*, 851–863. [CrossRef]
7. Shah, K.; Khan, R.A. Study of solution to a toppled system of fractional differential equations with integral boundary conditions. *Int. J. Appl. Comput. Math.* **2017**, *3*, 2369–2388. [CrossRef]
8. Ahmad, B.; Nieto, J.J. Existence results for a coupled system of nonlinear fractional differential equations with three-point boundary conditions. *Comput. Math. Appl.* **2009**, *58*, 1838–1843. [CrossRef]
9. Agarwal, R.P.; Benchohra, M.; Hamani, S. A survey on existence results for boundary value problems of nonlinear fractional differential equations and inclusions. *Acta Appl. Math.* **2010**, *109*, 973–1033. [CrossRef]
10. Li, C.F.; Luo, X.N.; Zhou, Y. Existence of positive solutions of the boundary value problem for nonlinear fractional differential equations. *Comput. Math. Appl.* **2010**, *59*, 1363–1375. [CrossRef]
11. Rehman, M.; Khan, R.A. Existence and uniqueness of solutions for multi-point boundary value problems for fractional differential equations. *Appl. Math. Lett.* **2010**, *23*, 1038–1044. [CrossRef]
12. Zhong, W.; Lin, W. Nonlocal and multiple-point boundary value problem for fractional differential equations. *Comput. Math. Appl.* **2010**, *59*, 1345–1351. [CrossRef]
13. El-Shahed, M.; Shammakh, W.M. Existence of positive solutions for m-point boundary value problem for nonlinear fractional differential equation. In *Abstract and Applied Analysis*; Hindawi: Cairo, Egypt, 2011; Volume 2011, p. 986575.
14. Khan, R.A.; Shah, K. Existence and uniqueness of solutions to fractional order multi point boundary value problems. *Commun. Appl. Anal.* **2015**, *19*, 515–526.
15. Khan, R.A. Three-point boundary value problems for higher order nonlinear fractional differential equations. *J. Appl. Math. Inform.* **2013**, *31*, 221–228. [CrossRef]
16. EL-Sayed, A.M.A.; Taher, E.O.B. Positive solutions for a nonlocal multi-point boundary-value problem of fractional and second order. *Electr. J. Differ. Equ.* **2013**, *64*, 1–8.
17. Yang, A.; Ge, W. Positive solutions of multi-point boundary value problems of nonlinear fractional differential equation at resonance. *J. Korea Soc. Math. Educ. Ser. B Pure Appl. Math.* **2009**, *16*, 181–193.
18. Jankowski, T. Differential equations with integral boundary condition. *J. Comput. Appl. Math.* **2002**, *147*, 2274–2280. [CrossRef]
19. Nanware, J.A.; Dhaigude, D.B. Existence and uniqueness of solution of differential equations of fractional order with integral boundary conditions. *J. Nonlinear Sci. Appl.* **2014**, *7*, 246–254. [CrossRef]

20. Cabada, A.; Wang, G. Positive solution of nonlinear fractional differential equations with integral boundary value conditions. *J. Math. Anal. Appl.* **2012**, *389*, 403–411. [CrossRef]
21. Deng, W. Smoothness and stability of the solutions for nonlinear fractional differential equations. *Nonlinear Anal. TMA* **2010**, *72*, 1768–1777. [CrossRef]
22. Li, Y.; Chen, Y.; Podlubny, I. Mittag-Leffler stability of fractional order nonlinear dynamic systems. *Automatica* **2009**, *45*, 1965–1969. [CrossRef]
23. Li, Y.; Chen, Y.; Podlubny, I. Stability of fractional-order nonlinear dynamic systems: Lyapunov direct method and generalized Mittag-Leffler stability. *Comput. Math. Appl.* **2010**, *59*, 1810–1821. [CrossRef]
24. Ulam, S.M. *Problems in Modern Mathematics*; John Wiley and Sons: New York, NY, USA, 1940; Chapter 6.
25. Hyers, D.H. On the stability of linear functional equation. *Proc. Natl. Acad. Sci. USA* **1941**, *27*, 222–224. [CrossRef] [PubMed]
26. Jung, S.M. On the Hyers-Ulam stability of the functional equations that have the quadratic property. *J. Math. Anal. Appl.* **1998**, *222*, 126–137. [CrossRef]
27. Jung, S.M. Hyers-Ulam stability of linear differential equations of first order. *Appl. Math. Lett.* **2006**, *19*, 854–858. [CrossRef]
28. Obloza, M. Hyers stability of the linear differential equation. *Rocznik Nauk. Dydakt. Prace Mat.* **1993**, *13*, 259–270.
29. Krasnoselskii, M.A. *Positive Solutions of Operator Equations*; Noordhoff: Groningen, The Netherlands, 1964.
30. Urs C. Coupled fixed point theorem and applications to periodic boundary value problem. *Miskolic Math. Notes* **2013**, *14*, 323–333. [CrossRef]
31. Stanêk, S. The existence of positive solutions of singular fractional boundary value problems. *Comput. Math. Appl.* **2011**, *62*, 1379–1388. [CrossRef]
32. Shah, K.; Khan, R.A. Multiple positive solutions to a coupled systems of nonlinear fractional differential equations. *SpringerPlus* **2016**, *5*, 1–20. [CrossRef] [PubMed]
33. Agarwal, R.; Meehan, M.; Regan, D.O. *Fixed Points Theory and Applications*; Cambridge University Press: Cambridge, UK, 2004.
34. Guo, D.; Lakshmikantham, V. *Nonlinear Problems in Abstract Cones*; Academic Press: New York, NY, USA, 1988.
35. Shah, K.; Khan, R.A. Existence and uniqueness of positive solutions to a coupled system of nonlinear fractional order differential equations with anti periodic boundary conditions. *Differ. Equ. Appl.* **2015**, *7*, 245–262. [CrossRef]

9

A Note on Double Conformable Laplace Transform Method and Singular One Dimensional Conformable Pseudohyperbolic Equations

Hassan Eltayeb [1,*], Said Mesloub [1], Yahya T. Abdalla [2] and Adem Kılıçman [3]

1 Mathematics Department, College of Science, King Saud University, P.O. Box 2455, Riyadh 11451, Saudi Arabia; mesloub@ksu.edu.sa

2 Department of Mathematics, College of Science, Sudan University of Science and Technology, P.O. Box 407, Khartoum 11111, Sudan; amynt2005@gmail.com

3 Department of Mathematics and Institute for Mathematical Research, Universiti Putra Malaysia, Serdang 43400 UPM, Selangor, Malaysia; akilic@upm.edu.my

* Correspondence: hgadain@ksu.edu.sa

Abstract: The purpose of this article is to obtain the exact and approximate numerical solutions of linear and nonlinear singular conformable pseudohyperbolic equations and conformable coupled pseudohyperbolic equations through the conformable double Laplace decomposition method. Further, the numerical examples were provided in order to demonstrate the efficiency, high accuracy, and the simplicity of present method.

Keywords: conformable derivative; conformable partial derivative; conformable double Laplace decomposition method; conformable Laplace transform; singular one dimensional coupled Burgers' equation

---

1. Introduction

In recent years, many mathematicians have been studying and discussing the linear and nonlinear fractional differential equations (FDEs) which arise in various fields of physical sciences, as well as in engineering. These types of equations play a significant role and also help to develop mathematical tools in order to understand fractional modelling.

However, there are many different methods to obtain exact and approximate solutions of these kinds of equations. In [1], the author point out a major flaw in the so-called conformable calculus. Recently, many researchers have also paid much attention to study the numerical and exact methods for finding the solution of conformable differential equations. In [2], the authors proposed so-called conformable derivatives. In [3], the conformable heat equation was studied. Similarly, in [4], the nonlinear conformable problems were also studied. The authors in [5] discussed the concepts underlying the formulation of operators capable of being interpreted as fractional derivatives or fractional integrals. In a very short period of time, many mathematicians became interested and provided mathematical models related to conformable derivatives, for the details we refer reader to see [6–9]. In [10,11], the conformable derivatives were applied to some problems in mechanics, and in [12] total frational derivative and directional fractional derivative of functions of several variables were studied.

In order to solve the conformable derivatives, the single Laplace transform method was first introduced and used in [13]. In [14], the idea was extended to the conformable double Laplace transform. In [15], the modified Laplace transform was applied to solve some ordinary differential equations in the frame work of a certain type generalized fractional derivatives. The authors in [16] applied the double Laplace decomposition method to solve singular linear and nonlinear one-dimensional pseudohyperbolic equations.

In this present research, the main objective is to solve linear and nonlinear singular pseudohyperbolic equations by using the conformable double Laplace transform decomposition method, which is a combination of the conformable double Laplace transformation and decomposition method.

2. Properties of Conformable Derivative and Conformable Double Laplace Transform

In this part, we present some background about the nature of the conformable Laplace transform. In the following example, we present the conformable partial derivatives of certain functions as follows.

Example 1. *Let $\mu, \nu \in (0,1]$ and $a, b, m, n, \lambda, \mu \in R$, then the conformable derivative follows*

$$\begin{aligned}
\frac{\partial^{\mu}}{\partial x^{\mu}}\left(au\left(x,t\right)+bv\left(x,t\right)\right) &= a\frac{\partial^{\mu}u}{\partial x^{\mu}}+b\frac{\partial^{\mu}u}{\partial x^{\mu}},\\
\frac{\partial^{\mu+\nu}}{\partial x^{\mu}\partial t^{\nu}}\left(x^{\mu}t^{\lambda}\right) &= \mu\lambda x^{\kappa-\mu}t^{\lambda-\nu},\\
\frac{\partial^{\mu}}{\partial x^{\mu}}\left(e^{\lambda\frac{x^{\mu}}{\mu}+\frac{\tau t^{\nu}}{\nu}}\right) &= \lambda e^{\lambda\frac{x^{\mu}}{\mu}+\frac{\tau t^{\nu}}{\nu}},\\
\frac{\partial^{\nu}}{\partial t^{\nu}}\left(e^{\lambda\frac{x^{\mu}}{\mu}+\frac{\tau t^{\nu}}{\nu}}\right) &= \tau e^{\lambda\frac{x^{\mu}}{\mu}+\frac{\tau t^{\nu}}{\nu}},\\
\frac{\partial^{\nu}}{\partial t^{\nu}}\left(\frac{x^{\mu}}{\mu}\right)^{n}\left(\frac{t^{\nu}}{\nu}\right)^{m} &= m\left(\frac{x^{\mu}}{\mu}\right)^{n}\left(\frac{t^{\nu}}{\nu}\right)^{m-1},\\
\frac{\partial^{\mu}}{\partial x^{\mu}}\left(\frac{x^{\mu}}{\mu}\right)^{n}\left(\frac{t^{\nu}}{\nu}\right) &= n\left(\frac{x^{\mu}}{\mu}\right)^{n-1}\left(\frac{t^{\nu}}{\nu}\right),\\
\frac{\partial^{\nu}}{\partial t^{\nu}}\left(\sin\left(\frac{x^{\mu}}{\mu}\right)\sin\left(\frac{t^{\nu}}{\nu}\right)\right) &= \sin\left(\frac{x^{\mu}}{\mu}\right)\cos\left(\frac{t^{\nu}}{\nu}\right),\\
\frac{\partial^{\mu}}{\partial x^{\mu}}\left(\sin a\left(\frac{x^{\mu}}{\mu}\right)\sin\left(\frac{t^{\nu}}{\nu}\right)\right) &= a\cos\left(\frac{x^{\mu}}{\mu}\right)\sin\left(\frac{t^{\nu}}{\nu}\right).
\end{aligned}$$

Next we recall the conformable single and double Laplace transforms, see [14,17], repectively.

Definition 1. *Let $f : [0,\infty) \to \mathbb{R}$ be a real valued function. The conformable single Laplace transform of f is defined by*

$$L_t^{\nu}\left(f\left(\frac{t^{\nu}}{\nu}\right)\right)=\int_0^{\infty}e^{-s\frac{t^{\nu}}{\nu}}f\left(\frac{t^{\nu}}{\nu}\right)t^{\nu-1}dt.$$

Similarly, if we let $u\left(\frac{x^{\mu}}{\mu},\frac{t^{\nu}}{\nu}\right)$ be a piecewise continuous function on $[0,\infty)\times[0,\infty)$ of exponential order and for some $a, b \in \mathbb{R}$,

$$\sup\left\{\frac{x^{\mu}}{\mu},\frac{t^{\nu}}{\nu}\right\}>0\text{, and }\frac{\left|u\left(\frac{x^{-}}{-},\frac{t^{\circ}}{\circ}\right)\right|}{e^{a\frac{x^{-}}{-}+b\frac{t^{\circ}}{\circ}}}\leq 1.$$

Then the conformable double Laplace transform is defined by

$$L_x^{\mu}L_t^{\nu}\left(u\left(\frac{x^{\mu}}{\mu},\frac{t^{\nu}}{\nu}\right)\right)=F_{\mu,\nu}\left(p,s\right)=\int_0^{\infty}\int_0^{\infty}e^{-p\frac{x^{\mu}}{\mu}-s\frac{t^{\nu}}{\nu}}u\left(\frac{x^{\mu}}{\mu},\frac{t^{\nu}}{\nu}\right)t^{\nu-1}x^{\mu-1}dtdx \tag{1}$$

where $p, s \in \mathbb{C}$, $0 < \mu, \nu \leq 1$ and integrals by means of conformable integrals with respect to x and t, respectively.

Further, the first and second order partial derivatives of the conformable double Laplace transform with respect to $\frac{x^{\mu}}{\mu}$ are given by

$$L_x^{\mu} L_t^{\nu} \left[\frac{\partial^{\mu} u}{\partial x^{\mu}} \right] = pU_{\mu,\nu}(p,s) - U_{\nu}(0,s), \tag{2}$$

$$L_x^{\mu} L_t^{\nu} \left(\frac{\partial^{2\mu} u}{\partial x^{2\mu}} \right) = p^2 U_{\mu,\nu}(p,s) - pU_{\nu}(0,s) - L_t^{\nu} \left(\frac{\partial^{\mu}}{\partial x^{\mu}} u \left(0, \frac{t^{\nu}}{\nu}\right) \right). \tag{3}$$

Similarly, with respect to $\frac{t^{\nu}}{\nu}$ they are given by

$$L_x^{\mu} L_t^{\nu} \left(\frac{\partial^{\nu} u}{\partial t^{\nu}} \right) = sU_{\mu,\nu}(p,s) - U_{\mu}(p,0), \tag{4}$$

$$L_x^{\mu} L_t^{\nu} \left(\frac{\partial^{2\nu} u}{\partial t^{2\nu}} \right) = s^2 U_{\mu,\nu}(p,s) - sU_{\mu}(p,0) - L_x^{\mu} \left(\frac{\partial^{\nu}}{\partial t^{\nu}} u \left(\frac{x^{\mu}}{\mu}, 0\right) \right). \tag{5}$$

In the following examples we state some conformable Laplace transforms of certain functions which are useful in this to Examples 3, 4, and 5.

Example 2. *In this example we calculate the conformable double Laplace for certain functions*

1. $L_x^{\mu} \left[(\frac{x^{\mu}}{\mu})^2 \right] = L_x \left[x^2 \right] = \frac{2!}{p^3}.$
2. $L_x^{\mu} L_t^{\nu} \left[(\frac{x^{\mu}}{\mu})^3 \sin(\tau \frac{t^{\nu}}{\nu}) \right] = L_x L_t \left[(x)^3 \sin(t) \right] = \frac{3!}{p^4} \frac{1}{s^2+1}.$
3. $L_x^{\mu} L_t^{\nu} \left[(\frac{x^{\mu}}{\mu}) \cos(\frac{t^{\nu}}{\nu}) \right] = L_x L_t \left[(x \cos(t) \right] = \frac{1}{p^2} \frac{s}{s^2+1}.$
4. $L_x^{\mu} L_t^{\nu} \left[4 \left(\frac{t^{\nu}}{\nu} \right) - 4 \sin \left(\frac{t^{\nu}}{\nu} \right) \right] = L_x L_t \left[(4t - 4 \sin(t) \right] = \frac{4}{ps^2(s^2+1)}.$
5. $L_x^{\mu} L_t^{\nu} \left[(4 - 4 \cos \left(\frac{t^{\nu}}{\nu} \right) \right] = L_x L_t \left[(4 - 4 \cos(t) \right] = \frac{4s}{ps^2(s^2+1)}.$

The next result generalizes the conformable double Laplace transform, see [14].

Theorem 1. *Let* $0 < \mu, \nu \leq 1$ *and* $m, n \in \mathbb{N}$ *such that* $u\left(\frac{x^{\mu}}{\mu}, \frac{t^{\nu}}{\nu}\right) \in C^l\left(\mathbb{R}^+ \times \mathbb{R}^+\right)$, *and* $l = \max(m,n)$. *Further, we also let the conformable Laplace transforms of the functions be denoted by* $u\left(\frac{x^{\mu}}{\mu}, \frac{t^{\nu}}{\nu}\right)$, $\frac{\partial^{m\mu} u}{\partial x^{m\mu}}$, and $\frac{\partial^{n\nu} u}{\partial t^{n\nu}}$. *Then*

$$\begin{aligned} L_x^{\mu} L_t^{\nu} \left(\frac{\partial^{m\mu}}{\partial x^{m\mu}} u \left(\frac{x^{\mu}}{\mu}, \frac{t^{\nu}}{\nu} \right) \right) &= p^m U_{\mu,\nu}(p,s) - p^{m-1} U_{\nu}(0,s) \\ &\quad - \sum_{i=1}^{m-1} p^{m-1-i} L_t^{\nu} \left(\frac{\partial^{i\mu}}{\partial x^{i\mu}} u \left(0, \frac{t^{\nu}}{\nu}\right) \right) \end{aligned}$$

$$\begin{aligned} L_x^{\mu} L_t^{\nu} \left(\frac{\partial^{n\nu}}{\partial t^{n\nu}} u \left(\frac{x^{\mu}}{\mu}, \frac{t^{\nu}}{\nu} \right) \right) &= s^n U_{\mu,\nu}(p,s) - s^{n-1} U_{\mu}(p,0) \\ &\quad - \sum_{j=1}^{n-1} s^{n-1-j} L_x^{\mu} \left(\frac{\partial^{j\nu}}{\partial t^{j\nu}} u \left(\frac{x^{\mu}}{\mu}, 0\right) \right) \end{aligned}$$

where

$$\frac{\partial^{m\mu}}{\partial x^{m\mu}} u \left(\frac{x^{\mu}}{\mu}, \frac{t^{\nu}}{\nu} \right) \text{ and } \frac{\partial^{n\nu}}{\partial t^{n\nu}} u \left(\frac{x^{\mu}}{\mu}, \frac{t^{\nu}}{\nu} \right)$$

denotes m, n *times conformable derivatives of function* $u(x,t)$ *respectively.*

Theorem 2. *If the conformable double Laplace transform of the conformable derivatives $\frac{\partial^{\nu} u}{\partial t^{\nu}}$ is given by Equation (4), then the double Laplace transforms of*

$$\left(\frac{x^{\mu}}{\mu}\right)^{n} \frac{\partial^{\nu}}{\partial t^{\nu}} f\left(\frac{x^{\mu}}{\mu}, \frac{t^{\nu}}{\nu}\right) \text{ and } \frac{x^{\mu}}{\mu} g\left(\frac{x^{\mu}}{\mu}, \frac{t^{\nu}}{\nu}\right)$$

are given by

$$(-1)^{n} \frac{d^{n}}{dp^{n}}\left(L_{x}^{\mu} L_{t}^{\nu}\left[f\left(\frac{x^{\mu}}{\mu}, \frac{t^{\nu}}{\nu}\right)\right]\right) = L_{x}^{\mu} L_{t}^{\nu}\left[\left(\frac{x^{\mu}}{\mu}\right)^{n} f\left(\frac{x^{\mu}}{\mu}, \frac{t^{\nu}}{\nu}\right)\right], \quad (6)$$

$$(-1)^{n} \frac{d^{n}}{dp^{n}}\left(L_{x}^{\mu} L_{t}^{\nu}\left[\frac{\partial^{\nu} f}{\partial t^{\nu}}\right]\right) = L_{x}^{\mu} L_{t}^{\nu}\left[\left(\frac{x^{\mu}}{\mu}\right)^{n} \frac{\partial^{\nu} f}{\partial t^{\nu}}\right] \quad (7)$$

and where $n = 1, 2, 3, \ldots$.

Proof. Using the definition of conformable double Laplace transform for Equation (6), we get

$$L_{x}^{\mu} L_{t}^{\nu}\left[f(\frac{x^{\mu}}{\mu}, \frac{t^{\nu}}{\nu})\right] = \int_{0}^{\infty}\int_{0}^{\infty} e^{-p\frac{x^{\mu}}{\mu} - s\frac{t^{\nu}}{\nu}} f\left(\frac{x^{\mu}}{\mu}, \frac{t^{\nu}}{\nu}\right) t^{\nu-1} x^{\mu-1} dt\, dx, \quad (8)$$

by taking the ^{n}th derivative with respect to p for both sides of Equation (8), we have

$$\begin{aligned}\frac{d^{n}}{dp^{n}}\left(L_{x}^{\mu} L_{t}^{\nu}\left[f(\frac{x^{\mu}}{\mu}, \frac{t^{\nu}}{\nu})\right]\right) &= \int_{0}^{\infty}\int_{0}^{\infty} \frac{d^{n}}{dp^{n}}\left(e^{-p\frac{x^{\mu}}{\mu} - s\frac{t^{\nu}}{\nu}} f\left(\frac{x^{\mu}}{\mu}, \frac{t^{\nu}}{\nu}\right)\right) t^{\nu-1} x^{\mu-1} dt dx \\ &= (-1)^{n} \int_{0}^{\infty}\int_{0}^{\infty} \left(\frac{x^{\mu}}{\mu}\right)^{n} e^{-p\frac{x^{\mu}}{\mu} - s\frac{t^{\nu}}{\nu}} t^{\nu-1} x^{\mu-1} f\left(\frac{x^{\mu}}{\mu}, \frac{t^{\nu}}{\nu}\right) dt\, dx \\ &= (-1)^{n} L_{x}^{\mu} L_{t}^{\nu}\left[\left(\frac{x^{\mu}}{\mu}\right)^{n} f\left(\frac{x^{\mu}}{\mu}, \frac{t^{\nu}}{\nu}\right)\right],\end{aligned}$$

and further we obtain

$$(-1)^{n} \frac{d^{n}}{dp^{n}}\left(L_{x}^{\mu} L_{t}^{\nu}\left[f\left(\frac{x^{\mu}}{\mu}, \frac{t^{\nu}}{\nu}\right)\right]\right) = L_{x}^{\mu} L_{t}^{\nu}\left[\left(\frac{x^{\mu}}{\mu}\right)^{n} f\left(\frac{x^{\mu}}{\mu}, \frac{t^{\nu}}{\nu}\right)\right].$$

Similarly, we can prove the Equation (7). □

3. Conformable Derivatives Double Laplace Transform Decomposition Method Applied to Singular Pseudohyperbolic Equation

The main aim of this section is to discuss the applicability of the conformable double Laplace transform decomposition method (CDLDM) for the linear and nonlinear singular pseudohyperbolic equation. The pseudo-hyperbolic equations arise, for example, in the description of the electron diffusion processes in a plate, and they also arise in hydrodynamics in the study of fluid motion with an alternating viscosity. In this study we define the conformable double Laplace transform of the function $u\left(\frac{x^{\mu}}{\mu}, \frac{t^{\nu}}{\nu}\right)$ by $U_{\mu,\nu}(p,s)$. To illustrate the idea of our method, let us suggest here two important problems.

The first problem:

Consider the linear pseudohyperbolic equations

$$\frac{\partial^{2\nu} u}{\partial t^{2\nu}} - \frac{\mu}{x^{\mu}} \frac{\partial^{\mu}}{\partial x^{\mu}}\left(\frac{x^{\mu}}{\mu} \frac{\partial^{\mu} u}{\partial x^{\mu}}\right) - \frac{\mu}{x^{\mu}} \frac{\partial^{\mu+\nu}}{\partial x^{\mu} \partial t^{\nu}}\left(\frac{x^{\mu}}{\mu} \frac{\partial^{\mu} u}{\partial x^{\mu}}\right) = f\left(\frac{x^{\mu}}{\mu}, \frac{t^{\nu}}{\nu}\right), \quad x, t > 0 \quad (9)$$

subject to the condition

$$u\left(x,0\right)=f_1\left(\frac{x^{\mu}}{\mu}\right),\quad \frac{\partial^{\nu}u\left(x,0\right)}{\partial t^{\nu}}=f_2\left(\frac{x^{\mu}}{\mu}\right), \tag{10}$$

where $f\left(\frac{x^{\mu}}{\mu},\frac{t^{\nu}}{\nu}\right)$, $f_1\left(\frac{x^{\mu}}{\mu}\right)$, and $f_2\left(\frac{x^{\mu}}{\mu}\right)$ are source term and initial conditions, respectively.

The method:

In order to obtain the solution of Equation (9) by using conformable double Laplace transform decomposition methods, we applying the following steps:

Step 1: Multiplying both sides of Equation (9) by the term $\frac{x^{\mu}}{\mu}$, we have

$$\frac{x^{\mu}}{\mu}\frac{\partial^{2\nu}u}{\partial t^{2\nu}}-\frac{\partial^{\mu}}{\partial x^{\mu}}\left(\frac{x^{\mu}}{\mu}\frac{\partial^{\mu}u}{\partial x^{\mu}}\right)-\frac{\partial^{\mu+\nu}}{\partial x^{\mu}\partial t^{\nu}}\left(\frac{x^{\mu}}{\mu}\frac{\partial^{\mu}u}{\partial x^{\mu}}\right)=\frac{x^{\mu}}{\mu}f\left(\frac{x^{\mu}}{\mu},\frac{t^{\nu}}{\nu}\right), \tag{11}$$

Step 2: Applying conformable double Laplace transform for Equation (11) we get

$$\begin{aligned} L_x^{\mu}L_t^{\nu}\left[\frac{x^{\mu}}{\mu}\frac{\partial^{2\nu}u}{\partial t^{2\nu}}\right] &= L_x^{\mu}L_t^{\nu}\left[\frac{\partial^{\mu}}{\partial x^{\mu}}\left(\frac{x^{\mu}}{\mu}\frac{\partial^{\mu}u}{\partial x^{\mu}}\right)+\frac{\partial^{\mu+\nu}}{\partial x^{\mu}\partial t^{\nu}}\left(\frac{x^{\mu}}{\mu}\frac{\partial^{\mu}u}{\partial x^{\mu}}\right)\right] \\ &\quad +L_x^{\mu}L_t^{\nu}\left[\frac{x^{\mu}}{\mu}f\left(\frac{x^{\mu}}{\mu},\frac{t^{\nu}}{\nu}\right)\right], \end{aligned} \tag{12}$$

Step 3: On using Equations (5)–(7), we obtain

$$-\frac{d}{dp}\left[s^2U_{\mu,\nu}(p,s)-sU_{\mu}\left(p,0\right)-\frac{\partial^{\nu}u\left(p,0\right)}{\partial t^{\nu}}\right]=L_x^{\mu}L_t^{\nu}\left[\Psi\right]-\frac{d}{dp}\left[L_x^{\mu}L_t^{\nu}\left(f\left(\frac{x^{\mu}}{\mu},\frac{t^{\nu}}{\nu}\right)\right)\right], \tag{13}$$

where the conformable Laplace transforms of $u\left(\frac{x^{\mu}}{\mu},0\right)$ and $\frac{\partial^{\nu}u\left(\frac{x^{\mu}}{\mu},0\right)}{\partial t^{\nu}}$ are denoted by

$$U_{\mu}\left(p,0\right)=F_1\left(p,0\right),\quad \frac{\partial^{\nu}u\left(p,0\right)}{\partial t^{\nu}}=F_2\left(p,0\right)$$

respectively, and

$$\Psi=\frac{\partial^{\mu}}{\partial x^{\mu}}\left(\frac{x^{\mu}}{\mu}\frac{\partial^{\mu}u}{\partial x^{\mu}}\right)+\frac{\partial^{\mu+\nu}}{\partial x^{\mu}\partial t^{\nu}}\left(\frac{x^{\mu}}{\mu}\frac{\partial^{\mu}u}{\partial x^{\mu}}\right),$$

using given initial condition Equation (13) becomes

$$\frac{d}{dp}\left[U_{\mu,\nu}(p,s)\right]=\frac{1}{s}\frac{d}{dp}F_1\left(p,0\right)+\frac{1}{s^2}\frac{d}{dp}F_2\left(p,0\right)-\frac{1}{s^2}L_x^{\mu}L_t^{\nu}\left[\Psi\right]+\frac{1}{s^2}\frac{d}{dp}F_{\mu,\nu}\left(p,s\right). \tag{14}$$

Step 4: By applying the integral for both sides of Equation (14), from 0 to p with respect to p, where p is transform of the variable $\frac{x^{\mu}}{\mu}$, we have

$$U_{\mu,\nu}(p,s)=\frac{F_1\left(p,0\right)}{s}+\frac{F_2\left(p,0\right)}{s^2}-\frac{1}{s^2}\int_0^p L_x^{\mu}L_t^{\nu}\left[\Psi\right]dp+\frac{1}{s^2}F_{\mu,\nu}\left(p,s\right), \tag{15}$$

where $F_{\mu,\nu}\left(p,s\right)$, $F_1\left(p,0\right)$, and $F_2\left(p,0\right)$ are conformable Laplace transforms of the functions $f\left(\frac{x^{\mu}}{\mu},\frac{t^{\nu}}{\nu}\right)$, $f_1\left(\frac{x^{\mu}}{\mu}\right)$, and $f_2\left(\frac{x^{\mu}}{\mu}\right)$, respectively.

Step 5: By taking the inverse conformable double Laplace transform for Equation (15), we can compute the solution $u\left(\frac{x^{\mu}}{\mu},\frac{t^{\nu}}{\nu}\right)$ as follows

$$u\left(\frac{x^{\mu}}{\mu},\frac{t^{\nu}}{\nu}\right)=f_1\left(\frac{x^{\mu}}{\mu}\right)+\frac{t^{\nu}}{\nu}f_2\left(\frac{x^{\mu}}{\mu}\right)+L_p^{-1}L_s^{-1}\left[\frac{1}{s^2}F_{\mu,\nu}(p,s)\right]-L_p^{-1}L_s^{-1}\left[\frac{1}{s^2}\int_0^p L_x^{\mu}L_t^{\nu}[\Psi]\,dp\right] \tag{16}$$

where $L_p^{-1}L_s^{-1}$ indicates the double inverse conformable derivatives double Laplace transform. Here, we assume that the double inverse Laplace transform with respect to p and s exists for each term in the right hand side of Equation (16).

Step 6: The conformable double Laplace transform decomposition method (CDLDM) defines the solutions $u(\frac{x^{\mu}}{\mu},\frac{t^{\nu}}{\nu})$ with the help of infinite series as:

$$u\left(\frac{x^{\mu}}{\mu},\frac{t^{\nu}}{\nu}\right)=\sum_{n=0}^{\infty}u_n\left(\frac{x^{\mu}}{\mu},\frac{t^{\nu}}{\nu}\right). \tag{17}$$

By substituting Equation (17) into Equation (16), we obtain

$$\begin{aligned}\sum_{n=0}^{\infty}u_n\left(\frac{x^{\mu}}{\mu},\frac{t^{\nu}}{\nu}\right) &= f_1\left(\frac{x^{\mu}}{\mu}\right)+\frac{t^{\nu}}{\nu}f_2\left(\frac{x^{\mu}}{\mu}\right)+L_p^{-1}L_s^{-1}\left[\frac{1}{s^2}F_{\mu,\nu}(p,s)\right]\\ &\quad -L_p^{-1}L_s^{-1}\left[\frac{1}{s^2}\int_0^p L_x^{\mu}L_t^{\nu}\left[\frac{\partial^{\mu}}{\partial x^{\mu}}\left(\frac{x^{\mu}}{\mu}\frac{\partial^{\mu}}{\partial x^{\mu}}\left(\sum_{n=0}^{\infty}u_n\right)\right)\right]dp\right]\\ &\quad -L_p^{-1}L_s^{-1}\left[\frac{1}{s^2}\int_0^p L_x^{\mu}L_t^{\nu}\left[\left(\frac{\partial^{\mu+\nu}}{\partial x^{\mu}\partial t^{\nu}}\left(\frac{x^{\mu}}{\mu}\frac{\partial^{\mu}}{\partial x^{\mu}}\sum_{n=0}^{\infty}u_n\right)\right)\right]dp\right].\end{aligned} \tag{18}$$

The zeroth component u_0, as suggested by Adomian method, is always identified by the given initial condition and the source term $L_p^{-1}L_s^{-1}\left[\frac{1}{s^2}F_{\mu,\nu}(p,s)\right]$, both of which are assumed to be known. Accordingly, we set

$$u_0\left(\frac{x^{\mu}}{\mu},\frac{t^{\nu}}{\nu}\right)=f_1\left(\frac{x^{\mu}}{\mu}\right)+\frac{t^{\nu}}{\nu}f_2\left(\frac{x^{\mu}}{\mu}\right)+L_p^{-1}L_s^{-1}\left[\frac{1}{s^2}F_{\mu,\nu}(p,s)\right].$$

The other components $u_{k+1}, k\geq 0$ are given by using the relation

$$\begin{aligned}u_{k+1}\left(\frac{x^{\mu}}{\mu},\frac{t^{\nu}}{\nu}\right) &= -L_p^{-1}L_s^{-1}\left[\frac{1}{s^2}\int_0^p L_x^{\mu}L_t^{\nu}\left[\frac{\partial^{\mu}}{\partial x^{\mu}}\left(\frac{x^{\mu}}{\mu}\frac{\partial^{\mu}u_k}{\partial x^{\mu}}\right)\right]dp\right]\\ &\quad - L_p^{-1}L_s^{-1}\left[\frac{1}{s^2}\int_0^p L_x^{\mu}L_t^{\nu}\left[\frac{\partial^{\mu+\nu}}{\partial x^{\mu}\partial t^{\nu}}\left(\frac{x^{\mu}}{\mu}\frac{\partial^{\mu}u_k}{\partial x^{\mu}}\right)\right]dp\right],\end{aligned} \tag{19}$$

the first few components from the last recursive relation are, at $k=0$,

$$\begin{aligned}u_1\left(\frac{x^{\mu}}{\mu},\frac{t^{\nu}}{\nu}\right) &= -L_p^{-1}L_s^{-1}\left[\frac{1}{s^2}\int_0^p L_x^{\mu}L_t^{\nu}\left[\frac{\partial^{\mu}}{\partial x^{\mu}}\left(\frac{x^{\mu}}{\mu}\frac{\partial^{\mu}u_0}{\partial x^{\mu}}\right)\right]dp\right]\\ &\quad - L_p^{-1}L_s^{-1}\left[\frac{1}{s^2}\int_0^p L_x^{\mu}L_t^{\nu}\left[\frac{\partial^{\mu+\nu}}{\partial x^{\mu}\partial t^{\nu}}\left(\frac{x^{\mu}}{\mu}\frac{\partial^{\mu}u_0}{\partial x^{\mu}}\right)\right]dp\right],\end{aligned}$$

at $k = 1$

$$\begin{aligned} u_2\left(\frac{x^\mu}{\mu}, \frac{t^\nu}{\nu}\right) &= -L_p^{-1}L_s^{-1}\left[\frac{1}{s^2}\int_0^p L_x^\mu L_t^\nu\left[\frac{\partial^\mu}{\partial x^\mu}\left(\frac{x^\mu}{\mu}\frac{\partial^\mu u_1}{\partial x^\mu}\right)\right]dp\right] \\ &- L_p^{-1}L_s^{-1}\left[\frac{1}{s^2}\int_0^p L_x^\mu L_t^\nu\left[\frac{\partial^{\mu+\nu}}{\partial x^\mu \partial t^\nu}\left(\frac{x^\mu}{\mu}\frac{\partial^\mu u_1}{\partial x^\mu}\right)\right]dp\right], \end{aligned}$$

at $k = 2$

$$\begin{aligned} u_3\left(\frac{x^\mu}{\mu}, \frac{t^\nu}{\nu}\right) &= -L_p^{-1}L_s^{-1}\left[\frac{1}{s^2}\int_0^p L_x^\mu L_t^\nu\left[\frac{\partial^\mu}{\partial x^\mu}\left(\frac{x^\mu}{\mu}\frac{\partial^\mu u_2}{\partial x^\mu}\right)\right]dp\right] \\ &- L_p^{-1}L_s^{-1}\left[\frac{1}{s^2}\int_0^p L_x^\mu L_t^\nu\left[\frac{\partial^{\mu+\nu}}{\partial x^\mu \partial t^\nu}\left(\frac{x^\mu}{\mu}\frac{\partial^\mu u_2}{\partial x^\mu}\right)\right]dp\right], \end{aligned}$$

etc. The important terms used in infinite series depend on the problems and may be three terms or four terms, etc.

In order to give a clear overview of this method, we present the following example:

Example 3. *Consider singular conformable derivatives in one dimensional pseudohyperbolic equations with the indicated initial condition*

$$\begin{aligned} \frac{\partial^{2\nu} u}{\partial t^{2\nu}} - \frac{\mu}{x^\mu}\frac{\partial^\mu}{\partial x^\mu}\left(\frac{x^\mu}{\mu}\frac{\partial^\mu u}{\partial x^\mu}\right) - \frac{\mu}{x^\mu}\frac{\partial^{\mu+\nu}}{\partial x^\mu \partial t^\nu}\left(\frac{x^\mu}{\mu}\frac{\partial^\mu u}{\partial x^\mu}\right) &= -\left(\frac{x^\mu}{\mu}\right)^2 \sin\left(\frac{t^\nu}{\nu}\right) \\ & -4\sin\left(\frac{t^\nu}{\nu}\right) - 4\cos\left(\frac{t^\nu}{\nu}\right), \end{aligned} \tag{20}$$

and

$$u\left(\frac{x^\mu}{\mu}, 0\right) = 0, \quad \frac{\partial u\left(\frac{x^\mu}{\mu}, 0\right)}{\partial t} = \left(\frac{x^\mu}{\mu}\right)^2. \tag{21}$$

By using the aforesaid method subject to the initial condition, we have

$$\begin{aligned} \frac{dU_{\mu,\nu}(p,s)}{dp} &= -\frac{1}{s^2}L_x^\mu L_t^\nu\left[\frac{\mu}{x^\mu}\frac{\partial^\mu}{\partial x^\mu}\left(\frac{x^\mu}{\mu}\frac{\partial^\mu u}{\partial x^\mu}\right) + \frac{\mu}{x^\mu}\frac{\partial^{\mu+\nu}}{\partial x^\mu \partial t^\nu}\left(\frac{x^\mu}{\mu}\frac{\partial^\mu u}{\partial x^\mu}\right)\right] \\ &+ \frac{3!}{p^4 s^2 (s^2+1)} + \frac{4}{p^2 s^2 (s^2+1)} + \frac{4}{p^2 s (s^2+1)} - \frac{3!}{p^4 s^2}, \end{aligned} \tag{22}$$

taking the integral for Equation (22), from 0 to p with respect to p, we get

$$\begin{aligned} U_{\mu,\nu}(p,s) &= -\frac{1}{s^2}\int_0^p L_x^\mu L_t^\nu\left[\frac{\partial^\mu}{\partial x^\mu}\left(\frac{x^\mu}{\mu}\frac{\partial^\mu u}{\partial x^\mu}\right) + \frac{\partial^{\mu+\nu}}{\partial x^\mu \partial t^\nu}\left(\frac{x^\mu}{\mu}\frac{\partial^\mu u}{\partial x^\mu}\right)\right]dp \\ & -\frac{2}{p^3 s^2 (s^2+1)} - \frac{4}{p s^2 (s^2+1)} - \frac{4}{p s (s^2+1)} + \frac{2!}{p^3 s^2}. \end{aligned} \tag{23}$$

Employing the inverse conformable derivatives double Laplace transform to Equation (23), we get

$$\begin{aligned} u\left(\frac{x^\mu}{\mu}, \frac{t^\nu}{\nu}\right) &= -L_p^{-1}L_s^{-1}\left[\frac{1}{s^2}\int_0^p L_x^\mu L_t^\nu\left[\frac{\partial^\mu}{\partial x^\mu}\left(\frac{x^\mu}{\mu}\frac{\partial^\mu u}{\partial x^\mu}\right) + \frac{\partial^{\mu+\nu}}{\partial x^\mu \partial t^\nu}\left(\frac{x^\mu}{\mu}\frac{\partial^\mu u}{\partial x^\mu}\right)\right]dp\right] \\ &+ \left(\frac{x^\mu}{\mu}\right)^2 \sin\left(\frac{t^\nu}{\nu}\right) - 4\left(\frac{t^\nu}{\nu}\right) + 4\sin\left(\frac{t^\nu}{\nu}\right) + 4\cos\left(\frac{t^\nu}{\nu}\right) - 4, \end{aligned} \tag{24}$$

by substituting Equation (17) into Equation (24), we obtain:

$$\begin{aligned}\sum_{n=0}^{\infty} u_n\left(\frac{x^\mu}{\mu},\frac{t^\nu}{\nu}\right) &= -L_p^{-1}L_s^{-1}\left[\frac{1}{s^2}\int_0^p L_x^\mu L_t^\nu\left[\frac{\partial^\mu}{\partial x^\mu}\left(\frac{x^\mu}{\mu}\frac{\partial^\mu}{\partial x^\mu}\sum_{n=0}^{\infty}u_n\right)\right]dp\right]\\ &\quad -L_p^{-1}L_s^{-1}\left[\frac{1}{s^2}\int_0^p L_x^\mu L_t^\nu\left[\frac{\partial^{\mu+\nu}}{\partial x^\mu \partial t^\nu}\left(\frac{x^\mu}{\mu}\frac{\partial^\mu}{\partial x^\mu}\sum_{n=0}^{\infty}u_n\right)\right]dp\right]\\ &\quad +\left(\frac{x^\mu}{\mu}\right)^2\sin\left(\frac{t^\nu}{\nu}\right)-4\left(\frac{t^\nu}{\nu}\right)+4\sin\left(\frac{t^\nu}{\nu}\right)+4\cos\left(\frac{t^\nu}{\nu}\right)-4.\end{aligned}$$

By applying the conformable double Laplace transform decomposition method, we obtain

$$u_0=\left(\frac{x^\mu}{\mu}\right)^2\sin\left(\frac{t^\nu}{\nu}\right)-4\left(\frac{t^\nu}{\nu}\right)+4\sin\left(\frac{t^\nu}{\nu}\right)+4\cos\left(\frac{t^\nu}{\nu}\right)-4,$$

eventually, we have the general recursive relation, given by

$$\begin{aligned}u_{k+1}\left(\frac{x^\mu}{\mu},\frac{t^\nu}{\nu}\right) &= -L_p^{-1}L_s^{-1}\left[\frac{1}{s^2}\int_0^p L_x^\mu L_t^\nu\left[\frac{\partial^\mu}{\partial x^\mu}\left(\frac{x^\mu}{\mu}\frac{\partial^\mu u_k}{\partial x^\mu}\right)\right]dp\right]\\ &\quad -L_p^{-1}L_s^{-1}\left[\frac{1}{s^2}\int_0^p L_x^\mu L_t^\nu\left[\frac{\partial^{\mu+\nu}}{\partial x^\mu\partial t^\nu}\left(\frac{x^\mu}{\mu}\frac{\partial^\mu u_k}{\partial x^\mu}\right)\right]dp\right],\end{aligned}$$

where $k \geq 0$, therefore

$$\begin{aligned}u_1 &= -L_p^{-1}L_s^{-1}\left[\frac{1}{s^2}\int_0^p L_x^\mu L_t^\nu\left[\frac{\partial^\mu}{\partial x^\mu}\left(\frac{x^\mu}{\mu}\frac{\partial^\mu}{\partial x^\mu}u_0\right)+\frac{\partial^{\mu+\nu}}{\partial x^\mu\partial t^\nu}\left(\frac{x^\mu}{\mu}\frac{\partial^\mu}{\partial x^\mu}u_0\right)\right]dp\right]\\ u_1 &= -L_p^{-1}L_s^{-1}\left[\frac{1}{s^2}\int_0^p L_x^\mu L_t^\nu\left[4\left(\frac{x^\mu}{\mu}\right)\sin\left(\frac{t^\nu}{\nu}\right)+4\left(\frac{x^\mu}{\mu}\right)\cos\left(\frac{t^\nu}{\nu}\right)\right]dp\right]\\ &= -L_p^{-1}L_s^{-1}\left[\frac{1}{s^2}\int_0^p\left[\frac{4}{p^2(s^2+1)}+\frac{4s}{p^2(s^2+1)}\right]dp\right],\\ u_1 &= L_p^{-1}L_s^{-1}\left[\frac{1}{s^2}\frac{4}{p(s^2+1)}+\frac{4s}{p(s^2+1)}\right]=L_s^{-1}\left[\frac{4}{s^2(s^2+1)}+\frac{4s}{s^2(s^2+1)}\right],\end{aligned}$$

by using partial fractional and inverse Laplace transform with respect to s, we have

$$\begin{aligned}u_1 &= L_s^{-1}\left[\frac{4}{s^2(s^2+1)}+\frac{4s}{s^2(s^2+1)}\right]\\ &= L_s^{-1}\left[\frac{4}{s^2}-\frac{4}{s^2+1}+\frac{4}{s}-\frac{4s}{s^2+1}\right]\\ &= 4\left(\frac{t^\nu}{\nu}\right)-4\sin\left(\frac{t^\nu}{\nu}\right)+4-4\cos\left(\frac{t^\nu}{\nu}\right),\end{aligned}$$

and

$$\begin{aligned}u_2 &= -L_p^{-1}L_s^{-1}\left[\frac{1}{s^2}\int_0^p L_x^\mu L_t^\nu\left[\frac{\partial^\mu}{\partial x^\mu}\left(\frac{x^\mu}{\mu}\frac{\partial^\mu}{\partial x^\mu}u_1\right)+\frac{\partial^{\mu+\nu}}{\partial x^\mu\partial t^\nu}\left(\frac{x^\mu}{\mu}\frac{\partial^\mu}{\partial x^\mu}u_1\right)\right]dp\right]\\ u_2 &= -L_p^{-1}L_s^{-1}\left[\frac{1}{s^2}\int_0^p L_x^\mu L_t^\nu\,[0+0]\,dp\right]=0.\end{aligned}$$

In the view of the above equations, the series solution is given by

$$\begin{aligned}\sum_{n=0}^{\infty} u_n &= u_0 + u_1 + u_2 + ... \\ &= \left(\frac{x^\mu}{\mu}\right)^2 \sin\left(\frac{t^\nu}{\nu}\right) - 4\left(\frac{t^\nu}{\nu}\right) + 4\sin\left(\frac{t^\nu}{\nu}\right) + 4\cos\left(\frac{t^\nu}{\nu}\right) - 4 \\ &\quad +4\left(\frac{t^\nu}{\nu}\right) - 4\sin\left(\frac{t^\nu}{\nu}\right) + 4 - 4\cos\left(\frac{t^\nu}{\nu}\right) + 0 + 0 + \dots\end{aligned}$$

Hence, the exact solution of Equation (20) is given by:

$$u\left(\frac{x^\mu}{\mu}, \frac{t^\nu}{\nu}\right) = \left(\frac{x^\mu}{\mu}\right)^2 \sin\left(\frac{t^\nu}{\nu}\right).$$

Second problem: Consider the following general form of the nonlinear singular pseudohyperbolic equations in one dimension of the form:

$$\frac{\partial^{2\nu} u}{\partial t^{2\nu}} - \frac{\mu}{x^\mu}\frac{\partial^\mu}{\partial x^\mu}\left(\frac{x^\mu}{\mu}\frac{\partial^\mu u}{\partial x^\mu}\right) - \frac{\mu}{x^\mu}\frac{\partial^{\mu+\nu}}{\partial x^\mu \partial t^\nu}\left(\frac{x^\mu}{\mu}\frac{\partial^\mu u}{\partial x^\mu}\right) - a\left(\frac{x^\mu}{\mu}\right) u \frac{\partial^\mu u}{\partial x^\mu} + u^2 = f\left(\frac{x^\mu}{\mu}, \frac{t^\nu}{\nu}\right) \tag{25}$$

with initial condition

$$u\left(\frac{x^\mu}{\mu}, 0\right) = g_1\left(\frac{x^\mu}{\mu}\right), \quad \frac{\partial^\nu u\left(\frac{x^\mu}{\mu}, 0\right)}{\partial t^\nu} = g_2\left(\frac{x^\mu}{\mu}\right), \tag{26}$$

where the functions $a(\frac{x^\mu}{\mu})$ are arbitrary. In order to obtain the solution of Equation (25), we use the following steps:

First step: By multiplying Equation (25) by $\frac{x^\mu}{\mu}$ and taking conformable double Laplace transform, we have

$$\begin{aligned} L_x^\mu L_t^\nu \left[\frac{x^\mu}{\mu}\frac{\partial^{2\nu} u}{\partial t^{2\nu}}\right] &= L_x^\mu L_t^\nu \left[\frac{\partial^\mu}{\partial x^\mu}\left(\frac{x^\mu}{\mu}\frac{\partial^\mu u}{\partial x^\mu}\right) + \frac{\partial^{\mu+\nu}}{\partial x^\mu \partial t^\nu}\left(\frac{x^\mu}{\mu}\frac{\partial^\mu u}{\partial x^\mu}\right)\right] \\ &+ L_x^\mu L_t^\nu \left[a(x)\frac{x^\mu}{\mu}\frac{x^\mu}{\mu} u \frac{\partial^\mu u}{\partial x^\mu} - \frac{x^\mu}{\mu} u^2\right] + L_x^\mu L_t^\nu \left[\frac{x^\mu}{\mu} f\left(\frac{x^\mu}{\mu}, \frac{t^\nu}{\nu}\right)\right], \end{aligned} \tag{27}$$

where conformable Laplace transform of $u\left(\frac{x^\mu}{\mu}, 0\right)$ and $\frac{\partial^\nu u\left(\frac{x^\mu}{\mu}, 0\right)}{\partial t^\nu}$ are given by

$$U_\mu(p, 0) = G_1(p, 0), \quad \frac{\partial^\nu u(p, 0)}{\partial t^\nu} = G_2(p, 0). \tag{28}$$

Second step: Applying Equations (5)–(28) into Equation (27), one can get that

$$\frac{d}{dp}\left[U_{\mu,\nu}(p, s)\right] = \frac{1}{s}\frac{d}{dp} G_1(p, 0) + \frac{1}{s^2}\frac{d}{dp} G_2(p, 0) - \frac{1}{s^2} L_x^\mu L_t^\nu [\Phi] + \frac{1}{s^2}\frac{d}{dp} F_{\mu,\nu}(p, s), \tag{29}$$

where,

$$\Phi = \frac{\partial^\mu}{\partial x^\mu}\left(\frac{x^\mu}{\mu}\frac{\partial^\mu u}{\partial x^\mu}\right) + \frac{\partial^{\mu+\nu}}{\partial x^\mu \partial t^\nu}\left(\frac{x^\mu}{\mu}\frac{\partial^\mu u}{\partial x^\mu}\right) + a(x)\frac{x^\mu}{\mu}\frac{x^\mu}{\mu} u \frac{\partial^\mu u}{\partial x^{2\mu}} - \frac{x^\mu}{\mu} u^2.$$

Third step: By taking the integral for Equation (29), from 0 to p with respect to p, where p is a transform of $\frac{x^{\mu}}{\mu}$, we have

$$U_{\mu,\nu}(p,s) = \frac{G_1(p,0)}{s} + \frac{vG_2(p,0)}{s^2} - \frac{1}{s^2}\int_0^p L_x^{\mu} L_t^{\nu}[\Phi]\,dp + F_{\mu,\nu}(p,s). \tag{30}$$

Fourth step: Using CFDLDM, the solution can be written in the infinite series as in Equation (17). By using the inverse Laplace transformation to Equation (30), we obtain.

$$\begin{aligned} u\left(\frac{x^{\mu}}{\mu},\frac{t^{\nu}}{\nu}\right) &= g_1\left(\frac{x^{\mu}}{\mu}\right) + \frac{t^{\nu}}{\nu} g_2\left(\frac{x^{\mu}}{\mu}\right) + L_p^{-1}L_s^{-1}\left[\frac{1}{s^2}F_{\mu,\nu}(p,s)\right] \\ &\quad - L_p^{-1}L_s^{-1}\left[\frac{1}{s^2}\int_0^p L_x^{\mu}L_t^{\nu}[\Phi]\,dp\right] \end{aligned} \tag{31}$$

furthermore, the nonlinear terms $u\frac{\partial^{\mu} u}{\partial x^{\mu}}$ and u^2 can be defined by:

$$u^2 = N_1 = \sum_{n=0}^{\infty} A_n, \quad u\frac{\partial^{\mu} u}{\partial x^{\mu}} = N_2 = \sum_{n=0}^{\infty} B_n. \tag{32}$$

We have a few terms of the Adomian polynomials for A_n and B_n that are denoted by

$$A_n = \frac{1}{n!}\left(\frac{d^n}{d\lambda^n}\left[N_1 \sum_{i=0}^{\infty}(\lambda^n u_n)\right]\right)_{\lambda=0}, \tag{33}$$

and

$$B_n = \frac{1}{n!}\left(\frac{d^n}{d\lambda^n}\left[N_2 \sum_{i=0}^{\infty}(\lambda^n u_n)\right]\right)_{\lambda=0}, \tag{34}$$

where $n = 0, 1, 2, ...$ By putting Equations (33)–(32) into Equation (31), we get

$$\begin{aligned} \sum_{n=0}^{\infty} u_n\left(\frac{x^{\mu}}{\mu},\frac{t^{\nu}}{\nu}\right) &= f_1\left(\frac{x^{\mu}}{\mu}\right) + \frac{t^{\nu}}{\nu} f_2\left(\frac{x^{\mu}}{\mu}\right) + L_p^{-1}L_s^{-1}\left[\frac{1}{s^2}F_{\mu,\nu}(p,s)\right] \\ &\quad - L_p^{-1}L_s^{-1}\left[\frac{1}{s^2}\int_0^p L_x^{\mu}L_t^{\nu}\left[\frac{\partial^{\mu}}{\partial x^{\mu}}\left(\frac{x^{\mu}}{\mu}\frac{\partial^{\mu}}{\partial x^{\mu}}\left(\sum_{n=0}^{\infty} u_n\right)\right)\right]dp\right] \\ &\quad - L_p^{-1}L_s^{-1}\left[\frac{1}{s^2}\int_0^p L_x^{\mu}L_t^{\nu}\left[\left(\frac{x^{\mu}}{\mu}\frac{\partial^{\mu+\nu}}{\partial x^{\mu}\partial t^{\nu}}\left(\sum_{n=0}^{\infty} u_n\left(\frac{x^{\mu}}{\mu},\frac{t^{\nu}}{\nu}\right)\right)\right)\right]dp\right] \\ &\quad - L_p^{-1}L_s^{-1}\left[\frac{1}{s^2}\int_0^p L_x^{\mu}L_t^{\nu}\left[\frac{x^{\mu}}{\mu}\left(\sum_{n=0}^{\infty} A_n\right)\right]dp\right] \\ &\quad + L_p^{-1}L_s^{-1}\left[\frac{1}{s^2}\int_0^p L_x^{\mu}L\left[a(x)\frac{x^{\mu}}{\mu}\sum_{n=0}^{\infty} B_n\right]dp\right], \end{aligned} \tag{35}$$

the few components of the Adomian polynomials of Equations (33) and (34) are given as follows

$$\begin{aligned} A_0 &= u_0^2, \\ A_1 &= 2u_0u_1, \\ A_2 &= 2u_0u_2 + u_1^2 \\ A_3 &= 2u_0u_3 + 2uu_2, \end{aligned} \tag{36}$$

and

$$\begin{aligned} B_0 &= u_0\frac{\partial^\mu u_0}{\partial x^\mu}, \\ B_1 &= u_0\frac{\partial^\mu u_1}{\partial x^\mu}+u_1\frac{\partial^\mu u_0}{\partial x^\mu}, \\ B_2 &= u_0\frac{\partial^\mu u_2}{\partial x^\mu}+u_1\frac{\partial^\mu u_1}{\partial x^\mu}+u_2\frac{\partial^\mu u_0}{\partial x^\mu}, \\ B_3 &= u_0\frac{\partial^\mu u_3}{\partial x^\mu}+u_1\frac{\partial^\mu u_2}{\partial x^\mu}+u_2\frac{\partial^\mu u_1}{\partial x^\mu}+u_3\frac{\partial^\mu u_0}{\partial x^\mu}. \end{aligned} \tag{37}$$

Hence, the zeroth component u_0 from Equation (35) is given by

$$u_0\left(\frac{x^\mu}{\mu},\frac{t^\nu}{\nu}\right)=f_1\left(\frac{x^\mu}{\mu}\right)+\frac{t^\nu}{\nu}f_2\left(\frac{x^\mu}{\mu}\right)+L_p^{-1}L_s^{-1}\left[\frac{1}{s^2}F_{\mu,\nu}(p,s)\right] \tag{38}$$

and

$$\begin{aligned} u_{k+1}\left(\frac{x^\mu}{\mu},\frac{t^\nu}{\nu}\right) &= -L_p^{-1}L_s^{-1}\left[\frac{1}{s^2}\int_0^p L_x^\mu L_t^\nu\left[\frac{\partial^\mu}{\partial x^\mu}\left(\frac{x^\mu}{\mu}\frac{\partial^\mu u_k}{\partial x^\mu}\right)\right]dp\right] \\ &-L_p^{-1}L_s^{-1}\left[\frac{1}{s^2}\int_0^p L_x^\mu L_t^\nu\left[\left(\frac{x^\mu}{\mu}\frac{\partial^{\mu+\nu}u_k}{\partial x^\mu\partial t^\nu}\right)\right]dp\right] \\ &+L_p^{-1}L_s^{-1}\left[\frac{1}{s^2}\int_0^p L_x^\mu L_t^\nu\left[a(x)\frac{x^\mu}{\mu}A_k-\frac{x^\mu}{\mu}B_k\right]dp\right], \end{aligned} \tag{39}$$

where $k \geq 0$.

Example 4. *Consider the nonlinear singular pseudohyperbolic equation in one dimensional is governed by*

$$\frac{\partial^{2\nu}u}{\partial t^{2\nu}}-\frac{\mu}{x^\mu}\frac{\partial^\mu}{\partial x^\mu}\left(\frac{x^\mu}{\mu}\frac{\partial^\mu u}{\partial x^\mu}\right)-\frac{\mu}{x^\mu}\frac{\partial^{\mu+\nu}}{\partial x^\mu\partial t^\nu}\left(\frac{x^\mu}{\mu}\frac{\partial^\mu u}{\partial x^\mu}\right)-\frac{1}{2}\frac{x^\mu}{\mu}u\frac{\partial^\mu u}{\partial x^\mu}+u^2=\left(\frac{x^\mu}{\mu}\right)^2e^{-\frac{t^\nu}{\nu}}, \tag{40}$$

subject to the following initial conditions

$$u\left(\frac{x^\mu}{\mu},0\right)=\left(\frac{x^\mu}{\mu}\right)^2,\quad \frac{\partial^\nu u\left(\frac{x^\mu}{\mu},0\right)}{\partial t^\nu}=-\left(\frac{x^\mu}{\mu}\right)^2. \tag{41}$$

The conformable double Laplace transform decomposition method leads to the following scheme

$$u_0\left(\frac{x^\mu}{\mu},\frac{t^\nu}{\nu}\right)=\left(\frac{x^\mu}{\mu}\right)^2e^{-\frac{t^\nu}{\nu}},$$

and

$$\begin{aligned} u_1\left(\frac{x^\mu}{\mu},\frac{t^\nu}{\nu}\right) &= -L_p^{-1}L_s^{-1}\left[\frac{1}{s^2}\int_0^p L_x^\mu L_t^\nu\left[\frac{\partial^\mu}{\partial x^\mu}\left(\frac{x^\mu}{\mu}\frac{\partial^\mu u_0}{\partial x^\mu}+\frac{x^\mu}{\mu}\frac{\partial^{\mu+\nu}u_0}{\partial x^\mu\partial t^\nu}\right)\right]dp\right] \\ &+L_p^{-1}L_s^{-1}\left[\frac{1}{s^2}\int_0^p L_x^\mu L_t^\nu\left[\frac{x^\mu}{\mu}u_0\frac{\partial^\mu u_0}{\partial x^\mu}-\frac{x^\mu}{\mu}u_0^2\right]dp\right] \\ &= -L_p^{-1}L_s^{-1}\left[\frac{1}{s^2}\int_0^p L_x^\mu L_t^\nu\left[4\left(\frac{x^\mu}{\mu}\right)e^{-\frac{t^\nu}{\nu}}-4\left(\frac{x^\mu}{\mu}\right)e^{-\frac{t^\nu}{\nu}}\right]dp\right] \\ &\quad -L_p^{-1}L_s^{-1}\left[\frac{1}{s^2}\int_0^p L_x^\mu L_t^\nu\left[\left(\frac{x^\mu}{\mu}\right)^5e^{-\frac{t^\nu}{\nu}}-\left(\frac{x^\mu}{\mu}\right)^5e^{-\frac{t^\nu}{\nu}}\right]dp\right] \\ &= 0, \end{aligned}$$

proceeding in a similar manner, we have

$$u_2\left(\frac{x^\mu}{\mu},\frac{t^\nu}{\nu}\right)=0,\ u_3\left(\frac{x^\mu}{\mu},\frac{t^\nu}{\nu}\right)=0,\ u_4\left(\frac{x^\mu}{\mu},\frac{t^\nu}{\nu}\right)=0,\dots$$

so that the solution $u\left(\frac{x^\mu}{\mu},\frac{t^\nu}{\nu}\right)$ *is given by*

$$\begin{aligned} u\left(\frac{x^\mu}{\mu},\frac{t^\nu}{\nu}\right) &= \sum_{n=0}^{\infty} u_n\left(\frac{x^\mu}{\mu},\frac{t^\nu}{\nu}\right)=u_0+u_1+u_2+\dots \\ &= \left(\frac{x^\mu}{\mu}\right)^2 e^{-\frac{t^\nu}{\nu}} \end{aligned}$$

and hence the conformable solution is given by

$$u\left(\frac{x^\mu}{\mu},\frac{t^\nu}{\nu}\right)=\left(\frac{x^\mu}{\mu}\right)^2 e^{-\frac{t^\nu}{\nu}}. \tag{42}$$

By substituting $\mu=1$ *and* $\nu=1$ *into Equation (42), the solution becomes*

$$\psi(x,t)=x^2e^{-t}.$$

Conformable double Laplace transform method and Singular conformable coupled pseudohyperbolic equation.

In this section, conformable double Laplace decomposition method is considered for the one-dimensional conformable derivatives coupled pseudohyperbolic equation since the method is much simpler and more efficient in the study of linear equations.

The thrid problem: Let us consider the conformable derivatives coupled pseudohyperbolic equations

$$\begin{aligned} \frac{\partial^{2\nu}u}{\partial t^{2\nu}}-\frac{\mu}{x^\mu}\frac{\partial^\mu}{\partial x^\mu}\left(\frac{x^\mu}{\mu}\frac{\partial^\mu}{\partial x^\mu}u\right)-\frac{\mu}{x^\mu}\frac{\partial^{\mu+\nu}}{\partial x^\mu\partial t^\nu}\left(\frac{x^\mu}{\mu}\frac{\partial^\mu}{\partial x^\mu}u\right)+\zeta v &= f\left(\frac{x^\mu}{\mu},\frac{t^\nu}{\nu}\right) \\ \frac{\partial^{2\nu}v}{\partial t^{2\nu}}-\frac{\mu}{x^\mu}\frac{\partial^\mu}{\partial x^\mu}\left(\frac{x^\mu}{\mu}\frac{\partial^\mu}{\partial x^\mu}v\right)-\frac{\mu}{x^\mu}\frac{\partial^{\mu+\nu}}{\partial x^\mu\partial t^\nu}\left(\frac{x^\mu}{\mu}\frac{\partial^\mu}{\partial x^\mu}v\right)+\zeta u &= g\left(\frac{x^\mu}{\mu},\frac{t^\nu}{\nu}\right), \end{aligned} \tag{43}$$

subject to

$$u\left(\frac{x^\mu}{\mu},0\right)=f_1\left(\frac{x^\mu}{\mu}\right),\ \frac{\partial^\nu u\left(\frac{x^\mu}{\mu},0\right)}{\partial t^\nu}=f_2\left(\frac{x^\mu}{\mu}\right) \text{ and } v\left(\frac{x^\mu}{\mu},0\right)=g_1(x),\ \frac{\partial^\nu u\left(\frac{x^\mu}{\mu},0\right)}{\partial t^\nu}=g_2\left(\frac{x^\mu}{\mu}\right) \tag{44}$$

where the linear terms $\frac{\mu}{x^\mu}\frac{\partial^\mu}{\partial x^\mu}\left(\frac{x^\mu}{\mu}\frac{\partial^\mu}{\partial x^\mu}\right)$ are the so-called conformable Bessel operators. Here, $f\left(\frac{x^\mu}{\mu},\frac{t^\nu}{\nu}\right)$, $g\left(\frac{x^\mu}{\mu},\frac{t^\nu}{\nu}\right)$, $f_1\left(\frac{x^\mu}{\mu}\right)$, $f_2\left(\frac{x^\mu}{\mu}\right)$, $g_1\left(\frac{x^\mu}{\mu}\right)$, and $g_2\left(\frac{x^\mu}{\mu}\right)$ are given functions, ζ is the coupling parameter. One can obtain the solution of Equation (43), by using the following steps.

(1): Multiply both sides of Equation (43) by $\frac{x^\mu}{\mu}$, we have

$$\begin{aligned} \frac{x^\mu}{\mu}\frac{\partial^{2\nu}u}{\partial t^{2\nu}}-\frac{\partial^\mu}{\partial x^\mu}\left(\frac{x^\mu}{\mu}\frac{\partial^\mu}{\partial x^\mu}u\right)-\frac{\partial^{\mu+\nu}}{\partial x^\mu\partial t^\nu}\left(\frac{x^\mu}{\mu}\frac{\partial^\mu}{\partial x^\mu}u\right)+\zeta\frac{x^\mu}{\mu}v &= \frac{x^\mu}{\mu}f\left(\frac{x^\mu}{\mu},\frac{t^\nu}{\nu}\right) \\ \frac{x^\mu}{\mu}\frac{\partial^{2\nu}v}{\partial t^{2\nu}}-\frac{\partial^\mu}{\partial x^\mu}\left(\frac{x^\mu}{\mu}\frac{\partial^\mu}{\partial x^\mu}v\right)-\frac{\partial^{\mu+\nu}}{\partial x^\mu\partial t^\nu}\left(\frac{x^\mu}{\mu}\frac{\partial^\mu}{\partial x^\mu}v\right)+\zeta\frac{x^\mu}{\mu}u &= \frac{x^\mu}{\mu}g\left(\frac{x^\mu}{\mu},\frac{t^\nu}{\nu}\right). \end{aligned} \tag{45}$$

(2): We apply conformable double Laplace transform on both sides of Equation (45) and single conformable Laplace transform for Equation (44), we get

$$\begin{aligned} L_x^{\mu} L_t^{\nu}\left[\frac{x^{\mu}}{\mu}\frac{\partial^{2\nu} u}{\partial t^{2\nu}}\right] &= L_x^{\mu} L_t^{\nu}\left[\frac{\partial^{\mu}}{\partial x^{\mu}}\left(\frac{x^{\mu}}{\mu}\frac{\partial^{\mu}}{\partial x^{\mu}}u\right)+\frac{\partial^{\mu+\nu}}{\partial x^{\mu}\partial t^{\nu}}\left(\frac{x^{\mu}}{\mu}\frac{\partial^{\mu}}{\partial x^{\mu}}u\right)-\zeta\frac{x^{\mu}}{\mu}v+\frac{x^{\mu}}{\mu}f\left(\frac{x^{\mu}}{\mu},\frac{t^{\nu}}{\nu}\right)\right], \\ L_x^{\mu} L_t^{\nu}\left[\frac{x^{\mu}}{\mu}\frac{\partial^{2\nu} v}{\partial t^{2\nu}}\right] &= L_x^{\mu} L_t^{\nu}\left[\frac{\partial^{\mu}}{\partial x^{\mu}}\left(\frac{x^{\mu}}{\mu}\frac{\partial^{\mu}}{\partial x^{\mu}}v\right)+\frac{\partial^{\mu+\nu}}{\partial x^{\mu}\partial t^{\nu}}\left(\frac{x^{\mu}}{\mu}\frac{\partial^{\mu}}{\partial x^{\mu}}v\right)-\zeta\frac{x^{\mu}}{\mu}u+\frac{x^{\mu}}{\mu}g\left(\frac{x^{\mu}}{\mu},\frac{t^{\nu}}{\nu}\right)\right], \end{aligned} \tag{46}$$

on using theorem 1 and theorem 2, we obtain

$$\begin{aligned} \frac{d}{dp}\left[U_{\mu,\nu}(p,s)\right] &= \frac{1}{s}\frac{d}{dp}F_1(p,0)+\frac{1}{s^2}\frac{d}{dp}F_2(p,0) \\ &\quad \frac{1}{s^2}L_x^{\mu} L_t^{\nu}\left[\frac{\partial^{\mu}}{\partial x^{\mu}}\left(\frac{x^{\mu}}{\mu}\frac{\partial^{\mu}}{\partial x^{\mu}}u\right)+\frac{\partial^{\mu+\nu}}{\partial x^{\mu}\partial t^{\nu}}\left(\frac{x^{\mu}}{\mu}\frac{\partial^{\mu}}{\partial x^{\mu}}u\right)-\zeta\frac{x^{\mu}}{\mu}v\right] \\ &\quad +\frac{1}{s^2}\frac{d}{dp}\left(L_x^{\mu} L_t^{\nu}\left[f\left(\frac{x^{\mu}}{\mu},\frac{t^{\nu}}{\nu}\right)\right]\right), \\ \frac{d}{dp}\left[U_{\mu,\nu}(p,s)\right] &= \frac{1}{s}\frac{d}{dp}G_1(p,0)+\frac{1}{s^2}\frac{d}{dp}G_2(p,0) \\ &\quad -\frac{1}{s^2}L_x^{\mu} L_t^{\nu}\left[\frac{\partial^{\mu}}{\partial x^{\mu}}\left(\frac{x^{\mu}}{\mu}\frac{\partial^{\mu}}{\partial x^{\mu}}v\right)+\frac{\partial^{\mu+\nu}}{\partial x^{\mu}\partial t^{\nu}}\left(\frac{x^{\mu}}{\mu}\frac{\partial^{\mu}}{\partial x^{\mu}}v\right)-\zeta\frac{x^{\mu}}{\mu}u\right] \\ &\quad +\frac{1}{s^2}\frac{d}{dp}\left(L_x^{\mu} L_t^{\nu}\left[g\left(\frac{x^{\mu}}{\mu},\frac{t^{\nu}}{\nu}\right)\right]\right). \end{aligned} \tag{47}$$

(3): By integrating both sides of Equation (47) from 0 to p with respect to p, we have

$$\begin{aligned} U_{\mu,\nu}(p,s) &= \frac{F_1(p,0)}{s}+\frac{F_2(p,0)}{s^2} \\ &\quad -\frac{1}{s^2}\int_0^p L_x^{\mu} L_t^{\nu}\left[\frac{\partial^{\mu}}{\partial x^{\mu}}\left(\frac{x^{\mu}}{\mu}\frac{\partial^{\mu}}{\partial x^{\mu}}u\right)+\frac{\partial^{\mu+\nu}}{\partial x^{\mu}\partial t^{\nu}}\left(\frac{x^{\mu}}{\mu}\frac{\partial^{\mu}}{\partial x^{\mu}}u\right)-\zeta\frac{x^{\mu}}{\mu}v\right]dp \\ &\quad +\frac{1}{s^2}\int_0^p\left(\frac{d}{dp}\left(L_x^{\mu} L_t^{\nu}\left[f\left(\frac{x^{\mu}}{\mu},\frac{t^{\nu}}{\nu}\right)\right]\right)\right)dp, \\ V_{\mu,\nu}(p,s) &= \frac{G_1(p,0)}{s}+\frac{G_2(p,0)}{s^2} \\ &\quad -\frac{1}{s^2}\int_0^p L_x^{\mu} L_t^{\nu}\left[\frac{\partial^{\mu}}{\partial x^{\mu}}\left(\frac{x^{\mu}}{\mu}\frac{\partial^{\mu}}{\partial x^{\mu}}v\right)+\frac{\partial^{\mu+\nu}}{\partial x^{\mu}\partial t^{\nu}}\left(\frac{x^{\mu}}{\mu}\frac{\partial^{\mu}}{\partial x^{\mu}}v\right)-\zeta\frac{x^{\mu}}{\mu}u\right]dp \\ &\quad +\frac{1}{s^2}\int_0^p\left(\frac{d}{dp}\left(L_x^{\mu} L_t^{\nu}\left[g\left(\frac{x^{\mu}}{\mu},\frac{t^{\nu}}{\nu}\right)\right]\right)\right)dp, \end{aligned} \tag{48}$$

where $F_1(p,0), F_2(p,0), G_1(p,0),$ and $G_2(p,0)$ are conformable Laplace transform of the functions $f_1\left(\frac{x^{\mu}}{\mu}\right), f_2\left(\frac{x^{\mu}}{\mu}\right), g_1\left(\frac{x^{\mu}}{\mu}\right)$, and $g_2\left(\frac{x^{\mu}}{\mu}\right)$, respectively. By applying double inverse Laplace transform for Equation (48), we have

$$\begin{aligned} u\left(\frac{x^{\mu}}{\mu},\frac{t^{\nu}}{\nu}\right) &= f_1\left(\frac{x^{\mu}}{\mu}\right)+\frac{t^{\nu}}{\nu}f_2\left(\frac{x^{\mu}}{\mu}\right)+L_p^{-1}L_s^{-1}\left[\frac{1}{s^2}\int_0^p\left(\frac{d}{dp}\left(L_x^{\mu} L_t^{\nu}\left[f\left(\frac{x^{\mu}}{\mu},\frac{t^{\nu}}{\nu}\right)\right]\right)\right)dp\right] \\ &\quad -L_p^{-1}L_s^{-1}\left[\frac{1}{s^2}\int_0^p L_x^{\mu} L_t^{\nu}\left[\frac{\partial^{\mu}}{\partial x^{\mu}}\left(\frac{x^{\mu}}{\mu}\frac{\partial^{\mu}}{\partial x^{\mu}}u\right)\right]dp\right] \\ &\quad -L_p^{-1}L_s^{-1}\left[\frac{1}{s^2}\int_0^p L_x^{\mu} L_t^{\nu}\left[\frac{\partial^{\mu+\nu}}{\partial x^{\mu}\partial t^{\nu}}\left(\frac{x^{\mu}}{\mu}\frac{\partial^{\mu}}{\partial x^{\mu}}u\right)-\zeta\frac{x^{\mu}}{\mu}v\right]dp\right] \end{aligned} \tag{49}$$

and

$$\begin{aligned} v\left(\frac{x^{\mu}}{\mu},\frac{t^{\nu}}{\nu}\right) &= g_1\left(\frac{x^{\mu}}{\mu}\right)+\frac{t^{\nu}}{\nu}g_2\left(\frac{x^{\mu}}{\mu}\right)+L_p^{-1}L_s^{-1}\left[\frac{1}{s^2}\int_0^p\left(\frac{d}{dp}\left(L_x^{\mu}L_t^{\nu}\left[g\left(\frac{x^{\mu}}{\mu},\frac{t^{\nu}}{\nu}\right)\right]\right)\right)dp\right] \\ & -L_p^{-1}L_s^{-1}\left[\frac{1}{s^2}\int_0^p L_x^{\mu}L_t^{\nu}\left[\frac{\partial^{\mu}}{\partial x^{\mu}}\left(\frac{x^{\mu}}{\mu}\frac{\partial^{\mu}}{\partial x^{\mu}}v\right)\right]dp\right] \\ & -L_p^{-1}L_s^{-1}\left[\frac{1}{s^2}\int_0^p L_x^{\mu}L_t^{\nu}\left[\frac{\partial^{\mu+\nu}}{\partial x^{\mu}\partial t^{\nu}}\left(\frac{x^{\mu}}{\mu}\frac{\partial^{\mu}}{\partial x^{\mu}}v\right)-\zeta\frac{x^{\mu}}{\mu}u\right]dp\right]. \end{aligned} \tag{50}$$

The conformable double Laplace decomposition methods represent the solutions of Equation (43), by the infinite series

$$u\left(\frac{x^{\mu}}{\mu},\frac{t^{\nu}}{\nu}\right)=\sum_{n=0}^{\infty}u_n\left(\frac{x^{\mu}}{\mu},\frac{t^{\nu}}{\nu}\right),\quad v\left(\frac{x^{\mu}}{\mu},\frac{t^{\nu}}{\nu}\right)=\sum_{n=0}^{\infty}v_n\left(\frac{x^{\mu}}{\mu},\frac{t^{\nu}}{\nu}\right). \tag{51}$$

By substituting Equation (51) into Equations (49) and (50), we get

$$\begin{aligned} \sum_{n=0}^{\infty}u_n\left(\frac{x^{\mu}}{\mu},\frac{t^{\nu}}{\nu}\right) &= f_1\left(\frac{x^{\mu}}{\mu}\right)+\frac{t^{\nu}}{\nu}f_2\left(\frac{x^{\mu}}{\mu}\right)+L_p^{-1}L_s^{-1}\left[\frac{1}{s^2}\int_0^p\left(\frac{d}{dp}\left(L_x^{\mu}L_t^{\nu}\left[f\left(\frac{x^{\mu}}{\mu},\frac{t^{\nu}}{\nu}\right)\right]\right)\right)dp\right] \\ & -L_p^{-1}L_s^{-1}\left[\frac{1}{s^2}\int_0^p\left(L_x^{\mu}L_t^{\nu}\left[\frac{\partial^{\mu}}{\partial x^{\mu}}\left(\frac{x^{\mu}}{\mu}\frac{\partial^{\mu}}{\partial x^{\mu}}\left(\sum_{n=0}^{\infty}u_n\right)\right)\right]\right)dp\right] \\ & -L_p^{-1}L_s^{-1}\left[\frac{1}{s^2}\int_0^p\left(L_x^{\mu}L_t^{\nu}\left[\frac{\partial^{\mu+\nu}}{\partial x^{\mu}\partial t^{\nu}}\left(\frac{x^{\mu}}{\mu}\frac{\partial^{\mu}}{\partial x^{\mu}}\left(\sum_{n=0}^{\infty}u_n\right)\right)\right]\right)dp\right] \\ & +L_p^{-1}L_s^{-1}\left[\frac{1}{s^2}L_xL_t\left[\zeta\frac{x^{\mu}}{\mu}\sum_{n=0}^{\infty}v_n\right]\right], \end{aligned} \tag{52}$$

and

$$\begin{aligned} \sum_{n=0}^{\infty}v_n\left(\frac{x^{\mu}}{\mu},\frac{t^{\nu}}{\nu}\right) &= g_1\left(\frac{x^{\mu}}{\mu}\right)+\frac{t^{\nu}}{\nu}g_2\left(\frac{x^{\mu}}{\mu}\right)+L_p^{-1}L_s^{-1}\left[\frac{1}{s^2}\int_0^p\left(\frac{d}{dp}\left(L_x^{\mu}L_t^{\nu}\left[g\left(\frac{x^{\mu}}{\mu},\frac{t^{\nu}}{\nu}\right)\right]\right)\right)dp\right] \\ & -L_p^{-1}L_s^{-1}\left[\frac{1}{s^2}\int_0^p\left(L_x^{\mu}L_t^{\nu}\left[\frac{\partial^{\mu}}{\partial x^{\mu}}\left(\frac{x^{\mu}}{\mu}\frac{\partial^{\mu}}{\partial x^{\mu}}\left(\sum_{n=0}^{\infty}v_n\right)\right)\right]\right)dp\right] \\ & -L_p^{-1}L_s^{-1}\left[\frac{1}{s^2}\int_0^p\left(L_x^{\mu}L_t^{\nu}\left[\frac{\partial^{\mu+\nu}}{\partial x^{\mu}\partial t^{\nu}}\left(\frac{x^{\mu}}{\mu}\frac{\partial^{\mu}}{\partial x^{\mu}}\left(\sum_{n=0}^{\infty}v_n\right)\right)\right]\right)dp\right] \\ & +L_p^{-1}L_s^{-1}\left[\frac{1}{s^2}L_xL_t\left[\zeta\frac{x^{\mu}}{\mu}\sum_{n=0}^{\infty}u_n\right]\right]. \end{aligned} \tag{53}$$

Our method suggests that the zeroth components u_0 and v_0 are identified by the initial conditions and from source terms as follows

$$\begin{aligned} u_0 &= f_1\left(\frac{x^{\mu}}{\mu}\right)+\frac{t^{\nu}}{\nu}f_2\left(\frac{x^{\mu}}{\mu}\right)+L_p^{-1}L_s^{-1}\left[\frac{1}{s}\int_0^p\left(\frac{d}{dp}\left(L_x^{\mu}L_t^{\nu}\left[f\left(\frac{x^{\mu}}{\mu},\frac{t^{\nu}}{\nu}\right)\right]\right)\right)dp\right], \\ v_0 &= g_1\left(\frac{x^{\mu}}{\mu}\right)+\frac{t^{\nu}}{\nu}g_2\left(\frac{x^{\mu}}{\mu}\right)+L_p^{-1}L_s^{-1}\left[\frac{1}{s}\int_0^p\left(\frac{d}{dp}\left(L_x^{\mu}L_t^{\nu}\left[g\left(\frac{x^{\mu}}{\mu},\frac{t^{\nu}}{\nu}\right)\right]\right)\right)dp\right]. \end{aligned} \tag{54}$$

The remaining terms are given by

$$\begin{aligned} u_{k+1} = & -L_p^{-1}L_s^{-1}\left[\frac{1}{s^2}\int_0^p\left(L_x^{\mu}L_t^{\nu}\left[\frac{\partial^{\mu}}{\partial x^{\mu}}\left(\frac{x^{\mu}}{\mu}\frac{\partial^{\mu}}{\partial x^{\mu}}u_k\right)\right]\right)dp\right] \\ & -L_p^{-1}L_s^{-1}\left[\frac{1}{s^2}\int_0^p\left(L_x^{\mu}L_t^{\nu}\left[\frac{\partial^{\mu+\nu}}{\partial x^{\mu}\partial t^{\nu}}\left(\frac{x^{\mu}}{\mu}\frac{\partial^{\mu}}{\partial x^{\mu}}u_k\right)\right]\right)dp\right] \\ & +L_p^{-1}L_s^{-1}\left[\frac{1}{s^2}L_xL_t\left[\zeta\,\frac{x^{\mu}}{\mu}v_k\right]\right], \end{aligned} \tag{55}$$

and

$$\begin{aligned} v_{k+1} = & -L_p^{-1}L_s^{-1}\left[\frac{1}{s^2}\int_0^p\left(L_x^{\mu}L_t^{\nu}\left[\frac{\partial^{\mu}}{\partial x^{\mu}}\left(\frac{x^{\mu}}{\mu}\frac{\partial^{\mu}}{\partial x^{\mu}}v_k\right)\right]\right)dp\right] \\ & -L_p^{-1}L_s^{-1}\left[\frac{1}{s^2}\int_0^p\left(L_x^{\mu}L_t^{\nu}\left[\frac{\partial^{\mu+\nu}}{\partial x^{\mu}\partial t^{\nu}}\left(\frac{x^{\mu}}{\mu}\frac{\partial^{\mu}}{\partial x^{\mu}}v_k\right)\right]\right)dp\right] \\ & +L_p^{-1}L_s^{-1}\left[\frac{1}{s^2}L_xL_t\left[\zeta\,\frac{x^{\mu}}{\mu}u_k\right]\right]. \end{aligned} \tag{56}$$

Here, we assume that the double inverse Laplace transform with respect to p and s exists for each term in the right hand side of the above equations.

To illustrate our method for solving the conformable derivatives coupled pseudohyperbolic equations, we will consider the following example:

Example 5. *Consider the following homogeneous form of a conformable derivatives coupled pseudohyperbolic equation*

$$\begin{aligned} \frac{\partial^{2\nu}v}{\partial t^{2\nu}}-\frac{\mu}{x^{\mu}}\frac{\partial^{\mu}}{\partial x^{\mu}}\left(\frac{x^{\mu}}{\mu}\frac{\partial^{\mu}}{\partial x^{\mu}}u\right)-\frac{\mu}{x^{\mu}}\frac{\partial^{\mu+\nu}}{\partial x^{\mu}\partial t^{\nu}}\left(\frac{x^{\mu}}{\mu}\frac{\partial^{\mu}}{\partial x^{\mu}}u\right)-v &= 0 \\ \frac{\partial^{2\nu}v}{\partial t^{2\nu}}-\frac{\mu}{x^{\mu}}\frac{\partial^{\mu}}{\partial x^{\mu}}\left(\frac{x^{\mu}}{\mu}\frac{\partial^{\mu}}{\partial x^{\mu}}v\right)-\frac{\mu}{x^{\mu}}\frac{\partial^{\mu+\nu}}{\partial x^{\mu}\partial t^{\nu}}\left(\frac{x^{\mu}}{\mu}\frac{\partial^{\mu}}{\partial x^{\mu}}v\right)-u &= 0, \end{aligned} \tag{57}$$

with initial condition

$$\begin{aligned} u\left(\frac{x^{\mu}}{\mu},0\right) &= \left(\frac{x^{\mu}}{\mu}\right)^2, \quad \frac{\partial^{\nu}u\left(\frac{x^{\mu}}{\mu},0\right)}{\partial t^{\nu}}=-\left(\frac{x^{\mu}}{\mu}\right)^2 \\ v\left(\frac{x^{\mu}}{\mu},0\right) &= \left(\frac{x^{\mu}}{\mu}\right)^2, \quad \frac{\partial^{\nu}v\left(\frac{x^{\mu}}{\mu},0\right)}{\partial t^{\nu}}=-\left(\frac{x^{\mu}}{\mu}\right)^2. \end{aligned} \tag{58}$$

By applying above method for Equations (57) and (58), we obtain

$$\begin{aligned} \sum_{n=0}^{\infty}u_n\left(\frac{x^{\mu}}{\mu},\frac{t^{\nu}}{\nu}\right)=\left(\frac{x^{\mu}}{\mu}\right)^2 & -\left(\frac{x^{\mu}}{\mu}\right)^2\frac{t^{\nu}}{\nu} \\ & -L_p^{-1}L_s^{-1}\left[\frac{1}{s^2}\int_0^p\left(L_x^{\mu}L_t^{\nu}\left[\frac{\partial^{\mu}}{\partial x^{\mu}}\left(\frac{x^{\mu}}{\mu}\frac{\partial^{\mu}}{\partial x^{\mu}}\left(\sum_{n=0}^{\infty}u_n\right)\right)\right]\right)dp\right] \\ & -L_p^{-1}L_s^{-1}\left[\frac{1}{s^2}\int_0^p\left(L_x^{\mu}L_t^{\nu}\left[\frac{\partial^{\mu+\nu}}{\partial x^{\mu}\partial t^{\nu}}\left(\frac{x^{\mu}}{\mu}\frac{\partial^{\mu}}{\partial x^{\mu}}\left(\sum_{n=0}^{\infty}u_n\right)\right)\right]\right)dp\right] \\ & -L_p^{-1}L_s^{-1}\left[\frac{1}{s^2}L_xL_t\left[\frac{x^{\mu}}{\mu}\sum_{n=0}^{\infty}v_n\right]\right], \end{aligned} \tag{59}$$

and

$$\begin{aligned}\sum_{n=0}^{\infty} v_n\left(\frac{x^\mu}{\mu},\frac{t^\nu}{\nu}\right) &= \left(\frac{x^\mu}{\mu}\right)^2-\left(\frac{x^\mu}{\mu}\right)^2\frac{t^\nu}{\nu}\\
&-L_p^{-1}L_s^{-1}\left[\frac{1}{s^2}\int_0^p\left(L_x^\mu L_t^\nu\left[\frac{\partial^\mu}{\partial x^\mu}\left(\frac{x^\mu}{\mu}\frac{\partial^\mu}{\partial x^\mu}\left(\sum_{n=0}^{\infty}v_n\right)\right)\right]\right)dp\right]\\
&-L_p^{-1}L_s^{-1}\left[\frac{1}{s^2}\int_0^p\left(L_x^\mu L_t^\nu\left[\frac{\partial^{\mu+\nu}}{\partial x^\mu\partial t^\nu}\left(\frac{x^\mu}{\mu}\frac{\partial^\mu}{\partial x^\mu}\left(\sum_{n=0}^{\infty}v_n\right)\right)\right]\right)dp\right]\\
&-L_p^{-1}L_s^{-1}\left[\frac{1}{s^2}L_xL_t\left[\zeta\,\frac{x^\mu}{\mu}\sum_{n=0}^{\infty}u_n\right]\right].\end{aligned} \tag{60}$$

By applying equations Equations (54)–(56), we have

$$\begin{aligned}
u_0 &= \left(\frac{x^\mu}{\mu}\right)^2-\left(\frac{x^\mu}{\mu}\right)^2\frac{t^\nu}{\nu},\quad v_0=\left(\frac{x^\mu}{\mu}\right)^2-\left(\frac{x^\mu}{\mu}\right)^2\frac{t^\nu}{\nu}\\
u_1 &= -L_p^{-1}L_s^{-1}\left[\frac{1}{s^2}L_xL_t\left[\frac{\partial^\mu}{\partial x^\mu}\left(\frac{x^\mu}{\mu}\frac{\partial^\mu}{\partial x^\mu}u_0\right)+\frac{\partial^{\mu+\nu}}{\partial x^\mu\partial t^\nu}\left(\frac{x^\mu}{\mu}\frac{\partial^\mu}{\partial x^\mu}u_0\right)+\frac{x^\mu}{\mu}v_0\right]\right]\\
&= -\frac{2}{3}\left(\frac{t^\nu}{\nu}\right)^3+\frac{1}{2}\left(\frac{x^\mu}{\mu}\right)^2\left(\frac{t^\nu}{\nu}\right)^2-\frac{1}{6}\left(\frac{x^\mu}{\mu}\right)^2\left(\frac{t^\nu}{\nu}\right)^3,\\
v_1 &= -L_p^{-1}L_s^{-1}\left[\frac{1}{s^2}L_xL_t\left[\frac{\partial^\mu}{\partial x^\mu}\left(\frac{x^\mu}{\mu}\frac{\partial^\mu}{\partial x^\mu}v_0\right)+\frac{\partial^{\mu+\nu}}{\partial x^\mu\partial t^\nu}\left(\frac{x^\mu}{\mu}\frac{\partial^\mu}{\partial x^\mu}v_0\right)+\frac{x^\mu}{\mu}u_0\right]\right]\\
&= -\frac{2}{3}\left(\frac{t^\nu}{\nu}\right)^3+\frac{1}{2}\left(\frac{x^\mu}{\mu}\right)^2\left(\frac{t^\nu}{\nu}\right)^2-\frac{1}{6}\left(\frac{x^\mu}{\mu}\right)^2\left(\frac{t^\nu}{\nu}\right)^3,\\
u_2 &= -L_p^{-1}L_s^{-1}\left[\frac{1}{s^2}L_xL_t\left[\frac{\partial^\mu}{\partial x^\mu}\left(\frac{x^\mu}{\mu}\frac{\partial^\mu}{\partial x^\mu}u_1\right)+\frac{\partial^{\mu+\nu}}{\partial x^\mu\partial t^\nu}\left(\frac{x^\mu}{\mu}\frac{\partial^\mu}{\partial x^\mu}u_1\right)+\frac{x^\mu}{\mu}v_1\right]\right]\\
&= \frac{2}{3}\left(\frac{t^\nu}{\nu}\right)^3+\frac{1}{24}\left(\frac{x^\mu}{\mu}\right)^2\left(\frac{t^\nu}{\nu}\right)^4-\frac{1}{120}\left(\frac{x^\mu}{\mu}\right)^2\left(\frac{t^\nu}{\nu}\right)^5,\\
v_2 &= -L_p^{-1}L_s^{-1}\left[\frac{1}{s^2}L_xL_t\left[\frac{\partial^\mu}{\partial x^\mu}\left(\frac{x^\mu}{\mu}\frac{\partial^\mu}{\partial x^\mu}v_1\right)+\frac{\partial^{\mu+\nu}}{\partial x^\mu\partial t^\nu}\left(\frac{x^\mu}{\mu}\frac{\partial^\mu}{\partial x^\mu}v_1\right)+\frac{x^\mu}{\mu}u_1\right]\right]\\
&= \frac{2}{3}\left(\frac{t^\nu}{\nu}\right)^3+\frac{1}{24}\left(\frac{x^\mu}{\mu}\right)^2\left(\frac{t^\nu}{\nu}\right)^4-\frac{1}{120}\left(\frac{x^\mu}{\mu}\right)^2\left(\frac{t^\nu}{\nu}\right)^5,
\end{aligned}$$

and

$$\begin{aligned}
u_3 &= -L_p^{-1}L_s^{-1}\left[\frac{1}{s^2}L_xL_t\left[\frac{\partial^\mu}{\partial x^\mu}\left(\frac{x^\mu}{\mu}\frac{\partial^\mu}{\partial x^\mu}u_2\right)+\frac{\partial^{\mu+\nu}}{\partial x^\mu\partial t^\nu}\left(\frac{x^\mu}{\mu}\frac{\partial^\mu}{\partial x^\mu}u_2\right)+\frac{x^\mu}{\mu}v_2\right]\right]\\
&= \frac{1}{15}\left(\frac{t^\nu}{\nu}\right)^5-\frac{1}{1260}\left(\frac{t^\nu}{\nu}\right)^7+\frac{1}{720}\left(\frac{x^\mu}{\mu}\right)^2\left(\frac{t^\nu}{\nu}\right)^6-\frac{1}{5040}\left(\frac{x^\mu}{\mu}\right)^2\left(\frac{t^\nu}{\nu}\right)^7,\\
v_3 &= -L_p^{-1}L_s^{-1}\left[\frac{1}{s}L_xL_t\left[\frac{\partial^\mu}{\partial x^\mu}\left(\frac{x^\mu}{\mu}\frac{\partial^\mu}{\partial x^\mu}v_2\right)+\frac{\partial^{\mu+\nu}}{\partial x^\mu\partial t^\nu}\left(\frac{x^\mu}{\mu}\frac{\partial^\mu}{\partial x^\mu}v_2\right)+\frac{x^\mu}{\mu}u_2\right]\right]\\
&= \frac{1}{15}\left(\frac{t^\nu}{\nu}\right)^5-\frac{1}{1260}\left(\frac{t^\nu}{\nu}\right)^7+\frac{1}{720}\left(\frac{x^\mu}{\mu}\right)^2\left(\frac{t^\nu}{\nu}\right)^6-\frac{1}{5040}\left(\frac{x^\mu}{\mu}\right)^2\left(\frac{t^\nu}{\nu}\right)^7,
\end{aligned}$$

and so on for other components. Using Equation (51), our required solutions are given below

$$u\left(\frac{x^\mu}{\mu},\frac{t^\nu}{\nu}\right) = u_0+u_1+u_2+u_3+\ldots = \left(1-\frac{t^\nu}{\nu}+\frac{\left(\frac{t^\nu}{\nu}\right)^2}{2!}-\frac{\left(\frac{t^\nu}{\nu}\right)^3}{3!}+\frac{\left(\frac{t^\nu}{\nu}\right)^4}{4!}-\ldots\right)\left(\frac{x^\mu}{\mu}\right)^2$$

$$v\left(\frac{x^\mu}{\mu},\frac{t^\nu}{\nu}\right) = v_0+v_1+v_2+v_3+\ldots = \left(1-\frac{t^\nu}{\nu}+\frac{\left(\frac{t^\nu}{\nu}\right)^2}{2!}-\frac{\left(\frac{t^\nu}{\nu}\right)^3}{3!}+\frac{\left(\frac{t^\nu}{\nu}\right)^4}{4!}-\ldots\right)\left(\frac{x^\mu}{\mu}\right)^2$$

and hence the exact solution becomes

$$u\left(\frac{x^\mu}{\mu},\frac{t^\nu}{\nu}\right)=\left(\frac{x^\mu}{\mu}\right)^2 e^{-\frac{t^\nu}{\nu}},\quad v\left(\frac{x^\mu}{\mu},\frac{t^\nu}{\nu}\right)=\left(\frac{x^\mu}{\mu}\right)^2 e^{-\frac{t^\nu}{\nu}}.$$

By taking $\mu=1$ and $\nu=1$, the conformable solution becomes

$$u(x,t)=x^2e^{-t},\quad v(x,t)=x^2e^{-t}.$$

4. Numerical Result

In this section, we shall illustrate the accuracy and effciency of the double conformable Laplace transform method by numerical results of $u(x,t)$ for the exact solution when $\mu=\nu=1$, and approximate solutions when μ and ν taken different fractional values in Equations (20) and (40), which are depicted through Figures 1–4, respectively.

The three dimensional surface in Figure 1 shows the exact solution of Equation (20) in standard form of singular pseudohyperbolic equation at $\mu=\nu=1$. Figure 2 compares the approximate solutions of Equation (20) when $t=\frac{\pi}{2}$. In Figure 2a, the numerical solution at $0<\mu=\nu\leq 1$, in this case $u(x,t)$, increases hastily at fractional derivative decrease, Figure 2b shows the solution at $\mu=0.99$ and $\nu=0.95, 0.90, 0.85$ and we see $u(x,t)$ increasing regularly when ν decreases, and in Figure 2c we can observe $u(x,t)$ increasing slowly at $\mu=0.95, 0.90, 0.85$ and $\nu=0.99$ when μ decreases.

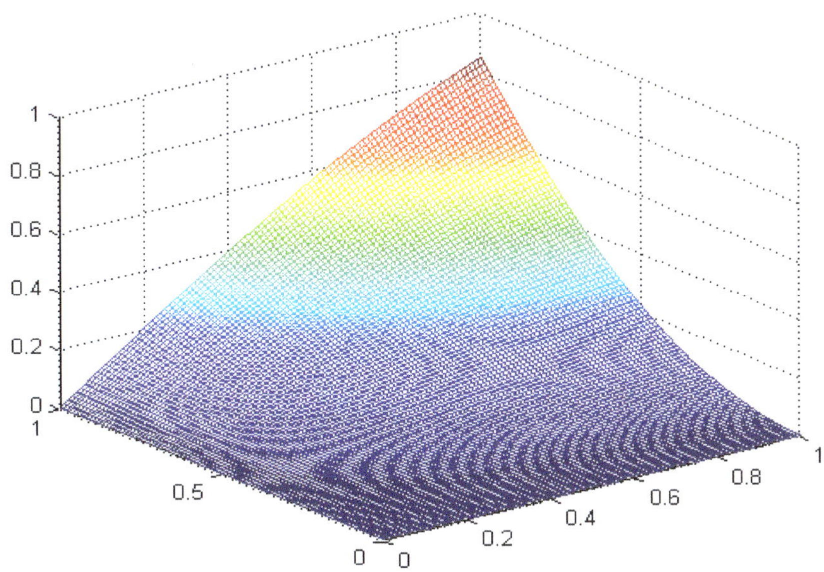

Figure 1. The Exact Solutions $u(x,t)$ for Equation (20) when $\mu=\nu=1$.

Similarly, the exact solution and approximate solution of Equation (40) were demonstrated in Figures 3 and 4 when $t = 1$. In the case $\nu = \mu = 1$, we get the exact solution of a singular pseudohyperbolic equation, as seen in Figure 3. Figure 4 shows the approximate solution of Equation (40) with different values of μ and ν. Figure 4a gives plots of the behavior of Equation (40) when $0 < \mu = \nu \leq 1$, in this case the function $u(x,t)$ increases quickly, and in Figure 4b we have obtained the solution for the values of $\mu = 0.99$ and different values of $0 < \nu \leq 1$, in this case the function $u(x,t)$ increases gradually, and Figure 4c gives the behavior of Equation (40) at $\nu = 0.99$ and different values of μ, in this case the function $u(x,t)$ increasing tardily.

It is clear from the solutions of Equations (20) and (40) that the conformable double Laplace decomposition method has good agreement with the exact solutions of the problems. The fractional-order solution of these two problems and exact solution of integer order problems are equal at $0 < \mu = \nu \leq 1$, in this case we have no error.

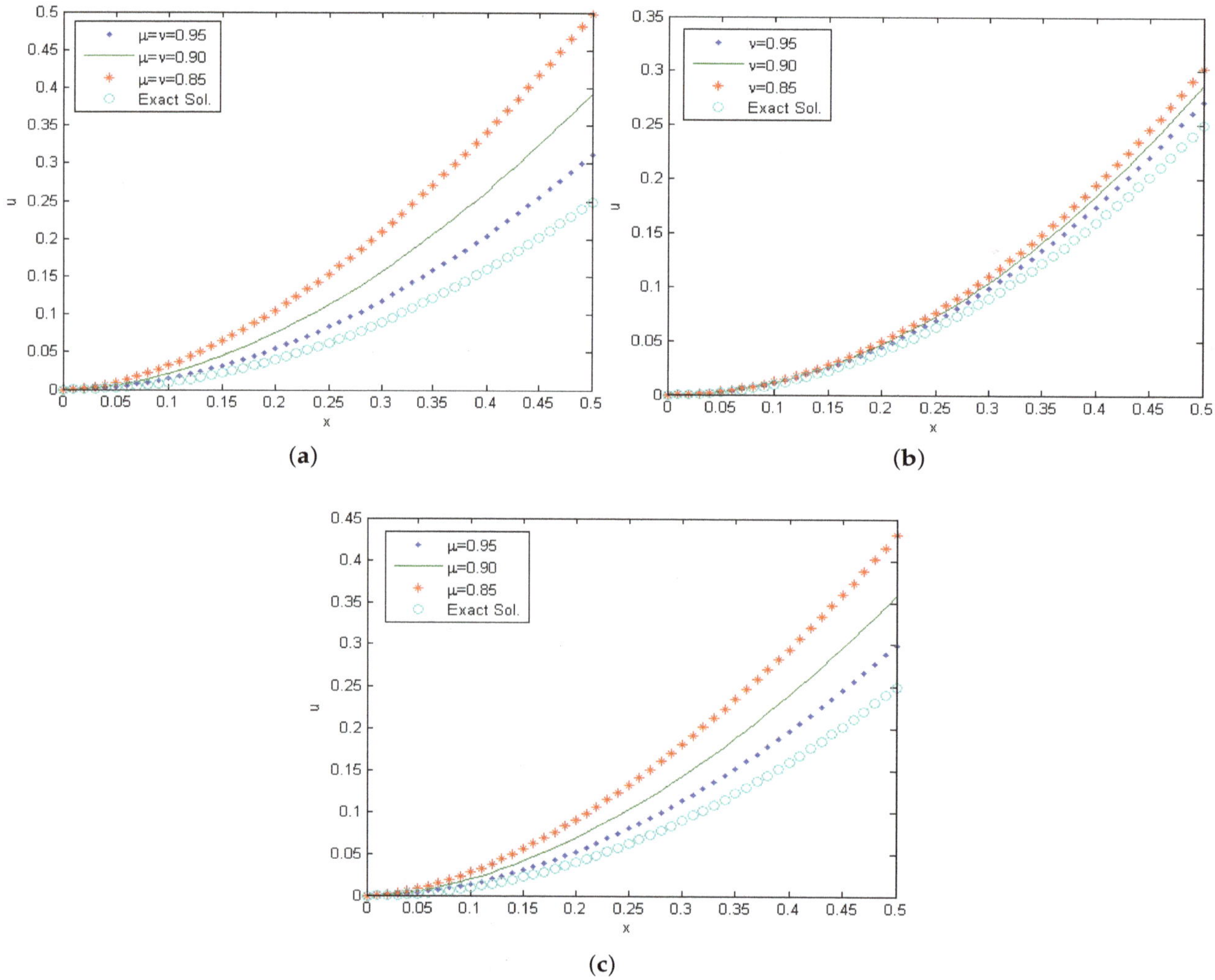

Figure 2. The solutions $u(x,t)$ for Equation (20) for different values of μ and ν when $t = \frac{\pi}{2}$. (**a**) Plot solutions $u(x,t)$ for Equation (20) at $\mu = \nu$. (**b**) Plot solutions $u(x,t)$ for Equation (20) when $\mu = 0.99$ and different values of ν. (**c**) Plot solutions $u(x,t)$ for Equation (20) for different values of μ at $\nu = 0.99$.

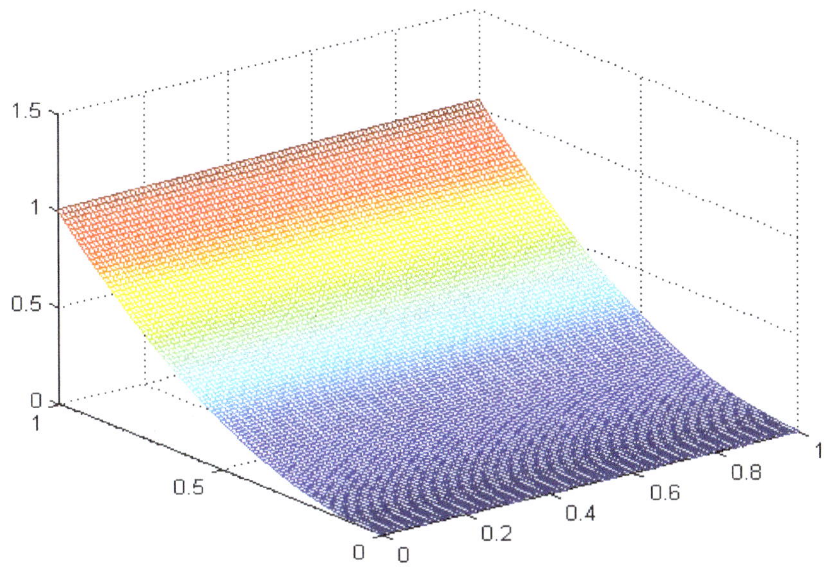

Figure 3. The Exact Solutions $u(x,t)$ for Equation (40) when $\mu = \nu = 1$.

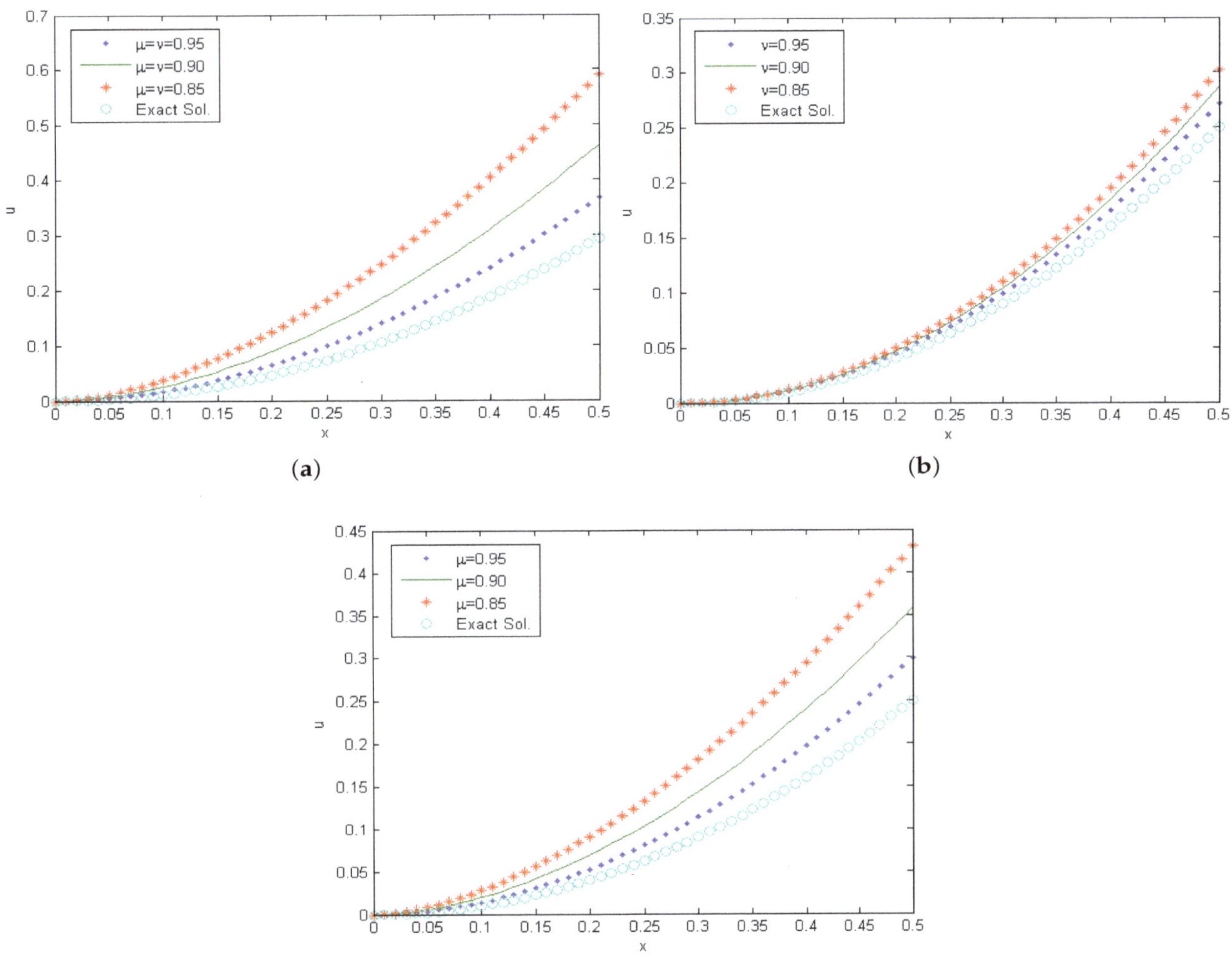

Figure 4. The solutions $u(x,t)$ for Equation (40) for different values of μ and ν when $t = 1$. (**a**) Plot solutions $u(x,t)$ for Equation (40) at $\mu = \nu$. (**b**) Plot solutions $u(x,t)$ for Equation (40) when $\mu = 0.99$ and different values of ν. (**c**) Plot solutions $u(x,t)$ for Equation (40) for different values of μ at $\nu = 0.99$.

5. Conclusions

In the present work we have studied singular linear and nonlinear pseudohyperbolic equations by employing the conformable double Laplace transform decomposition method (CDLDM), and we obtain analytic solutions when $\mu = \nu = 1$ and numerical solutions for different fractional values. Further, we also studied singular coupled pseudohyperbolic equations. It is clear that the solutions of Equations (20) and (40) were obtained as infinite series by using the conformable double Laplace decomposition method and they are in good agreement with the exact solutions of the problems. We have provided three different examples in order to demonstrate the efficiency, high accuracy, and the simplicity of the present method. Further, we plot the exact solutions, as well as the numerical solutions, in Figures 1–4, and we can easily see the efficieny of and agreement among the solutions.

Author Contributions: Conceptualization, H.E. and Y.T.; Data curation, H.E.; Formal analysis, H.E. and Y.T.; Methodology, H.E.; Supervision, H.E.; Validation, S.M.; Visualization, H.E. and A.K.; Writing, H.E. draft preparation, H.E.; Writing review and editing, A.K.

References

1. Abdelhakim, A.A. The flaw in the conformable calculus: It is conformable because it is not fractional. *Fract. Calcul. Appl. Anal.* **2019**, *22*, 242–254. [CrossRef]
2. Khalil, R.; Al Horani, M.; Yousef, A.; Sababheh, M. A new definition of fractional derivative. *J. Comput. Appl. Math.* **2014**, *264*, 65–70. [CrossRef]
3. Çenesiz, Y.; Baleanu, D.; Kurt, A.; Tasbozan, O. New exact solutions of burgers' type equations with conformable derivative. *Waves Random Complex Media* **2017**, *27*, 103–116. [CrossRef]
4. Korkmaz, A.; Hosseini, K. Exact solutions of a nonlinear conformable time-fractional parabolic equation with exponential nonlinearity using reliable methods. *Opt. Quantum Electron.* **2017**, *49*, 278. [CrossRef]
5. Ortigueira, M.D.; Machado, J.T. What is a fractional derivative? *J. Comput. Phys.* **2015**, *293*, 4–13. [CrossRef]
6. Aminikhah, H.; Sheikhani, A.R.; Rezazadeh, H. Sub-equation method for the fractional regularized long-wave equations with conformable fractional derivatives. *Sci. Iranica Trans. B Mech. Eng.* **2016**, *23*, 1048. [CrossRef]
7. Chung, W.S. Fractional newton mechanics with conformable fractional derivative. *J. Comput. Appl. Math.* **2015**, *290*, 150–158. [CrossRef]
8. Eslami, M.; Rezazadeh, H. The first integral method for wu–zhang system with conformable time-fractional derivative. *Calcolo* **2016**, *53*, 475–485. [CrossRef]
9. Ünal, E.; Gökdoğan, A. Solution of conformable fractional ordinary differential equations via differential transform method. *Optik* **2017**, *128*, 264–273. [CrossRef]
10. Rahimi, Z.; Rezazadeh, G.; Sumelka, W.; Yang, X. A study of critical point instability of micro and nano beams under a distributed variable-pressure force in the framework of the inhomogeneous non-linear nonlocal theory. *Arch. Mech.* **2017**, *69*, 413–433.
11. Rahimi, Z.; Sumelka, W.; Yang, X.-J. A new fractional nonlocal model and its application in free vibration of timoshenko and euler-bernoulli beams. *Eur. Phys. J. Plus* **2017**, *132*, 479. [CrossRef]
12. Tallafha, A.; Al Hihi, S. Total and directional fractional derivatives. *Int. J. Pure Appl. Math.* **2016**, *107*, 1037–1051. [CrossRef]
13. Hashemi, M. Invariant subspaces admitted by fractional differential equations with conformable derivatives. *Chaos Solitons Fractals* **2018**, *107*, 161–169. [CrossRef]
14. Özkan, O.; Kurt, A. On conformable double laplace transform. *Opt. Quantum Electron.* **2018**, *50*, 103. [CrossRef]
15. Jarad, F.; Abdeljawad, T. A modified laplace transform for certain generalized fractional operators. *Results Nonlinear Anal.* **2018**, *1*, 88–98.
16. Eltayeb, H.; Mesloub, S.; Kılıçman, A. Application of double laplace decomposition method to solve a singular one-dimensional pseudohyperbolic equation. *Adv. Mech. Eng.* **2017**, *9*, 1687814017716638. [CrossRef]
17. Eroğlu, B.; Avcı, D.; Özdemir, N. Optimal control problem for a conformable fractional heat conduction equation. *Acta Phys. Pol. A* **2017**, *132*, 658–662. [CrossRef]

10

Positive Solutions for a System of Fractional Integral Boundary Value Problems of Riemann–Liouville type Involving Semipositone Nonlinearities

Youzheng Ding [1], Jiafa Xu [2] and Zhengqing Fu [3,*]

1 School of Science, Shandong Jianzhu University, Jinan 250101, China; dingyz@sdjzu.edu.cn
2 School of Mathematical Sciences, Chongqing Normal University, Chongqing 401331, China; 20150028@cqnu.edu.cn
3 College of Mathematics and System Sciences, Shandong University of Science and Technology, Qingdao 266590, China
* Correspondence: skd992050@sdust.edu.cn

Abstract: In this work by the index of fixed point and matrix theory, we discuss the positive solutions for the system of Riemann–Liouville type fractional boundary value problems

$$D^{\alpha}_{0+}u(t) + f_1(t, u(t), v(t), w(t)) = 0, t \in (0,1),$$

$$D^{\alpha}_{0+}v(t) + f_2(t, u(t), v(t), w(t)) = 0, t \in (0,1),$$

$$D^{\alpha}_{0+}w(t) + f_3(t, u(t), v(t), w(t)) = 0, t \in (0,1),$$

$$u(0) = u'(0) = \cdots = u^{(n-2)}(0) = 0, D^{p}_{0+}u(t)|_{t=1} = \int_0^1 h(t)D^{q}_{0+}u(t)dt,$$

$$v(0) = v'(0) = \cdots = v^{(n-2)}(0) = 0, D^{p}_{0+}v(t)|_{t=1} = \int_0^1 h(t)D^{q}_{0+}v(t)dt,$$

$$w(0) = w'(0) = \cdots = w^{(n-2)}(0) = 0, D^{p}_{0+}w(t)|_{t=1} = \int_0^1 h(t)D^{q}_{0+}w(t)dt,$$

where $\alpha \in (n-1, n]$ with $n \in \mathbb{N}$, $n \geq 3$, $p, q \in \mathbb{R}$ with $p \in [1, n-2]$, $q \in [0, p]$, D^{α}_{0+} is the α order Riemann–Liouville type fractional derivative, and $f_i (i = 1,2,3) \in C([0,1] \times \mathbb{R}^+ \times \mathbb{R}^+ \times \mathbb{R}^+, \mathbb{R})$ are semipositone nonlinearities.

Keywords: Riemann–Liouville type fractional problem; positive solutions; the index of fixed point; matrix theory

---

1. Introduction

In this work the positive solutions for the system of fractional boundary value problems involving Riemann–Liouville type are considered:

$$\begin{cases} D^{\alpha}_{0+}u(t) + f_1(t, u(t), v(t), w(t)) = 0, t \in (0,1), \\ D^{\alpha}_{0+}v(t) + f_2(t, u(t), v(t), w(t)) = 0, t \in (0,1), \\ D^{\alpha}_{0+}w(t) + f_3(t, u(t), v(t), w(t)) = 0, t \in (0,1), \\ u(0) = u'(0) = \cdots = u^{(n-2)}(0) = 0, D^{p}_{0+}u(t)|_{t=1} = \int_0^1 h(t)D^{q}_{0+}u(t)dt, \\ v(0) = v'(0) = \cdots = v^{(n-2)}(0) = 0, D^{p}_{0+}v(t)|_{t=1} = \int_0^1 h(t)D^{q}_{0+}v(t)dt, \\ w(0) = w'(0) = \cdots = w^{(n-2)}(0) = 0, D^{p}_{0+}w(t)|_{t=1} = \int_0^1 h(t)D^{q}_{0+}w(t)dt, \end{cases} \quad (1)$$

where D_{0+}^{α} is the α order Riemann–Liouville type fractional derivative, the constants α, p, q, n, and the functions h, $f_i (i = 1, 2, 3)$ satisfy the assumptions

(C0) $n \in \mathbb{N}, n \geq 3, \alpha \in (n-1, n], p \in [1, n-2], q \in [0, p]$,

(C1) there exists h with $h(t) \geq 0 (\not\equiv 0)$ on $[0, 1]$ such that $A := \frac{\Gamma(\alpha)}{\Gamma(\alpha-p)} - \frac{\Gamma(\alpha)}{\Gamma(\alpha-q)} \int_0^1 h(t) t^{\alpha-q-1} dt > 0$,

(C2) $f_i (i = 1, 2, 3) \in C([0, 1] \times \mathbb{R}^+ \times \mathbb{R}^+ \times \mathbb{R}^+, \mathbb{R})$, and there is a $M > 0$ such that

$$f_i(t, x_1, x_2, x_3) \geq -M, \text{ for } (t, x_1, x_2, x_3) \in [0, 1] \times \mathbb{R}^+ \times \mathbb{R}^+ \times \mathbb{R}^+, i = 1, 2, 3.$$

Fractional calculus theory shows undoubted advantages in aerodynamics, electrodynamics in complex medium, the theory of control, signal and image processing, rheology, and many other issues, see the books [1–3]. The study of such kind of problems has received considerable attention in the previous studies, see for instance [4–79] and the references therein.

In [4] by the fixed point theorem of Guo–Krasnosel'skii, the authors discussed the positive solutions for the multi-point Riemann–Liouville fractional boundary value problems

$$\begin{cases} D_{0+}^{\alpha} u(t) + \lambda f(t, u(t)) = 0, t \in (0, 1), \\ u(0) = u'(0) = \cdots = u^{(n-2)}(0) = 0, \\ D_{0+}^{p} u(t)|_{t=1} = \sum_{i=1}^{m} a_i D_{0+}^{q} u(t)|_{t=\xi_i} \end{cases} \tag{2}$$

where f is a sign-changing nonlinearity. In [5], the authors studied the multiple positive solutions for the problem (2) ($\lambda = 1$), where f is a sign-changing nonlinearity, and permits singularities on t and u. In [6], by means of the index of fixed point, the authors researched the positive solutions for the boundary value problems of Hadamard fractional equations

$$\begin{cases} -{}^{H}D^{\alpha} u(t) = f(t, u(t)), \quad t \in [1, e], \\ u(1) = \delta u(1) = \delta u(e) = 0, \end{cases} \tag{3}$$

where f is a sign-changing nonlinearity, and may grow superlinearly and sublinearly at ∞.

The fractional-order equations in systems have also been widely investigated in the literature, see for example [52–79]. In [52], the authors studied the system of Hadamard fractional integral boundary value problems

$$\begin{cases} {}^{H}D^{\beta} u(t) + f_1(t, u(t), v(t)) = 0, \quad 1 < t < e, \\ {}^{H}D^{\beta} v(t) + f_2(t, u(t), v(t)) = 0, \quad 1 < t < e, \\ u(1) = v(1) = u'(1) = v'(1) = 0, \\ u(e) = \int_1^e h(s) v(s) \frac{ds}{s}, \\ v(e) = \int_1^e g(s) u(s) \frac{ds}{s}, \end{cases} \tag{4}$$

where the nonlinearities $f_i (i = 1, 2) \in C([1, e] \times \mathbb{R}^+ \times \mathbb{R}^+, \mathbb{R}^+)$.

In [53], by means of the alternative of Leray–Schauder, the authors obtained the uniqueness and existence of solutions for the system of fractional integral boundary value problems

$$\begin{cases} D^{\alpha} x(t) = f(t, x(t), y(t), D^{\gamma} y(t)), t \in [0, T], \\ D^{\beta} y(t) = g(t, x(t), D^{\delta} x(t), y(t)), t \in [0, T], \end{cases} \tag{5}$$

with the integral boundary conditions

$$\begin{cases} x(0) = h(y), \int_0^T y(s) ds = \mu_1 x(\eta), \\ y(0) = \phi(x), \int_0^T x(s) ds = \mu_2 y(\xi), \end{cases}$$

where $D^{\alpha}, D^{\beta}, D^{\delta}, D^{\gamma}$ are the fractional derivatives of Caputo type.

In [54], the authors studied the positive solutions of the abstract fractional semipositone differential system with integral boundary conditions, which arises from HIV infection models

$$\begin{cases} D_{0+}^{\alpha}u(t) + \lambda f\left(t, u(t), D_{0+}^{\beta}u(t), v(t)\right) = 0, \\ D_{0+}^{\gamma}v(t) + \lambda g(t, u(t)) = 0,\ 0 < t < 1, \\ D_{0+}^{\beta}u(0) = D_{0+}^{\beta+1}u(0) = 0,\ D_{0+}^{\beta}u(1) = \int_0^1 D_{0+}^{\beta}u(s)dA(s), \\ v(0) = v'(0) = 0,\ v(1) = \int_0^1 v(s)dB(s), \end{cases} \tag{6}$$

where f, g are the semipositone nonlinearities (so-called semipositone problems), which originally modeled nonlinear phenomena of chemical reactions by Dutch chemist Aris [80]. For some relevant work, we refer the reader to [4–7,71–75].

Motivated by the works aforementioned, in this work we use the index of fixed point and nonnegative matrix theory to study the positive solutions for the system of Riemann–Liouville type fractional boundary value problems (1). We first transform our problem into the equivalent system of Hammerstein type integral equations, and establish some nonnegative operator equations. Then, using some superlinear and sublinear conditions for our nonlinearities, we obtain two existence theorems. Finally, we offer two examples to explain our main theorems.

2. Preliminaries

Now, we offer the definition of the $\alpha(> 0)$ order Riemann–Liouville type fractional derivative, which is given by

$$D_{0+}^{\alpha}f(t) = \frac{1}{\Gamma(n-\alpha)}\left(\frac{\mathrm{d}}{\mathrm{d}t}\right)^n \int_0^t (t-s)^{n-\alpha-1}f(s)ds,$$

where $f : (0, +\infty) \to (-\infty, +\infty)$ is a continuous function, and $n = [\alpha] + 1$. For more materials, we refer to the books [1–3].

Lemma 1. *lSuppose that (C0)–(C1) hold. Let $f \in C[0, 1]$, then the problem*

$$\begin{cases} D_{0+}^{\alpha}u(t) + f(t) = 0, t \in (0, 1), \\ u(0) = u'(0) = \cdots = u^{(n-2)}(0) = 0, D_{0+}^{p}u(t)|_{t=1} = \int_0^1 h(t)D_{0+}^{q}u(t)dt, \end{cases} \tag{7}$$

has a solution, which can take the form

$$u(t) = \int_0^1 G(t, s)f(s)ds,$$

where

$$G(t, s) = g_1(t, s) + \frac{t^{\alpha-1}}{A}\int_0^1 h(t)g_2(t, s)dt,$$

and

$$g_1(t, s) = \frac{1}{\Gamma(\alpha)}\begin{cases} t^{\alpha-1}(1-s)^{\alpha-p-1} - (t-s)^{\alpha-1}, & 0 \le s \le t \le 1, \\ t^{\alpha-1}(1-s)^{\alpha-p-1}, & 0 \le t \le s \le 1, \end{cases} \tag{8}$$

$$g_2(t, s) = \frac{1}{\Gamma(\alpha-q)}\begin{cases} t^{\alpha-q-1}(1-s)^{\alpha-p-1} - (t-s)^{\alpha-q-1}, & 0 \le s \le t \le 1, \\ t^{\alpha-q-1}(1-s)^{\alpha-p-1}, & 0 \le t \le s \le 1. \end{cases} \tag{9}$$

Proof. Using similar arguments in ([4], [Lemma 1 and 2]), we have

$$u(t) = c_1t^{\alpha-1} + c_2t^{\alpha-2} + \cdots + c_nt^{\alpha-n} - \int_0^t \frac{(t-s)^{\alpha-1}}{\Gamma(\alpha)}f(s)ds,$$

where $c_i \in \mathbb{R}, i = 1, 2, ..., n$. Note that $u(0) = u'(0) = \cdots = u^{(n-2)}(0) = 0$, and thus $c_2 = \cdots = c_n = 0$. Consequently, we get

$$u(t) = c_1 t^{\alpha-1} - \int_0^t \frac{(t-s)^{\alpha-1}}{\Gamma(\alpha)} f(s)ds.$$

Therefore, we find

$$D_{0+}^p u(t) = c_1 \frac{\Gamma(\alpha)}{\Gamma(\alpha-p)} t^{\alpha-p-1} - I_{0+}^{\alpha-p} f(t),\ D_{0+}^q u(t) = c_1 \frac{\Gamma(\alpha)}{\Gamma(\alpha-q)} t^{\alpha-q-1} - I_{0+}^{\alpha-q} f(t).$$

Using the condition $D_{0+}^p u(t)|_{t=1} = \int_0^1 h(t) D_{0+}^q u(t)dt$, we have

$$c_1 \frac{\Gamma(\alpha)}{\Gamma(\alpha-p)} - \frac{1}{\Gamma(\alpha-p)} \int_0^1 (1-s)^{\alpha-p-1} f(s)ds = c_1 \frac{\Gamma(\alpha)}{\Gamma(\alpha-q)} \int_0^1 h(t) t^{\alpha-q-1} dt - \frac{1}{\Gamma(\alpha-q)} \int_0^1 h(t) \int_0^t (t-s)^{\alpha-q-1} f(s)dsdt.$$

Solving this equation, we obtain

$$c_1 = \frac{1}{A\Gamma(\alpha-p)} \int_0^1 (1-s)^{\alpha-p-1} f(s)ds - \frac{1}{A\Gamma(\alpha-q)} \int_0^1 h(t) \int_0^t (t-s)^{\alpha-q-1} f(s)dsdt.$$

As a result, we get

$$\begin{aligned}
u(t) &= \frac{1}{A\Gamma(\alpha-p)} \int_0^1 t^{\alpha-1}(1-s)^{\alpha-p-1} f(s)ds - \frac{t^{\alpha-1}}{A\Gamma(\alpha-q)} \int_0^1 h(t) \int_0^t (t-s)^{\alpha-q-1} f(s)dsdt - \int_0^t \frac{(t-s)^{\alpha-1}}{\Gamma(\alpha)} f(s)ds \\
&= \frac{1}{\Gamma(\alpha)} \int_0^1 t^{\alpha-1}(1-s)^{\alpha-p-1} f(s)ds - \int_0^t \frac{(t-s)^{\alpha-1}}{\Gamma(\alpha)} f(s)ds + \left[\frac{1}{A\Gamma(\alpha-p)} - \frac{1}{\Gamma(\alpha)}\right] \int_0^1 t^{\alpha-1}(1-s)^{\alpha-p-1} f(s)ds \\
&\quad - \frac{t^{\alpha-1}}{A\Gamma(\alpha-q)} \int_0^1 h(t) \int_0^t (t-s)^{\alpha-q-1} f(s)dsdt \\
&= \int_0^1 g_1(t,s) f(s)ds + \frac{t^{\alpha-1}}{A\Gamma(\alpha-q)} \left[\int_0^1 \int_0^1 h(t) t^{\alpha-q-1}(1-s)^{\alpha-p-1} f(s)dsdt - \int_0^1 h(t) \int_0^t (t-s)^{\alpha-q-1} f(s)dsdt\right] \\
&= \int_0^1 g_1(t,s) f(s)ds + \frac{t^{\alpha-1}}{A} \int_0^1 \int_0^1 h(t) g_2(t,s) dt f(s)ds \\
&= \int_0^1 G(t,s) f(s)ds.
\end{aligned}$$

□

Lemma 2. *(see ([4], [Lemma 3])). Suppose that (C0) holds. The functions $g_i (i = 1, 2)$ have the properties*
(i) $g_i \in C([0,1] \times [0,1], \mathbb{R}^+)$, and $g_i(t,s) > 0$ for $t, s \in (0,1)$, $i = 1, 2$,
(ii) $t^{\alpha-1}\widetilde{\varphi}(s) \le g_1(t,s) \le \widetilde{\varphi}(s)$ for all $t, s \in [0,1]$, where

$$\widetilde{\varphi}(s) = \frac{(1-s)^{\alpha-p-1}(1-(1-s)^p)}{\Gamma(\alpha)}, s \in [0,1],$$

(iii) $g_1(t,s) \le \frac{t^{\alpha-1}(1-s)^{\alpha-p-1}}{\Gamma(\alpha)}$, $t, s \in [0,1]$.

Lemma 3. *Suppose that (C0)–(C1) hold. The Green's function G has the properties*
(i) $G \in C([0,1] \times [0,1], \mathbb{R}^+)$, and $G(t,s) > 0$ for $t, s \in (0,1)$,
(ii) $t^{\alpha-1}\varphi(s) \le G(t,s) \le \varphi(s), \forall t, s \in [0,1]$, where

$$\varphi(s) = \widetilde{\varphi}(s) + \frac{1}{A} \int_0^1 h(t) g_2(t,s) dt, s \in [0,1],$$

(iii) $G(t,s) \le t^{\alpha-1}\left[\frac{(1-s)^{\alpha-p-1}}{\Gamma(\alpha)} + \frac{1}{A}\int_0^1 h(t) g_2(t,s)dt\right], \forall t, s \in [0,1]$.
This is a direct result of Lemma 2, so we omit its proof.

Lemma 4. *Let* $\kappa_1 = \int_0^1 t^{\alpha-1}\varphi(t)dt$, $\kappa_2 = \int_0^1 \varphi(t)dt$. *Then we have the following inequalities*

$$\kappa_1\varphi(s) \le \int_0^1 G(t,s)\varphi(t)dt \le \kappa_2\varphi(s),\ \forall s \in [0,1]. \tag{10}$$

From Lemma 3(ii), we easily obtain (10).

Next we will consider the problem

$$\begin{cases} D_{0+}^{\alpha}u(t) + \widetilde{f}(t,u(t)) = 0, t \in (0,1), \\ u(0) = u'(0) = \cdots = u^{(n-2)}(0) = 0, D_{0+}^{p}u(t)|_{t=1} = \int_0^1 h(t)D_{0+}^{q}u(t)dt, \end{cases} \tag{11}$$

where $\widetilde{f}$ satisfies the condition

(C2)′ $\widetilde{f} \in C([0,1] \times \mathbb{R}^+, \mathbb{R})$, and there is a $M > 0$ such that

$$\widetilde{f}(t,x_1) \ge -M, \text{ for } (t,x_1) \in [0,1] \times \mathbb{R}^+.$$

Lemma 5. *Suppose that (C0)–(C1) and (C2)′. Then the problem* (11) *is equivalent to*

$$u(t) = \int_0^1 G(t,s)\widetilde{f}(s,u(s))ds, \tag{12}$$

where G is defined in Lemma 1.

Now, we take care of the following auxiliary problem associated with (11):

$$\begin{cases} D_{0+}^{\alpha}u(t) + \widetilde{F}(t,u(t)-z(t)) = 0, t \in (0,1), \\ u(0) = u'(0) = \cdots = u^{(n-2)}(0) = 0, D_{0+}^{p}u(t)|_{t=1} = \int_0^1 h(t)D_{0+}^{q}u(t)dt, \end{cases} \tag{13}$$

where $\widetilde{F}(t,x_1) = \begin{cases} \widetilde{f}(t,x_1) + M, & t \in [0,1], x_1 \ge 0, \\ \widetilde{f}(t,0) + M, & t \in [0,1], x_1 < 0, \end{cases}$ *and* $z(t) = M\int_0^1 G(t,s)ds$, *for* $t \in [0,1]$. *Then* $\widetilde{F}$ *is nonnegative continuous on* $[0,1] \times \mathbb{R}^+$, *and from Lemma 5 we have* (13) *is equivalent to*

$$u(t) = \int_0^1 G(t,s)\widetilde{F}(s,u(s)-z(s))ds, \tag{14}$$

where G is as in Lemma 1.

Lemma 6. *(i) If* (11) *has a positive solution* u^*, *then* (13) *has a solution* $u^* + z$.

(ii) If u^* *is a solution for* (13), *and* $u^*(t) \ge z(t)$ *for* $t \in [0,1]$, *then* $u^* - z$ *is a positive solution for* (11).

Proof. Note that z satisfies the fractional boundary value problem

$$\begin{cases} D_{0+}^{\alpha}z(t) + M = 0, t \in (0,1), \\ z(0) = z'(0) = \cdots = z^{(n-2)}(0) = 0, D_{0+}^{p}z(t)|_{t=1} = \int_0^1 h(t)D_{0+}^{q}z(t)dt. \end{cases} \tag{15}$$

Substituting $u^* + z$ into (13), we have

$$D_{0+}^{\alpha}(u^*+z)(t) + \widetilde{F}(t,u^*(t)+z(t)-z(t)) = 0 \Longrightarrow D_{0+}^{\alpha}u^*(t) + D_{0+}^{\alpha}(z)(t) + \widetilde{f}(t,u^*(t)) + M = 0.$$

Using $D_{0+}^{\alpha}(z)(t) = -M$, we have $D_{0+}^{\alpha}u^*(t) + \widetilde{f}(t,u^*(t)) = 0$, and note that u^*, z satisfy the boundary conditions in (11), (15), we obtain Lemma 6(i) holds.

Next, substituting $u^* - z$ into (11), and using $D_{0+}^{\alpha}(z)(t) = -M$ we have

$$D_{0+}^{\alpha}(u^*-z)(t) + \widetilde{f}(t,u^*(t)-z(t)) = 0 \Longrightarrow D_{0+}^{\alpha}u^*(t) - D_{0+}^{\alpha}z(t) + \widetilde{f}(t,u^*(t)-z(t)) = 0,$$

and

$$D_{0+}^{\alpha}u^*(t)+\widetilde{F}(t,u^*(t)-z(t))=0.$$

Note that u^*,z satisfy the boundary conditions in (13), (15), we obtain Lemma 6(ii) holds.

Lemma 6 implies that we only need to seek the solution u^* for (13), which is greater than z, we can obtain the positive solution u^*-z for (11).

Let $E:=C[0,1]$, $\|u\|:=\max_{t\in[0,1]}|u(t)|$, $P:=\{u\in E:u(t)\geq 0,\forall t\in[0,1]\}$, $P_0=\{u\in P:u(t)\geq t^{\alpha-1}\|u\|,\forall t\in[0,1]\}$. Then $(E,\|\cdot\|)$ is a real Banach space, and P,P_0 are cones on E. Note that the relations between (13) and (14), we let an operator $T:P\to P$ as follows:

$$(Tu)(t)=\int_0^1 G(t,s)\widetilde{F}(s,u(s)-z(s))ds,\ \text{for } u\in P,t\in[0,1].$$

From the continuity of $G,\widetilde{F}$ we obtain $T:P\to P$ is a completely continuous operator, and if there exists $\overline{u}\in P\backslash\{0\}$ such that $T\overline{u}=\overline{u}$, then this $\overline{u}$ is a positive solution for (13). □

Lemma 7. $T(P)\subset P_0$.

By Lemma 3(ii) we can easily obtain this conclusion, so we omit its proof.

Note that if $\overline{u}$ is a positive fixed point of T, from Lemma 7 we have $\overline{u}\in P_0$. Moreover, when

$$\|\overline{u}\|\geq\widetilde{M}=M\int_0^1\left[\frac{(1-s)^{\alpha-p-1}}{\Gamma(\alpha)}+\frac{1}{A}\int_0^1 h(t)g_2(t,s)dt\right]ds>0,$$

we have

$$\begin{aligned}\overline{u}(t)-z(t)&\geq t^{\alpha-1}\|\overline{u}\|-M\int_0^1 G(t,s)ds\\&\geq t^{\alpha-1}\|\overline{u}\|-M\int_0^1 t^{\alpha-1}\left[\frac{(1-s)^{\alpha-p-1}}{\Gamma(\alpha)}+\frac{1}{A}\int_0^1 h(t)g_2(t,s)dt\right]ds\\&\geq 0.\end{aligned}$$

Then from Lemma 6 we have $\overline{u}-z$ is a positive solution for (11). *Therefore, we only need to study the positive fixed point u^* for T, which the norm is greater than $\widetilde{M}$, then u^*-z is a positive solution for* (11).

In the following two lemmas, we let X be a real Banach space and P a cone on X.

Lemma 8. *(see [81]). Let $\Omega\subset X$ be a bounded open set, and $T:\overline{\Omega}\cap P\to P$ a continuous compact operator. If there exists $\mu_0\in P\backslash\{0\}$ such that*

$$u-Tu\neq\lambda\mu_0,\forall\lambda\geq 0,u\in\partial\Omega\cap P,$$

then $i(T,\Omega\cap P,P)=0$, where i is the index of fixed point on P.

Lemma 9. *(see [81]). Let $\Omega\subset X$ be a bounded open set with $0\in\Omega$, and $T:\overline{\Omega}\cap P\to P$ a continuous compact operator. If*

$$u-\lambda Tu\neq 0,\forall\lambda\in[0,1],u\in\partial\Omega\cap P,$$

then $i(T,\Omega\cap P,P)=1$.

In what follows, in order to build our main theorems, we need to introduce some basic knowledge for nonnegative matrices, for more details see [82,83].

Definition 1. *Let $\mathcal{M}$ be a real matrix. If all elements of $\mathcal{M}$ are nonnegative, then $\mathcal{M}$ is called to be nonnegative.*

Definition 2. *A real square matrix* $\mathcal{M} = (m_{ij})_{n\times n}$ *is called* $\mathbb{R}^n_+$*-monotone, if for every column vector* $x \in \mathbb{R}^n, \mathcal{M}x \in \mathbb{R}^n_+ \Longrightarrow x \in \mathbb{R}^n_+$.

Lemma 10. *A real square matrix* $\mathcal{M}$ *is* $\mathbb{R}^n_+$*-monotone* $\Longleftrightarrow$ $\det \mathcal{M} \neq 0$, *and* $\mathcal{M}^{-1}$ *is nonnegative.*

Remark 1. *Note that our boundary condition at* $t = 1$ *is integral and generalizes multi-point fractional boundary conditions. However, our problem (7) can be considered as a perturbation of the two-point boundary value problem*

$$\begin{cases} D^{\alpha}_{0+}u(t) + f(t) = 0, t \in (0,1), \\ u(0) = u'(0) = \cdots = u^{(n-2)}(0) = D^{p}_{0+}u(t)|_{t=1} = 0, \end{cases} \tag{16}$$

which is equivalent to

$$u(t) = \int_0^1 g_1(t,s)f(s)ds,$$

where g_1 *is defined by (8). Therefore, our method, by making good use of the original Green's function for the problem (16), will dispense with constructing a new Green's function, in contrast to some papers dealing with multi-point boundary value problems. For example, in [50] the author studied the problem*

$$\begin{cases} D^{\alpha}_{0+}u(t) + f(t,u(t)) = 0, 0 < t < 1, \\ u(0) = 0, \beta u(\eta) = u(1), \end{cases} \tag{17}$$

where $\alpha \in (1,2], \beta\eta^{\alpha-1}, \eta \in (0,1)$. *The author obtained the Green's function associated with (17) is*

$$G_{Bai}(t,s) = \begin{cases} \frac{[t(1-s)]^{\alpha-1} - \beta t^{\alpha-1}(\eta-s)^{\alpha-1} - (t-s)^{\alpha-1}(1-\beta\eta^{\alpha-1})}{(1-\beta\eta^{\alpha-1})\Gamma(\alpha)}, & 0 \le s \le t \le 1, s \le \eta, \\ \frac{[t(1-s)]^{\alpha-1} - (t-s)^{\alpha-1}(1-\beta\eta^{\alpha-1})}{(1-\beta\eta^{\alpha-1})\Gamma(\alpha)}, & 0 < \eta \le s \le t \le 1, \\ \frac{[t(1-s)]^{\alpha-1} - \beta t^{\alpha-1}(\eta-s)^{\alpha-1}}{(1-\beta\eta^{\alpha-1})\Gamma(\alpha)}, & 0 \le t \le s \le \eta < 1, \\ \frac{[t(1-s)]^{\alpha-1}}{(1-\beta\eta^{\alpha-1})\Gamma(\alpha)}, & 0 \le t \le s \le 1, \eta \le s. \end{cases} \tag{18}$$

This function is very complicated. However, we note that this function can be expressed by

$$G_{Bai}(t,s) = g_{Bai}(t,s) + \frac{\beta t^{\alpha-1}}{1-\beta\eta^{\alpha-1}} g_{Bai}(\eta,s),$$

$$g_{Bai}(t,s) = \frac{1}{\Gamma(\alpha)} \begin{cases} [t(1-s)]^{\alpha-1} - (t-s)^{\alpha-1}, & 0 \le s \le t \le 1, \\ [t(1-s)]^{\alpha-1} & 0 \le t \le s \le 1, \end{cases}$$

where g_{Bai} *is the Green's function for the problem*

$$\begin{cases} D^{\alpha}_{0+}u(t) + f(t,u(t)) = 0, 0 < t < 1, \\ u(0) = u(1) = 0. \end{cases} \tag{19}$$

Compared with G_{Bai}, g_{Bai} *is much simpler.*

3. Main Results

From the discission of Section 2, we can define the operators $T_i (i = 1,2,3) : P \times P \times P \to P$ and $T : P \times P \times P \to P \times P \times P$ as follows:

$$T_i(u,v,w)(t) = \int_0^1 G(t,s)F_i(s,u(s)-z(s),v(s)-z(s),w(s)-z(s))ds,$$
$$T(u,v,w)(t) = (T_1,T_2,T_3)(u,v,w)(t), \text{ for } t \in [0,1],$$

where $F_i(t,x_1,x_2,x_3) = \begin{cases} f_i(t,x_1,x_2,x_3)+M, & t\in[0,1], \text{ for } x_i\geq 0, i=1,2,3, \\ f_i(t,0,0,0)+M, & t\in[0,1], \text{ for else cases.}\end{cases}$ Consequently, if there exists $(\overline{u},\overline{v},\overline{w})$ is a positive fixed pint of T with $\|\overline{u}\|,\|\overline{v}\|,\|\overline{w}\|\geq \widetilde{M}$, then we obtain $(\overline{u}-z,\overline{v}-z,\overline{w}-z)$ is a positive solution for (1).

Now, we list our assumptions for $F_i(i=1,2,3)$:

(C3) There exist $a_{ji},b_{ji}\geq 0$ and $l_j>0(i,j=1,2,3)$ such that

$$\begin{pmatrix} F_1(t,x_1,x_2,x_3) \\ F_2(t,x_1,x_2,x_3) \\ F_3(t,x_1,x_2,x_3)\end{pmatrix} \geq \begin{pmatrix} a_{11}x_1+a_{12}x_2+a_{13}x_3-l_1 \\ a_{21}x_1+a_{22}x_2+a_{23}x_3-l_2 \\ a_{31}x_1+a_{32}x_2+a_{33}x_3-l_3\end{pmatrix}, \forall(t,x_1,x_2,x_3)\in[0,1]\times\mathbb{R}^+\times\mathbb{R}^+\times\mathbb{R}^+,$$

and the matrix $\mathcal{M}_1$ is a $\mathbb{R}^3_+$-monotone matrix, where

$$\mathcal{M}_1 = \begin{pmatrix} \kappa_1 a_{11}-1 & \kappa_1 a_{12} & \kappa_1 a_{13} \\ \kappa_1 a_{21} & \kappa_1 a_{22}-1 & \kappa_1 a_{23} \\ \kappa_1 a_{31} & \kappa_1 a_{32} & \kappa_1 a_{33}-1\end{pmatrix}.$$

(C4) There exists $Q_i(t)$ in $[0,1]$ such that

$$\int_0^1 \varphi(t)Q_i(t)dt<\widetilde{M}, \text{ and } F_i(t,x_1,x_2,x_3)\leq Q_i(t), \forall(t,x_1,x_2,x_3)\in[0,1]\times\left[0,\widetilde{M}\right]^3, i=1,2,3.$$

(C5) There exist $\widetilde{a}_{ji},\widetilde{b}_{ji}\geq 0$ and $\widetilde{l}_j>0(i,j=1,2,3)$ such that

$$\begin{pmatrix} F_1(t,x_1,x_2,x_3) \\ F_2(t,x_1,x_2,x_3) \\ F_3(t,x_1,x_2,x_3)\end{pmatrix} \leq \begin{pmatrix} \widetilde{a}_{11}x_1+\widetilde{a}_{12}x_2+\widetilde{a}_{13}x_3+\widetilde{l}_1 \\ \widetilde{a}_{21}x_1+\widetilde{a}_{22}x_2+\widetilde{a}_{23}x_3+\widetilde{l}_2 \\ \widetilde{a}_{31}x_1+\widetilde{a}_{32}x_2+\widetilde{a}_{33}x_3+\widetilde{l}_3\end{pmatrix}, \forall(t,x_1,x_2,x_3)\in[0,1]\times\mathbb{R}^+\times\mathbb{R}^+\times\mathbb{R}^+,$$

and the matrix $\mathcal{M}_2$ is a $\mathbb{R}^3_+$-monotone matrix, where

$$\mathcal{M}_2 = \begin{pmatrix} 1-\kappa_2\widetilde{a}_{11} & -\kappa_2\widetilde{a}_{12} & -\kappa_2\widetilde{a}_{13} \\ -\kappa_2\widetilde{a}_{21} & 1-\kappa_2\widetilde{a}_{22} & -\kappa_2\widetilde{a}_{23} \\ -\kappa_2\widetilde{a}_{31} & -\kappa_2\widetilde{a}_{32} & 1-\kappa_2\widetilde{a}_{33}\end{pmatrix}.$$

(C6) There exists $\widetilde{Q}_i(t)$ in $[0,1]$, and $t_0\in(0,1)$ such that

$$\int_0^1 \varphi(t)\widetilde{Q}_i(t)dt>\widetilde{M}t_0^{1-\alpha}, \text{ and } F_i(t,x_1,x_2,x_3)\geq \widetilde{Q}_i(t), \forall(t,x_1,x_2,x_3)\in[1,e]\times\left[0,\widetilde{M}\right]^3, i=1,2,3.$$

Let $B_\rho=\{u\in P:\|u\|<\rho\}$ for $\rho>0$ in the sequel. Then we easily have $\partial B_\rho=\{u\in P:\|u\|=\rho\}$, $\overline{B}_\rho=\{u\in P:\|u\|\leq\rho\}$.

Theorem 1. *Suppose that (C0)–(C4) hold. Then* (1) *has a positive solution.*

Proof. We first show that:

$$(u,v,w)\neq T(u,v,w)+\lambda(\phi_1,\phi_2,\phi_3), \text{ for } u,v,w\in\partial B_{R_1}\cap P,\ \lambda\geq 0, \tag{20}$$

where $\phi_i(i=1,2,3)$ are given elements in cone P_0, and $R_1>\widetilde{M}$. Argument by contrary, there exists $u,v,w\in\partial B_{R_1}\cap P$ and $\lambda_0\geq 0$ such that

$$(u,v,w)=T(u,v,w)+\lambda_0(\phi_1,\phi_2,\phi_3), \text{ for } u,v,w\in\partial B_{R_1}\cap P,\ \lambda\geq 0. \tag{21}$$

This implies that

$$\begin{pmatrix} u(t) \\ v(t) \\ w(t) \end{pmatrix} = \begin{pmatrix} T_1(u,v,w)(t)+\lambda_0\phi_1(t) \\ T_2(u,v,w)(t)+\lambda_0\phi_2(t) \\ T_3(u,v,w)(t)+\lambda_0\phi_3(t) \end{pmatrix} \geq \begin{pmatrix} \int_0^1 G(t,s)F_1(s,u(s)-z(s),v(s)-z(s),w(s)-z(s))ds \\ \int_0^1 G(t,s)F_2(s,u(s)-z(s),v(s)-z(s),w(s)-z(s))ds \\ \int_0^1 G(t,s)F_3(s,u(s)-z(s),v(s)-z(s),w(s)-z(s))ds \end{pmatrix}.$$

Note that Lemma 7 we have

$$u,v,w \in P_0. \tag{22}$$

From (C3) we have

$$\begin{pmatrix} u(t) \\ v(t) \\ w(t) \end{pmatrix} \geq \begin{pmatrix} \int_0^1 G(t,s)(a_{11}(u(s)-z(s))+a_{12}(v(s)-z(s))+a_{13}(w(s)-z(s))-l_1)ds \\ \int_0^1 G(t,s)(a_{21}(u(s)-z(s))+a_{22}(v(s)-z(s))+a_{23}(w(s)-z(s))-l_2)ds \\ \int_0^1 G(t,s)(a_{31}(u(s)-z(s))+a_{32}(v(s)-z(s))+a_{33}(w(s)-z(s))-l_3)ds \end{pmatrix}.$$

Multiplying by $\varphi(t)$ for the above both sides, and integrating on $[0,1]$, by Lemma 4 we get

$$\begin{pmatrix} \int_0^1 u(t)\varphi(t)dt \\ \int_0^1 v(t)\varphi(t)dt \\ \int_0^1 w(t)\varphi(t)dt \end{pmatrix} \geq \begin{pmatrix} \int_0^1 \kappa_1\varphi(t)(a_{11}(u(t)-z(t))+a_{12}(v(t)-z(t))+a_{13}(w(t)-z(t)))dt - l_1\kappa_2^2 \\ \int_0^1 \kappa_1\varphi(t)(a_{21}(u(t)-z(t))+a_{22}(v(t)-z(t))+a_{23}(w(t)-z(t)))dt - l_2\kappa_2^2 \\ \int_0^1 \kappa_1\varphi(t)(a_{31}(u(t)-z(t))+a_{32}(v(t)-z(t))+a_{33}(w(t)-z(t)))dt - l_3\kappa_2^2 \end{pmatrix}.$$

Consequently, we find

$$\begin{pmatrix} \kappa_1a_{11}-1 & \kappa_1a_{12} & \kappa_1a_{13} \\ \kappa_1a_{21} & \kappa_1a_{22}-1 & \kappa_1a_{23} \\ \kappa_1a_{31} & \kappa_1a_{32} & \kappa_1a_{33}-1 \end{pmatrix} \begin{pmatrix} \int_0^1 u(t)\varphi(t)dt \\ \int_0^1 v(t)\varphi(t)dt \\ \int_0^1 w(t)\varphi(t)dt \end{pmatrix} \leq \begin{pmatrix} \kappa_1(a_{11}+a_{12}+a_{13})\int_0^1 \varphi(t)z(t)dt + l_1\kappa_2^2 \\ \kappa_1(a_{21}+a_{22}+a_{23})\int_0^1 \varphi(t)z(t)dt + l_2\kappa_2^2 \\ \kappa_1(a_{31}+a_{32}+a_{33})\int_0^1 \varphi(t)z(t)dt + l_3\kappa_2^2 \end{pmatrix}$$

$$\leq \begin{pmatrix} \kappa_1(a_{11}+a_{12}+a_{13})M\kappa_2^2 + l_1\kappa_2^2 \\ \kappa_1(a_{21}+a_{22}+a_{23})M\kappa_2^2 + l_2\kappa_2^2 \\ \kappa_1(a_{31}+a_{32}+a_{33})M\kappa_2^2 + l_3\kappa_2^2 \end{pmatrix}.$$

Therefore, we obtain

$$\begin{pmatrix} \int_0^1 u(t)\varphi(t)dt \\ \int_0^1 v(t)\varphi(t)dt \\ \int_0^1 w(t)\varphi(t)dt \end{pmatrix} \leq \begin{pmatrix} \kappa_1 a_{11}-1 & \kappa_1 a_{12} & \kappa_1 a_{13} \\ \kappa_1 a_{21} & \kappa_1 a_{22}-1 & \kappa_1 a_{23} \\ \kappa_1 a_{31} & \kappa_1 a_{32} & \kappa_1 a_{33}-1 \end{pmatrix}^{-1} \begin{pmatrix} \kappa_1(a_{11}+a_{12}+a_{13})M\kappa_2^2+l_1\kappa_2^2 \\ \kappa_1(a_{21}+a_{22}+a_{23})M\kappa_2^2+l_2\kappa_2^2 \\ \kappa_1(a_{31}+a_{32}+a_{33})M\kappa_2^2+l_3\kappa_2^2 \end{pmatrix}$$

$$= \frac{1}{\Delta_1}\begin{pmatrix} (\kappa_1 a_{22}-1)(\kappa_1 a_{33}-1)-\kappa_1^2 a_{23}a_{32} & \kappa_1^2 a_{13}a_{32}-\kappa_1 a_{12}(\kappa_1 a_{33}-1) & \kappa_1^2 a_{12}a_{23}-\kappa_1 a_{13}(\kappa_1 a_{22}-1) \\ \kappa_1^2 a_{23}a_{31}-\kappa_1 a_{21}(\kappa_1 a_{33}-1) & (\kappa_1 a_{11}-1)(\kappa_1 a_{33}-1)-\kappa_1^2 a_{13}a_{31} & \kappa_1^2 a_{13}a_{21}-\kappa_1 a_{23}(\kappa_1 a_{11}-1) \\ \kappa_1^2 a_{21}a_{32}-\kappa_1 a_{31}(\kappa_1 a_{22}-1) & \kappa_1^2 a_{12}a_{31}-\kappa_1 a_{32}(\kappa_1 a_{11}-1) & (\kappa_1 a_{11}-1)(\kappa_1 a_{22}-1)-\kappa_1^2 a_{12}a_{21} \end{pmatrix}$$

$$\cdot \begin{pmatrix} \kappa_1(a_{11}+a_{12}+a_{13})M\kappa_2^2+l_1\kappa_2^2 \\ \kappa_1(a_{21}+a_{22}+a_{23})M\kappa_2^2+l_2\kappa_2^2 \\ \kappa_1(a_{31}+a_{32}+a_{33})M\kappa_2^2+l_3\kappa_2^2 \end{pmatrix},$$

where

$$\Delta_1 = \det\begin{pmatrix} \kappa_1 a_{11}-1 & \kappa_1 a_{12} & \kappa_1 a_{13} \\ \kappa_1 a_{21} & \kappa_1 a_{22}-1 & \kappa_1 a_{23} \\ \kappa_1 a_{31} & \kappa_1 a_{32} & \kappa_1 a_{33}-1 \end{pmatrix}.$$

As a result of this, there exist $\mathcal{N}_i > 0 (i = 1, 2, 3)$ such that

$$\begin{pmatrix} \int_0^1 u(t)\varphi(t)dt \\ \int_0^1 v(t)\varphi(t)dt \\ \int_0^1 w(t)\varphi(t)dt \end{pmatrix} \leq \begin{pmatrix} \mathcal{N}_1 \\ \mathcal{N}_2 \\ \mathcal{N}_3 \end{pmatrix},$$

where $\mathcal{N}_1 = \frac{1}{\Delta_1}[((\kappa_1 a_{22}-1)(\kappa_1 a_{33}-1)-\kappa_1^2 a_{23}a_{32})(\kappa_1(a_{11}+a_{12}+a_{13})M\kappa_2^2+l_1\kappa_2^2)+(\kappa_1^2 a_{13}a_{32}-\kappa_1 a_{12}(\kappa_1 a_{33}-1))(\kappa_1(a_{21}+a_{22}+a_{23})M\kappa_2^2+l_2\kappa_2^2)+(\kappa_1^2 a_{12}a_{23}-\kappa_1 a_{13}(\kappa_1 a_{22}-1))(\kappa_1(a_{31}+a_{32}+a_{33})M\kappa_2^2+l_3\kappa_2^2)]$, $\mathcal{N}_2 = \frac{1}{\Delta_1}[(\kappa_1^2 a_{23}a_{31}-\kappa_1 a_{21}(\kappa_1 a_{33}-1))(\kappa_1(a_{11}+a_{12}+a_{13})M\kappa_2^2+l_1\kappa_2^2)+((\kappa_1 a_{11}-1)(\kappa_1 a_{33}-1)-\kappa_1^2 a_{13}a_{31})(\kappa_1(a_{21}+a_{22}+a_{23})M\kappa_2^2+l_2\kappa_2^2)+(\kappa_1^2 a_{13}a_{21}-\kappa_1 a_{23}(\kappa_1 a_{11}-1))(\kappa_1(a_{31}+a_{32}+a_{33})M\kappa_2^2+l_3\kappa_2^2)]$, $\mathcal{N}_3 = \frac{1}{\Delta_1}[(\kappa_1^2 a_{21}a_{32}-\kappa_1 a_{31}(\kappa_1 a_{22}-1))(\kappa_1(a_{11}+a_{12}+a_{13})M\kappa_2^2+l_1\kappa_2^2)+(\kappa_1^2 a_{12}a_{31}-\kappa_1 a_{32}(\kappa_1 a_{11}-1))(\kappa_1(a_{21}+a_{22}+a_{23})M\kappa_2^2+l_2\kappa_2^2)+((\kappa_1 a_{11}-1)(\kappa_1 a_{22}-1)-\kappa_1^2 a_{21}a_{21})(\kappa_1(a_{31}+a_{32}+a_{33})M\kappa_2^2+l_3\kappa_2^2)]$.

Note that (22), we have

$$\begin{pmatrix} \|u\| \\ \|v\| \\ \|w\| \end{pmatrix} \leq \begin{pmatrix} \mathcal{N}_1\kappa_1^{-1} \\ \mathcal{N}_2\kappa_1^{-1} \\ \mathcal{N}_3\kappa_1^{-1} \end{pmatrix}.$$

Therefore, we can choose $R_1 > \max\{\widetilde{M}, \mathcal{N}_1\kappa_1^{-1}, \mathcal{N}_2\kappa_1^{-1}, \mathcal{N}_3\kappa_1^{-1}\}$ such that when $u, v, w \in \partial B_{R_1} \cap P$, (21) is not satisfied. This also indicates that (20) holds for $u, v, w \in \partial B_{R_1} \cap P$, and Lemma 8 indicates that

$$i(T, B_{R_1} \cap (P \times P \times P), P \times P \times P) = 0. \tag{23}$$

On the other hand, we prove that

$$(u, v, w) \neq \lambda T(u, v, w), \text{ for } u, v, w \in \partial B_{\widetilde{M}} \cap P, \lambda \in [0, 1]. \tag{24}$$

If this claim is not true, there exist $u, v, w \in \partial B_{\widetilde{M}} \cap P, \lambda_1 \in [0,1]$ such that

$$(u, v, w) = \lambda_1 T(u, v, w).$$

This implies that

$$\|u\| \leq \|T_1(u, v, w)\|, \ \|v\| \leq \|T_2(u, v, w)\|, \text{ and } \|w\| \leq \|T_3(u, v, w)\|.$$

However, from (C4) we have

$$\begin{aligned} T_1(u, v, w)(t) &= \int_0^1 G(t, s) F_1(s, u(s) - z(s), v(s) - z(s), w(s) - z(s)) ds \\ &\leq \int_0^1 \varphi(s) Q_1(s) \frac{ds}{s} \\ &< \widetilde{M}. \end{aligned}$$

Note that by (C4), $\|u\| = \widetilde{M}$. Hence, we obtain $\|T_1(u, v, w)\| < \|u\|$. Similarly, $\|T_2(u, v, w)\| < \|v\|$ and $\|T_3(u, v, w)\| < \|w\|$. This has a contradiction. Hence (24) holds. By Lemma 9 we get

$$i(T, B_{\widetilde{M}} \cap (P \times P \times P), P \times P \times P) = 1. \tag{25}$$

By use of (23) and (25) we can calculate

$$\begin{aligned} &i(T, (B_{R_1} \backslash \overline{B}_{\widetilde{M}}) \cap (P \times P \times P), P \times P \times P) \\ &\quad = i(T, B_{R_1} \cap (P \times P \times P), P \times P \times P) - i(T, B_{\widetilde{M}} \cap (P \times P \times P), P \times P \times P) \\ &\quad = -1. \end{aligned}$$

Therefore, T has a fixed point (u^*, v^*, w^*) on $(B_{R_1} \backslash \overline{B}_{\widetilde{M}}) \cap (P \times P \times P)$. Consequently, $(u^* - z, v^* - z, w^* - z)$ is a positive solution for (1), i.e., (1) has a positive solution. □

Theorem 2. *Suppose that (C0)–(C2), (C5)–(C6) hold. Then* (1) *has a positive solution.*

Proof. We first claim that:

$$(u, v, w) \neq \lambda T(u, v, w), \text{ for } u, v, w \in \partial B_{R_2} \cap P, \lambda \in [0,1], \tag{26}$$

where $R_2 > \widetilde{M}$. If this claim does not hold, there exist $u, v, w \in \partial B_{R_2} \cap P, \lambda_2 \in [0,1]$ such that

$$(u, v, w) = \lambda_2 T(u, v, w). \tag{27}$$

This indicates that

$$\begin{pmatrix} u(t) \\ v(t) \\ w(t) \end{pmatrix} = \begin{pmatrix} \lambda_2 T_1(u, v, w)(t) \\ \lambda_2 T_2(u, v, w)(t) \\ \lambda_2 T_3(u, v, w)(t) \end{pmatrix} \leq \begin{pmatrix} \int_0^1 G(t, s) F_1(s, u(s) - z(s), v(s) - z(s), w(s) - z(s)) ds \\ \int_0^1 G(t, s) F_2(s, u(s) - z(s), v(s) - z(s), w(s) - z(s)) ds \\ \int_0^1 G(t, s) F_3(s, u(s) - z(s), v(s) - z(s), w(s) - z(s)) ds \end{pmatrix}.$$

Using Lemma 7, we know $u, v, w \in P_0$. By virtue of (C5), we obtain

$$\begin{pmatrix} u(t) \\ v(t) \\ w(t) \end{pmatrix} \le \begin{pmatrix} \int_0^1 G(t,s)(\widetilde{a}_{11}(u(s)-z(s)) + \widetilde{a}_{12}(v(s)-z(s)) + \widetilde{a}_{13}(w(s)-z(s)) + \widetilde{l}_1)ds \\ \int_0^1 G(t,s)(\widetilde{a}_{21}(u(s)-z(s)) + \widetilde{a}_{22}(v(s)-z(s)) + \widetilde{a}_{23}(w(s)-z(s)) + \widetilde{l}_2)ds \\ \int_0^1 G(t,s)(\widetilde{a}_{31}(u(s)-z(s)) + \widetilde{a}_{32}(v(s)-z(s)) + \widetilde{a}_{33}(w(s)-z(s)) + \widetilde{l}_3)ds \end{pmatrix}.$$

Multiplying by $\varphi(t)$, and integrating over $[0,1]$, Lemma 4 enables us to get

$$\begin{pmatrix} \int_0^1 u(t)\varphi(t)dt \\ \int_0^1 v(t)\varphi(t)dt \\ \int_0^1 w(t)\varphi(t)dt \end{pmatrix} \le \begin{pmatrix} \int_0^1 \kappa_2\varphi(t)(\widetilde{a}_{11}(u(t)-z(t)) + \widetilde{a}_{12}(v(t)-z(t)) + \widetilde{a}_{13}(w(t)-z(t)) + \widetilde{l}_1)dt \\ \int_0^1 \kappa_2\varphi(t)(\widetilde{a}_{21}(u(t)-z(t)) + \widetilde{a}_{22}(v(t)-z(t)) + \widetilde{a}_{23}(w(t)-z(t)) + \widetilde{l}_2)dt \\ \int_0^1 \kappa_2\varphi(t)(\widetilde{a}_{31}(u(t)-z(t)) + \widetilde{a}_{32}(v(t)-z(t)) + \widetilde{a}_{33}(w(t)-z(t)) + \widetilde{l}_3)dt \end{pmatrix}$$

$$\le \begin{pmatrix} \int_0^1 \kappa_2\varphi(t)(\widetilde{a}_{11}u(t) + \widetilde{a}_{12}v(t) + \widetilde{a}_{13}w(t))dt + \widetilde{l}_1\kappa_2^2 \\ \int_0^1 \kappa_2\varphi(t)(\widetilde{a}_{21}u(t) + \widetilde{a}_{22}v(t) + \widetilde{a}_{23}w(t))dt + \widetilde{l}_2\kappa_2^2 \\ \int_0^1 \kappa_2\varphi(t)(\widetilde{a}_{31}u(t) + \widetilde{a}_{32}v(t) + \widetilde{a}_{33}w(t))dt + \widetilde{l}_3\kappa_2^2 \end{pmatrix}.$$

Therefore, we find

$$\begin{pmatrix} 1-\kappa_2\widetilde{a}_{11} & -\kappa_2\widetilde{a}_{12} & -\kappa_2\widetilde{a}_{13} \\ -\kappa_2\widetilde{a}_{21} & 1-\kappa_2\widetilde{a}_{22} & -\kappa_2\widetilde{a}_{23} \\ -\kappa_2\widetilde{a}_{31} & -\kappa_2\widetilde{a}_{32} & 1-\kappa_2\widetilde{a}_{33} \end{pmatrix} \begin{pmatrix} \int_0^1 u(t)\varphi(t)dt \\ \int_0^1 v(t)\varphi(t)dt \\ \int_0^1 w(t)\varphi(t)dt \end{pmatrix} \le \begin{pmatrix} \widetilde{l}_1\kappa_2^2 \\ \widetilde{l}_2\kappa_2^2 \\ \widetilde{l}_3\kappa_2^2 \end{pmatrix},$$

and

$$\begin{pmatrix} \int_0^1 u(t)\varphi(t)dt \\ \int_0^1 v(t)\varphi(t)dt \\ \int_0^1 w(t)\varphi(t)dt \end{pmatrix} \le \begin{pmatrix} 1-\kappa_2\widetilde{a}_{11} & -\kappa_2\widetilde{a}_{12} & -\kappa_2\widetilde{a}_{13} \\ -\kappa_2\widetilde{a}_{21} & 1-\kappa_2\widetilde{a}_{22} & -\kappa_2\widetilde{a}_{23} \\ -\kappa_2\widetilde{a}_{31} & -\kappa_2\widetilde{a}_{32} & 1-\kappa_2\widetilde{a}_{33} \end{pmatrix}^{-1} \begin{pmatrix} \widetilde{l}_1\kappa_2^2 \\ \widetilde{l}_2\kappa_2^2 \\ \widetilde{l}_3\kappa_2^2 \end{pmatrix}$$

$$= \frac{1}{\Delta_2} \begin{pmatrix} (1-\kappa_2\widetilde{a}_{22})(1-\kappa_2\widetilde{a}_{33}) - \kappa_2^2\widetilde{a}_{23}\widetilde{a}_{32} & \kappa_2^2\widetilde{a}_{13}\widetilde{a}_{32} + \kappa_2\widetilde{a}_{12}(1-\kappa_2\widetilde{a}_{33}) & \kappa_2^2\widetilde{a}_{12}\widetilde{a}_{23} + \kappa_2\widetilde{a}_{13}(1-\kappa_2\widetilde{a}_{22}) \\ \kappa_2^2\widetilde{a}_{23}\widetilde{a}_{31} + \kappa_2\widetilde{a}_{21}(1-\kappa_2\widetilde{a}_{33}) & (1-\kappa_2\widetilde{a}_{11})(1-\kappa_2\widetilde{a}_{33}) - \kappa_2^2\widetilde{a}_{13}\widetilde{a}_{31} & \kappa_2^2\widetilde{a}_{13}\widetilde{a}_{21} + \kappa_2\widetilde{a}_{23}(1-\kappa_2\widetilde{a}_{11}) \\ \kappa_2^2\widetilde{a}_{21}\widetilde{a}_{32} + \kappa_2\widetilde{a}_{31}(1-\kappa_2\widetilde{a}_{22}) & \kappa_2^2\widetilde{a}_{12}\widetilde{a}_{31} + \kappa_2\widetilde{a}_{32}(1-\kappa_2\widetilde{a}_{11}) & (1-\kappa_2\widetilde{a}_{11})(1-\kappa_2\widetilde{a}_{22}) - \kappa_2^2\widetilde{a}_{12}\widetilde{a}_{21} \end{pmatrix} \cdot \begin{pmatrix} \widetilde{l}_1\kappa_2^2 \\ \widetilde{l}_2\kappa_2^2 \\ \widetilde{l}_3\kappa_2^2 \end{pmatrix},$$

where

$$\Delta_2 = \det \begin{pmatrix} 1-\kappa_2\widetilde{a}_{11} & -\kappa_2\widetilde{a}_{12} & -\kappa_2\widetilde{a}_{13} \\ -\kappa_2\widetilde{a}_{21} & 1-\kappa_2\widetilde{a}_{22} & -\kappa_2\widetilde{a}_{23} \\ -\kappa_2\widetilde{a}_{31} & -\kappa_2\widetilde{a}_{32} & 1-\kappa_2\widetilde{a}_{33} \end{pmatrix}.$$

Hence, there exist $\mathcal{N}_i > 0 (i = 4,5,6)$ such that

$$\begin{pmatrix} \int_0^1 u(t)\varphi(t)dt \\ \int_0^1 v(t)\varphi(t)dt \\ \int_0^1 w(t)\varphi(t)dt \end{pmatrix} \le \begin{pmatrix} \mathcal{N}_4 \\ \mathcal{N}_5 \\ \mathcal{N}_6 \end{pmatrix},$$

where $\mathcal{N}_4 = \frac{\kappa_2^2}{\Delta_2}[\widetilde{l}_1((1-\kappa_2\widetilde{a}_{22})(1-\kappa_2\widetilde{a}_{33}) - \kappa_2^2\widetilde{a}_{23}\widetilde{a}_{32}) + \widetilde{l}_2(\kappa_2^2\widetilde{a}_{13}\widetilde{a}_{32} + \kappa_2\widetilde{a}_{12}(1-\kappa_2\widetilde{a}_{33})) + \widetilde{l}_3(\kappa_2^2\widetilde{a}_{12}\widetilde{a}_{23} + \kappa_2\widetilde{a}_{13}(1-\kappa_2\widetilde{a}_{22}))]$, $\mathcal{N}_5 = \frac{\kappa_2^2}{\Delta_2}[\widetilde{l}_1(\kappa_2^2\widetilde{a}_{23}\widetilde{a}_{31} + \kappa_2\widetilde{a}_{21}(1-\kappa_2\widetilde{a}_{33})) + \widetilde{l}_2((1-\kappa_2\widetilde{a}_{11})(1-\kappa_2\widetilde{a}_{33}) - \kappa_2^2\widetilde{a}_{13}\widetilde{a}_{31}) + \widetilde{l}_3(\kappa_2^2\widetilde{a}_{13}\widetilde{a}_{21} + \kappa_2\widetilde{a}_{23}(1-\kappa_2\widetilde{a}_{11}))]$, $\mathcal{N}_6 = \frac{\kappa_2^2}{\Delta_2}[\widetilde{l}_1(\kappa_2^2\widetilde{a}_{21}\widetilde{a}_{32} + \kappa_2\widetilde{a}_{31}(1-\kappa_2\widetilde{a}_{22})) + \widetilde{l}_2(\kappa_2^2\widetilde{a}_{12}\widetilde{a}_{31} + \kappa_2\widetilde{a}_{32}(1-\kappa_2\widetilde{a}_{11})) + \widetilde{l}_3((1-\kappa_2\widetilde{a}_{11})(1-\kappa_2\widetilde{a}_{22}) - \kappa_2^2\widetilde{a}_{12}\widetilde{a}_{21})]$. Note that $u, v, w \in P_0$, we have

$$\begin{pmatrix} \|u\| \\ \|v\| \\ \|w\| \end{pmatrix} \le \begin{pmatrix} \mathcal{N}_4\kappa_1^{-1} \\ \mathcal{N}_5\kappa_1^{-1} \\ \mathcal{N}_6\kappa_1^{-1} \end{pmatrix}.$$

Therefore, we can choose $R_2 > \max\{\widetilde{M}, \mathcal{N}_4\kappa_1^{-1}, \mathcal{N}_5\kappa_1^{-1}, \mathcal{N}_6\kappa_1^{-1}\}$ such that when $u, v, w \in \partial B_{R_2} \cap P$, (27) is not satisfied. This also indicates that (26) holds for $u, v, w \in \partial B_{R_2} \cap P$, and by Lemma 9 we get

$$i(T, B_{R_2} \cap (P \times P \times P), P \times P \times P) = 1. \tag{28}$$

On the other hand, we prove that

$$(u, v, w) \neq T(u, v, w) + \lambda(\widetilde{\phi}_1, \widetilde{\phi}_2, \widetilde{\phi}_3), \text{ for } u, v, w \in \partial B_{\widetilde{M}} \cap P, \ \forall \lambda \ge 0, \tag{29}$$

where $\widetilde{\phi}_i \in P (i = 1,2,3)$ are fixed elements. Otherwise, there exist $u, v, w \in \partial B_{\widetilde{M}} \cap P, \lambda_3 \ge 0$ such that

$$(u, v, w) = T(u, v, w) + \lambda_3(\widetilde{\phi}_1, \widetilde{\phi}_2, \widetilde{\phi}_3).$$

This implies that

$$\begin{pmatrix} \|u\| \\ \|v\| \\ \|w\| \end{pmatrix} \ge \begin{pmatrix} \|T_1(u,v,w)\| \\ \|T_2(u,v,w)\| \\ \|T_3(u,v,w)\| \end{pmatrix}. \tag{30}$$

However, from (C6) we have

$$\begin{aligned} T_i(u, v, w)(t_0) &= \int_0^1 G(t_0, s)F_i(s, u(s) - z(s), v(s) - z(s), w(s) - z(s))ds \\ &\ge t_0^{\alpha-1} \int_0^1 \varphi(s)\widetilde{Q}_i(s)ds \\ &> \widetilde{M}, i = 1,2,3. \end{aligned}$$

Note that from (C6), we have $\|u\| = \widetilde{M}$. Hence, we obtain

$$\begin{pmatrix} \|T_1(u,v,w)\| \\ \|T_2(u,v,w)\| \\ \|T_3(u,v,w)\| \end{pmatrix} \ge \begin{pmatrix} T_1(u,v,w)(t_0) \\ T_2(u,v,w)(t_0) \\ T_3(u,v,w)(t_0) \end{pmatrix} > \begin{pmatrix} \|u\| \\ \|v\| \\ \|w\| \end{pmatrix}.$$

This has a contradiction with (30), and thus (29) holds. By Lemma 8 we find

$$i(T, B_{\widetilde{M}} \cap (P \times P \times P), P \times P \times P) = 0. \tag{31}$$

From (28) and (31) we can calculate

$$\begin{aligned} &i(T,(B_{R_2} \backslash \overline{B}_{\widetilde{M}}) \cap (P \times P \times P), P \times P \times P) \\ &= i(T, B_{R_2} \cap (P \times P \times P), P \times P \times P) - i(T, B_{\widetilde{M}} \cap (P \times P \times P), P \times P \times P) \\ &= 1. \end{aligned}$$

Therefore T has a fixed point (u^*, v^*, w^*) on $(B_{R_2} \backslash \overline{B}_{\widetilde{M}}) \cap (P \times P \times P)$. Therefore, $(u^* - z, v^* - z, w^* - z)$ is a positive solution for (1), i.e., (1) has a positive solution.

Let $n = 4, \alpha = 3.5, p = 1.5, q = 0.5$, and $h(t) = t, t \in [0,1]$. Then we have $A = 2.91$, and $\int_0^1 h(t)g_2(t,s)dt = \frac{5}{24}s - \frac{1}{4}s^2 + \frac{1}{24}s^4, s \in [0,1]$. This implies that (C0)–(C1) hold. Moreover, we can calculate

$$\kappa_1 = 0.017, \kappa_2 = 0.075, \widetilde{M} = 0.16M.$$

□

Example 1. *Let $\kappa_1 a_{11} - 1 = \kappa_1 a_{22} - 1 = \kappa_1 a_{33} - 1 = \kappa_1$, and we have $a_{11} = a_{22} = a_{33} = \frac{\kappa_1 + 1}{\kappa_1} = 59.82$. Moreover, we take the matrix*

$$\begin{pmatrix} a_{11} & a_{12} & a_{13} \\ a_{21} & a_{22} & a_{23} \\ a_{31} & a_{32} & a_{33} \end{pmatrix} = \begin{pmatrix} 59.82 & 0 & 0 \\ 0 & 59.82 & 0 \\ 0 & 0 & 59.82 \end{pmatrix},$$

and

$$\begin{pmatrix} F_1(t,x_1,x_2,x_3) \\ F_2(t,x_1,x_2,x_3) \\ F_3(t,x_1,x_2,x_3) \end{pmatrix} = \begin{pmatrix} 2M(9.57M)^{-\gamma_1}(a_{11}x_1 + a_{12}x_2 + a_{13}x_3)^{\gamma_1} \\ 1.8M(9.57M)^{-\gamma_2}(a_{21}x_1 + a_{22}x_2 + a_{23}x_3)^{\gamma_2} \\ 1.5M(9.57M)^{-\gamma_3}(a_{31}x_1 + a_{32}x_2 + a_{33}x_3)^{\gamma_3} \end{pmatrix}, \forall (t,x_1,x_2,x_3) \in [0,1] \times \mathbb{R}^+ \times \mathbb{R}^+ \times \mathbb{R}^+,$$

where $\gamma_i > 1 (i = 1,2,3)$. Note that

$$\mathcal{M}_1 = \begin{pmatrix} \kappa_1 a_{11} - 1 & \kappa_1 a_{12} & \kappa_1 a_{13} \\ \kappa_1 a_{21} & \kappa_1 a_{22} - 1 & \kappa_1 a_{23} \\ \kappa_1 a_{31} & \kappa_1 a_{32} & \kappa_1 a_{33} - 1 \end{pmatrix} = \begin{pmatrix} \kappa_1 & 0 & 0 \\ 0 & \kappa_1 & 0 \\ 0 & 0 & \kappa_1 \end{pmatrix}.$$

Hence, $\mathcal{M}_1$ is a $\mathbb{R}^3_+$-monotone matrix. Furthermore, for all $t \in [0,1]$ we have

$$\begin{aligned} \liminf_{a_{11}x_1+a_{12}x_2+a_{13}x_3 \to +\infty} \frac{F_1(x_1,x_2,x_3)}{a_{11}x_1 + a_{12}x_2 + a_{13}x_3} &= \liminf_{a_{11}x_1+a_{12}x_2+a_{13}x_3 \to +\infty} \frac{2M(9.57M)^{-\gamma_1}(a_{11}x_1 + a_{12}x_2 + a_{13}x_3)^{\gamma_1}}{a_{11}x_1 + a_{12}x_2 + a_{13}x_3} = +\infty, \\ \liminf_{a_{21}x_1+a_{22}x_2+a_{23}x_3 \to +\infty} \frac{F_2(x_1,x_2,x_3)}{a_{21}x_1 + a_{22}x_2 + a_{23}x_3} &= \liminf_{a_{21}x_1+a_{22}x_2+a_{23}x_3 \to +\infty} \frac{1.8M(9.57M)^{-\gamma_2}(a_{21}x_1 + a_{22}x_2 + a_{23}x_3)^{\gamma_2}}{a_{21}x_1 + a_{22}x_2 + a_{23}x_3} = +\infty, \\ \liminf_{a_{31}x_1+a_{32}x_2+a_{33}x_3 \to +\infty} \frac{F_3(x_1,x_2,x_3)}{a_{31}x_1 + a_{32}x_2 + a_{33}x_3} &= \liminf_{a_{31}x_1+a_{32}x_2+a_{33}x_3 \to +\infty} \frac{1.5M(9.57M)^{-\gamma_3}(a_{31}x_1 + a_{32}x_2 + a_{33}x_3)^{\gamma_3}}{a_{31}x_1 + a_{32}x_2 + a_{33}x_3} = +\infty. \end{aligned}$$

On the other hand, if $(t,x_1,x_2,x_3) \in [0,1] \times [0, 0.16M]^3$, we have

$$F_1 \leq 2M, \ F_2 \leq 1.8M, F_3 \leq 1.5M.$$

If we choose $Q_1(t) \equiv 2M, Q_2(t) \equiv 1.8M, Q_3(t) \equiv 1.5M$ *for* $t \in [0,1]$, *and we have*

$$\int_0^1 \varphi(t)Q_i(t)dt \le \int_0^1 \varphi(t)Q_1(t)dt = 2\kappa_2 M = 0.15M < \widetilde{M}, i = 1,2,3.$$

Therefore, (C3)–(C4) hold.

Example 2. *Let* $t_0 = 0.5$, $\widetilde{Q}_1(t) = 13M, \widetilde{Q}_2(t) = 14M, \widetilde{Q}_3(t) = 15M$ *for* $t \in [0,1]$, *and*

$$\begin{pmatrix} \widetilde{a}_{11} & \widetilde{a}_{12} & \widetilde{a}_{13} \\ \widetilde{a}_{21} & \widetilde{a}_{22} & \widetilde{a}_{23} \\ \widetilde{a}_{31} & \widetilde{a}_{32} & \widetilde{a}_{33} \end{pmatrix} = \begin{pmatrix} 2 & 5 & 3 \\ 8 & 3 & 4 \\ 6 & 3 & 4 \end{pmatrix},$$

and

$$\begin{pmatrix} F_1(t,x_1,x_2,x_3) \\ F_2(t,x_1,x_2,x_3) \\ F_3(t,x_1,x_2,x_3) \end{pmatrix} = \begin{pmatrix} 13Me^{1.6M}e^{-2x_1-5x_2-3x_3} \\ 14Me^{2.4M}e^{-8x_1-3x_2-4x_3} \\ 15Me^{2.08M}e^{-6x_1-3x_2-4x_3} \end{pmatrix}, \forall (t,x_1,x_2,x_3) \in [0,1]\times\mathbb{R}^+\times\mathbb{R}^+\times\mathbb{R}^+.$$

Then if $(t,x_1,x_2,x_3) \in [0,1]\times[0,0.16M]^3$, *we have* $F_1 \ge 13M, F_2 \ge 14M, F_3 \ge 15M$, *and* $\int_0^1 \varphi(t)\widetilde{Q}_i(t)dt \ge \int_0^1 \varphi(t)\widetilde{Q}_1(t)dt = 13\kappa_2 M > 0.16 \times 5.6569M$.

On the other hand, we can calculate

$$\det \mathcal{M}_2 = \det \begin{pmatrix} 1-0.075\times 2 & -0.075\times 5 & -0.075\times 3 \\ -0.075\times 8 & 1-0.075\times 3 & -0.075\times 4 \\ -0.075\times 6 & -0.075\times 3 & 1-0.075\times 4 \end{pmatrix} = 0.0868,$$

and

$$\mathcal{M}_2^{-1} = \frac{1}{0.0868}\begin{pmatrix} 0.475 & 0.313 & 0.287 \\ 0.555 & 0.494 & 0.39 \\ 0.484 & 0.36 & 0.434 \end{pmatrix}.$$

Consequently, $\mathcal{M}_2$ *is a* $\mathbb{R}^3_+$*-monotone matrix. Furthermore, for all* $t \in [0,1]$ *we have*

$$\limsup_{\widetilde{a}_{11}x_1+\widetilde{a}_{12}x_2+\widetilde{a}_{13}x_3\to+\infty} \frac{F_1(t,x_1,x_2,x_3)}{\widetilde{a}_{11}x_1+\widetilde{a}_{12}x_2+\widetilde{a}_{13}x_3} = \limsup_{\widetilde{a}_{11}x_1+\widetilde{a}_{12}x_2+\widetilde{a}_{13}x_3\to+\infty} \frac{13Me^{1.6M}e^{-\widetilde{a}_{11}x_1-\widetilde{a}_{12}x_2-\widetilde{a}_{13}x_3}}{\widetilde{a}_{11}x_1+\widetilde{a}_{12}x_2+\widetilde{a}_{13}x_3} = 0,$$
$$\limsup_{\widetilde{a}_{21}x_1+\widetilde{a}_{22}x_2+\widetilde{a}_{23}x_3\to+\infty} \frac{F_2(t,x_1,x_2,x_3)}{\widetilde{a}_{21}x_1+\widetilde{a}_{22}x_2+\widetilde{a}_{23}x_3} = \limsup_{\widetilde{a}_{21}x_1+\widetilde{a}_{22}x_2+\widetilde{a}_{23}x_3\to+\infty} \frac{14Me^{2.4M}e^{-\widetilde{a}_{21}x_1-\widetilde{a}_{22}x_2-\widetilde{a}_{23}x_3}}{\widetilde{a}_{21}x_1+\widetilde{a}_{22}x_2+\widetilde{a}_{23}x_3} = 0,$$
$$\limsup_{\widetilde{a}_{31}x_1+\widetilde{a}_{32}x_2+\widetilde{a}_{33}x_3\to+\infty} \frac{F_3(t,x_1,x_2,x_3)}{\widetilde{a}_{31}x_1+\widetilde{a}_{32}x_2+\widetilde{a}_{33}x_3} = \limsup_{\widetilde{a}_{31}x_1+\widetilde{a}_{32}x_2+\widetilde{a}_{33}x_3\to+\infty} \frac{15Me^{2.08M}e^{-\widetilde{a}_{31}x_1-\widetilde{a}_{32}x_2-\widetilde{a}_{33}x_3}}{\widetilde{a}_{31}x_1+\widetilde{a}_{32}x_2+\widetilde{a}_{33}x_3} = 0.$$

As a result, (C5)–(C6) hold.

4. Conclusions

In this paper, we utilize the index of fixed point to research the positive solutions for the system of Riemann–Liouville type fractional boundary value problems (1). We first investigate corresponding operator equations for (1), and then establish some coupling behaviors for our nonlinearities $f_i (i = 1,2,3)$ by virtue of nonnegative matrix theory, which ensure that our nonlinearities can grow superlinearly and sublinearly at ∞.

Author Contributions: Conceptualization, Y.D. and J.X.; methodology, J.X.; software, Y.D.; validation, Y.D., J.X. and Z.F.; formal analysis, J.X.; investigation, Y.D.; resources, J.X.; data curation, J.X.; writing—original draft preparation, Y.D.; writing—review and editing, J.X.; visualization, J.X.; supervision, J.X. and Z.F.; project administration, J.X.; funding acquisition, J.X.

References

1. Kilbas, A.A.; Srivastava, H.M.; Trujillo, J.J. *Theory and Applications of Fractional Differential Equations, Volume 204 of North-Holland Mathematics Studies*; Elsevier: Amsterdam, The Netherlands, 2006.
2. Podlubny, I. *Fractional Differential Equations, Volume 198 of Mathematics in Science and Engineering*; Academic Press: San Diego, CA, USA, 1999.
3. Samko, S.G.; Kilbas, A.A.; Marichev, O.I. *Fractional Integrals and Derivatives, Theory and Applications*; Gordon and Breach: Zurich, Switzerland, 1993.
4. Henderson, J.; Luca, R. Existence of positive solutions for a singular fractional boundary value problem. *Nonlinear Anal. Model. Control* **2017**, *22*, 99–114. [CrossRef]
5. Zhang, X.; Shao, Z.; Zhong, Q.; Zhao, Z. Triple positive solutions for semipositone fractional differential equations m-point boundary value problems with singularities and p-q-order derivatives. *Nonlinear Anal. Model. Control* **2018**, *23*, 889–903. [CrossRef]
6. Zhang, K.; Fu, Z. Solutions for a class of Hadamard fractional boundary value problems with sign-changing nonlinearity. *J. Funct. Spaces* **2019**, *2019*, 9046472. [CrossRef]
7. Pu, R.; Zhang, X.; Cui, Y.; Li, P.; Wang, W. Positive solutions for singular semipositone fractional differential equation subject to multipoint boundary conditions. *J. Funct. Spaces* **2017**, *2017*, 5892616. [CrossRef]
8. Wei, Z.; Li, Q.; Che, J. Initial value problems for fractional differential equations involving Riemann–Liouville sequential fractional derivative. *J. Math. Anal. Appl.* **2010**, *367*, 260–272. [CrossRef]
9. Wei, Z.; Dong, W.; Che, J. Periodic boundary value problems for fractional differential equations involving a Riemann–Liouville fractional derivative. *Nonlinear Anal. Theory Methods Appl.* **2010**, *73*, 3232–3238. [CrossRef]
10. Shu, X.; Lai, Y.; Chen, Y. The existence of mild solutions for impulsive fractional partial differential equations. *Nonlinear Anal. Theory Methods Appl.* **2011**, *74*, 2003–2011. [CrossRef]
11. Baleanu, D.; Rezapour, S.; Mohammadi, M. Some existence results on nonlinear fractional differential equations. *Philos. Trans. A* **2013**, *371*, 20120144. [CrossRef]
12. Samet, B.; Aydi, H. On some inequalities involving Caputo fractional derivatives and applications to special means of real numbers. *Mathematics* **2018**, *6*, 193. [CrossRef]
13. Yue, Z.; Zou, Y. New uniqueness results for fractional differential equation with dependence on the first order derivative. *Adv. Differ. Equ.* **2019**, *2019*, 38. [CrossRef]
14. Zou, Y.; He, G. The existence of solutions to integral boundary value problems of fractional differential equations at resonance. *J. Funct. Spaces* **2017**, *2017*, 2785937. [CrossRef]
15. Zou, Y.; He, G. On the uniqueness of solutions for a class of fractional differential equations. *Appl. Math. Lett.* **2017**, *74*, 68–73. [CrossRef]
16. Zou, Y. Positive solutions for a fractional boundary value problem with a perturbation term. *J. Funct. Spaces* **2018**, *2018*, 9070247. [CrossRef]
17. Cui, Y. Uniqueness of solution for boundary value problems for fractional differential equations. *Appl. Math. Lett.* **2016**, *51*, 48–54. [CrossRef]
18. Cui, Y.; Ma, W.; Sun, Q.; Su, X. New uniqueness results for boundary value problem of fractional differential equation. *Nonlinear Anal. Model. Control* **2018**, *23*, 31–39. [CrossRef]
19. Meng, S.; Cui, Y. The extremal solution to conformable fractional differential equations involving integral boundary condition. *Mathematics* **2019**, *7*, 186. [CrossRef]
20. Meng, S.; Cui, Y. Multiplicity results to a conformable fractional differential equations involving integral boundary condition. *Complexity* **2019**, *2019*, 8402347. [CrossRef]
21. Cui, Y.; Sun, Q.; Su, X. Monotone iterative technique for nonlinear boundary value problems of fractional order $p \in (2,3]$. *Adv. Differ. Equ.* **2017**, *2017*, 248. [CrossRef]
22. Zhang, X.; Wu, J.; Liu, L.; Wu, Y.; Cui, Y. Convergence analysis of iterative scheme and error estimation of positive solution for a fractional differential equation. *Math. Model. Anal.* **2018**, *23*, 611–626. [CrossRef]

23. Wu, J.; Zhang, X.; Liu, L.; Wu, Y.; Cui, Y. The convergence analysis and error estimation for unique solution of a p-Laplacian fractional differential equation with singular decreasing nonlinearity. *Bound. Value Probl.* **2018**, *2018*, 82. [CrossRef]
24. He, J.; Zhang, X.; Liu, L.; Wu, Y.; Cui, Y. Existence and asymptotic analysis of positive solutions for a singular fractional differential equation with nonlocal boundary conditions. *Bound. Value Probl.* **2018**, *2018*, 189. [CrossRef]
25. He, J.; Zhang, X.; Liu, L.; Wu, Y.; Cui, Y. A singular fractional Kelvin-Voigt model involving a nonlinear operator and their convergence properties. *Bound. Value Probl.* **2019**, *2019*, 112. [CrossRef]
26. Zhong, Q.; Zhang, X.; Lu, X.; Fu, Z. Uniqueness of successive positive solution for nonlocal singular higher-order fractional differential equations involving arbitrary derivatives. *J. Funct. Spaces* **2018**, *2018*, 6207682. [CrossRef]
27. Wang, F.; Cui, Y. Unbounded solutions to abstract boundary value problems of fractional differential equations on a half line. *Math. Methods Appl. Sci.* **2019**. [CrossRef]
28. Wang, Y.; Liu, Y.; Cui, Y. Infinitely many solutions for impulsive fractional boundary value problem with p-Laplacian. *Bound. Value Probl.* **2018**, *2018*, 94. [CrossRef]
29. Zuo, M.; Hao, X.; Liu, L.; Cui, Y. Existence results for impulsive fractional integro-differential equation of mixed type with constant coefficient and antiperiodic boundary conditions. *Bound. Value Probl.* **2017**, *2017*, 161. [CrossRef]
30. Sun, Q.; Ji, H.; Cui, Y. Positive solutions for boundary value problems of fractional differential equation with integral boundary conditions. *J. Funct. Spaces* **2018**, *2018*, 6461930. [CrossRef]
31. Sun, Q.; Meng, S.; Cui, Y. Existence results for fractional order differential equation with nonlocal Erdélyi-Kober and generalized Riemann–Liouville type integral boundary conditions at resonance. *Adv. Differ. Equ.* **2018**, *2018*, 243. [CrossRef]
32. Ma, W.; Cui, Y. The eigenvalue problem for Caputo type fractional differential equation with Riemann–Stieltjes integral boundary conditions. *J. Funct. Spaces* **2018**, *2018*, 2176809. [CrossRef]
33. Zhang, K.; Wang, J.; Ma, W. Solutions for integral boundary value problems of nonlinear Hadamard fractional differential equations. *J. Funct. Spaces* **2018**, *2018*, 2193234. [CrossRef]
34. Ma, W.; Meng, S.; Cui, Y. Resonant integral boundary value problems for Caputo fractional differential equations. *Math. Probl. Eng.* **2018**, *2018*, 5438592. [CrossRef]
35. Wang, G.; Bai, Z.; Zhang, L. Successive iterations for unique positive solution of a nonlinear fractional q-integral boundary value problem. *J. Appl. Anal. Comput.* **2019**, *9*, 1204–1215.
36. Jia, M.; Li, L.; Liu, X.; Song, J.; Bai, Z. A class of nonlocal problems of fractional differential equations with composition of derivative and parameters. *Adv. Differ. Equ.* **2019**, *2019*, 280. [CrossRef]
37. Zhang, W.; Bai, Z.; Sun, S. Extremal solutions for some periodic fractional differential equations. *Adv. Differ. Equ.* **2016**, *2016*, 179. [CrossRef]
38. Sheng, K.; Zhang, W.; Bai, Z. Positive solutions to fractional boundary-value problems with p-Laplacian on time scales. *Bound. Value Probl.* **2018**, *2018*, 70. [CrossRef]
39. Dong, X.; Bai, Z.; Zhang, S. Positive solutions to boundary value problems of p-Laplacian with fractional derivative. *Bound. Value Probl.* **2017**, *2017*, 5. [CrossRef]
40. Tian, Y.; Sun, S.; Bai, Z. Positive solutions of fractional differential equations with p-Laplacian. *J. Funct. Spaces* **2017**, *2017*, 3187492. [CrossRef]
41. Song, Q.; Bai, Z. Positive solutions of fractional differential equations involving the Riemann–Stieltjes integral boundary condition. *Adv. Differ. Equ.* **2018**, *2018*, 183. [CrossRef]
42. He, L.; Dong, X.; Bai, Z.; Chen, B. Solvability of some two-point fractional boundary value problems under barrier strip conditions. *J. Funct. Spaces* **2017**, *2017*, 1465623. [CrossRef]
43. Song, Q.; Dong, X.; Bai, Z.; Chen, B. Existence for fractional Dirichlet boundary value problem under barrier strip conditions. *J. Nonlinear Sci. Appl.* **2017**, *10*, 3592–3598. [CrossRef]
44. Zhao, Y.; Hou, X.; Sun, Y.; Bai, Z. Solvability for some class of multi-order nonlinear fractional systems. *Adv. Differ. Equ.* **2019**, *2019*, 23. [CrossRef]
45. Bai, Z.; Dong, X.; Yin, C. Existence results for impulsive nonlinear fractional differential equation with mixed boundary conditions. *Bound. Value Probl.* **2016**, *2016*, 63. [CrossRef]
46. Zhai, C.; Li, P.; Li, H. Single upper-solution or lower-solution method for Langevin equations with two fractional orders. *Adv. Differ. Equ.* **2018**, *2018*, 360. [CrossRef]

47. Fu, Z.; Bai, S.; O'Regan, D.; Xu, J. Nontrivial solutions for an integral boundary value problem involving Riemann–Liouville fractional derivatives. *J. Inequal. Appl.* **2019**, *2019*, 104. [CrossRef]
48. Zhang, K.; O'Regan, D.; Xu, J.; Fu, Z. Nontrivial solutions for a higher order nonlinear fractional boundary value problem involving Riemann–Liouville fractional derivatives. *J. Funct. Spaces* **2019**, *2019*, 2381530. [CrossRef]
49. Fazli, H.; Nieto, J.J.; Bahrami, F. On the existence and uniqueness results for nonlinear sequential fractional differential equations. *Appl. Comput. Math.* **2018**, *17*, 36–47.
50. Bai, Z. On positive solutions of a nonlocal fractional boundary value problem. *Nonlinear Anal.* **2010**, *72*, 916–924. [CrossRef]
51. Jiang, J.; O'Regan, D.; Xu, J.; Cui, Y. Positive solutions for a Hadamard fractional p-Laplacian three-point boundary value problem. *Mathematics* **2019**, *7*, 439. [CrossRef]
52. Jiang, J.; O'Regan, D.; Xu, J.; Fu, Z. Positive solutions for a system of nonlinear Hadamard fractional differential equations involving coupled integral boundary conditions. *J. Inequal. Appl.* **2019**, *2019*, 204. [CrossRef]
53. Ahmad, B.; Ntouyas, S.K.; Alsaedi, A. On a coupled system of fractional differential equations with coupled nonlocal and integral boundary conditions. *Chaos Solitons Fractals* **2016**, *83*, 234–241. [CrossRef]
54. Wang, Y.; Liu, L.; Zhang, X.; Wu, Y. Positive solutions of an abstract fractional semipositone differential system model for bioprocesses of HIV infection. *Appl. Math. Comput.* **2015**, *258*, 312–324. [CrossRef]
55. Aljoudi, S.; Ahmad, B.; Nieto, J.J.; Alsaedi, A. A coupled system of Hadamard type sequential fractional differential equations with coupled strip conditions. *Chaos Solitons Fractals* **2016**, *91*, 39–46. [CrossRef]
56. Henderson, J.; Luca, R.; Tudorache, A. On a system of fractional differential equations with coupled integral boundary conditions. *Fract. Calc. Appl. Anal.* **2015**, *18*, 361–386. [CrossRef]
57. Ahmad, B.; Luca, R. Existence of solutions for a system of fractional differential equations with coupled nonlocal boundary conditions. *Fract. Calc. Appl. Anal.* **2018**, *21*, 423–441. [CrossRef]
58. Henderson, J.; Luca, R. Positive solutions for a system of coupled fractional boundary value problems. *Lith. Math. J.* **2018**, *58*, 15–32. [CrossRef]
59. Ahmad, B.; Nieto, J.J. Existence results for a coupled system of nonlinear fractional differential equations with three-point boundary conditions. *Comput. Math. Appl.* **2009**, *58*, 1838–1843. [CrossRef]
60. Ali, A.; Shah, K.; Jarad, F.; Gupta, V.; Abdeljawad, T. Existence and stability analysis to a coupled system of implicit type impulsive boundary value problems of fractional-order differential equations. *Adv. Differ. Equ.* **2019**, *2019*, 101. [CrossRef]
61. Mahmudov, N.I.; Bawaneh, S.; Al-Khateeb, A. On a coupled system of fractional differential equations with four point integral boundary conditions. *Mathematics* **2019**, *7*, 279. [CrossRef]
62. Zhao, Y.; Chen, H.; Xu, C. Nontrivial solutions for impulsive fractional differential equations via Morse theory. *Appl. Math. Comput.* **2017**, *307*, 170–179. [CrossRef]
63. Zhang, X.; Liu, L.; Wu, Y.; Zou, Y. Existence and uniqueness of solutions for systems of fractional differential equations with Riemann–Stieltjes integral boundary condition. *Adv. Differ. Equ.* **2018**, *2018*, 204. [CrossRef]
64. Zhang, X.; Liu, L.; Zou, Y. Fixed-point theorems for systems of operator equations and their applications to the fractional differential equations. *J. Funct. Spaces* **2018**, *2018*, 7469868. [CrossRef]
65. Hao, X.; Wang, H.; Liu, L.; Cui, Y. Positive solutions for a system of nonlinear fractional nonlocal boundary value problems with parameters and p-Laplacian operator. *Bound. Value Probl.* **2017**, *2017*, 182. [CrossRef]
66. Qi, T.; Liu, Y.; Zou, Y. Existence result for a class of coupled fractional differential systems with integral boundary value conditions. *J. Nonlinear Sci. Appl.* **2017**, *10*, 4034–4045. [CrossRef]
67. Qi, T.; Liu, Y.; Cui, Y. Existence of solutions for a class of coupled fractional differential systems with nonlocal boundary conditions. *J. Funct. Spaces* **2017**, *2017*, 6703860. [CrossRef]
68. Zhang, Y. Existence results for a coupled system of nonlinear fractional multi-point boundary value problems at resonance. *J. Inequal. Appl.* **2018**, *2018*, 198. [CrossRef] [PubMed]
69. Li, H.; Zhang, J. Positive solutions for a system of fractional differential equations with two parameters. *J. Funct. Spaces* **2018**, *2018*, 1462505. [CrossRef]
70. Zhai, C.; Wang, W.; Li, H. A uniqueness method to a new Hadamard fractional differential system with four-point boundary conditions. *J. Inequal. Appl.* **2018**, *2018*, 207. [CrossRef]
71. Cheng, W.; Xu, J.; Cui, Y.; Ge, Q. Positive solutions for a class of fractional difference systems with coupled boundary conditions. *Adv. Differ. Equ.* **2019**, *2019*, 249. [CrossRef]

72. Cheng, W.; Xu, J.; Cui, Y. Positive solutions for a system of nonlinear semipositone fractional q-difference equations with q-integral boundary conditions. *J. Nonlinear Sci. Appl.* **2017**, *10*, 4430–4440. [CrossRef]
73. Xu, J.; Goodrich, C.S.; Cui, Y. Positive solutions for a system of first-order discrete fractional boundary value problems with semipositone nonlinearities. *Rev. R. Acad. Cienc. Exactas Fís. Nat. Ser. A Mat.* **2019**, *113*, 1343–1358. [CrossRef]
74. Qiu, X.; Xu, J.; O'Regan, D.; Cui, Y. Positive solutions for a system of nonlinear semipositone boundary value problems with Riemann–Liouville fractional derivatives. *J. Funct. Spaces* **2018**, *2018*, 7351653. [CrossRef]
75. Chen, C.; Xu, J.; O'Regan, D.; Fu, Z. Positive solutions for a system of semipositone fractional difference boundary value problems. *J. Funct. Spaces* **2018**, *2018*, 6835028. [CrossRef]
76. Wang, F.; Cui, Y.; Zhou, H. Solvability for an infinite system of fractional order boundary value problems. *Ann. Funct. Anal.* **2019**, *10*, 395–411. [CrossRef]
77. Wang, F.; Cui, Y. Positive solutions for an infinite system of fractional order boundary value problems. *Adv. Differ. Equ.* **2019**, *2019*, 169. [CrossRef]
78. Riaz, U.; Zada, A.; Ali, Z.; Ahmad, M.; Xu, J.; Fu, Z. Analysis of nonlinear coupled systems of impulsive fractional differential equations with Hadamard derivatives. *Math. Probl. Eng.* **2019**, *2019*, 5093572. [CrossRef]
79. Riaz, U.; Zada, A.; Ali, Z.; Cui, Y.; Xu, J. Analysis of coupled systems of implicit impulsive fractional differential equations involving Hadamard derivatives. *Adv. Differ. Equ.* **2019**, *2019*, 226. [CrossRef]
80. Aris, R. *Introduction to the Analysis of Chemical Reactors*; Prentice Hall: Englewood Cliffs, NJ, USA, 1965.
81. Guo, D.; Lakshmikantham, V. *Nonlinear Problems in Abstract Cones*; Academic Press: Orlando, FL, USA, 1988.
82. Yang, Z.; Zhang, Z. Positive solutions for a system of nonlinear singular Hammerstein integral equations via nonnegative matrices and applications. *Positivity* **2012**, *16*, 783–800. [CrossRef]
83. Abraham, B.; Plemmons, R.J. *Nonnegative Matrices in the Mathematical Sciences*; Academic Press: New York, NY, USA, 1979.

11

A Mollification Regularization Method for the Inverse Source Problem for a Time Fractional Diffusion Equation

Le Dinh Long [1], **Yong Zhou** [2,3], **Tran Thanh Binh** [4] **and Nguyen Can** [5,*]

1 Faculty of Natural Sciences, Thu Dau Mot University, Thu Dau Mot City 820000, Binh Duong Province, Vietnam; ledinhlong@tdmu.edu.vn
2 Faculty of Information Technology, Macau University of Science and Technology, Macau 999078, China; yzhou@xtu.edu.cn
3 Faculty of Mathematics and Computational Science, Xiangtan University, Xiangtan 411105, China
4 Institute of Research and Development, Duy Tan University, Da Nang 550000, Vietnam; tranthanhbinh6@duytan.edu.vn
5 Applied Analysis Research Group, Faculty of Mathematics and Statistics, Ton Duc Thang University, Ho Chi Minh City 700000, Vietnam
* Correspondence: nguyenhuucan@tdtu.edu.vn

Abstract: We consider a time-fractional diffusion equation for an inverse problem to determine an unknown source term, whereby the input data is obtained at a certain time. In general, the inverse problems are ill-posed in the sense of Hadamard. Therefore, in this study, we propose a mollification regularization method to solve this problem. In the theoretical results, the error estimate between the exact and regularized solutions is given by a priori and a posteriori parameter choice rules. Besides, the proposed regularized methods have been verified by a numerical experiment.

Keywords: time-fractional diffusion equation; inverse problem; ill-posed problem; convergence estimates

---

1. Introduction

In this work, we study an inverse source problem for the time-fractional diffusion equation in a infinite domain as follows:

$$\begin{cases} \dfrac{\partial^\beta u(x,t)}{\partial t^\beta} = u_{xx}(x,t) + \phi(t) f(x), \quad (x,t) \in \mathbb{R} \times (0,T], \\ u(x,0) = 0, x \in \mathbb{R}, \\ u(x,T) = g(x), \quad x \in \mathbb{R}, \end{cases} \tag{1}$$

where the fractional derivative $\frac{\partial^\beta u}{\partial t^\beta}$ is the Caputo derivative of order β $(0 < \beta < 1)$ as defined by

$$\frac{d^\beta f(t)}{dt^\beta} = \frac{1}{\Gamma(1-\beta)} \int_0^t \frac{df(s)}{ds} \frac{ds}{(t-s)^\beta}, \tag{2}$$

and $\Gamma(\cdot)$ denotes the standard Gamma function.

The biggest motivation for developing the problem (1) is the inverse problems for the heat equation; we recover the unknown source function under different assumptions on the smoothness of input data, which were proposed by Igor Malyshev in Reference [1]. The inverse problems of the restoration of a source function in the heat equation with the classical derivative are studied by many researchers, that is, Geng [2] and Shidfar [3].

The mathematical model (1) arising in control theory, physical, generalized voltage divider, elasticity and the model of neurons in biology is studied in References [4–6].

According to our search, the fractional inverse source problems (1) are the subject of very few works, for example, Sakamoto et al. [7] used the data $u(x_0, t)(x_0 \in \mathbb{R})$ to determine $\phi(t)$ once $f(x)$ was given, where the authors obtained a Lipschitz stability for $\phi(t)$. In Problem (1) for a one-dimensional problem with special coefficients, Wei et al. [8] used the Fourier truncation method to solve an inverse source problem with $\phi(t) = 1$. In Reference [9], using the mollification regularization method, Yang and Fu determined the inverse spatial-dependent heat source problem. In Reference [10], Wei and Wang considered a modified quasi boundary value regularization method for identifying this problem. In Reference [11], using the quasi-reversibility regularization method, Yang and his group identified the unknown source for a time fractional diffusion equation. In Reference [12], with the quasi-reversibility regularization method, Wei and her group investigated a space-dependent source for the time fractional diffusion equation. Actually, to our knowledge, in the case $\phi(t)$, dependent on time, the results of the inverse source problem for the time-fractional diffusion equation still has a limited achievement, if $\phi(t) \neq 0$, we know Huy and his group investigated this problem by way of the Tikhonov regularization method, see Reference [13]. In these regularization methods, the priori parameter choice rule depends on the noise level and the priori bound. But in practice, to know exactly this is very difficult. In the above research, by using Morozov's Discrepancy Principle for choosing the regularization parameter in Tikhonov regularization, the authors show error estimation for both the priori choice rule parameter and the posteriori choice rule parameter.

In this paper, we use the mollification method to solve the inverse source problem. Instead of receiving the correct input data, we only get the approximate input data. We assume that the measured data in functions couple $(g_\varepsilon(x) \in \mathscr{L}_2(\mathbb{R}), \phi_\varepsilon(t) \in C[0,T])$ satisfies

$$\|g - g_\varepsilon\|_{\mathscr{L}^2(\mathbb{R})} \leq \varepsilon, \ \ \|\phi - \phi_\varepsilon\|_{C[0,T]} \leq \varepsilon, \tag{3}$$

where the constant $\varepsilon > 0$ represents a noise level. It is known that the inverse source problem mentioned above is ill-posed in the sense of Hadamard, that is, a solution of this problem (1) does not always exist, if the solution does exist, it is not dependent continuously on the given data, meaning that the error of the initial data is small, the error of the solution will be large. This makes trouble for the numerical solution; here a regularization is required. The Fourier transform of a function $\mathcal{F}$ is defined by

$$\widehat{\mathcal{F}}(\xi) = \frac{1}{\sqrt{2\pi}} \int_{\mathbb{R}} e^{-i\xi x} \mathcal{F}(x) dx. \tag{4}$$

We imposed an a priori bound on the input data, that is,

$$\|\mathcal{F}\|_{\mathbb{H}^k(\mathbb{R})} \leq \mathcal{M}, \ \ k > 0, \tag{5}$$

where $\mathcal{M} \geq 0$ is a constant, $\|\cdot\|_{\mathbb{H}^k(\mathbb{R})}$ denotes the norm in Sobolev space $\mathbb{H}^k(\mathbb{R})$ is defined

$$\mathbb{H}^k(\mathbb{R}) = \left\{ \mathcal{F} \in \mathscr{L}_2(\mathbb{R}), \|\mathcal{F}\|_k \leq \infty \right\}, \ \ and \ \ \|\mathcal{F}\|_{\mathbb{H}^k(\mathbb{R})} = \left(\int_{\mathbb{R}} \left| (1+\xi^2)^{k/2} \widehat{\mathcal{F}}(\xi) \right|^2 d\xi \right)^{\frac{1}{2}}. \tag{6}$$

The outline of this paper is divided into the following sections: Section 2 gives some auxiliary results. In Section 3, by the priori bound assumption of the exact solution and the priori parameter choice rule, we present the convergence rate. In Section 4, we show the convergence rate between the exact and regularized solutions under the posteriori parameter choice rule. Next, a numerical example is proposed to show the illustration of the results in theory in Section 5. Finally, a conclusion is given in Section 6.

2. Some Auxiliary Results

Before showing some lemmas, we recall the Mittag-Leffler function which is defined by

$$E_{\beta,\kappa}(z) = \sum_{k=0}^{\infty} \frac{z^k}{\Gamma(\beta k + \kappa)}, \quad z \in \mathbb{C}, \tag{7}$$

where $\beta > 0$ and $\kappa \in \mathbb{R}$ are arbitrary constant. In Reference [14], the properties of the Mittag-Leffler function are discussed. Hereby, we present the following Lemmas of the Mittag-Leffler function which can be found in Reference ([14], Chapter 1).

Lemma 1. *Let $0 < \beta_0 < \beta_1 < 1$. Then there exist the constants $\overline{B}_1, \overline{B}_2, \overline{B}_3$ depending only on β_0, β_1 such that for all $\beta \in [\beta_0, \beta_1]$,*

$$\frac{\overline{B}_1}{\Gamma(1-\beta)} \frac{1}{1-x} \leq E_{\beta,1}(x) \leq \frac{\overline{B}_2}{\Gamma(1-\beta)} \frac{1}{1-x}, \quad E_{\beta,\alpha}(x) \leq \frac{\overline{B}_3}{1-x}, \quad \forall x \leq 0, \forall \alpha \in \mathbb{R}. \tag{8}$$

These estimates are uniform for all $\beta \in [\beta_0, \beta_1]$·

Lemma 2. *(see Reference [7]) For $0 < \beta < 1$, we have:*

$$E_{\beta,\beta}(-\zeta) \geq 0, \quad \zeta \geq 0.$$

Proof. As for the proof, see Miller and Samko [15]. □

Lemma 3. *(see Reference [7]) For $\xi > 0$, $\alpha > 0$ and a positive integer $n \in \mathbb{N}$, we have:*

$$\begin{aligned} &\frac{d^n}{dt^n} E_{\beta,1}(-\xi^2 t^\beta) = -\xi^2 t^{\beta-n} E_{\beta,\beta-n+1}(-\xi^2 t^\beta), \quad t > 0, \\ &\frac{d}{dt}(t E_{\beta,2}(-\xi^2 t^\beta)) = E_{\beta,1}(-\xi^2 t^\beta), \quad t \geq 0. \end{aligned} \tag{9}$$

Lemma 4. *(see Reference [7]) By Lemma 2 and Lemma 3, we have*

$$\begin{aligned} \int_0^\varrho \left| t^{\gamma-1} E_{\beta,\beta}(-\xi^2 t^\gamma) \right| dt &= \int_0^\varrho t^{\gamma-1} E_{\beta,\beta}(-\xi^2 t^\gamma) dt \\ &= -\frac{1}{\xi^2} \int_0^\varrho \frac{d}{dt} E_{\beta,1}(-\xi^2 t^\gamma) dt = \frac{1}{\xi^2}\Big(1 - E_{\alpha,1}(-\xi^2 \varrho^\alpha)\Big), \quad \varrho > 0. \end{aligned} \tag{10}$$

Lemma 5. *(see Reference [16]) For $0 < \alpha < 1$, $\xi \in \mathbb{R}$, the following inequalities hold:*

$$\sup_{\xi \in \mathbb{R}} \left| (1+\xi^2)^{-k} (1 - e^{-\frac{\alpha^2 \xi^2}{4}}) \right| \leq \max\left\{\alpha^{2k}, \alpha^2\right\}. \tag{11}$$

Proof. The proof can be found in Reference [9]. □

Lemma 6. *Let $\beta \in (0,1)$ and $\xi \in \mathbb{R}$, the following estimate holds*

$$\left(\int_0^T s^{\beta-1}E_{\beta,\beta}(-\xi^2 s^\beta)ds\right)^{-1} = \frac{\xi^2}{1-E_{\beta,1}(-\xi^2T^\beta)} \leq \begin{cases} \dfrac{\xi^2}{1-E_{\beta,1}(-T^\beta)}, & \text{if } |\xi| \geq 1, \\ \dfrac{1}{1-E_{\beta,1}(-T^\beta)}, & \text{if } |\xi| < 1. \end{cases} \tag{12}$$

Proof. If $|\xi| \geq 1$ then since $E_{\beta,1}(-y)$ for $0<\beta<1$ is a decreasing function for $y>0$, we get $E_{\beta,1}(-\xi^2T^\beta) \leq E_{\beta,1}(-T^\beta)$. Whereupon

$$\left(\int_0^T s^{\beta-1}E_{\beta,\beta}(-\xi^2 s^\beta)ds\right)^{-1} = \frac{\xi^2}{1-E_{\beta,1}(-\xi^2T^\beta)} \leq \frac{\xi^2}{1-E_{\beta,1}(-T^\beta)}, \quad \text{for } |\xi| \geq 1. \tag{13}$$

If $|\xi| \leq 1$ then since $E_{\beta,\beta}(-y)$ with $0<\beta<1$ is a decreasing function for $y>0$, we get $E_{\beta,\beta}(-\xi^2s^\beta) \geq E_{\beta,\beta}(-s^\beta)$, so

$$\left(\int_0^T s^{\beta-1}E_{\beta,\beta}(-\xi^2 s^\beta)ds\right)^{-1} \leq \left(\int_0^T s^{\beta-1}E_{\beta,\beta}(-s^\beta)ds\right)^{-1} = \frac{1}{1-E_{\beta,1}(-T^\beta)}, \quad \text{for } |\xi| \leq 1. \tag{14}$$

□

Lemma 7. *For $\alpha \in (0,1)$ and $\xi \in \mathbb{R}$, from Lemma 6, one has:*

$$\begin{aligned} \frac{1}{\left(\int_0^T s^{\beta-1}E_{\beta,\beta}(-\xi^2 s^\beta)ds\right)e^{\frac{\alpha^2\xi^2}{4}}} &= \frac{\xi^2}{\left(1-E_{\beta,1}(-\xi^2T^\beta)\right)e^{\frac{\alpha^2\xi^2}{4}}} \\ &\leq \begin{cases} \dfrac{\frac{\xi^2\alpha^2}{4}}{\left(1-E_{\beta,1}(-T^\beta)\right)e^{\frac{\alpha^2\xi^2}{4}}} \leq \left(\dfrac{4}{\alpha^2}\right)\left(\dfrac{1}{(1-E_{\beta,1}(-T^\beta))}\right), & \text{if } |\xi| \geq 1, \\ \dfrac{1}{\left(1-E_{\beta,1}(-T^\beta)\right)e^{\frac{\alpha^2\xi^2}{4}}} \leq \left(\dfrac{4}{\alpha^2}\right)\left(\dfrac{1}{1-E_{\beta,1}(-T^\beta)}\right), & \text{if } |\xi| < 1. \end{cases} \end{aligned} \tag{15}$$

This gives

$$\frac{1}{\left(\int_0^T s^{\beta-1}E_{\beta,\beta}(-\xi^2 s^\beta)ds\right)e^{\frac{\alpha^2\xi^2}{4}}} \leq \left(\frac{4}{\alpha^2}\right)\left(\frac{1}{1-E_{\beta,1}(-T^\beta)}\right). \tag{16}$$

3. The Priori Parameter Choice

Next, the error estimate of the mollification regularization method will be derived under the priori parameter choice rule in this section. We consider the Gauss function

$$\rho_\alpha(x) := \frac{1}{\alpha\sqrt{\pi}}e^{-\frac{x^2}{\alpha^2}}, \tag{17}$$

as the mollifer kernel, where α is a positive constant.

We define an operator K_α as

$$K_\alpha f(x) := \rho_\alpha f(x) := \int_{\mathbb{R}} \rho_\alpha(t) f(x-t)dt = \int_{\mathbb{R}} \rho_\alpha(x-t) f(t)dt, \tag{18}$$

for $f(x) \in \mathscr{L}_2(\mathbb{R})$. The original ill-posed problem is replaced by a new problem of searching its approximation $f_{\varepsilon,\alpha}(x)$ which is defined by

$$f_{\varepsilon,\alpha}(x) := \frac{1}{\sqrt{2\pi}} \int_{\mathbb{R}} e^{ix\xi} \rho_\alpha \widehat{f_\varepsilon(\xi)} d\xi, \tag{19}$$

The Inverse Source Problem

By using the Fourier transform, the problem (1) is formulated in the following frequency space

$$\begin{cases} \dfrac{\partial^\beta \widehat{u}(\xi,t)}{\partial t^\beta} + \xi^2 \widehat{u}(\xi,t) = \phi(t)\widehat{f}(\xi), & (\xi,t) \in \mathbb{R} \times (0,T], \\ \widehat{u}(\xi,0) = 0, & \xi \in \mathbb{R}, \\ \widehat{u}(\xi,T) = \widehat{g}(\xi), & \xi \in \mathbb{R}. \end{cases} \tag{20}$$

From the equation and the initial value in (20), we obtain

$$\widehat{u}(\xi,t) = \int_0^t (t-s)^{\beta-1} E_{\beta,\beta}(-\xi^2(t-s)^\beta)\phi(s)\widehat{f}(\xi)ds. \tag{21}$$

Or equivalently,

$$u(x,t) = \frac{1}{\sqrt{2\pi}} \int_{\mathbb{R}} e^{i\xi x} \Big(\int_0^t (t-s)^{\beta-1} E_{\beta,\beta}(-\xi^2(t-s)^\beta)\phi(s)ds \Big) \widehat{f}(\xi) d\xi. \tag{22}$$

Set

$$\mathcal{D}_\beta(\xi, t-s) = (t-s)^{\beta-1} E_{\beta,\beta}(-\xi^2(t-s)^\beta).$$

And $\widehat{u}(\xi,T) = \widehat{g}(\xi)$ in (20), one has

$$\widehat{f}(\xi) = \frac{\widehat{g}(\xi)}{\int_0^T \mathcal{D}_\beta(\xi, T-s)\phi(s)ds}. \tag{23}$$

Using the inverse Fourier transform, then we obtain the formula of the source function f

$$f(x) = \frac{1}{\sqrt{2\pi}} \int_{\mathbb{R}} e^{i\xi x} \frac{\widehat{g}(\xi)}{\int_0^T \mathcal{D}_\beta(\xi, T-s)\phi(s)ds} d\xi. \tag{24}$$

On the other hand, if $\phi(t)$ is bounded by $\inf_{t\in[0,T]} |\phi(t)| \le \phi(t) \le \sup_{t\in[0,T]} |\phi(t)| = \|\phi\|_{C[0,T]}$, we have $\Big(\int_0^T \mathcal{D}_\beta(\xi, T-s)\phi(s)ds \Big)^{-1}$ can be written then $\frac{1}{\inf_{t\in[0,T]}|\phi(t)|} \frac{\xi^2}{(1-E_{\beta,1}(-\xi^2 T^\beta))}$. The unbounded function $\frac{\xi^2}{\left(1-E_{\beta,1}(-\xi^2 T^\beta)\right)}$ can be seen as an amplification when $\xi \to \infty$. From now on, putting $\inf_{t\in[0,T]} |\phi(t)| = \mathcal{A}_0$, $\inf_{t\in[0,T]} |\phi_\varepsilon(t)| = \mathcal{A}_1$, $\sup_{t\in[0,T]} |\phi(t)| = \|\phi\|_{C[0,T]} = \Phi$. From (19) with α is a regularization parameter and α depends on ε, we found the regularized solution

$$\widehat{f_{\varepsilon,\alpha}}(\xi) = \frac{\widehat{g_\varepsilon}(\xi)}{\left(\int\limits_0^T \mathcal{D}_\beta(\xi, T-s)\phi_\varepsilon(s)ds\right) e^{\frac{\alpha^2\xi^2}{4}}}. \tag{25}$$

Using inverse Fourier transform, we get

$$f_{\varepsilon,\alpha}(x) = \frac{1}{\sqrt{2\pi}} \int\limits_{\mathbb{R}} \frac{\widehat{g_\varepsilon}(\xi)}{\left(\int\limits_0^T \mathcal{D}_\beta(\xi, T-s)\phi_\varepsilon(s)ds\right) e^{\frac{\alpha^2\xi^2}{4}}} e^{i\xi x} d\xi. \tag{26}$$

The main conclusion of this section are given below.

Theorem 1. *Let $f(x)$, given by (24), be the exact solution of (1) with exact data $g \in \mathscr{L}_2(\mathbb{R})$, and $f_{\varepsilon,\alpha}(x)$ is approximation solution of $f(x)$ with measured data $g_\varepsilon \in \mathscr{L}_2(\mathbb{R})$. Then we obtain*

a. If $0 < k < 1$, and choosing $\alpha(\varepsilon) = \left(\frac{\varepsilon}{\mathcal{M}}\right)^{\frac{1}{2(k+1)}}$, we have a convergence estimate

$$\|f(.) - f_{\varepsilon,\alpha}(.)\|_{\mathscr{L}_2(\mathbb{R})} \le \varepsilon^{\frac{k}{k+1}} \mathcal{M}^{\frac{1}{k+1}} \left(\max\left\{1, \left(\frac{\varepsilon}{\mathcal{M}}\right)^{\frac{1-k}{k+1}}\right\} + \mathcal{R}(\mathcal{A}_0, \mathcal{A}_1, \widehat{g})\right). \tag{27}$$

b. If $k > 1$, by choosing $\alpha(\varepsilon) = \left(\frac{\varepsilon}{\mathcal{M}}\right)^{\frac{1}{4}}$, we have a convergence estimate

$$\|f(.) - f_{\varepsilon,\alpha}(.)\|_{\mathscr{L}_2(\mathbb{R})} \le \varepsilon^{\frac{1}{2}} \mathcal{M}^{\frac{1}{2}} \left(1 + \mathcal{R}(\mathcal{A}_0, \mathcal{A}_1, \widehat{g})\right), \tag{28}$$

in which

$$\mathcal{R}(\mathcal{A}_0, \mathcal{A}_1, \widehat{g}) = \frac{4}{(1 - E_{\beta,1}(-T^\beta))} \left(\frac{1}{\mathcal{A}_1} + \frac{\|\widehat{g}\|_{\mathscr{L}_2(\mathbb{R})}}{\mathcal{A}_1\mathcal{A}_0}\right). \tag{29}$$

Proof. From (24) and (26), by the Parseval formula, the triangle inequality, we obtain

$$\begin{aligned} \|f(.) - f_{\varepsilon,\alpha}(.)\|_{\mathscr{L}_2(\mathbb{R})} &= \|\widehat{f}(.) - \widehat{f_{\varepsilon,\alpha}}(.)\|_{\mathscr{L}_2(\mathbb{R})} \\ &= \left\| \frac{\widehat{g}(\xi)}{\int\limits_0^T \mathcal{D}_\beta(\xi, T-s)\phi(s)ds} - \frac{\widehat{g_\varepsilon}(\xi)}{\left(\int\limits_0^T \mathcal{D}_\beta(\xi, T-s)\phi_\varepsilon(s)ds\right) e^{\frac{\alpha^2\xi^2}{4}}} \right\|_{\mathscr{L}_2(\mathbb{R})} \\ &= \|\mathcal{I}_1\|_{\mathscr{L}_2(\mathbb{R})} + \|\mathcal{I}_2\|_{\mathscr{L}_2(\mathbb{R})} + \|\mathcal{I}_3\|_{\mathscr{L}_2(\mathbb{R})}, \end{aligned} \tag{30}$$

in which

$$\mathcal{I}_1 = \frac{\widehat{g}(\xi)}{\int\limits_0^T \mathcal{D}_\beta(\xi, T-s)\phi(s)ds} - \frac{\widehat{g}(\xi)}{\Big(\int\limits_0^T \mathcal{D}_\beta(\xi, T-s)\phi(s)ds\Big)e^{\frac{\alpha^2\xi^2}{4}}},$$

$$\mathcal{I}_2 = \frac{\widehat{g}(\xi)}{\Big(\int\limits_0^T \mathcal{D}_\beta(\xi, T-s)\phi_\varepsilon(s)ds\Big)e^{\frac{\alpha^2\xi^2}{4}}} - \frac{\widehat{g}_\varepsilon(\xi)}{\Big(\int\limits_0^T \mathcal{D}_\beta(\xi, T-s)\phi_\varepsilon(s)ds\Big)e^{\frac{\alpha^2\xi^2}{4}}},$$

$$\mathcal{I}_3 = \frac{\widehat{g}(\xi)}{\Big(\int\limits_0^T \mathcal{D}_\beta(\xi, T-s)\phi(s)ds\Big)e^{\frac{\alpha^2\xi^2}{4}}} - \frac{\widehat{g}(\xi)}{\Big(\int\limits_0^T \mathcal{D}_\beta(\xi, T-s)\phi_\varepsilon(s)ds\Big)e^{\frac{\alpha^2\xi^2}{4}}}. \tag{31}$$

Next, we estimate the error by diving it into three steps as follows
Step 1: Estimate for $\|\mathcal{I}_1\|^2_{\mathscr{L}_2(\mathbb{R})}$, we have

$$\begin{aligned}
\|\mathcal{I}_1\|^2_{\mathscr{L}_2(\mathbb{R})} &= \left\| \frac{\widehat{g}(\xi)}{\Big(\int\limits_0^T \mathcal{D}_\beta(\xi, T-s)\phi(s)ds\Big)} (1 - e^{-\frac{\alpha^2\xi^2}{4}}) \right\|^2_{\mathscr{L}_2(\mathbb{R})} \\
&= \left\| (1+\xi^2)^{-k}(1 - e^{-\frac{\alpha^2\xi^2}{4}})(1+\xi^2)^k \widehat{f}(\xi) \right\|^2_{\mathscr{L}_2(\mathbb{R})} \\
&\le \sup_{\xi\in\mathbb{R}} \left| (1+\xi^2)^{-k}(1 - e^{-\frac{\alpha^2\xi^2}{4}}) \right|^2 \|f\|^2_{H^k(\mathbb{R})} \le \mathcal{M}^2 \max\left\{\alpha^{4k}, \alpha^4\right\}.
\end{aligned}$$

Hence,

$$\|\mathcal{I}_1\|_{\mathscr{L}_2(\mathbb{R})} \le \mathcal{M} \max\left\{\alpha^{2k}, \alpha^2\right\}. \tag{32}$$

Step 2: Estimate for $\|\mathcal{I}_2\|^2_{\mathscr{L}_2(\mathbb{R})}$, we get

$$\begin{aligned}
\|\mathcal{I}_2\|^2_{\mathscr{L}_2(\mathbb{R})} &= \left\| \frac{\widehat{g}(\xi)}{\Big(\int\limits_0^T \mathcal{D}_\beta(\xi, T-s)\phi_\varepsilon(s)ds\Big)e^{\frac{\alpha^2\xi^2}{4}}} - \frac{\widehat{g}_\varepsilon(\xi)}{\Big(\int\limits_0^T \mathcal{D}_\beta(\xi, T-s)\phi_\varepsilon(s)ds\Big)e^{\frac{\alpha^2\xi^2}{4}}} \right\|^2_{\mathscr{L}_2(\mathbb{R})} \\
&\le \mathcal{A}_1^{-2} \|\widehat{g}(\xi) - \widehat{g}_\varepsilon(\xi)\|^2_{\mathscr{L}_2(\mathbb{R})} \sup_{\xi\in\mathbb{R}} \left| \Big(\int\limits_0^T (T-s)^{\beta-1} E_{\beta,\beta}(-\xi^2(T-s)^\beta)ds\Big) e^{\frac{\alpha^2\xi^2}{4}} \right|^{-2} \\
&\le \mathcal{A}_1^{-2} \|\widehat{g}(\xi) - \widehat{g}_\varepsilon(\xi)\|^2_{\mathscr{L}_2(\mathbb{R})} \sup_{\xi\in\mathbb{R}} \left| \frac{\xi^2}{(1 - E_{\beta,1}(-\xi^2 T^\beta)e^{\frac{\alpha^2\xi^2}{4}}} \right|^2 \\
&\le \Big(\frac{16\varepsilon^2}{\alpha^4}\Big)\Big(\mathcal{A}_1(1 - E_{\beta,1}(-T^\beta))\Big)^{-2}.
\end{aligned} \tag{33}$$

Hence, we conclude that

$$\|\mathcal{I}_2\|_{\mathscr{L}_2(\mathbb{R})} \le \Big(\frac{4\varepsilon}{\alpha^2}\Big)\Big(\mathcal{A}_1(1 - E_{\beta,1}(-T^\beta))\Big)^{-1}. \tag{34}$$

Step 3: Estimate for $\|\mathcal{I}_3\|^2_{\mathscr{L}_2(\mathbb{R})}$, we have

$$\begin{aligned}\|\mathcal{I}_3\|^2_{\mathscr{L}_2(\mathbb{R})} &= \left\| \frac{\widehat{g}(\xi)}{\Big(\int_0^T \mathcal{D}_\beta(\xi, T-s)\phi(s)ds\Big)e^{\frac{\alpha^2\xi^2}{4}}} - \frac{\widehat{g}(\xi)}{\Big(\int_0^T \mathcal{D}_\beta(\xi, T-s)\phi_\varepsilon(s)ds\Big)e^{\frac{\alpha^2\xi^2}{4}}} \right\|^2_{\mathscr{L}_2(\mathbb{R})} \\ &= \left\| \frac{\widehat{g}(\xi)}{e^{\frac{\alpha^2\xi^2}{4}}} \frac{\int_0^T \mathcal{D}_\beta(\xi, T-s)\Big(\phi_\varepsilon(s)-\phi(s)\Big)ds}{\Big(\int_0^T \mathcal{D}_\beta(\xi, T-s)\phi_\varepsilon(s)ds\Big)\Big(\int_0^T \mathcal{D}_\beta(\xi, T-s)\phi(s)ds\Big)} \right\|^2_{\mathscr{L}_2(\mathbb{R})}. \end{aligned} \tag{35}$$

From (35), we get

$$\begin{aligned}\|\mathcal{I}_3\|^2_{\mathscr{L}_2(\mathbb{R})} &= \mathcal{A}_1^{-2}\|\phi_\varepsilon-\phi\|^2_{C[0,T]} \left\| \frac{\widehat{g}(\xi)}{\Big(\int_0^T \mathcal{D}_\beta(\xi, T-s)\phi(s)ds\Big)e^{\frac{\alpha^2\xi^2}{4}}}\right\|^2_{\mathscr{L}_2(\mathbb{R})} \\ &\le \mathcal{A}_1^{-2}\|\phi_\varepsilon-\phi\|^2_{C[0,T]} \left\| \frac{\widehat{g}(\xi)}{\Big(\int_0^T \mathcal{D}_\beta(\xi, T-s)\phi(s)ds\Big)e^{\frac{\alpha^2\xi^2}{4}}}\right\|^2_{\mathscr{L}_2(\mathbb{R})} \\ &\le (\mathcal{A}_0\mathcal{A}_1)^{-2}\|\phi_\varepsilon-\phi\|^2_{C[0,T]} \left| \frac{\xi^2}{(1-E_{\beta,1}(-\xi^2T^\beta))\,e^{\frac{\alpha^2\xi^2}{4}}}\right|^2 \int_{\mathbb{R}} |\widehat{g}(\xi)|^2 d\xi \\ &\le \left(\frac{16}{\alpha^4}\right)\Big(\mathcal{A}_0\mathcal{A}_1(1-E_{\beta,1}(-T^\beta))\Big)^{-2}\|\phi_\varepsilon-\phi\|^2_{C[0,T]}\int_{\mathbb{R}} |\widehat{g}(\xi)|^2 d\xi. \end{aligned} \tag{36}$$

Hence,

$$\|\mathcal{I}_3\|_{\mathscr{L}_2(\mathbb{R})} \le \left(\frac{4\varepsilon}{\alpha^2}\right)\Big(\mathcal{A}_0\mathcal{A}_1(1-E_{\beta,1}(-T^\beta))\Big)^{-1}\|\widehat{g}\|_{\mathscr{L}_2(\mathbb{R})}. \tag{37}$$

Combining (32), (34) and (37), we received

(a) If $0 \le k \le 1$ by choosing $\alpha(\varepsilon) = \left(\frac{\varepsilon}{\mathcal{M}}\right)^{\frac{1}{2(k+1)}}$, we have a convergent estimation:

$$\|f(.) - f_{\varepsilon,\alpha}(.)\|_{\mathscr{L}_2(\mathbb{R})} \text{ is of order } \varepsilon^{\frac{k}{k+1}}. \tag{38}$$

(b) If $k > 1$, by choosing $\alpha(\varepsilon) = \left(\frac{\varepsilon}{\mathcal{M}}\right)^{\frac{1}{4}}$, we have a convergent estimation:

$$\|f(.) - f_{\varepsilon,\alpha}(.)\|_{\mathscr{L}_2(\mathbb{R})} \text{ is of order } \varepsilon^{\frac{1}{2}}. \tag{39}$$

□

4. The Discrepancy Principle

Now, we present the posteriori regularization parameter choice rule. The most general of the posteriori rules is the Morozov discrepancy principle [17]. Choosing the regularization parameter α as the solution of the equation

$$l(\alpha) = \left\| e^{-\frac{\alpha^2\xi^2}{4}} \widehat{g}_\varepsilon(\xi) - \widehat{g}_\varepsilon(\xi) \right\|_{\mathscr{L}_2(\mathbb{R})} = \varepsilon + \eta\left(\log\left(\log\left(\frac{T}{\varepsilon}\right)\right)\right)^{-1}, \tag{40}$$

where $\eta > 1$ is a constant.

Remark 1. *To ensure the existence and uniqueness, we can choose η such that*

$$0 < \varepsilon + \eta\left(\log\left(\log\left(\frac{T}{\varepsilon}\right)\right)\right)^{-1} < \|\widehat{g}_\varepsilon\|_{\mathscr{L}_2(\mathbb{R})}.$$

To establish the existence and uniqueness of the solution of Equation (40), we consider the following lemmas

Lemma 8. *If $\varepsilon > 0$ then there holds:*

(a) *$l(\alpha)$ is a continous function.*
(b) $\lim_{\alpha\to 0^+} l(\alpha) = 0$.
(c) $\lim_{\alpha\to+\infty} l(\alpha) = \|\widehat{g}_\varepsilon\|_{\mathscr{L}_2(\mathbb{R})}$.
(d) *$l(\alpha)$ is a strictly increasing function.*

The proof is very easy and we omit it here.

Lemma 9. *The following inequality holds:*

$$\|e^{-\frac{\alpha^2\xi^2}{4}} \widehat{g}_\varepsilon(\xi) - \widehat{g}(\xi)\|_{\mathscr{L}_2(\mathbb{R})} \le 2\varepsilon + \eta\left(\log\left(\log\left(\frac{T}{\varepsilon}\right)\right)\right)^{-1}. \tag{41}$$

Proof. Applying the triangle inequality and (40), we have

$$\begin{aligned}
\|e^{-\frac{\alpha^2\xi^2}{4}} \widehat{g}_\varepsilon(\xi) - \widehat{g}(\xi)\|_{\mathscr{L}_2(\mathbb{R})} &\le \|e^{-\frac{\alpha^2\xi^2}{4}} \widehat{g}_\varepsilon(\xi) - \widehat{g}_\varepsilon(\xi)\|_{\mathscr{L}_2(\mathbb{R})} + \|\widehat{g}_\varepsilon(\xi) - \widehat{g}(\xi)\|_{\mathscr{L}_2(\mathbb{R})} \\
&\le \|e^{-\frac{\alpha^2\xi^2}{4}} \widehat{g}_\varepsilon(\xi) - \widehat{g}_\varepsilon(\xi)\|_{\mathscr{L}_2(\mathbb{R})} + \|\widehat{g}_\varepsilon(\xi) - \widehat{g}(\xi)\|_{\mathscr{L}_2(\mathbb{R})} \\
&\le 2\varepsilon + \eta\left(\log\left(\log\left(\frac{T}{\varepsilon}\right)\right)\right)^{-1}.
\end{aligned} \tag{42}$$

□

Lemma 10. *For any $0 \neq \xi \in \mathbb{R}$, let $s, t \in [0, T]$ such that $0 \le s \le t \le T$, making the substitution ξ^2 and using the inequality: $\frac{\overline{B}_3 s^{\beta-1}}{1+\xi^2 s^\beta} \le \overline{B}_3 s^{\beta-1}$, we have the following estimate*

$$\int_0^T \mathcal{D}_\beta(\xi, T-s)ds = \int_0^T (T-s)^{\beta-1} E_{\beta,\beta}(-\xi^2(T-s)^\beta)ds \le \frac{\overline{B}_3 T^\beta}{\beta}. \tag{43}$$

Lemma 11. *If α is the solution of Equation (40), then the following inequality also holds:*

$$\frac{4}{\alpha^2} \le \frac{\mathcal{H}_\beta(\overline{B}_3, T, \Phi, \mathcal{M})}{\eta}\left(\log\left(\log\left(\frac{T}{\varepsilon}\right)\right)\right). \tag{44}$$

whereby $\mathcal{M} \geq \|f\|_{\mathbb{H}^k(\mathbb{R})}$.

Proof. Due to (40), we receive

$$
\begin{aligned}
&\varepsilon+\eta\Big(\log\Big(\log\Big(\frac{T}{\varepsilon}\Big)\Big)\Big)^{-1} = \Big\|(1-e^{-\frac{\alpha^2\xi^2}{4}})\widehat{g_\varepsilon}(\xi)\Big\|_{\mathscr{L}_2(\mathbb{R})} \\
&\leq \Big\|(1-e^{-\frac{\alpha^2\xi^2}{4}})\widehat{g_\varepsilon}(\xi)-(1-e^{-\frac{\alpha^2\xi^2}{4}})\widehat{g}(\xi)+(1-e^{-\frac{\alpha^2\xi^2}{4}})\widehat{g}(\xi)\Big\|_{\mathscr{L}_2(\mathbb{R})} \\
&\leq \varepsilon+\Big\|(1-e^{-\frac{\alpha^2\xi^2}{4}})(1+\xi^2)^{-k}\Big(\int_0^T \mathcal{D}_\beta(\xi,T-s)\phi(s)ds\Big)(1+\xi^2)^k\widehat{f}(\xi)\Big\|_{\mathscr{L}_2(\mathbb{R})} \\
&\leq \varepsilon+\sup_{\xi\in\mathbb{R}}\Big|(1-e^{-\frac{\alpha^2\xi^2}{4}})(1+\xi^2)^{-k}\Big(\int_0^T \mathcal{D}_\beta(\xi,T-s)\phi(s)ds\Big)\Big|\mathcal{M} \\
&\leq \varepsilon+\frac{\alpha^2}{4}\mathcal{H}_\beta(\overline{B}_3,T,\Phi,\mathcal{M}),
\end{aligned}
\tag{45}
$$

whereby

$$
\mathcal{H}_\beta(\overline{B}_3,T,\Phi,\mathcal{M}) = (\beta)^{-1}\Phi\overline{B}_3T^\beta\mathcal{M}. \tag{46}
$$

So

$$
\frac{4}{\alpha^2} \leq \frac{\mathcal{H}_\beta(\overline{B}_3,T,\Phi,\mathcal{M})}{\eta}\log\Big(\log\Big(\frac{T}{\varepsilon}\Big)\Big). \tag{47}
$$

□

Lemma 12. *For* $0<\alpha<1$, *using the Lemma 7, the following inequality holds:*

$$
\sup_{\xi\in\mathbb{R}}\Big|\Big(\frac{\xi^2}{1-E_{\beta,1}(-\xi^2T^\beta)}\Big)^{k+1}e^{-\frac{\alpha^2\xi^2}{4}}\Big| \leq \Big(\frac{k+1}{(1-E_{\beta,1}(-T^\beta))}\Big)^{k+1}\Big(\frac{4}{\alpha^2}\Big)^{k+1}. \tag{48}
$$

The proof is similar to Lemma 7 and we omit it here.

Next, the main results of this section are shown under Theorem.

Theorem 2. *Assume the condition* $\|g_\varepsilon - g\| \leq \varepsilon$ *where* $\|.\|$ *denotes the* $\mathscr{L}_2(\mathbb{R})$*-norm with* $\varepsilon > 0$ *is a noise level and the condition* (5) *holds, then there holds the following error estimate*

$$
\begin{aligned}
&\|f(.)-f_{\varepsilon,\alpha}(.)\|_{\mathscr{L}_2(\mathbb{R})} = \|\widehat{f}(.)-\widehat{f_{\varepsilon,\alpha}}(.)\|_{\mathscr{L}_2(\mathbb{R})} \\
&\leq \Big(\varepsilon\Big(\log\Big(\log\Big(\frac{T}{\varepsilon}\Big)\Big)\Big)^{k+1}\Big(\frac{\mathcal{L}_\beta(k,T)\ \mathcal{H}_\beta(\overline{B}_3,T,\Phi)}{\mathcal{A}_0\eta}\Big)^{k+1}\mathcal{M}^k \\
&+\frac{\Phi}{\mathcal{A}_0^{k+1}}\Big(\frac{1}{1-E_{\beta,1}(-T^\beta)}\Big)^k\Big)^{\frac{1}{k+1}}\mathcal{M}^{\frac{1}{k+1}}.\Big(2\varepsilon+\eta\Big(\log\Big(\log\Big(\frac{T}{\varepsilon}\Big)\Big)\Big)^{-1}\Big)^{\frac{k}{k+1}} \\
&+\Big(\varepsilon\log\Big(\log\Big(\frac{T}{\varepsilon}\Big)\Big)\Big)\Big(\frac{\mathcal{H}_\beta(\overline{B}_3,T,\Phi,\mathcal{M})}{\eta\mathcal{A}_0(1-E_{\beta,1}(-T^\beta))}+\Big(\frac{\mathcal{H}_\beta(\overline{B}_3,T,\Phi,\mathcal{M})\|\widehat{g}\|_{\mathscr{L}_2(\mathbb{R})}}{\eta(1-E_{\beta,1}(-T^\beta))\mathcal{A}_0\mathcal{A}_1}\Big)\Big).
\end{aligned}
\tag{49}
$$

Proof. By the Parseval formula, we get

$$\begin{aligned}
\|f(.) - f_{\varepsilon,\alpha}(.)\|_{\mathscr{L}_2(\mathbb{R})} &= \|\widehat{f}(.) - \widehat{f_{\varepsilon,\alpha}}(.)\|_{\mathscr{L}_2(\mathbb{R})} \\
&\leq \left\| \frac{e^{-\frac{\alpha^2\xi^2}{4}}\widehat{g}_\varepsilon(\xi) - \widehat{g}(\xi)}{\int_0^T \mathcal{D}_\beta(\xi, T-s)\phi(s)ds} \right\|_{\mathscr{L}_2(\mathbb{R})} \\
&+ \left\| \frac{\widehat{g}_\varepsilon(\xi)}{\left(\int_0^T \mathcal{D}_\beta(\xi, T-s)\phi(s)ds\right)e^{\frac{\alpha^2\xi^2}{4}}} - \frac{\widehat{g}(\xi)}{\left(\int_0^T \mathcal{D}_\beta(\xi, T-s)\phi(s)ds\right)e^{\frac{\alpha^2\xi^2}{4}}} \right\|_{\mathscr{L}_2(\mathbb{R})} \\
&+ \left\| \frac{\widehat{g}(\xi)}{\left(\int_0^T \mathcal{D}_\beta(\xi, T-s)\phi(s)ds\right)e^{\frac{\alpha^2\xi^2}{4}}} - \frac{\widehat{g}(\xi)}{\left(\int_0^T \mathcal{D}_\beta(\xi, T-s)\phi_\varepsilon(s)ds\right)e^{\frac{\alpha^2\xi^2}{4}}} \right\|_{\mathscr{L}_2(\mathbb{R})} \\
&\leq \|\mathcal{J}_1\|_{\mathscr{L}_2(\mathbb{R})} + \|\mathcal{J}_2\|_{\mathscr{L}_2(\mathbb{R})} + \|\mathcal{J}_3\|_{\mathscr{L}_2(\mathbb{R})},
\end{aligned} \tag{50}$$

We can divide the proof into three steps as follows:
Step 1: Estimate for $\|\mathcal{J}_1\|^2_{\mathscr{L}_2(\mathbb{R})}$, using the Hölder inequality, we obtain

$$\begin{aligned}
\|\mathcal{J}_1\|^2_{\mathscr{L}_2(\mathbb{R})} &= \frac{1}{\left(\int_0^T \mathcal{D}_\beta(\xi, T-s)\phi(s)ds\right)^2} \left\| e^{-\frac{\alpha^2\xi^2}{4}}\widehat{g}_\varepsilon(\xi) - \widehat{g}(\xi) \right\|^2_{\mathscr{L}_2(\mathbb{R})} \\
&\leq \left\| \frac{\xi^2}{\mathcal{A}_0(1 - E_{\beta,1}(-\xi^2 T^\beta))} \left(e^{-\frac{\alpha^2\xi^2}{4}}\widehat{g}_\varepsilon(\xi) - \widehat{g}(\xi)\right) \right\|^2_{\mathscr{L}_2(\mathbb{R})} \\
&\leq (\mathcal{C}_1^2)^{\frac{1}{k+1}} \times (\mathcal{C}_2^2)^{\frac{k}{k+1}},
\end{aligned} \tag{51}$$

whereby

$$\begin{aligned}
\mathcal{C}_1^2 &= \left(\int_{\mathbb{R}} \left(\left(\frac{\xi^2}{\mathcal{A}_0(1 - E_{\beta,1}(-\xi^2 T^\beta))} \right)^2 \left(e^{-\frac{\alpha^2\xi^2}{4}}\widehat{g}_\varepsilon(\xi) - \widehat{g}(\xi)\right)^{\frac{2}{k+1}} \right)^{k+1} d\xi \right), \\
\mathcal{C}_2^2 &= \left(\int_{\mathbb{R}} \left(\left(e^{-\frac{\alpha^2\xi^2}{4}}\widehat{g}_\varepsilon(\xi) - \widehat{g}(\xi)\right)^{\frac{2k}{k+1}} \right)^{k+1} d\xi \right).
\end{aligned} \tag{52}$$

From (52), we can check that $(\mathcal{C}_2^2)^{\frac{k}{k+1}}$ as follows

$$\begin{aligned}
(\mathcal{C}_2^2)^{\frac{k}{k+1}} &= \left(\int_{\mathbb{R}} \left(e^{-\frac{\alpha^2\xi^2}{4}}\widehat{g}_\varepsilon(\xi) - \widehat{g}(\xi)\right)^2 d\xi \right)^{\frac{k}{k+1}} \\
&= \left\| e^{-\frac{\alpha^2\xi^2}{4}}\widehat{g}_\varepsilon(\xi) - \widehat{g}(\xi) \right\|^{\frac{2k}{k+1}}_{\mathscr{L}_2(\mathbb{R})} \leq \left(2\varepsilon + \eta\left(\log\left(\log\left(\frac{T}{\varepsilon}\right)\right)\right)^{-1} \right)^{\frac{2k}{k+1}}.
\end{aligned} \tag{53}$$

On the other hand, we deduce

$$
\begin{aligned}
(\mathcal{C}_1^2)^{\frac{1}{k+1}} &\leq \left(\int_{-\infty}^{+\infty} \Big(\frac{\xi^2}{\mathcal{A}_0(1 - E_{\beta,1}(-\xi^2 T^\beta))} \Big)^{2(k+1)} \Big(e^{-\frac{\alpha^2 \xi^2}{4}} \widehat{g}_\varepsilon(\xi) - \widehat{g}(\xi) \Big)^2 d\xi \right)^{\frac{1}{k+1}} \\
&= \left\| \Big(\frac{\xi^2}{\mathcal{A}_0(1 - E_{\beta,1}(-\xi^2 T^\beta))} \Big)^{k+1} \Big(e^{-\frac{\alpha^2 \xi^2}{4}} \widehat{g}_\varepsilon(\xi) - \widehat{g}(\xi) \Big) \right\|_{\mathscr{L}_2(\mathbb{R})}^{\frac{2}{k+1}} \\
&\leq \left(\left\| \Big(\frac{\xi^2}{\mathcal{A}_0(1 - E_{\beta,1}(-\xi^2 T^\beta))} \Big)^{k+1} e^{-\frac{\alpha^2 \xi^2}{4}} \Big(\widehat{g}_\varepsilon(\xi) - \widehat{g}(\xi) \Big) \right\|_{\mathscr{L}_2(\mathbb{R})} \right. \\
&\quad \left. + \left\| \Big(\frac{\xi^2}{\mathcal{A}_0(1 - E_{\beta,1}(-\xi^2 T^\beta))} \Big)^{k+1} \Big(e^{-\frac{\alpha^2 \xi^2}{4}} \widehat{g}(\xi) - \widehat{g}(\xi) \Big) \right\|_{\mathscr{L}_2(\mathbb{R})} \right)^{\frac{2}{k+1}}. \qquad (54)
\end{aligned}
$$

To estimate $\mathcal{C}_1$, we give two Lemmas as follows:

Lemma 13. *Assume that the condition $\|\widehat{g}_\varepsilon(\xi) - \widehat{g}(\xi)\|_{\mathscr{L}_2(\mathbb{R})} \leq \varepsilon$ holds. Then we have the following estimate*

$$
\begin{aligned}
&\left\| \Big(\frac{\xi^2}{\mathcal{A}_0(1 - E_{\beta,1}(-\xi^2 T^\beta))} \Big)^{k+1} e^{-\frac{\alpha^2 \xi^2}{4}} \Big(\widehat{g}_\varepsilon(\xi) - \widehat{g}(\xi) \Big) \right\|_{\mathscr{L}_2(\mathbb{R})} \\
&\leq \varepsilon \left(\log \left(\log \left(\frac{T}{\varepsilon} \right) \right) \right)^{k+1} \left(\frac{\mathcal{L}_\beta(k,T)\, \mathcal{H}_\beta(\overline{B}_3, T, \Phi, \mathcal{M})}{\mathcal{A}_0 \eta} \right)^{k+1}. \qquad (55)
\end{aligned}
$$

Proof. Using the Lemma 12 and setting $\mathcal{L}_\beta(k,T) = \Big(\frac{k+1}{(1-E_{\beta,1}(-T^\beta))} \Big)$, we get

$$
\begin{aligned}
&\left\| \Big(\frac{\xi^2}{\mathcal{A}_0(1 - E_{\beta,1}(-\xi^2 T^\beta))} \Big)^{k+1} e^{-\frac{\alpha^2 \xi^2}{4}} \Big(\widehat{g}_\varepsilon(\xi) - \widehat{g}(\xi) \Big) \right\|_{\mathscr{L}_2(\mathbb{R})} \\
&\leq \varepsilon \Big(\frac{4}{\alpha^2} \Big)^{k+1} \left(\frac{k+1}{\mathcal{A}_0(1 - E_{\beta,1}(-T^\beta))} \right)^{k+1} \\
&\leq \varepsilon \left(\log \left(\log \left(\frac{T}{\varepsilon} \right) \right) \right)^{k+1} \left(\frac{\mathcal{L}_\beta(k,T)\, \mathcal{H}_\beta(\overline{B}_3, T, \Phi, \mathcal{M})}{\mathcal{A}_0 \eta} \right)^{k+1} \qquad (56)
\end{aligned}
$$

in which $\mathcal{H}_\beta(\overline{B}_3, T, \Phi, M)$ is defined in Lemma 11. □

Lemma 14. *Let $\xi \in \mathbb{R}$ and exist $\mathcal{M}$ is a positive constant such that $\mathcal{M} \geq \|f\|_{H^k(\mathbb{R})}$, we get*

$$
\left\| \Big(\frac{\xi^2}{\mathcal{A}_0(1 - E_{\beta,1}(-\xi^2 T^\beta))} \Big)^{k+1} \Big(e^{-\frac{\alpha^2 \xi^2}{4}} \widehat{g}(\xi) - \widehat{g}(\xi) \Big) \right\|_{\mathscr{L}_2(\mathbb{R})} \leq \frac{\Phi}{\mathcal{A}_0^{k+1}} \left(\frac{1}{1 - E_{\beta,1}(-T^\beta)} \right)^k \mathcal{M}. \qquad (57)
$$

Proof. Applying the Lemma 4, we receive

$$\begin{aligned}
&\Big\|\Big(\frac{\xi^2}{\mathcal{A}_0(1-E_{\beta,1}(-\xi^2T^\beta))}\Big)^{k+1}\Big(e^{-\frac{\alpha^2\xi^2}{4}}\widehat{g}(\xi)-\widehat{g}(\xi)\Big)\Big\|_{\mathscr{L}_2(\mathbb{R})}\\
&=\Bigg\|\Big(\frac{\xi^2(1-e^{-\frac{\alpha^2\xi^2}{4}})^{\frac{1}{k+1}}}{\mathcal{A}_0(1-E_{\beta,1}(-\xi^2T^\beta))}\Big)^{k+1}(1+\xi^2)^{-k}(1+\xi^2)^k\widehat{f}(\xi)\int_0^T\mathcal{D}_\beta(\xi,T-s)\phi(s)ds\Bigg\|_{\mathscr{L}_2(\mathbb{R})}\\
&\le\frac{\Phi}{\mathcal{A}_0}\sup_{\xi\in\mathbb{R}}\Big|\Big(\frac{\xi^2}{\mathcal{A}_0(1-E_{\beta,1}(-\xi^2T^\beta))}\Big)^k\frac{(1-e^{-\frac{\alpha^2\xi^2}{4}})}{(1+\xi^2)^k}\Big|\mathcal{M}\\
&\le\frac{\Phi}{\mathcal{A}_0^{k+1}}\sup_{\xi\in\mathbb{R}}\Big|\Big(\frac{\xi^2}{(1+\xi^2)(1-E_{\beta,1}(-\xi^2T^\beta))}\Big)^k(1-e^{-\frac{\alpha^2\xi^2}{4}})\Big|\mathcal{M}\\
&\le\frac{\Phi}{\mathcal{A}_0^{k+1}}\left(\frac{1}{1-E_{\beta,1}(-T^\beta)}\right)^k\mathcal{M}. \qquad (58)
\end{aligned}$$

□

Combining (54), (56) and (58), we have estimate $(\mathcal{C}_1^2)^{\frac{1}{k+1}}$ as follows

$$\begin{aligned}
(\mathcal{C}_1^2)^{\frac{1}{k+1}}\le&\left(\varepsilon\left(\log\left(\log\left(\frac{T}{\varepsilon}\right)\right)\right)^{k+1}\left(\frac{\mathcal{L}_\beta(k,T)\,\mathcal{H}_\beta(\overline{B}_3,T,\Phi)}{\mathcal{A}_0\eta}\right)^{k+1}\mathcal{M}^k\right.\\
&\left.+\frac{\Phi}{\mathcal{A}_0^{k+1}}\left(\frac{1}{1-E_{\beta,1}(-T^\beta)}\right)^k\right)^{\frac{2}{k+1}}\mathcal{M}^{\frac{2}{k+1}}. \qquad (59)
\end{aligned}$$

From (51) to (59), so

$$\begin{aligned}
\|\mathcal{J}_1\|_{\mathscr{L}_2(\mathbb{R})}\le&\left(\varepsilon\left(\log\left(\log\left(\frac{T}{\varepsilon}\right)\right)\right)^{k+1}\left(\frac{\mathcal{L}_\beta(k,T)\,\mathcal{H}_\beta(\overline{B}_3,T,\Phi)}{\mathcal{A}_0\eta}\right)^{k+1}\mathcal{M}^k\right.\\
&\left.+\frac{\Phi}{\mathcal{A}_0^{k+1}}\left(\frac{1}{1-E_{\beta,1}(-T^\beta)}\right)^k\right)^{\frac{1}{k+1}}\mathcal{M}^{\frac{1}{k+1}}.\left(2\varepsilon+\eta\left(\log\left(\log\left(\frac{T}{\varepsilon}\right)\right)\right)^{-1}\right)^{\frac{k}{k+1}}. \qquad (60)
\end{aligned}$$

Step 2: Estimate for $\|\mathcal{J}_2\|^2_{\mathscr{L}_2(\mathbb{R})}$, we have

$$\begin{aligned}
\|\mathcal{J}_2\|^2_{\mathscr{L}_2(\mathbb{R})}&\le\left\|\frac{(\widehat{g}_\varepsilon(\xi)-\widehat{g}(\xi))}{\Big(\int_0^T\mathcal{D}_\beta(\xi,T-s)\phi(s)ds\Big)e^{\frac{\alpha^2\xi^2}{4}}}\right\|^2_{\mathscr{L}_2(\mathbb{R})}\\
&=\left\|\frac{\xi^2}{\mathcal{A}_0\Big(1-E_{\beta,1}(-\xi^2T^\beta)\Big)}e^{-\frac{\alpha^2\xi^2}{4}}\Big(\widehat{g}_\varepsilon(\xi)-\widehat{g}(\xi)\Big)\right\|^2_{\mathscr{L}_2(\mathbb{R})}\\
&\le\left(\frac{16}{\alpha^4}\right)\frac{\|\widehat{g}_\varepsilon-\widehat{g}\|^2_{\mathscr{L}_2(\mathbb{R})}}{\mathcal{A}_0^2}\left(\frac{1}{1-E_{\beta,1}(-T^\beta)}\right)^2. \qquad (61)
\end{aligned}$$

Applying the Lemmas 11 and 12 in case $k=0$, we know that

$$\|\mathcal{J}_2\|^2_{\mathscr{L}_2(\mathbb{R})}\le\left(\varepsilon\log\left(\log\left(\frac{T}{\varepsilon}\right)\right)\right)^2\left(\frac{\mathcal{H}_\beta(\overline{B}_3,T,\Phi,\mathcal{M})}{\eta\mathcal{A}_0(1-E_{\beta,1}(-T^\beta))}\right)^2. \qquad (62)$$

Hence, we conclude that

$$\|\mathcal{J}_2\|_{\mathscr{L}_2(\mathbb{R})} \leq \left(\varepsilon \log\left(\log\left(\frac{T}{\varepsilon}\right)\right)\right)\left(\frac{\mathcal{H}_\beta(\overline{B}_3, T, \Phi, \mathcal{M})}{\eta \mathcal{A}_0(1-E_{\beta,1}(-T^\beta))}\right). \tag{63}$$

Step 3: Estimate for $\|\mathcal{J}_3\|^2_{\mathscr{L}_2(\mathbb{R})}$, we have :

$$\|\mathcal{J}_3\|^2_{\mathscr{L}_2(\mathbb{R})} \leq \left\| \frac{\widehat{g}(\xi)}{\left(\int_0^T \mathcal{D}_\beta(\xi, T-s)\phi(s)ds\right)e^{\frac{\alpha^2\xi^2}{4}}} - \frac{\widehat{g}(\xi)}{\left(\int_0^T \mathcal{D}_\beta(\xi, T-s)\phi_\varepsilon(s)ds\right)e^{\frac{\alpha^2\xi^2}{4}}} \right\|^2_{\mathscr{L}_2(\mathbb{R})}$$

$$= \left\| \frac{\left(e^{-\frac{\alpha^2\xi^2}{4}}\widehat{g}(\xi)\right)\int_0^T \mathcal{D}_\beta(\xi, T-s)(\phi_\varepsilon(s)-\phi(s))ds}{\left(\int_0^T \mathcal{D}_\beta(\xi, T-s)\phi(s)ds\right)\left(\int_0^T \mathcal{D}_\beta(\xi, T-s)\phi_\varepsilon(s)ds\right)} \right\|^2_{\mathscr{L}_2(\mathbb{R})}. \tag{64}$$

From (64), it gives

$$\|\mathcal{J}_3\|^2_{\mathscr{L}_2(\mathbb{R})} \leq \left\| \frac{\|\phi_\varepsilon - \phi\|_{C[0,T]} \int_0^T \mathcal{D}_\beta(\xi, T-s)ds}{\mathcal{A}_0\mathcal{A}_1\left(\int_0^T \mathcal{D}_\beta(\xi, T-s)ds\right)^2}\left(e^{-\frac{\alpha^2\xi^2}{4}}\widehat{g}(\xi)\right) \right\|^2_{\mathscr{L}_2(\mathbb{R})}$$

$$\leq \frac{\|\phi - \phi_\varepsilon\|^2_{C[0,T]}}{\mathcal{A}_0^2\mathcal{A}_1^2}\left\| \frac{\xi^2}{(1-E_{\beta,1}(-\xi^2T^\beta))}e^{-\frac{\alpha^2\xi^2}{4}}\widehat{g}(\xi) \right\|^2_{\mathscr{L}_2(\mathbb{R})}. \tag{65}$$

Applying Lemma 12 with $k = 0$ and Lemma 11 , we know that

$$\begin{aligned}
\|\mathcal{J}_3\|^2_{\mathscr{L}_2(\mathbb{R})} &\leq \left(\frac{16}{\alpha^4}\right)\left(\frac{1}{1-E_{\beta,1}(-T^\beta)}\right)^2 \frac{\|\phi - \phi_\varepsilon\|^2_{C[0,T]}}{\mathcal{A}_0^2\mathcal{A}_1^2}\|\widehat{g}\|^2_{\mathscr{L}_2(\mathbb{R})} \\
&\leq \left(\varepsilon \log\left(\log\left(\frac{T}{\varepsilon}\right)\right)\right)^2\left(\frac{\mathcal{H}_\beta(\overline{B}_3, T, \Phi, \mathcal{M})\|\widehat{g}\|_{\mathscr{L}_2(\mathbb{R})}}{\eta(1-E_{\beta,1}(-T^\beta))\mathcal{A}_0\mathcal{A}_1}\right)^2.
\end{aligned} \tag{66}$$

Therefore,

$$\|\mathcal{J}_3\|_{\mathscr{L}_2(\mathbb{R})} \leq \left(\varepsilon \log\left(\log\left(\frac{T}{\varepsilon}\right)\right)\right)\left(\frac{\mathcal{H}_\beta(\overline{B}_3, T, \Phi, \mathcal{M})\|\widehat{g}\|_{\mathscr{L}_2(\mathbb{R})}}{\eta(1-E_{\beta,1}(-T^\beta))\mathcal{A}_0\mathcal{A}_1}\right). \tag{67}$$

Combining (60), (63) and (67), we get:

$$
\begin{aligned}
&\|f(.)-f_{\varepsilon,\alpha}(.)\|_{\mathscr{L}_2(\mathbb{R})} = \|\widehat{f}(.)-\widehat{f_{\varepsilon,\alpha}}(.)\|_{\mathscr{L}_2(\mathbb{R})}\\
&\le \Bigg(\varepsilon\Big(\log\Big(\log\Big(\frac{T}{\varepsilon}\Big)\Big)^{k+1}\Big(\frac{\mathcal{L}_\beta(k,T)\,\mathcal{H}_\beta(\overline{B}_3,T,\Phi)}{\mathcal{A}_0\eta}\Big)^{k+1}\mathcal{M}^k\\
&+\frac{\Phi}{\mathcal{A}_0^{k+1}}\Big(\frac{1}{1-E_{\beta,1}(-T^\beta)}\Big)^k\Bigg)^{\frac{1}{k+1}}\mathcal{M}^{\frac{1}{k+1}}.\Big(2\varepsilon+\eta\Big(\log\Big(\log\Big(\frac{T}{\varepsilon}\Big)\Big)\Big)^{-1}\Big)^{\frac{k}{k+1}}\\
&+\Big(\varepsilon\log\Big(\log\Big(\frac{T}{\varepsilon}\Big)\Big)\Big)\Big(\frac{\mathcal{H}_\beta(\overline{B}_3,T,\Phi,\mathcal{M})}{\eta\mathcal{A}_0(1-E_{\beta,1}(-T^\beta))}+\Big(\frac{\mathcal{H}_\beta(\overline{B}_3,T,\Phi,\mathcal{M})\|\widehat{g}\|_{\mathscr{L}_2(\mathbb{R})}}{\eta(1-E_{\beta,1}(-T^\beta))\mathcal{A}_0\mathcal{A}_1}\Big)\Big).
\end{aligned}
\tag{68}
$$

Nohing that

$$\lim_{\varepsilon\to 0}\varepsilon\Big(\log\Big(\log\Big(\frac{T}{\varepsilon}\Big)\Big)\Big)=0,\quad \lim_{\varepsilon\to 0}\varepsilon\Big(\log\Big(\log\Big(\frac{T}{\varepsilon}\Big)\Big)\Big)^{k+1}=0. \tag{69}$$

Combining (68) and (69), we conclude that

$$\|f(.)-f_{\varepsilon,\alpha}(.)\|_{\mathscr{L}_2(\mathbb{R})} = \|\widehat{f}(.)-\widehat{f_{\varepsilon,\alpha}}(.)\|_{\mathscr{L}_2(\mathbb{R})} \to 0, \text{as } \varepsilon\to 0. \tag{70}$$

The proof of Theorem 2 is completed. □

5. Numerical Experiments

In this section, in order to illustrate the usefulness of the proposed methods, we consider the numerical examples intended. We carry out numerically above regularization methods to verify our proposed methods. The numerical examples with $T = 1$, and $\beta = 0.4$, $\beta = 0.95$ are shown in this section, respectively. In the following, we give an example which had the exact expression of the solutions $(u(x,t), f(x))$. We use the computations in Matlab codes which are given by Podlubny [18] for computing the generalized Mittag-Leffler function and the accuracy control in computing is 10^{-10}. We will do the numerical tests with $x \in [-7,7]$ and $\eta = 1.1$. The couple of $(\phi_\varepsilon, g_\varepsilon)$, which are determined below, play as measured data with a random noise as follows:

$$\phi_\varepsilon(\cdot) = \phi(\cdot)+\varepsilon\,(2\text{rand}(.)-1),\ g_\varepsilon(\cdot) = g(\cdot)+\varepsilon\,(2\text{rand}(.)-1). \tag{71}$$

Following Reference [9], we know the function $rand(\cdot)$ generates arrays of random numbers whose elements are normally distributed with mean 0, variance $\sigma^2 = 1$ and standard deviation $\sigma = 1$, and it gives rand(size(.)) and rand(size(.)) returns an array of random entries that is the same size as g and ϕ, respectively. We can easily verify the validity of the inequality:

$$\|\phi_\varepsilon-\phi\|_{C[0,T]}\le\varepsilon,\ \|g_\varepsilon-g\|_{\mathscr{L}_2(\mathbb{R})}\le\varepsilon. \tag{72}$$

In this example, we consider particularly a one-dimensional case of the problem (1) for f is an exact data function.

$$
\begin{cases}
\dfrac{\partial^\beta u(x,t)}{\partial t^\beta} = u_{xx}(x,t)+\phi(t)f(x),\ \ (x,t)\in\mathbb{R}\times(0,T],\\
u(x,0)=0, x\in\mathbb{R},\\
u(x,1)=g(x),\ \ x\in\mathbb{R}\cdot
\end{cases}
\tag{73}
$$

In this example, we choose the following solution

$$u(x,t) = \Big(E_{\beta,1}(t^\beta)-E_{\beta,1}(-t^\beta)\Big)\sin\Big(\frac{x}{2}\Big). \tag{74}$$

Then a simple computation yields

$$\phi(t) = \frac{5}{4} E_{\beta,1}(t^\beta) + \frac{3}{4} E_{\beta,1}(-t^\beta). \tag{75}$$

and $f(x) = \sin\left(\frac{x}{2}\right)$. Moreover, we have $u(x,0) = u_0(x) = 0$ and

$$u(x,1) = u_1(x) = g(x) = \Big(E_{\beta,1}(1) - E_{\beta,1}(-1)\Big) \sin\left(\frac{x}{2}\right). \tag{76}$$

Next, for computing the integral in the latter equality, see Reference [19], we use the fact that

$$\int_0^x u^{\kappa-1} E_{\beta,\kappa}(yu^\beta)(x-u)^{\beta-1} E_{\beta,\beta}(z(x-u)^\beta)du = \frac{yE_{\beta,\kappa+\beta}(yx^\beta) - zE_{\beta,\beta+\kappa}(zx^\beta)}{y-z} x^{\beta+\kappa+1}. \tag{77}$$

From $\phi_\varepsilon(.) = \phi(.) + \varepsilon\Big(2\text{rand}(.) - 1\Big)$, we have

$$\begin{aligned} \int_0^1 s^{\beta-1} E_{\beta,\beta}(-\xi^2 s^\beta)\phi_\varepsilon(1-s)ds &= \int_0^1 s^{\beta-1} E_{\beta,\beta}(-\xi^2 s^\beta)\phi(1-s)ds \\ &\quad + \varepsilon\,(2\text{rand}(.) - 1) \int_0^1 s^{\beta-1} E_{\beta,\beta}(-\xi^2 s^\beta)ds. \end{aligned} \tag{78}$$

Combining (72), (75) and (78), we have

$$\begin{aligned} \int_0^1 s^{\beta-1} E_{\beta,\beta}(-\xi^2 s^{\beta-1})\phi_\varepsilon(1-s)ds &= \frac{5}{4}\Big(\frac{E_{\beta,\beta+1}(1) + \xi^2 E_{\beta,\beta+1}(-\xi^2)}{1+\xi^2}\Big) \\ &\quad - \frac{3}{4}\Big(\frac{E_{\beta,\beta+1}(-1) - \xi^2 E_{\beta,\beta+1}(-\xi^2)}{-1+\xi^2}\Big) \\ &\quad + \frac{\varepsilon\Big(2\text{rand}(.) - 1\Big)}{\xi^2}\Big(1 - E_{\beta,1}(-\xi^2)\Big). \end{aligned} \tag{79}$$

In general, the numerical methods referenced by References [20,21] are summarized in three steps as follows.

Step 1: Choose N to generate the spatial and temporal discretization in such a manner as:

$$x_i = i\Delta x,\ \Delta x = \frac{\pi}{N},\ i = \overline{0,N}. \tag{80}$$

Obviously, the higher value of N will provide numerical results that are more accurate and stable. Here we choose $N = 100$ is satisfied.

Step 2: Setting $f_{\varepsilon,\alpha}(x_i) = f^i_{\varepsilon,\alpha}$ and $f(x_i) = f^i$, constructing two vectors contained all discrete values of $f_{\varepsilon,\alpha}$ and f denoted by $\Lambda_{\varepsilon,\alpha}$ and Ψ, respectively.

$$\begin{aligned} &\Lambda_{\varepsilon,\alpha} = [f^0_{\varepsilon,\alpha}\ f^1_{\varepsilon,\alpha}\ \ldots\ f^Q_{\varepsilon,\alpha}] \in \mathbb{R}^{Q+1}, \\ &\Psi = [f^0\ f^1\ \ldots\ f^{Q-1}\ f^Q] \in \mathbb{R}^{Q+1}. \end{aligned} \tag{81}$$

Step 3: The estimation is defined:

Relative error estimation:

$$E_1 = \frac{\sqrt{\sum_{i=1}^{N} |f_{\varepsilon,\alpha}(x_i) - f(x_i)|^2_{\mathscr{L}_2(-7,7)}}}{\sqrt{\sum_{i=1}^{N} |f(x_i)|^2_{\mathscr{L}_2(-7,7)}}}. \tag{82}$$

Absolute error estimation:

$$E_2 = \sqrt{\frac{1}{N}\sum_{i=1}^{N} |f_{\varepsilon,\alpha}(x_i) - f(x_i)|^2_{\mathscr{L}_2(-7,7)}}. \tag{83}$$

Figure 1 shows a 2D figures between the sought and its regularized solutions for $N = 100$ and $\beta = 0.95$. All figures are presented with $\varepsilon = 0.1$, $\varepsilon = 0.01$ and $\varepsilon = 0.001$, respectively.

In Tables 1 and 2 of this example, we show the error estimation both the priori and the posteriori within case $N = 100$, that is, in Table 1 we show the error estimation for both the priori and the posteriori at $\beta = 0.95$ with $\varepsilon \in \{0.1, 0.01, 0.001\}$, respectively. In Table 2, we show the relative error estimation and absolute error estimation both the priori and the posteriori with $\varepsilon = 0.01$ with the different values of $\beta \in \{0.2, 0.4, 0.6, 0.8\}$, when ε is fixed and the mesh resolutions are increased, the regularized solution convergence is better than that of the exact solution. From observing the results from the tables and figures above, we conclude that when ε tends to zero, the regularized solution approaches the exact solution.

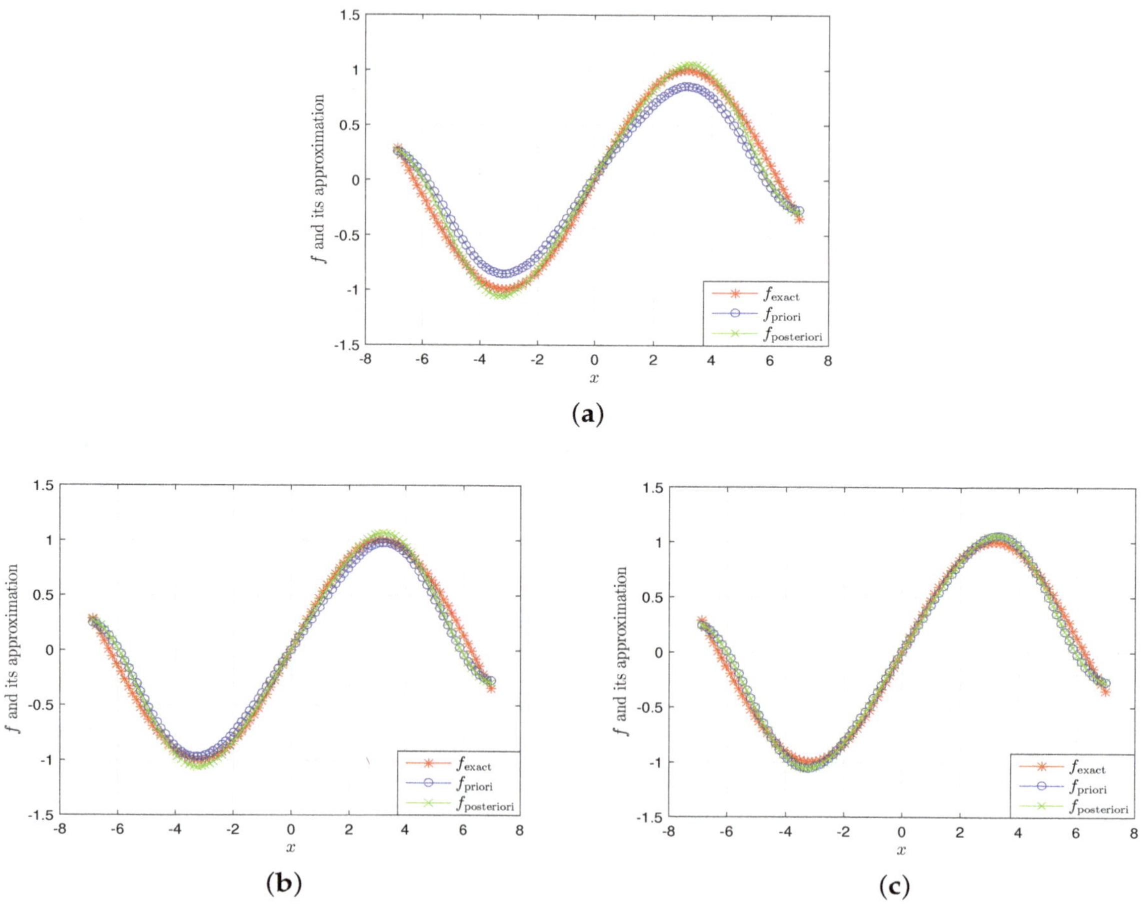

Figure 1. A comparison between the exact and regularized solutions for $k = 1$, $\beta = 0.95$ with $N = 100$. (**a**) $\varepsilon = 0.1$. (**b**) $\varepsilon = 0.01$. (**c**) $\varepsilon = 0.001$.

Table 1. The error estimation between the exact and regularized solutions of this example at $\beta = 0.95$ with $N = 100$.

ε	$E_1^{\beta_{pri}}$	$E_1^{\beta_{pos}}$	$E_2^{\beta_{pri}}$	$E_2^{\beta_{pos}}$
0.1	0.279660141830880	0.163452531664322	0.188256991900635	0.110030273632189
0.01	0.167130513450332	0.146077554813055	0.112506156619184	0.098334073898654
0.001	0.144054212078375	0.144599158066180	0.096972033479447	0.097338871212350

Table 2. The error estimation between the exact and regularized solutions with the different values of β, $\varepsilon = 0.01$ and $N = 100$.

β	$E_1^{\beta_{pri}}$	$E_1^{\beta_{pos}}$	$E_2^{\beta_{pri}}$	$E_2^{\beta_{pos}}$
0.2	0.156401672575436	0.176079016470940	0.078962919638416	0.092189970426402
0.4	0.146364358305196	0.165153671589525	0.073895354649786	0.086469770247512
0.6	0.136338164832119	0.153413164488168	0.068833404246973	0.080322774289912
0.8	0.124692172130227	0.140316883268202	0.062953661590221	0.073465933522836

6. Conclusions

In this study, by using the mollification regularization method, we solved the inverse problem and recovered the source term for time fractional diffusion equation with the time dependent coefficient. In the theoretical results, which we have shown, we obtained the error estimates of both a priori and a posteriori parameter choice rule methods based on a priori condition. In addition, in the numerical results, it shows that the regularized solutions are converged to the exact solution. Furthermore, it also shows that the smaller error of the input data, the better the convergence results.

Author Contributions: Project administration, Y.Z.; Resources, T.T.B.; Methodology, L.D.L.; Writing-review, editing and software, N.C.

References

1. Malyshev, I. An inverse source problem for heat equation. *J. Math. Anal. Appl.* **1989**, *142*, 206–218. [CrossRef]
2. Geng, F.; Lin, Y. Application of the variational iteration method to inverse heat source problems. *Comput. Math. Appl.* **2009**, *58*, 2098–2102. [CrossRef]
3. Shidfar, A.; Babaei, A.; Molabahrami, A. Solving the inverse problem of identifying an unknown source term in a parabolic equation. *Comput. Math. Appl.* **2010**, *60*, 1209–1213. [CrossRef]
4. Barkai, E.; Metzler, R.; Klafter, J. From continuous time random walks to the fractional Fokker-Planck equation. *Phys. Rev. E* **2000**, *61*, 132–138. [CrossRef] [PubMed]
5. Chaves, A. Fractional diffusion equation to describe Levy flights. *Phys. Lett. A* **1998**, *239*, 13–16. [CrossRef]
6. Gorenflo, R.; Mainardi, F.; Scalas, E.; Raberto, M. Fractional calculus and continuoustime finance, the diffusion limit. In *Mathematical Finance. Trends in Mathematics*; Birkhuser: Basel, Switzerland, 2001; pp. 171–180.
7. Sakamoto, K.; Yamamoto, M. Initial value/boundary value problems for fractional diffusion-wave equations and applications to some inverse problems. *J. Math. Anal. Appl.* **2011**, *382*, 426–447. [CrossRef]
8. Zhang, Z.Q.; Wei, T. Identifying an unknown source in time-fractional diffusion equation by a truncation method. *Appl. Math. Comput.* **2013**, *219*, 5972–5983. [CrossRef]
9. Yang, F.; Fu, C.L. The quasi-reversibility regularization method for identifying the unknown source for time fractional diffusion equation. *Appl. Math. Modell.* **2015**, *39*, 1500–1512. [CrossRef]
10. Wei, T.; Wang, J.G. A modified Quasi-Boundary Value method for an inverse source problem of the time fractional diffusion equation. *Appl. Numer. Math.* **2014**, *78*, 95–111. [CrossRef]
11. Yang, F.; Fu, C.L.; Li, X.X. A mollification regularization method for identifying the time-dependent heat source problem. *J. Engine Math.* **2016** *100*, 67–80. [CrossRef]

12. Wei, T.; Wang, J.G. Quasi-reversibility method to identify a space-dependent source for the time-fractional diffusion equation. *Appl. Math. Modell.* **2015**, *39*, 6139–6149. [CrossRef]
13. Nguyen, H.T.; Le, D.L.; Nguyen, V.T. Regularized solution of an inverse source problem for the time fractional diffusion equation. *Appl. Math. Modell.* **2016**, *40*, 8244–8264. [CrossRef]
14. Podlubny, I. *Fractional Differential Equations*; Academic Press: San Diego, CA, USA, 1999.
15. Samko, S.G.; Kilbas, A.A.; Marichev, O.I. *Fractional Integrals and Derivatives: Theory and Application*; Gordon and Breach: New York, NY, USA, 1993.
16. Yang, F.; Fu, C.L. A mollification regularization method for the inverse spatial-dependent heat source problem. *J. Comput. Appl. Math.* **2014**, *255*, 555–567. [CrossRef]
17. Kirsch, A. *An Introduction to the Mathematical Theory of Inverse Problem*; Applied Mathematical Sciences; Springer Science Business Media: Berlin/Heidelberg, Germany, 2011.
18. Podlubny, I.; Kacenak, M. Mittag-Leffler Function. The MATLAB Routine, 2006. Available online: http://www.mathworks.com/matlabcentral/fileexchange (accessed on 19 September 2019).
19. Mathai, A.M.; Haubold, H.J. Mittag Leffler function and Fractional Calculus. In *Special Functions for Applied Scientists*; Springer: New York, NY, USA, 2008; Chapter 2.
20. Meerschaert, M.; Tadjeran, C. Finite difference approximations for two-sided spacefractional partial differential equations. *Appl. Numer. Math.* **2006**, *56*, 80–90. [CrossRef]
21. Bu, W.; Liu, X.; Tang, Y.; Jiang, Y. Finite element multigrid method for multiterm time fractional advection-diffusion equations. *Int. J. Model. Simul. Sci. Comput.* **2015**, *6*, 1540001. [CrossRef]

12

On Neutral Functional Differential Inclusions involving Hadamard Fractional Derivatives

Bashir Ahmad [1,*], **Ahmed Alsaedi** [1], **Sotiris K. Ntouyas** [1,2] **and Hamed H. Al-Sulami** [1]

[1] Nonlinear Analysis and Applied Mathematics (NAAM)-Research Group, Department of Mathematics, Faculty of Science, King Abdulaziz University, P.O. Box 80203, Jeddah 21589, Saudi Arabia; aalsaedi@kau.edu.sa (A.A.); sntouyas@uoi.gr (S.K.N.); hhaalsalmi@kau.edu.sa (H.H.A.-S.)

[2] Department of Mathematics, University of Ioannina, 451 10 Ioannina, Greece

* Correspondence: bahmad@kau.edu.sa or bashirahmad_qau@yahoo.com

Abstract: We prove the existence of solutions for neutral functional differential inclusions involving Hadamard fractional derivatives by applying several fixed point theorems for multivalued maps. We also construct examples for illustrating the obtained results.

Keywords: functional fractional differential inclusions; Hadamard fractional derivative; existence; fixed point

---

1. Introduction

Fractional calculus has emerged as an important area of investigation in view of the application of its tools in scientific and engineering disciplines. Examples include bio-medical sciences, ecology, finance, reaction-diffusion systems, wave propagation, electromagnetics, viscoelasticity, material sciences, and so forth. Fractional-order operators give rise to more informative and realistic mathematical models in contrast to their integer-order counterparts. It has been due to the non-local nature of fractional-order operators, which enables us to gain insight into the hereditary behavior (past history) of the associated phenomena. For examples and recent development of the topic, see References [1,2] and the references cited therein.

Differential inclusions—known as generalization of differential equations and inequalities—are found to be of great utility in the study of dynamical systems, stochastic processes, optimal control theory, and so forth. One can find a detailed account of the topic in Reference [3]. In recent years, an overwhelming interest in the subject of fractional-order differential equations and inclusions has been shown, for instance, see References [4–14] and the references cited therein.

In Reference [15], the authors obtained some existence results for sequential neutral differential equations involving Hadamard derivatives:

$$\begin{cases} \mathcal{D}^{\alpha}[\mathcal{D}^{\beta}y(t) - g(t,y_t)] = f(t,y(t)), \ \ t \in J := [1,b], \\ y(t) = \phi(t), \ t \in [1-r,1], \qquad \mathcal{D}^{\beta}y(1) = \eta \in \mathbb{R}, \end{cases} \tag{1}$$

where $\mathcal{D}^{\alpha}, \mathcal{D}^{\beta}$ are the Hadamard fractional derivatives of order $0 < \alpha, \beta < 1$, respectively and $f, g : J \times \mathbb{R} \to \mathbb{R}$ are continuous functions, $J \subseteq \mathbb{R}$ and $\phi \in C([1-r,1], \mathbb{R})$.

In this paper, we cover the multivalued case of problem (1) and investigate the Hadamard type neutral fractional differential inclusions given by

$$\begin{cases} \mathcal{D}^{\alpha}[\mathcal{D}^{\beta}y(t) - g(t,y_t)] \in F(t,y(t)), \ \ t \in J := [1,b], \\ y(t) = \phi(t), \ t \in [1-r,1], \qquad \mathcal{D}^{\beta}y(1) = \eta \in \mathbb{R}, \end{cases} \tag{2}$$

where $F : J \times \mathbb{R} \to \mathcal{P}(\mathbb{R})$ is a multivalued map, $\mathcal{P}(\mathbb{R})$ represents the family of all nonempty subsets of $\mathbb{R}$, and the other quantities in (2) are the same as taken in (1). Here y_t is an element of the Banach space $C_r := C([-r,0],\mathbb{R})$ equipped with norm $\|\phi\|_C := \sup\{|\phi(\theta)| : -r \le \theta \le 0\}$, and is defined by $y_t(\theta) = y(t+\theta),\ \theta \in [-r,0]$, where y is a function defined on $[1-r,b]$ and $t \in J$. The standard fixed point theorems for multivalued maps are applied to establish the existence results for the problem (2).

The remaining content of the paper is composed as follows. In Section 2, we describe the necessary background material needed for our work. Section 3 deals with the main theorems. In Section 4, we construct illustrative examples for the obtained results.

2. Preliminaries

Let us begin this section with some necessary definitions of fractional calculus [1].

Definition 1. *For a function $h : [1,\infty) \to \mathbb{R}$, the Hadamard derivative of fractional order χ is defined by*

$$D^{\chi}h(t) = \frac{1}{\Gamma(n-\chi)}\left(t\frac{d}{dt}\right)^n \int_1^t \left(\log\frac{t}{s}\right)^{n-\chi-1}\frac{h(s)}{s}ds,\ \ n = [\chi]+1,$$

where $[\chi]$ denotes the integer part of the real number χ and $\log(\cdot) = \log_e(\cdot)$.

Definition 2. *The Hadamard fractional integral of order χ for a function h is defined as*

$$I^{\chi}h(t) = \frac{1}{\Gamma(\chi)}\int_1^t \left(\log\frac{t}{s}\right)^{\chi-1}\frac{h(s)}{s}ds,\ \ \chi > 0,$$

provided the integral exists.

Now we state a known result [15], which plays a key role in the forthcoming analysis.

Lemma 1 (Lemma 2.3 in [15]). *The function y is a solution of the problem*

$$\begin{cases} D^{\alpha}[D^{\beta}y(t) - g(t,y_t)] = f(t,y_t),\ t \in J := [1,b], \\ y(t) = \phi(t),\ t \in [1-r,1], \\ D^{\beta}y(1) = \eta \in \mathbb{R}, \end{cases} \tag{3}$$

if and only if

$$y(t) = \begin{cases} \phi(t), & \text{if } t \in [1-r,1], \\ \Big\{\phi(1) + (\eta - g(1,\phi))\dfrac{(\log t)^{\beta}}{\Gamma(\beta+1)} + \dfrac{1}{\Gamma(\alpha)}\displaystyle\int_1^t \left(\log\frac{t}{s}\right)^{\alpha-1}\frac{g(s,y_s)}{s}ds & \\ +\dfrac{1}{\Gamma(\alpha+\beta)}\displaystyle\int_1^t \left(\log\frac{t}{s}\right)^{\alpha+\beta-1}\frac{f(s,y_s)}{s}ds\Big\}, & \text{if } t \in J. \end{cases} \tag{4}$$

3. Existence Results

For a normed space $(X, \|\cdot\|)$, we define $\mathcal{P}_{cl}(X) = \{Y \in \mathcal{P}(X) : Y \text{ is closed}\}$, $\mathcal{P}_{cp}(X) = \{Y \in \mathcal{P}(X) : Y \text{ is compact}\}$, $\mathcal{P}_{cl,b}(X) = \{Y \in \mathcal{P}(X) : Y \text{ is closed and bounded}\}$, $\mathcal{P}_{cp,c}(X) = \{Y \in \mathcal{P}(X) : Y \text{ is compact and convex}\}$ and $\mathcal{P}_{b,cl,c}(X) = \{Y \in \mathcal{P}(X) : Y \text{ is bounded, closed and convex}\}$. In passing, we remark that a closed and bounded set in a metric space is not necessarily compact in general; however, it is true that a set in a metric space of real or complex numbers is compact if and only if it is closed and bounded.

For each $y \in C(J, \mathbb{R})$, define the set of selections of F by

$$S_{F,y} := \{\xi \in L^1(J, \mathbb{R}) : \xi(t) \in F(t, y(t)) \text{ on } J\}.$$

Denote by $C(J, \mathbb{R})$ the Banach space of all continuous functions from J into $\mathbb{R}$ endowed with the norm $\|y\| := \sup\{|y(t)| : t \in J\}$. $L^1(J, \mathbb{R})$ represents the space of functions $y : J \to \mathbb{R}$ such that $\|y\|_{L^1} = \int_1^b |y(t)|dt$.

Our first existence result deals with the case when F has convex values and is based on nonlinear alternative for Kakutani maps [16] with the assumption that the multivalued map F is Carathéodory.

Definition 3 (Granas, Dugundji [16]). *A multivalued map $F : J \times \mathbb{R} \to \mathcal{P}(\mathbb{R})$ is said to be Carathéodory if*

(i) $t \longmapsto F(t, x)$ is measurable for each $x \in \mathbb{R}$;
(ii) $x \longmapsto F(t, x)$ is upper semicontinuous for almost all $t \in J$.

Further a Carathéodory function F is called L^1-Carathéodory if

(iii) for each $\rho > 0$, there exists $\varphi_\rho \in L^1(J, \mathbb{R}^+)$ such that

$$\|F(t, x)\| = \sup\{|v| : v \in F(t, x)\} \le \varphi_\rho(t)$$

for all $x \in \mathbb{R}$ with $\|x\| \le \rho$ and for almost everywhere $t \in J$.

Theorem 1. *Assume that:*

(H_0) *there exists a non-negative constant $k < \Gamma(\alpha+1)(\log b)^{-\alpha}$ such that*

$$|g(t, u_1) - g(t, u_2)| \le k\|u_1 - u_2\|_C, \quad \text{for } t \in J \text{ and every } u_1, u_2 \in C_r.$$

(H_1) *$F : J \times \mathbb{R} \to \mathcal{P}_{cp,c}(\mathbb{R})$ is L^1-Carathéodory;*
(H_2) *there exists a continuous non-decreasing function $\Phi : [0, \infty) \to (0, \infty)$ and a function $p \in C(J, \mathbb{R}^+)$ such that*

$$\|F(t, x)\|_{\mathcal{P}} := \sup\{|y| : y \in F(t, x)\} \le p(t)\Phi(\|x\|) \text{ for each } (t, x) \in J \times \mathbb{R};$$

(H_3) *there exists a constant $\omega > 0$ such that*

$$\frac{\left(1 - \dfrac{k(\log b)^\alpha}{\Gamma(\alpha+1)}\right)\omega}{\|\phi\|_C + (|\eta| + k\|\phi\|_C + g_0)\dfrac{(\log b)^\beta}{\Gamma(\beta+1)} + \dfrac{g_0(\log b)^\alpha}{\Gamma(\alpha+1)} + \dfrac{\Phi(\omega)\|p\|}{\Gamma(\alpha+\beta+1)}(\log b)^{\alpha+\beta}} > 1,$$

where $g_0 = |g(1, 0)|$.

Then the problem (2) has at least one solution on $[1 - r, b]$.

Proof. Let us first transform the problem (2) into a fixed point problem by introducing an operator $\mathcal{V} : C([1-r,b],\mathbb{R}) \longrightarrow \mathcal{P}(C([1-r,b],\mathbb{R}))$ by

$$\mathcal{V}(y) = \left\{ \begin{array}{l} h \in C([1-r,b],\mathbb{R}) : \\ h(t) = \begin{cases} \phi(t), & \text{if } t \in [1-r,1], \\ \Big\{\phi(1) + (\eta - g(1,\phi))\dfrac{(\log t)^{\beta}}{\Gamma(\beta+1)} & \\ +\dfrac{1}{\Gamma(\alpha)}\displaystyle\int_1^t \left(\log\frac{t}{s}\right)^{\alpha-1}\frac{g(s,y_s)}{s}ds & \\ +\dfrac{1}{\Gamma(\alpha+\beta)}\displaystyle\int_1^t \left(\log\frac{t}{s}\right)^{\alpha+\beta-1}\frac{\xi(s)}{s}ds\Big\}, & \text{if } t \in J, \end{cases} \end{array} \right\} \tag{5}$$

for $\xi \in S_{F,x}$. It is obvious by Lemma 1 that the fixed points of the operator $\mathcal{V}$ are solutions of the problem (2).

We verify the hypothesis of nonlinear alternative for Kakutani maps [16] in several steps.

Step 1. *$\mathcal{V}(y)$ is convex for each $y \in C([1-r,b],\mathbb{R})$.* It directly follows from the fact that $S_{F,y}$ is convex (F has convex values).

Step 2. *$\mathcal{V}$ maps bounded sets (balls) into bounded sets in $C([1-r,b],\mathbb{R})$.* Let $B_\zeta = \{y \in C([1-r,b],\mathbb{R}) : \|y\|_{[1-r,b]} \leq \zeta\}$ be a bounded set in $C([1-r,b],\mathbb{R})$. Then, for each $h \in \mathcal{B}(y), y \in B_\zeta$, there exists $\xi \in S_{F,y}$ such that

$$\begin{aligned} h(t) &= \phi(1) + (\eta - g(1,\phi))\frac{(\log t)^{\beta}}{\Gamma(\beta+1)} + \frac{1}{\Gamma(\alpha)}\int_1^t \left(\log\frac{t}{s}\right)^{\alpha-1}\frac{g(s,y_s)}{s}ds \\ &\quad + \frac{1}{\Gamma(\alpha+\beta)}\int_1^t \left(\log\frac{t}{s}\right)^{\alpha+\beta-1}\frac{\xi(s)}{s}ds. \end{aligned}$$

Then, for $t \in J$, we have

$$\begin{aligned} |h(t)| &\leq \|\phi\|_C + (|\eta| + k\|\phi\|_C + g_0)\frac{(\log b)^{\beta}}{\Gamma(\beta+1)} + \frac{k\|y\|_{[1-r,b]} + g_0}{\Gamma(\alpha+1)}(\log b)^{\alpha} \\ &\quad + \frac{\Phi(\|y\|_{[1-r,b]})\|p\|}{\Gamma(\alpha+\beta+1)}(\log b)^{\alpha+\beta}. \end{aligned}$$

Thus,

$$\|h\| \leq \|\phi\|_C + (|\eta| + k\|\phi\|_C + g_0)\frac{(\log b)^{\beta}}{\Gamma(\beta+1)} + \frac{k\zeta + g_0}{\Gamma(\alpha+1)}(\log b)^{\alpha} + \frac{\Phi(\zeta)\|p\|}{\Gamma(\alpha+\beta+1)}(\log b)^{\alpha+\beta}.$$

Step 3. *$\mathcal{V}$ maps bounded sets into equicontinuous sets of $C([1-r,b],\mathbb{R})$.*

Let $t_1, t_2 \in J$ with $t_1 < t_2$ and $y \in B_\zeta$. Then, for each $h \in \mathcal{B}(y)$, we obtain

$$\begin{aligned} |h(t_2) - h(t_1)| &\leq \frac{|\eta| + k\|\phi\|_C + g_0}{\Gamma(\beta+1)}\left[(\log t_2)^{\beta} - (\log t_1)^{\beta}\right] \\ &\quad + \frac{k\zeta + g_0}{\Gamma(\alpha+\beta)}\int_1^{t_1}\left|\left(\log\frac{t_2}{s}\right)^{\alpha+\beta-1} - \left(\log\frac{t_1}{s}\right)^{\alpha+\beta-1}\right|\frac{ds}{s} \\ &\quad + \frac{k\zeta + g_0}{\Gamma(\alpha+\beta)}\int_{t_1}^{t_2}\left(\log\frac{t_2}{s}\right)^{\alpha+\beta-1}\frac{ds}{s} \\ &\quad + \frac{\Phi(\zeta)\|p\|}{\Gamma(\alpha+\beta)}\int_1^{t_1}\left|\left(\log\frac{t_2}{s}\right)^{\alpha+\beta-1} - \left(\log\frac{t_1}{s}\right)^{\alpha+\beta-1}\right|\frac{ds}{s} \end{aligned}$$

$$
\begin{aligned}
& +\frac{\Phi(\zeta)\|p\|}{\Gamma(\alpha+\beta)}\int_{t_1}^{t_2}\left(\log\frac{t_2}{s}\right)^{\alpha+\beta-1}\frac{ds}{s} \\
\leq\ & \frac{|\eta|+k\|\phi\|_C+g_0}{\Gamma(\beta+1)}\left[(\log t_2)^\beta-(\log t_1)^\beta\right] \\
& +\left\{\frac{k\zeta+g_0}{\Gamma(\alpha+\beta)}+\frac{\Phi(\zeta)\|p\|}{\Gamma(\alpha+\beta+1)}\right\}\left[\left|(\log t_2)^{\alpha+\beta}-(\log t_1)^{\alpha+\beta}\right|+|\log t_2/t_1|^{\alpha+\beta}\right],
\end{aligned}
$$

which tends to zero as $t_2-t_1\to 0$ independently of $y\in B_\zeta$. For the cases $t_1<t_2\leq 0$ and $t_1\leq 0\leq t_2$, the equicontinuity can be established in a similar manner. Thus, by Arzelá-Ascoli theorem [17], we deduce that $\mathcal{V}: C([1-r,b],\mathbb{R})\to\mathcal{P}(C([1-r,b],\mathbb{R}))$ is completely continuous.

Now we show that $\mathcal{V}$ has a closed graph. Then it will follow by the Proposition 1.2 in Reference [18] that $\mathcal{V}$ is upper semi-continuous, as it is already proved to be completely continuous.

Step 4. $\mathcal{V}$ has a closed graph. We need to show that $h_*\in\mathcal{V}(y_*)$ when $y_n\to x_*, h_n\in\mathcal{V}(y_n)$ and $h_n\to h_*$. Associated with $h_n\in\mathcal{V}(y_n)$, there exists $\xi_n\in S_{F,y_n}$ such that, for each $t\in J$,

$$
\begin{aligned}
h_n(t) &= \phi(1)+(\eta-g(1,\phi))\frac{(\log t)^\beta}{\Gamma(\beta+1)}+\frac{1}{\Gamma(\alpha)}\int_1^t\left(\log\frac{t}{s}\right)^{\alpha-1}\frac{g(s,y_s)}{s}ds \\
&\quad+\frac{1}{\Gamma(\alpha+\beta)}\int_1^t\left(\log\frac{t}{s}\right)^{\alpha+\beta-1}\frac{\xi_n(s)}{s}ds.
\end{aligned}
$$

Thus it suffices to show that there exists $\xi_*\in S_{F,y_*}$ such that, for each $t\in J$,

$$
\begin{aligned}
h_*(t) &= \phi(1)+(\eta-g(1,\phi))\frac{(\log t)^\beta}{\Gamma(\beta+1)}+\frac{1}{\Gamma(\alpha)}\int_1^t\left(\log\frac{t}{s}\right)^{\alpha-1}\frac{g(s,y_s)}{s}ds \\
&\quad+\frac{1}{\Gamma(\alpha+\beta)}\int_1^t\left(\log\frac{t}{s}\right)^{\alpha+\beta-1}\frac{\xi_*(s)}{s}ds.
\end{aligned}
$$

Let us introduce the linear operator $\Theta: L^1(J,\mathbb{R})\to C(J,\mathbb{R})$ given by

$$
\begin{aligned}
\xi\mapsto\Theta(\xi)(t) &= \phi(1)+(\eta-g(1,\phi))\frac{(\log t)^\beta}{\Gamma(\beta+1)}+\frac{1}{\Gamma(\alpha)}\int_1^t\left(\log\frac{t}{s}\right)^{\alpha-1}\frac{g(s,y_s)}{s}ds \\
&\quad+\frac{1}{\Gamma(\alpha+\beta)}\int_1^t\left(\log\frac{t}{s}\right)^{\alpha+\beta-1}\frac{\xi(s)}{s}ds.
\end{aligned}
$$

Notice that $\|h_n(t)-h_*(t)\|\to 0$, as $n\to\infty$. Therefore, it follows from a result dealing with the closed graph operators derived in Reference [19] that $\Theta\circ S_F$ is a closed graph operator. Further, we have $h_n(t)\in\Theta(S_{F,y_n})$. Since $y_n\to y_*$, we have

$$
\begin{aligned}
h(t) &= \phi(1)+(\eta-g(1,\phi))\frac{(\log t)^\beta}{\Gamma(\beta+1)}+\frac{1}{\Gamma(\alpha)}\int_1^t\left(\log\frac{t}{s}\right)^{\alpha-1}\frac{g(s,y_s)}{s}ds \\
&\quad+\frac{1}{\Gamma(\alpha+\beta)}\int_1^t\left(\log\frac{t}{s}\right)^{\alpha+\beta-1}\frac{\xi_*(s)}{s}ds,
\end{aligned}
$$

for some $\xi_*\in S_{F,y_*}$.

Step 5. We can find an open set $U\subseteq C([1-r,b],\mathbb{R})$ with $y\notin\nu\mathcal{V}(y)$ for any $\nu\in(0,1)$ and all $y\in\partial U$.

Let $\nu\in(0,1)$ and $y\in\nu\mathcal{V}(y)$. Then there exists $\xi\in L^1(J,\mathbb{R})$ with $\xi\in S_{F,y}$ such that for $t\in J$,

$$
|y(t)| \leq \|\phi\|_C+(|\eta|+k\|\phi\|_C+g_0)\frac{(\log b)^\beta}{\Gamma(\beta+1)}+\frac{k\|y\|_{[1-r,b]}+g_0}{\Gamma(\alpha+1)}(\log b)^\alpha
$$

$$+\frac{\Phi(\|y\|_{[1-r,b]})\|p\|}{\Gamma(\alpha+\beta+1)}(\log b)^{\alpha+\beta}, \ t\in J,$$

which implies that

$$\begin{aligned}\|y\|_{[1-r,b]}\left\{1-\frac{k(\log b)^{\alpha}}{\Gamma(\alpha+1)}\right\} &\le \|\phi\|_C+(|\eta|+k\|\phi\|_C+g_0)\frac{(\log b)^{\beta}}{\Gamma(\beta+1)} \\ &+\frac{g_0(\log b)^{\alpha}}{\Gamma(\alpha+1)}+\frac{\Phi(\|y\|_{[1-r,b]})\|p\|}{\Gamma(\alpha+\beta+1)}(\log b)^{\alpha+\beta}.\end{aligned}$$

Consequently

$$\frac{\left(1-\frac{k(\log b)^{\alpha}}{\Gamma(\alpha+1)}\right)\|y\|_{[1-r,b]}}{\|\phi\|_C+(|\eta|+k\|\phi\|_C+g_0)\frac{(\log b)^{\beta}}{\Gamma(\beta+1)}+\frac{g_0(\log b)^{\alpha}}{\Gamma(\alpha+1)}+\frac{\Phi(\|y\|_{[1-r,b]})\|p\|}{\Gamma(\alpha+\beta+1)}(\log b)^{\alpha+\beta}}\le 1.$$

By (H_3), there exists a real number ω such that $\|y\|_{[1-r,b]}\neq\omega$. Let us consider an open set

$$U=\{y\in C([1-r,b],\mathbb{R})) : \|y\|_{[1-r,b]}<\omega\},$$

with $\overline{U}=U\cup\partial U$. Notice that $\mathcal{V}:\overline{U}\to\mathcal{P}(C([1-r,b],\mathbb{R}))$ is compact and upper semi-continuous multivalued map with convex closed values. The choice of U implies that there does not exist any $y\in\partial U$ satisfying $y\in\nu\mathcal{V}(y)$ for some $\nu\in(0,1)$. In consequence, we deduce from the nonlinear alternative for Kakutani maps [16] that $\mathcal{V}$ has a fixed point $y\in\overline{U}$ which corresponds to a solution to the problem (2). This finishes the proof. □

In the following result, we make use of the nonlinear alternative for contractive maps ([20] Corollary 3.8) to show the existence of solutions for the problem (2).

Lemma 2. *(Nonlinear alternative [20]) Let D be a bounded neighborhood of $0\in X$, where X is a Banach space. Let $Z_1:X\to\mathcal{P}_{cp,c}(X)$ and $Z_2:\bar{D}\to\mathcal{P}_{cp,c}(X)$ be multivalued operators such that (a) Z_1 is contraction, and (b) Z_2 is upper semi-continuous and compact. Then, if $G=Z_1+Z_2$, either (i) G has a fixed point in $\overline{D}$ or (ii) there is a point $u\in\partial D$ and $\lambda\in(0,1)$ with $u\in\lambda G(u)$.*

Theorem 2. *If the conditions $(H_0)-(H_3)$ of Theorem 1 hold, then there exists at least one solution for the problem (2) on $[1-r,b]$.*

Proof. In order to verify the hypotheses of Lemma 2, we introduce the operator $\Psi_1: C([1-r,b],\mathbb{R})\longrightarrow C([1-r,b],\mathbb{R})$ by

$$\Psi_1 y(t)=\begin{cases}0, & \text{if } t\in[1-r,1],\\ (\eta-g(1,\phi))\frac{(\log t)^{\beta}}{\Gamma(\beta+1)}+\frac{1}{\Gamma(\alpha)}\int_1^t\left(\log\frac{t}{s}\right)^{\alpha-1}\frac{g(s,y_s)}{s}ds, & \text{if } t\in J.\end{cases} \tag{6}$$

and the multivalued operator $\Psi_2: C([1-r,b],\mathbb{R})\longrightarrow\mathcal{P}(C([1-r,b],\mathbb{R}))$ by

$$\Psi_2 y(t)=\left\{\begin{array}{l} h\in C([1-r,b],\mathbb{R}):\\ h(t)=\begin{cases}\phi(t), & \text{if } t\in[1-r,1],\\ \phi(1)+\frac{1}{\Gamma(\alpha+\beta)}\int_1^t\left(\log\frac{t}{s}\right)^{\alpha+\beta-1}\frac{\xi(s)}{s}ds, & \text{if } t\in J,\end{cases}\end{array}\right\} \tag{7}$$

for $\xi \in S_{F,y}$. Observe that $\mathcal{V} = \Psi_1 + \Psi_2$, where $\mathcal{V}$ is defined by (5). In the first step, it will be established that the operators Ψ_1 and Ψ_2 define the multivalued operators $\Psi_1, \Psi_2 : B_\theta \to \mathcal{P}_{cp,c}(C([1-r,b],\mathbb{R}))$, where $B_\theta = \{y \in C([1-r,b],\mathbb{R}) : \|y\|_{[1-r,b]} \leq \theta\}$ is a bounded set in $C([1-r,b],\mathbb{R})$. Let us show that Ψ_2 is compact-valued on B_θ. Observe that the operator Ψ_2 is equivalent to the composition $\mathcal{L} \circ S_F$, where $\mathcal{L}$ is the continuous linear operator on $L^1(J,\mathbb{R})$ into $C([1-r,b],\mathbb{R})$, defined by

$$\mathcal{L}(v)(t) = \phi(1) + \frac{1}{\Gamma(\alpha+\beta)} \int_1^t \left(\log\frac{t}{s}\right)^{\alpha+\beta-1} \frac{v(s)}{s} ds.$$

Let $y \in B_\theta$ be arbitrary and let $\{\xi_n\}$ be a sequence in $S_{F,y}$. Then it follows by the definition of $S_{F,y}$ that $\xi_n(t) \in F(t, y(t))$ for almost all $t \in J$. As $F(t,y(t))$ is compact for all $t \in J$, we have a convergent subsequence of $\{\xi_n(t)\}$ (we denote it by $\{\xi_n(t)\}$ again) that converges in measure to some $\xi(t) \in S_{F,y}$ for almost all $t \in J$. On the other hand, $\mathcal{L}$ is continuous, so $\mathcal{L}(\xi_n)(t) \to \mathcal{L}(\xi)(t)$ pointwise on J.

The convergence will be uniform once it is shown that $\{\mathcal{L}(\xi_n)\}$ is an equicontinuous sequence. For $t_1, t_2 \in J$ with $t_1 < t_2$, we have

$$\begin{aligned} |\mathcal{L}(\xi_n)(t_2) - \mathcal{L}(\xi_n)(t_1)| &\leq \frac{\Phi(\theta)\|p\|}{\Gamma(\alpha+\beta)} \int_1^{t_1} \left| \left(\log\frac{t_2}{s}\right)^{\alpha+\beta-1} - \left(\log\frac{t_1}{s}\right)^{\alpha+\beta-1} \right| \frac{ds}{s} \\ &\quad + \frac{\Phi(\theta)\|p\|}{\Gamma(\alpha+\beta)} \int_{t_1}^{t_2} \left(\log\frac{t_2}{s}\right)^{\alpha+\beta-1} \frac{ds}{s} \\ &\leq \frac{\Phi(\theta)\|p\|}{\Gamma(\alpha+\beta+1)} \left[\left|(\log t_2)^{\alpha+\beta} - (\log t_1)^{\alpha+\beta}\right| + |\log t_2/t_1|^{\alpha+\beta} \right] \to 0, \end{aligned}$$

as $t_2 \to t_1$, which shows that the sequence $\{\mathcal{L}(\xi_n)\}$ is equicontinuous. As a consequence of the Arzelá-Ascoli theorem, there exists a uniformly convergent subsequence of $\{\xi_n\}$ (we denote it again by $\{\xi_n\}$) such that $\mathcal{L}(\xi_n) \to \mathcal{L}(\xi)$. Noting that $\mathcal{L}(\xi) \in \mathcal{L}(S_{F,y})$, we deduce that $\mathcal{B}(y) = \mathcal{L}(S_{F,y})$ is compact for all $y \in B_\theta$. So $\Psi_2(y)$ is compact.

Now, we show that $\Psi_2(y)$ is convex for all $y \in C([1-r,b],\mathbb{R})$. Let $h_1, h_2 \in \Psi_2(y)$. We select $\xi_1, \xi_2 \in S_{F,y}$ such that

$$h_i(t) = \phi(1) + \frac{1}{\Gamma(\alpha+\beta)} \int_1^t \left(\log\frac{t}{s}\right)^{\alpha+\beta-1} \frac{\xi_i(s)}{s} ds, \quad i = 1, 2,$$

for almost all $t \in J$. Then

$$[\lambda h_1 + (1-\lambda)h_2](t) = \phi(1) + \frac{1}{\Gamma(\alpha+\beta)} \int_1^t \left(\log\frac{t}{s}\right)^{\alpha+\beta-1} \frac{[\lambda\xi_1(s) + (1-\lambda)\xi_2(s)]}{s} ds,$$

where $0 \leq \lambda \leq 1$. Since $S_{F,y}$ is convex (as F has convex values), $\lambda\xi_1(s) + (1-\lambda)\xi_2(s) \in S_{F,y}$. Thus $\lambda h_1 + (1-\lambda)h_2 \in \Psi_2(y)$, which shows that Ψ_2 is convex-valued.

On the other hand, it is easy to show that Ψ_1 is compact and convex-valued. Next we prove that Ψ_1 is a contraction on $C([1-r,b],\mathbb{R})$. For $y, z \in C([1-r,b],\mathbb{R})$, we have

$$\begin{aligned} |\Psi_1(y)(t) - \Psi_1(z)(t)| &\leq \frac{1}{\Gamma(\alpha)} \int_1^t \left(\log\frac{t}{s}\right)^{\alpha-1} \frac{|g(s,y_s) - g(s,z_s)|}{s} ds \\ &\leq \frac{k}{\Gamma(\alpha)} \int_1^t \left(\log\frac{t}{s}\right)^{\alpha-1} \frac{\|y_s - z_s\|_C}{s} ds \\ &\leq \frac{k(\log t)^\alpha}{\Gamma(\alpha+1)} \|y - z\|_{[1-r,b]}, \end{aligned}$$

which implies that $\|\Psi_1(y) - \Psi_1(z)\|_{[1-r,b]} \leq \frac{k(\log b)^\alpha}{\Gamma(\alpha+1)} \|y - z\|_{[1-r,b]}$. By the assumption (H_0), we conclude that Ψ_1 is a contraction.

As in the proof of Theorem 1, it can easily be shown that the operator Ψ_2 is compact and upper semi-continuous.

In view of the foregoing steps, we deduce that Ψ_1 and Ψ_2 satisfy the hypothesis of Lemma 2. So, from the conclusion of Lemma 2, either condition (i) or condition (ii) holds. We show that conclusion (ii) is not possible. If $y \in \lambda\Psi_1(y) + \lambda\Psi_2(y)$ for $\lambda \in (0,1)$, then there exist $\xi \in S_{F,y}$ such that

$$\begin{aligned} y(t) &= \lambda\Bigg(\phi(1) + (\eta - g(1,\phi))\frac{(\log t)^\beta}{\Gamma(\beta+1)} + \frac{1}{\Gamma(\alpha)}\int_1^t \left(\log\frac{t}{s}\right)^{\alpha-1}\frac{g(s,y_s)}{s}ds \\ &\quad + \frac{1}{\Gamma(\alpha+\beta)}\int_1^t \left(\log\frac{t}{s}\right)^{\alpha+\beta-1}\frac{\xi(s)}{s}ds\Bigg), \quad t \in J. \end{aligned}$$

By our assumptions, we can obtain

$$\begin{aligned} |y(t)| &\le \|\phi\|_C + [|\eta| + k\|\phi\|_C + g_0]\frac{(\log b)^\beta}{\Gamma(\beta+1)} + \frac{k\|y\|_{[1-r,b]} + g_0}{\Gamma(\alpha)}\int_1^t \left(\log\frac{t}{s}\right)^{\alpha-1}\frac{ds}{s} \\ &\quad + \frac{1}{\Gamma(\alpha+\beta)}\int_1^t \left(\log\frac{t}{s}\right)^{\alpha+\beta-1} p(s)\Phi(\|y_s\|_C)\frac{ds}{s} \\ &\le \|\phi\|_C + [|\eta| + k\|\phi\|_C + g_0]\frac{(\log b)^\beta}{\Gamma(\beta+1)} + \frac{k\|y\|_{[1-r,b]} + g_0}{\Gamma(\alpha+1)}(\log b)^\alpha \\ &\quad + \frac{\|p\|\Phi(\|y\|_{[1-r,b]})}{\Gamma(\alpha+\beta+1)}(\log b)^{\alpha+\beta}. \end{aligned}$$

Thus

$$\frac{\left(1 - \frac{k(\log b)^\alpha}{\Gamma(\alpha+1)}\right)\|y\|_{[1-r,b]}}{\|\phi\|_C + [|\eta| + k\|\phi\|_C + g_0]\frac{(\log b)^\beta}{\Gamma(\beta+1)} + \frac{g_0(\log b)^\alpha}{\Gamma(\alpha+1)} + \frac{\Phi(\|y\|_{[1-r,b]})\|p\|}{\Gamma(\alpha+\beta+1)}(\log b)^{\alpha+\beta}} \le 1. \tag{8}$$

If condition (ii) of Lemma 2 is satisfied, then there exists $\lambda \in (0,1)$ and $y \in \partial B_\omega$ with $y = \lambda\mathcal{V}(y)$. Then, y is a solution of (2) with $\|y\|_{[1-r,b]} = \omega$. Now, by the inequality (8), we get

$$\frac{\left(1 - \frac{k(\log b)^\alpha}{\Gamma(\alpha+1)}\right)\omega}{\|\phi\|_C + [|\eta| + k\|\phi\|_C + g_0]\frac{(\log b)^\beta}{\Gamma(\beta+1)} + \frac{g_0(\log b)^\alpha}{\Gamma(\alpha+1)} + \frac{\Phi(\omega)\|p\|}{\Gamma(\alpha+\beta+1)}(\log b)^{\alpha+\beta}} \le 1,$$

which contradicts (H_3). Hence, $\mathcal{V}$ has a fixed point on $[1-r,b]$ by Lemma 2, which implies that the problem (2) has a solution. The proof is complete. □

Our next result deals with the non-convex valued map in the problem (2) and is based on Covitz and Nadler's fixed point theorem [21] (*If* $N : X \to \mathcal{P}_{cl}(X)$ *is a contraction, then* $Fix N \neq \emptyset$, *where* X *is a metric space*).

For a metric space (X,d) induced from the normed space $(X; \|\cdot\|)$, it is argued in Reference [22] that $(\mathcal{P}_{cl,b}(X), H_d)$ is a metric space, where $H_d : \mathcal{P}(X) \times \mathcal{P}(X) \to \mathbb{R} \cup \{\infty\}$ is defined by $H_d(A,B) = \max\{\sup_{a\in A} d(a,B), \sup_{b\in B} d(A,b)\}$, $d(A,b) = \inf_{a\in A} d(a;b)$ and $d(a,B) = \inf_{b\in B} d(a;b)$.

Definition 4 (Granas, Dugundji [16]). *A multivalued operator* $N : X \to \mathcal{P}_{cl}(X)$ *is called*

(a) $\gamma-$*Lipschitz if and only if there exists* $\gamma > 0$ *such that*

$$H_d(N(x), N(y)) \le \gamma d(x,y) \text{ for each } x, y \in X;$$

(b) *a contraction if and only if it is* $\gamma-$*Lipschitz with* $\gamma < 1$.

Theorem 3. *Assume that* (H_0) *and the following conditions hold:*

(A_1) $F : J \times \mathbb{R} \to \mathcal{P}_{cp}(\mathbb{R})$ *is such that* $F(\cdot, y) : J \to \mathcal{P}_{cp}(\mathbb{R})$ *is measurable for each* $y \in \mathbb{R}$.
(A_2) $H_d(F(t,y), F(t,\bar{y})) \leq m(t)|y - \bar{y}|$ *for almost all* $t \in J$ *and* $y, \bar{y} \in \mathbb{R}$ *with* $m \in C(J, \mathbb{R}^+)$ *and* $d(0, F(t,0)) \leq m(t)$ *for almost all* $t \in J$.

Then there exists at least one solution for the problem (2) on J, *provided that*

$$\delta := \frac{k}{\Gamma(\alpha+1)}(\log b)^{\alpha} + \frac{\|m\|}{\Gamma(\alpha+\beta+1)}(\log b)^{\alpha+\beta} < 1. \tag{9}$$

Proof. Observe that the set $S_{F,y}$ is nonempty for each $y \in C(J, \mathbb{R})$ by the assumption (A_1). Therefore F has a measurable selection (see Theorem III.6 [23]). Next we consider the operator $\mathcal{V}$ given by (5) and verify that it satisfies the hypothesis of the Covitz and Nadler theorem [21]. We show that $\mathcal{V}(y) \in \mathcal{P}_{cl}(C(J, \mathbb{R}))$ for each $y \in C(J, \mathbb{R})$. Let $\{v_n\}_{n \geq 0} \in \mathcal{F}(y)$ be such that $v_n \to v$ $(n \to \infty)$ in $C(J, \mathbb{R})$. Then $v \in C(J, \mathbb{R})$ and we can find $\xi_n \in S_{F,y_n}$ such that, for each $t \in J$,

$$\begin{aligned} v_n(t) &= \phi(1) + (\eta - g(1,\phi))\frac{(\log t)^{\beta}}{\Gamma(\beta+1)} + \frac{1}{\Gamma(\alpha)}\int_1^t \left(\log\frac{t}{s}\right)^{\alpha-1}\frac{g(s,y_s)}{s}ds \\ &\quad + \frac{1}{\Gamma(\alpha+\beta)}\int_1^t \left(\log\frac{t}{s}\right)^{\alpha+\beta-1}\frac{\xi_n(s)}{s}ds. \end{aligned}$$

Since F has compact values, we pass onto a subsequence (if necessary) such that ξ_n converges to ξ in $L^1(J, \mathbb{R})$. So $\xi \in S_{F,y}$ and for each $t \in J$, we have

$$\begin{aligned} u_n(t) \to \xi(t) &= \phi(1) + (\eta - g(1,\phi))\frac{(\log t)^{\beta}}{\Gamma(\beta+1)} + \frac{1}{\Gamma(\alpha)}\int_1^t \left(\log\frac{t}{s}\right)^{\alpha-1}\frac{g(s,y_s)}{s}ds \\ &\quad + \frac{1}{\Gamma(\alpha+\beta)}\int_1^t \left(\log\frac{t}{s}\right)^{\alpha+\beta-1}\frac{\xi(s)}{s}ds. \end{aligned}$$

Hence, $v \in \mathcal{V}(y)$.

Next we prove that there exists $0 < \delta < 1$ (δ is defined by (9)) such that

$$H_d(\mathcal{V}(y), \mathcal{V}(\bar{y})) \leq \delta\|y - \bar{y}\| \text{ for each } y, \bar{y} \in C^2(J, \mathbb{R}).$$

Let $y, \bar{y} \in C^2(J, \mathbb{R})$ and $h_1 \in \mathcal{V}(y)$. Then there exists $\xi_1(t) \in F(t, y(t))$ such that, for each $t \in J$,

$$\begin{aligned} h_1(t) &= \phi(1) + (\eta - g(1,\phi))\frac{(\log t)^{\beta}}{\Gamma(\beta+1)} + \frac{1}{\Gamma(\alpha)}\int_1^t \left(\log\frac{t}{s}\right)^{\alpha-1}\frac{g(s,y_s)}{s}ds \\ &\quad + \frac{1}{\Gamma(\alpha+\beta)}\int_1^t \left(\log\frac{t}{s}\right)^{\alpha+\beta-1}\frac{\xi_1(s)}{s}ds. \end{aligned}$$

By (A_2), we have

$$H_d(F(t,y), F(t,\bar{y})) \leq m(t)|y(t) - \bar{y}(t)|.$$

So, there exists $v \in F(t, \bar{y}(t))$ such that

$$|\xi_1(t) - v(t)| \leq m(t)|y(t) - \bar{y}(t)|, \quad t \in J.$$

Define $V : J \to \mathcal{P}(\mathbb{R})$ by

$$V(t) = \{v \in \mathbb{R} : |\xi_1(t) - v(t)| \leq m(t)|y(t) - \bar{y}(t)|\}.$$

By Proposition III.4 in Reference [23], it follows that the multivalued operator $V(t) \cap F(t, \bar{y}(t))$ is measurable. So we can find a measurable selection $\xi_2(t)$ for V. So $\xi_2(t) \in F(t, \bar{y}(t))$ and satisfying $|\xi_1(t) - \xi_2(t)| \leq m(t)|y(t) - \bar{y}(t)|$ for each $t \in J$.

For each $t \in J$, we define

$$\begin{aligned} h_2(t) &= \phi(1) + (\eta - g(1,\phi))\frac{(\log t)^\beta}{\Gamma(\beta+1)} + \frac{1}{\Gamma(\alpha)}\int_1^t \left(\log\frac{t}{s}\right)^{\alpha-1}\frac{g(s,\bar{y}_s)}{s}ds \\ &+ \frac{1}{\Gamma(\alpha+\beta)}\int_1^t \left(\log\frac{t}{s}\right)^{\alpha+\beta-1}\frac{\xi_2(s)}{s}ds. \end{aligned}$$

Thus,

$$\begin{aligned} |h_1(t) - h_2(t)| &\leq \frac{1}{\Gamma(\alpha)}\int_1^t \left(\log\frac{t}{s}\right)^{\alpha-1}\frac{|g(s,y_s) - g(s,\bar{y}_s)|}{s}ds \\ &+ \frac{1}{\Gamma(\alpha+\beta)}\int_1^t \left(\log\frac{t}{s}\right)^{\alpha+\beta-1}\frac{|\xi_1(s) - \xi_2(s)|}{s}ds \\ &\leq \frac{k\|y-\bar{y}\|_{[1-r,b]}}{\Gamma(\alpha+1)}(\log b)^\alpha + \frac{\|m\|}{\Gamma(\alpha+\beta+1)}(\log b)^{\alpha+\beta}\|y-\bar{y}\|_{[1-r,b]}. \end{aligned}$$

Hence

$$\|h_1 - h_2\| \leq \left\{\frac{k}{\Gamma(\alpha+1)}(\log b)^\alpha + \frac{\|m\|}{\Gamma(\alpha+\beta+1)}(\log b)^{\alpha+\beta}\right\}\|y-\bar{y}\|_{[1-r,b]}.$$

On the other hand, interchanging the roles of y and $\bar{y}$ leads to

$$H_d(\mathcal{F}(y), \mathcal{F}(\bar{y})) \leq \left\{\frac{k}{\Gamma(\alpha+1)}(\log b)^\alpha + \frac{\|m\|}{\Gamma(\alpha+\beta+1)}(\log b)^{\alpha+\beta}\right\}\|y-\bar{y}\|_{[1-r,b]}.$$

So $\mathcal{V}$ is a contraction. Therefore, from the conclusion of Covitz and Nadler theorem [21], the operator $\mathcal{V}$ has a fixed point y which is indeed a solution of the problem (2). This finishes the proof. □

Finally, we prove an existence result by applying the multivalued version of Krasnoselskii's fixed point theorem [24], which is stated below.

Lemma 3 (Krasnoselskii [24]). *Let X be a Banach space, $Y \in \mathcal{P}_{b,cl,c}(X)$ and $W_1, W_2 : Y \to \mathcal{P}_{cp,c}(X)$ be multivalued operators satisfying the conditions: (i) $W_1 y + W_2 y \subset Y$ for all $y \in Y$; (ii) W_1 is contraction; and (iii) W_2 is upper semicontinuous and compact. Then there exists $y \in Y$ such that $y \in W_1 y + W_2 y$.*

Theorem 4. *Suppose that (H_0), (H_1) and the following assumption are satisfied*

(B_1) *there exists a function $q \in C([1,b], \mathbb{R}^+)$ such that*

$$\|F(t,u)\|_{\mathcal{P}} := \sup\{|y| : y \in F(t,u)\} \leq q(t), \text{ for each } (t,u) \in [1,b] \times C_r.$$

Then there exists at least one solution for the problem (2) on $[1-r,b]$.

Proof. Let us consider the operators Ψ_1 and Ψ_2 defined by (6) and (7) respectively. As in Theorem 2, one can show that $\Psi_1, \Psi_2 : B_\theta \to \mathcal{P}_{cp,c}(C([1-r,b],\mathbb{R}))$ are indeed multivalued operators, where $B_\theta = \{y \in C([1-r,b],\mathbb{R}) : \|y\|_{[1-r,b]} \leq \theta\}$ is a bounded set in $C([1-r,b],\mathbb{R})$. Moreover, Ψ_1 is a contraction on $C([1-r,b],\mathbb{R})$ and Ψ_2 is upper semi-continuous and compact.

Next we show that $\Psi_1(y)+\Psi_2(y)\subset B_\theta$ for all $y\in B_\theta$. Let $y\in B_\theta$ and suppose that

$$\theta\left(1-\frac{k(\log b)^\alpha}{\Gamma(\alpha+1)}\right)>\|\phi\|_C+\frac{[|\eta|+k\|\phi\|_C+g_0](\log b)^\beta}{\Gamma(\beta+1)}+\frac{g_0(\log b)^\alpha}{\Gamma(\alpha+1)}+\frac{\|q\|(\log b)^{\alpha+\beta}}{\Gamma(\alpha+\beta+1)}.$$

For $h\in\Psi_1,\Psi_2$ and $\xi\in S_{F,y}$, we have

$$\begin{aligned} h(t) &= \phi(1)+(\eta-g(1,\phi))\frac{(\log t)^\beta}{\Gamma(\beta+1)}+\frac{1}{\Gamma(\alpha)}\int_1^t\left(\log\frac{t}{s}\right)^{\alpha-1}\frac{g(s,y_s)}{s}ds \\ &\quad +\frac{1}{\Gamma(\alpha+\beta)}\int_1^t\left(\log\frac{t}{s}\right)^{\alpha+\beta-1}\frac{\xi(s)}{s}ds,\ \ t\in J. \end{aligned}$$

With the given assumptions, one can obtain

$$\begin{aligned} |h(t)| &\leq \|\phi\|_C+[|\eta|+k\|\phi\|_C+g_0]\frac{(\log b)^\beta}{\Gamma(\beta+1)}+\frac{k\|y\|_{[1-r,b]}+g_0}{\Gamma(\alpha+1)}(\log b)^\alpha \\ &\quad +\frac{\|q\|}{\Gamma(\alpha+\beta+1)}(\log b)^{\alpha+\beta}. \end{aligned}$$

Thus

$$\|h\| \leq \|\phi\|_C+[|\eta|+k\|\phi\|_C+g_0]\frac{(\log b)^\beta}{\Gamma(\beta+1)}+\frac{k\theta+g_0}{\Gamma(\alpha+1)}(\log b)^\alpha+\frac{\|q\|}{\Gamma(\alpha+\beta+1)}(\log b)^{\alpha+\beta}<\theta,$$

which means that $\Psi_1(y)+\Psi_2(y)\subset B_\theta$ for all $y\in B_\theta$.

Thus, the operators Ψ_1 and Ψ_2 satisfy the hypothesis of Lemma 3 and hence its conclusion implies that $y\in\mathcal{A}(y)+\mathcal{B}(y)$ in B_θ. Therefore the problem (2) has a solution in B_θ and the proof is finished. □

4. Examples

In this section, we demonstrate the application of our main results by considering the following Hadamard type neutral fractional differential inclusions:

$$D^{1/4}\Big(D^{2/3}y(t)-g(t,y_t)\Big)\in F(t,y_t),\ \ t\in J=[1,e], \tag{10}$$

$$y(t)=\phi(t),\ \ t\in[1/2,1],\ D^{2/3}y(1)=1/4. \tag{11}$$

Here $\alpha=1/4,\beta=2/3,r=1/2,b=e$,

$$F(t,y_t)=\left[\frac{\sqrt{3+\ln t}}{4}\sin(y_t),\ \frac{\sqrt{3}|y_t|^3}{8(1+|y_t|^3)}\sin(\pi t/2e)+\frac{1}{16}\right],$$

$$g(t,y_t)=\frac{1}{4+\ln t}\tan^{-1}(y_t)+\sin(\pi t/2),\ \ \phi(t)=\frac{1}{16\sqrt{\frac{3}{4}+t^2}}.$$

With the given data, it is easy to see that (H_0) is satisfied with $k<\Gamma(5/4)$, (H_2) is satisfied with $p(t)=\sqrt{3+\ln t}/4, \|p\|=1/2, \Phi(\|u\|_C)=\|u\|_C$ and (H_3) holds true for $M>7.05996548$ ($M_0=1.46447352, g_0=1$) with a particular choice of $k=1/4$. Thus all the conditions of Theorem 1 hold true. Hence the problem (10) and (11) has at least one solution on $[1/2,e]$ by the conclusion of Theorem 1. In a similar manner, one can check that the hypotheses of Theorem 2 hold with $M>1.71978641$ and consequently the conclusion of Theorem 2 applies to the problem (10) and (11).

In order to illustrate Theorem 3, let us take

$$F(t, y_t) = \left[0, \frac{\sqrt{15 + (\ln t)^2}}{8} \frac{|y_t|}{(1 + |y_t|)} + \frac{1}{4}\right] \quad (12)$$

in (10). Then $\|m\| = 1/2$ and from the condition (9), $\delta \approx 0.74950391 < 1$. Clearly the hypothesis of Theorem 3 is satisfied. Therefore, there exists at least one solution for the problem (10) and (11) with $F(t, y_t)$ given by (12) on $[1/2, e]$.

5. Conclusions

In this paper, we have derived several existence results for an initial value problem of neutral functional Hadamard-type fractional differential inclusions. In our first result (Theorem 1), we apply a nonlinear alternative for Kakutani multivalued maps to prove the existence of solutions for the problem at hand when the multivalued map F is assumed to be convex-valued. The nonlinear alternative for contractive maps is applied to prove the existence of solutions for the given problem in Theorem 2. In Theorem 3, we show the existence of solutions for the given problem involving non-convex valued maps with the aid of Covitz and Nadler's fixed point theorem. Our final existence result (Theorem 4) relies on the multivalued version of Krasnoselskii's fixed point theorem. In the nutshell, we have presented a comprehensive study of neutral functional Hadamard-type fractional differential inclusions by making use of different tools of fixed point theory for multivalued maps. In our future work, we plan to investigate the existence of solutions to an initial value problem for neutral functional fractional differential inclusions involving a combination of Caputo and Hadamard fractional derivatives.

Author Contributions: Conceptualization, B.A. and S.K.N.; methodology, A.A. and S.K.N.; validation, B.A., A.A., S.K.N. and H.H.A.-S.; formal analysis, A.A., B.A., and S.K.N.; writing—original draft preparation, S.K.N.; writing—review and editing, A.A., B.A. S.K.N. and H.H.A.-S.; project administration, B.A.; funding acquisition, B.A. and A.A.

Acknowledgments: This project was funded by the Deanship of Scientific Research (DSR), King Abdulaziz University, Jeddah, under grant no. (RG-39-130-38). The authors, therefore, acknowledge with thanks DSR technical and financial support. The authors also thank the reviewers for their constructive remarks on our work.

References

1. Kilbas, A.A.; Srivastava, H.M.; Trujillo, J.J. *Theory and Applications of Fractional Differential Equations*; North-Holland Mathematics Studies, 204; Elsevier Science B.V.: Amsterdam, The Netherlands, 2006.
2. Sabatier, J.; Agrawal, O.P.; Machado, J.A.T. (Eds.) *Advances in Fractional Calculus: Theoretical Developments and Applications in Physics and Engineering*; Springer: Dordrecht, The Netherlands, 2007.
3. Kisielewicz, M. *Stochastic Differential Inclusions and Applications*; Springer Optimization and Its Applications, 80; Springer: New York, NY, USA, 2013.
4. Agarwal, R.P.; Baleanu, D.; Hedayati, V.; Rezapour, S. Two fractional derivative inclusion problems via integral boundary condition. *Appl. Math. Comput.* **2015**, *257*, 205–212. [CrossRef]
5. Ahmad, B.; Ntouyas, S.K.; Alsaedi, A. New results for boundary value problems of Hadamard-type fractional differential inclusions and integral boundary conditions. *Bound. Value Probl.* **2013**, *2013*, 275. [CrossRef]
6. Ahmad, B.; Ntouyas, S.K.; Alsaedi, A. Existence theorems for nonlocal multivalued Hadamard fractional integro-differential boundary value problems. *J. Inequal. Appl.* **2014**, *2014*, 454. [CrossRef]
7. Ahmad, B.; Ntouyas, S.K. Existence results for a coupled system of Caputo type sequential fractional differential equations with nonlocal integral boundary conditions. *Appl. Math. Comput.* **2015**, *266*, 615–622. [CrossRef]

8. Ahmad, B.; Ntouyas, S.K. Nonlocal fractional boundary value problems with slit-strips boundary conditions. *Fract. Calc. Appl. Anal.* **2015**, *18*, 261–280. [CrossRef]
9. Balochian, S.; Nazari, M. Stability of particular class of fractional differential inclusion systems with input delay. *Control Intell. Syst.* **2014**, *42*, 279–283. [CrossRef]
10. Liu, X.; Liu, Z.; Fu, X. Relaxation in nonconvex optimal control problems described by fractional differential equations. *J. Math. Anal. Appl.* **2014**, *409*, 446–458. [CrossRef]
11. Ntouyas, S.K.; Etemad, S.; Tariboon, J. Existence results for multi-term fractional differential inclusions. *Adv. Difference Equ.* **2015**, *2015*, 140. [CrossRef]
12. Sun, J.; Yin, Q. Robust fault-tolerant full-order and reduced-order observer synchronization for differential inclusion chaotic systems with unknown disturbances and parameters. *J. Vib. Control* **2015**, *21*, 2134–2148. [CrossRef]
13. Wang, X.; Schiavone, P. Harmonic three-phase circular inclusions in finite elasticity. *Contin. Mech. Thermodyn.* **2015**, *27*, 739–747. [CrossRef]
14. Yukunthorn, W.; Ahmad, B.; Ntouyas, S.K.; Tariboon, J. On Caputo-Hadamard type fractional impulsive hybrid systems with nonlinear fractional integral conditions. *Nonlinear Anal. Hybrid Syst.* **2016**, *19*, 77–92. [CrossRef]
15. Ahmad, B.; Ntouyas, S.K. Existence and uniqueness of solutions for Caputo-Hadamard sequential fractional order neutral functional differential equations. *Electron. J. Differ. Equ.* **2017**, *36*, 1–11.
16. Granas, A.; Dugundji, J. *Fixed Point Theory*; Springer: New York, NY, USA, 2005.
17. Rudin, W. *Principles of Mathematical Analysis*, 3rd ed.; McGraw Hill: Singapore, 1976.
18. Deimling, K. *Multivalued Differential Equations*; Walter De Gruyter: Berlin, Germany; New York, NY, USA, 1992.
19. Lasota, A.; Opial, Z. An application of the Kakutani-Ky Fan theorem in the theory of ordinary differential equations. *Bull. Acad. Polon. Sci. Ser. Sci. Math. Astronom. Phys.* **1965**, *13*, 781–786.
20. Petryshyn, W.V.; Fitzpatric, P.M. A degree theory, fixed point theorems, and mapping theorems for multivalued noncompact maps. *Trans. Amer. Math. Soc.* **1974**, *194*, 1–25. [CrossRef]
21. Covitz, H.; Nadler, S.B., Jr. Multivalued contraction mappings in generalized metric spaces. *Israel J. Math.* **1970**, *8*, 5–11. [CrossRef]
22. Kisielewicz, M. *Differential Inclusions and Optimal Control*; Kluwer: Dordrecht, The Netherlands, 1991.
23. Castaing, C.; Valadier, M. *Convex Analysis and Measurable Multifunctions*; Lecture Notes in Mathematics 580; Springer-Verlag: Berlin/Heidelberg, Germany; New York, NY, USA, 1977.
24. Petrusel, A. Fixed points and selections for multivalued operators. *Semin. Fixed Point Theory-Cluj-Napoca* **2001**, *2*, 3–22.

13

Certain Fractional Proportional Integral Inequalities via Convex Functions

Gauhar Rahman [1], Kottakkaran Sooppy Nisar [2,*], Thabet Abdeljawad [3,4,5,*] and Samee Ullah [6]

1 Department of Mathematics, Shaheed Benazir Bhutto University, Sheringal, Upper Dir 18000, Pakistan; gauhar55uom@gmail.com
2 Department of Mathematics, College of Arts and Sciences, Prince Sattam bin Abdulaziz University, Wadi Aldawaser 11991, Saudi Arabia
3 Department of Mathematics and General Sciences, Prince Sultan University, Riyadh 12345, Saudi Arabia
4 Department of Medical Research, China Medical University, Taichung 40402, Taiwan
5 Department of Computer Science and Information Engineering, Asia University, Taichung 40402, Taiwan
6 Department of Mathematics, University of Malakand, Lower Dir 18800, Chakdara, Pakistan; sameeullah413@gmail.com
* Correspondence: n.sooppy@psau.edu.sa or ksnisar1@gmail.com (K.S.N.); tabdeljawad@psu.edu.sa (T.A.)

Abstract: The goal of this article is to establish some fractional proportional integral inequalities for convex functions by employing proportional fractional integral operators. In addition, we establish some classical integral inequalities as the special cases of our main findings.

Keywords: convex function; fractional integrals; proportional fractional integrals; inequalities; Qi inequality

---

1. Introduction

Integral inequalities play a vital role in the field of fractional differential equations. In the past few decades, researchers have paid their valuable consideration to this area. The significant developments in this area have been investigated, for example, [1–3], and [4] (cf. references cited therein). In [5], Ngo et al. established the following inequalities

$$\int_0^1 g^{\sigma+1}(t)dt \geq \int_0^1 t^{\sigma} g(t)dt \tag{1}$$

and

$$\int_0^1 g^{\sigma+1}(t)dt \geq \int_0^1 t g^{\sigma}(t)dt, \tag{2}$$

where $\sigma > 0$ and the positive continuous function g on $[0,1]$ such that

$$\int_x^1 g(t)dt \geq \int_x^1 tdt, x \in [0,1].$$

Later on, Liu et al. [6] established the following inequalities

$$\int_a^b g^{\sigma+\gamma}(t)dt \geq \int_a^b (t-a)^{\sigma} g^{\gamma}(t)dt, \tag{3}$$

where $\sigma > 0, \gamma > 0$, and the positive continuous g on $[a,b]$ is such that

$$\int_a^b g^\delta(t)dt \geq \int_a^b (t-a)^\delta dt, \delta = \min(1,\gamma), t \in [a,b].$$

Liu et al. [7] derived two theorems for integral inequalities as follows:

Theorem 1. *Suppose that the functions f_1 and g_1 are positive and continuous on $[a,b], (a < b)$ with $f_1 \leq g_1$ on $[a,b]$ such that the function $\frac{f_1}{g_1}, (g_1 \neq 0)$ is decreasing and the function f_1 is increasing. Assume that the function Φ is a convex with $\Phi(0) = 0$. Then, the following inequality holds*

$$\frac{\int_a^b f_1(t)dt}{\int_a^b g_1(t)dt} \geq \frac{\int_a^b \Phi\left(f_1(t)\right)dt}{\int_a^b \Phi\left(g_1(t)\right)dt}.$$

Theorem 2. *Suppose that the functions f_1, f_2, and f_3 be positive and continuous on $[a,b], (a < b)$ with $f_1 \leq f_2$ on $[a,b]$ such that the function $\frac{f_1}{f_2}, (f_2 \neq 0)$ is decreasing and the functions f_1 and f_3 are increasing. Assume that the function Φ is a convex with $\Phi(0) = 0$. Then, the following inequality holds*

$$\frac{\int_a^b f_1(t)dt}{\int_a^b f_2(t)dt} \geq \frac{\int_a^b \Phi\left(f_1(t)\right)f_3(t)dt}{\int_a^b \Phi\left(f_2(t)\right)f_3(t)dt}.$$

The inequalities in Equations (1)–(3) and their various generalizations have gained attention of the researchers [8–12].

Furthermore, the research of fractional integral inequalities is also of prominent importance. In [13,14], the authors presented some weighted Grüss type and new inequalities involving Riemann–Liouville (R-L) fractional integrals. In [15], Nisar et al. introduced many inequalities for extended gamma and confluent hypergeometric k-functions. Certain Gronwall inequalities for R-L and Hadamard k-fractional derivatives with applications are observed in [16]. The inequalities concerning the generalized (k,ρ)-fractional integral operators can be seen in [17].

The generalized fractional integral and Grüss type inequalities via generalized fractional integrals can be found in [18,19]. In [20], the authors examined the (k,s)-R-L fractional integral and its applications. In [21], the authors presented generalized Hermite–Hadamard type inequalities through fractional integral operators. Dahmani [22] introduced some classes of fractional integral inequalities by employing a family of n positive functions. Further the applications of fractional integral inequalities can be found [23,24].

In the last few decades, the researchers have paid their valuable consideration to the field of fractional calculus. This field has received more attention from various researchers due to its wide applications in various fields. In the growth of fractional calculus, researchers concentrate to develop several fractional integral operators and their applications in distinct fields (see, e.g., [25–33]). Zaher et al. [34] presented a new fractional nonlocal model.

Such types of these new fractional integral operators promote the future study to develop certain new approaches to unify the fractional operators and secure fractional integral inequalities. Especially, several striking inequalities, properties, and applicability for the fractional conformable integrals and derivatives are recently studied by various researchers. We refer the interesting readers to the works by [35–44], and [45]. The applications of conformable derivative can be found in [46–49] (cf. references cited therein).

2. Preliminaries

Jarad et al. [50] proposed the following left and right generalized proportional integral operators, which are sequentially defined by

$$\left({}_a\mathcal{J}^{\xi,\delta}f\right)(\tau)=\frac{1}{\delta^{\xi}\Gamma(\xi)}\int_a^{\tau}\exp[\frac{\delta-1}{\delta}(\tau-t)](\tau-t)^{\xi-1}f(t)dt,a<\tau \tag{4}$$

and

$$\left(\mathcal{J}_b^{\xi,\delta}f\right)(\tau)=\frac{1}{\delta^{\xi}\Gamma(\xi)}\int_{\tau}^{b}\exp[\frac{\delta-1}{\delta}(t-\tau)](t-\tau)^{\xi-1}f(t)dt,\tau<b, \tag{5}$$

where the proportional index $\delta\in(0,1]$ and $\xi\in\mathbb{C}$ with $Re(\xi)>0$ and $\Gamma(\tau)$ is the well-know gamma function defined by $\Gamma(\tau)=\int_0^{\infty}t^{\tau-1}e^{-t}dt$ [51–53].

Remark 1. *Setting $\delta=1$ in Equations (4) and (5), we obtain the following left and right R-L:*

$$\left({}_a\mathcal{J}^{\xi}f\right)(\tau)=\frac{1}{\Gamma(\xi)}\int_a^{\tau}(\tau-t)^{\xi-1}f(t)dt,a<\tau,$$

and

$$\left(\mathcal{J}_b^{\xi}f\right)(\tau)=\frac{1}{\Gamma(\xi)}\int_{\tau}^{b}(t-\tau)^{\xi-1}f(t)dt,\tau<b,$$

where $\xi\in\mathbb{C}$ with $Re(\xi)>0$.

Recently, the generalized proportional derivative, and integral operators are established and studied in [54,55]. Certain new classes of integral inequalities for a class of n ($n\in\mathbb{N}$) positive continuous and decreasing functions on $[a,b]$ via generalized proportional fractional integrals can be found in the work of Rahman et al. [56]. The generalized Hadamard proportional fractional integrals and certain inequalities for convex functions by employing were recently proposed by Rahman et al. [57]. The bounds of proportional integrals in the sense of another function can be found in the work of Rahman et al. [58].

3. Main Results

In this section, we establish proportional fractional integral inequalities for convex functions by employing proportional fractional integral operators.

Theorem 3. *Suppose that the functions f and g are positive and continuous on the interval $[a,b]$, $(a<b)$ and $f\leq g$ on $[a,b]$. If the function $\frac{f}{g}$, $(g\neq 0)$ is decreasing and the function f is increasing on $[a,b]$, then, for any convex function Φ with $\Phi(0)=0$, the following inequality satisfies the proportional fractional integral operator given by Equation (4)*

$$\frac{{}_a\mathcal{J}^{\xi,\delta}\left[f(\tau)\right]}{{}_a\mathcal{J}^{\xi,\delta}\left[g(\tau)\right]}\geq\frac{{}_a\mathcal{J}^{\xi,\delta}\left[\Phi(f(\tau))\right]}{{}_a\mathcal{J}^{\xi,\delta}\left[\Phi(g(\tau))\right]}, \tag{6}$$

where $\delta\in(0,1]$, $\xi\in\mathbb{C}$ with $Re(\xi)>0$.

Proof. Since Φ is convex function with $\Phi(0) = 0$, the function $\frac{f(\tau)}{\tau}$ is increasing. As f is increasing, the function $\frac{\Phi(f(\tau))}{f(\tau)}$ is also increasing. Obviously, $\frac{f(\tau)}{g(\tau)}$ is decreasing function. Thus, for all $\rho, \theta \in [a,b]$, we have

$$\left(\frac{\Phi(f(\rho))}{f(\rho)} - \frac{\Phi(f(\theta))}{f(\theta)}\right)\left(\frac{f(\theta)}{g(\theta)} - \frac{f(\rho)}{g(\rho)}\right) \geq 0.$$

It follows that

$$\frac{\Phi(f(\rho))}{f(\rho)}\frac{f(\theta)}{g(\theta)} + \frac{\Phi(f(\theta))}{f(\theta)}\frac{f(\rho)}{g(\rho)} - \frac{\Phi(f(\theta))}{f(\theta)}\frac{f(\theta)}{g(\theta)} - \frac{\Phi(f(\rho))}{f(\rho)}\frac{f(\rho)}{g(\rho)} \geq 0. \tag{7}$$

Multiplying Equation (7) by $g(\rho)g(\theta)$, we have

$$\frac{\Phi(f(\rho))}{f(\rho)}f(\theta)g(\rho) + \frac{\Phi(f(\theta))}{f(\theta)}f(\rho)g(\theta) - \frac{\Phi(f(\theta))}{f(\theta)}f(\theta)g(\rho) - \frac{\Phi(f(\rho))}{f(\rho)}f(\rho)g(\theta) \geq 0. \tag{8}$$

Multiplying Equation (8) by $\frac{1}{\delta^{\xi}\Gamma(\xi)}\exp[\frac{\delta-1}{\delta}(\tau-\rho)](\tau-\rho)^{\xi-1}$, and integrating with respect to ρ over $[a,\tau]$, $a < \tau \leq b$, we have

$$\begin{aligned}
&\frac{1}{\delta^{\xi}\Gamma(\xi)}\int_a^{\tau}\exp[\frac{\delta-1}{\delta}(\tau-\rho)](\tau-\rho)^{\xi-1}\frac{\Phi(f(\rho))}{f(\rho)}f(\theta)g(\rho)d\rho \\
&+\frac{1}{\delta^{\xi}\Gamma(\xi)}\int_a^{\tau}\exp[\frac{\delta-1}{\delta}(\tau-\rho)](\tau-\rho)^{\xi-1}\frac{\Phi(f(\theta))}{f(\theta)}f(\rho)g(\theta)d\rho \\
&-\frac{1}{\delta^{\xi}\Gamma(\xi)}\int_a^{\tau}\exp[\frac{\delta-1}{\delta}(\tau-\rho)](\tau-\rho)^{\xi-1}\frac{\Phi(f(\theta))}{f(\theta)}f(\theta)g(\rho)d\rho \\
&-\frac{1}{\delta^{\xi}\Gamma(\xi)}\int_a^{\tau}\exp[\frac{\delta-1}{\delta}(\tau-\rho)](\tau-\rho)^{\xi-1}\frac{\Phi(f(\rho))}{f(\rho)}f(\rho)g(\theta)d\rho \geq 0.
\end{aligned}$$

Then, it follows that

$$\begin{aligned}
&f(\theta)\,{}_a\mathcal{J}^{\xi,\delta}\left(\frac{\Phi(f(\tau))}{f(\tau)}g(\tau)\right) + \left(\frac{\Phi(f(\theta))}{f(\theta)}g(\theta)\right){}_a\mathcal{J}^{\xi,\delta}(f(\tau)) \\
&-\left(\frac{\Phi(f(\theta))}{f(\theta)}f(\theta)\right){}_a\mathcal{J}^{\xi,\delta}(g(\tau)) - g(\theta)\,{}_a\mathcal{J}^{\xi,\delta}\left(\frac{\Phi(f(\tau))}{f(\tau)}f(\tau)\right) \geq 0.
\end{aligned} \tag{9}$$

Again, multiplying both sides of Equation (9) by $\frac{1}{\delta^{\xi}\Gamma(\xi)}\exp[\frac{\delta-1}{\delta}(\tau-\theta)](\tau-\theta)^{\xi-1}$, and integrating the resultant inequality with respect to θ over $[a,\tau]$, $a < \tau \leq b$, we get

$$\begin{aligned}
&{}_a\mathcal{J}^{\xi,\delta}(f(\tau))\,{}_a\mathcal{J}^{\xi,\delta}\left(\frac{\Phi(f(\tau))}{f(\tau)}g(\tau)\right) + {}_a\mathcal{J}^{\xi,\delta}\left(\frac{\Phi(f(\tau))}{f(\tau)}g(\tau)\right){}_a\mathcal{J}^{\xi,\delta}(f(\tau)) \\
&\geq {}_a\mathcal{J}^{\xi,\delta}(g(\tau))\,{}_a\mathcal{J}^{\xi,\delta}(\Phi(f(\tau))) + {}_a\mathcal{J}^{\xi,\delta}(\Phi(f(\tau)))\,{}_a\mathcal{J}^{\xi,\delta}(g(\tau)).
\end{aligned}$$

It follows that

$$\frac{{}_a\mathcal{J}^{\xi,\delta}(f(\tau))}{{}_a\mathcal{J}^{\xi,\delta}(g(\tau))} \geq \frac{{}_a\mathcal{J}^{\xi,\delta}(\Phi(f(\tau)))}{{}_a\mathcal{J}^{\xi,\delta}\left(\frac{\Phi(f(\tau))}{f(\tau)}g(\tau)\right)}. \tag{10}$$

Now, since $f \leq g$ on $[a,b]$ and $\frac{\Phi(\tau)}{\tau}$ is an increasing function, for $\rho \in [a,\tau]$, $a < \tau \leq b$, we have

$$\frac{\Phi(f(\rho))}{f(\rho)} \leq \frac{\Phi(g(\rho))}{g(\rho)}. \tag{11}$$

Multiplying both sides of Equation (11) by $\frac{1}{\delta^{\xi}\Gamma(\xi)}\exp[\frac{\delta-1}{\delta}(\tau-\rho)](\tau-\rho)^{\xi-1}g(\rho)$ and integrating the resultant inequality with respect to ρ over $[a,\tau]$, $a<\tau\leq b$, we get

$$\frac{1}{\delta^{\xi}\Gamma(\xi)}\int_a^{\tau}\exp[\frac{\delta-1}{\delta}(\tau-\rho)](\tau-\rho)^{\xi-1}\frac{\Phi(f(\rho))}{f(\rho)}g(\rho)d\rho$$
$$\leq\frac{1}{\delta^{\xi}\Gamma(\xi)}\int_a^{\tau}\exp[\frac{\delta-1}{\delta}(\tau-\rho)](\tau-\rho)^{\xi-1}\frac{\Phi(g(\rho))}{g(\rho)}g(\rho)d\rho,$$

which, in view of Equation (4), can be written as

$$_a\mathcal{J}^{\xi,\delta}\left(\frac{\Phi(f(\tau))}{f(\tau)}g(\tau)\right)\leq {_a\mathcal{J}^{\xi,\delta}}\left(\Phi(g(\tau))\right). \tag{12}$$

Hence, from Equations (10) and (12), we get Equation (6). □

Remark 2. *Applying Theorem 3 for $\delta=1$, we get Theorem 3.1 proved by [59].*

Remark 3. *Applying Theorem 3 for $\xi=\delta=1$ and $x=b$, we get Theorem 1.*

Theorem 4. *Suppose that the functions f and g are positive and continuous on $[a,b]$, $(a<b)$ and $f\leq g$ on $[a,b]$. If the function $\frac{f}{g}$, $(g\neq 0)$ is decreasing and the function f is increasing on $[a,b]$, then, for any convex function Φ with $\Phi(0)=0$, the following inequality satisfies the proportional fractional integral operator given by Equation* (4)

$$\frac{_a\mathcal{J}^{\xi,\delta}\left[f(\tau)\right]\,{_a\mathcal{J}^{\lambda,\delta}}\left[\Phi(g(\tau))\right]+{_a\mathcal{J}^{\lambda,\delta}}\left[f(\tau)\right]\,{_a\mathcal{J}^{\xi,\delta}}\left[\Phi(g(\tau))\right]}{_a\mathcal{J}^{\xi,\delta}\left[g(\tau)\right]\,{_a\mathcal{J}^{\lambda,\delta}}\left[\Phi(f(\tau))\right]+{_a\mathcal{J}^{\lambda,\delta}}\left[g(\tau)\right]\,{_a\mathcal{J}^{\xi,\delta}}\left[\Phi(f(\tau))\right]}\geq 1,$$

where $\delta\in(0,1]$, $\xi,\lambda\in\mathbb{C}$ with $Re(\xi)>0$ and $Re(\lambda)>0$.

Proof. Since Φ is convex function with $\Phi(0)=0$, the function $\frac{f(\tau)}{\tau}$ is increasing. As f is increasing, the function $\frac{\Phi(f(\tau))}{f(\tau)}$ is also increasing. Clearly, the function $\frac{f(\tau)}{g(\tau)}$ is decreasing for all $\rho,\theta\in[a,\tau]$, $a<\tau\leq b$. Multiplying Equation (9) by $\frac{1}{\delta^{\lambda}\Gamma(\lambda)}\exp[\frac{\delta-1}{\delta}(\tau-\theta)](\tau-\theta)^{\lambda-1}$ and integrating the resultant inequality with respect to θ over $[a,\tau]$, $a<\tau\leq b$, we get

$$_a\mathcal{J}^{\lambda,\delta}(f(\tau))\,{_a\mathcal{J}^{\xi,\delta}}\left(\frac{\Phi(f(\tau))}{f(\tau)}g(\tau)\right)+{_a\mathcal{J}^{\lambda,\delta}}\left(\frac{\Phi(f(\tau))}{f(\tau)}g(\tau)\right){_a\mathcal{J}^{\xi,\delta}}\left(f(\tau)\right)$$
$$\geq {_a\mathcal{J}^{\xi,\delta}}\left(g(\tau)\right)\,{_a\mathcal{J}^{\lambda,\delta}}\left(\frac{\Phi(f(\tau)}{f(\tau)}f(\tau)\right)+{_a\mathcal{J}^{\xi,\delta}}\left(\frac{\Phi(f(\tau)}{f(\tau)}f(\tau)\right){_a\mathcal{J}^{\lambda,\delta}}\left(g(\tau)\right). \tag{13}$$

Now, since $f\leq g$ on $[a,b]$ and $\frac{\Phi(\tau)}{\tau}$ is an increasing function, for $\rho\in[a,\tau]$, $a<\tau\leq b$, we have

$$\frac{\Phi(f(\rho))}{f(\rho)}\leq\frac{\Phi(g(\rho))}{g(\rho)}. \tag{14}$$

Multiplying both sides of Equation (14) by $\frac{1}{\delta^{\xi}\Gamma(\xi)}\exp[\frac{\delta-1}{\delta}(\tau-\rho)](\tau-\rho)^{\xi-1}g(\rho)$ and integrating the resultant inequality with respect to ρ over $[a,\tau]$, $a<\tau\leq b$, we get

$$\frac{1}{\delta^{\xi}\Gamma(\xi)}\int_a^{\tau}\exp[\frac{\delta-1}{\delta}(\tau-\rho)](\tau-\rho)^{\xi-1}\frac{\Phi(f(\rho))}{f(\rho)}g(\rho)d\rho$$
$$\leq\frac{1}{\delta^{\xi}\Gamma(\xi)}\int_a^{\tau}\exp[\frac{\delta-1}{\delta}(\tau-\rho)](\tau-\rho)^{\xi-1}\frac{\Phi(g(\rho))}{g(\rho)}g(\rho)d\rho,$$

which, in view of Equation (4), can be written as

$$ {}_a\mathcal{J}^{\xi,\delta}\left(\frac{\Phi(f(\tau))}{f(\tau)}g(\tau)\right) \leq {}_a\mathcal{J}^{\xi,\delta}\left(\Phi(g(\tau))\right). \tag{15} $$

Similarly, one can obtain

$$ {}_a\mathcal{J}^{\lambda,\delta}\left(\frac{\Phi(f(\tau))}{f(\tau)}g(\tau)\right) \leq {}_a\mathcal{J}^{\lambda,\delta}\left(\Phi(g(\tau))\right). \tag{16} $$

Hence, from Equations (12), (13), (15), and (16), we get the desired result. □

Remark 4. *Setting $\xi = \lambda$, Theorem 4 leads to Theorem 3.*

Remark 5. *Applying Theorem 4 for $\delta = 1$, we get Theorem 3.3 proved by Dahmani [59].*

Theorem 5. *Suppose that the functions f, h, and g are positive and continuous on $[a,b], (a < b)$ and $f \leq h$ on $[a,b]$. If the function $\frac{f}{g}$ is decreasing and the functions f and h are increasing on $[a,b]$, then, for any convex function Φ with $\Phi(0) = 0$, the following inequality satisfies the proportional fractional integral operator given by Equation* (4)

$$ \frac{{}_a\mathcal{J}^{\xi,\delta}\left[f(\tau)\right]}{{}_a\mathcal{J}^{\xi,\delta}\left[g(\tau)\right]} \geq \frac{{}_a\mathcal{J}^{\xi,\delta}\left[\Phi(f(\tau))h(\tau)\right]}{{}_a\mathcal{J}^{\xi,\delta}\left[\Phi(g(\tau))h(\tau)\right]}, $$

where $\delta \in (0,1]$, $\xi \in \mathbb{C}$ with $Re(\xi) > 0$.

Proof. Since Φ is convex function such that $\Phi(0) = 0$, the function $\frac{\Phi(\tau)}{\tau}$ is increasing. As the function f is increasing, $\frac{\Phi(f(\tau))}{f(\tau)}$ is also increasing. Clearly, the function $\frac{f(\tau)}{g(\tau)}$ is decreasing for all $\rho, \theta \in [a,\tau], a < \tau \leq b$.

$$ \left(\frac{\Phi(f(\rho))}{f(\rho)}h(\rho) - \frac{\Phi(f(\theta))}{f(\theta)}h(\theta)\right)\left(f(\theta)g(\rho) - f(\rho)g(\theta)\right) \geq 0. $$

It follows that

$$ \frac{\Phi(f(\rho))h(\rho)}{f(\rho)}f(\theta)g(\rho) + \frac{\Phi(f(\theta))h(\theta)}{f(\theta)}f(\rho)g(\theta) - \frac{\Phi(f(\theta))h(\theta)}{f(\theta)}f(\theta)g(\rho) - \frac{\Phi(f(\rho))h(\rho)}{f(\rho)}f(\rho)g(\theta) \geq 0. \tag{17} $$

Multiplying Equation (17) by $\frac{1}{\delta^{\xi}\Gamma(\xi)}\exp[\frac{\delta-1}{\delta}(\tau-\rho)](\tau-\rho)^{\xi-1}$ and integrating the resultant inequality with respect to ρ over $[a,\tau], a < \tau \leq b$, we have

$$ \begin{aligned} &\frac{1}{\delta^{\xi}\Gamma(\xi)}\int_a^{\tau}\exp[\frac{\delta-1}{\delta}(\tau-\rho)](\tau-\rho)^{\xi-1}\frac{\Phi(f(\rho))}{f(\rho)}f(\theta)g(\rho)h(\rho)d\rho \\ &+\frac{1}{\delta^{\xi}\Gamma(\xi)}\int_a^{\tau}\exp[\frac{\delta-1}{\delta}(\tau-\rho)](\tau-\rho)^{\xi-1}\frac{\Phi(f(\theta))}{f(\theta)}f(\rho)g(\theta)h(\theta)d\rho \\ &-\frac{1}{\delta^{\xi}\Gamma(\xi)}\int_a^{\tau}\exp[\frac{\delta-1}{\delta}(\tau-\rho)](\tau-\rho)^{\xi-1}\frac{\Phi(f(\theta))}{f(\theta)}f(\theta)h(\theta)g(\rho)d\rho \\ &-\frac{1}{\delta^{\xi}\Gamma(\xi)}\int_a^{\tau}\exp[\frac{\delta-1}{\delta}(\tau-\rho)](\tau-\rho)^{\xi-1}\frac{\Phi(f(\rho))}{f(\rho)}f(\rho)h(\rho)g(\theta)d\rho \geq 0. \end{aligned} $$

It follows that

$$f(\theta)\,{}_a\mathcal{J}^{\xi,\delta}\left(\frac{\Phi(f(\tau))}{f(\tau)}g(\tau)h(\tau)\right)+\left(\frac{\Phi(f(\theta))}{f(\theta)}g(\theta)h(\theta)\right){}_a\mathcal{J}^{\xi,\delta}(f(\tau))$$
$$-\left(\frac{\Phi(f(\theta))}{f(\theta)}f(\theta)h(\theta)\right){}_a\mathcal{J}^{\xi,\delta}(g(\tau))-g(\theta)\,{}_a\mathcal{J}^{\xi,\delta}\left(\frac{\Phi(f(\tau))}{f(\tau)}f(\tau)h(\tau)\right)\geq 0. \tag{18}$$

Again, multiplying both sides of Equation (18) by $\frac{1}{\delta^{\xi}\Gamma(\xi)}\exp[\frac{\delta-1}{\delta}(\tau-\theta)](\tau-\theta)^{\xi-1}$ and integrating the resultant inequality with respect to θ over $[a,\tau]$, $a<\tau\leq b$, we get

$${}_a\mathcal{J}^{\xi,\delta}(f(\tau))\,{}_a\mathcal{J}^{\xi,\delta}\left(\frac{\Phi(f(\tau))}{f(\tau)}g(\tau)h(\tau)\right)+{}_a\mathcal{J}^{\xi,\delta}\left(\frac{\Phi(f(\tau))}{f(\tau)}g(\tau)h(\tau)\right){}_a\mathcal{J}^{\xi,\delta}\left(f(\tau)\right)$$
$$\geq{}_a\mathcal{J}^{\xi,\delta}\left(g(\tau)\right)\,{}_a\mathcal{J}^{\xi,\delta}\left(\Phi(f(\tau))h(\tau)\right)+{}_a\mathcal{J}^{\xi,\delta}\left(\Phi(f(\tau))h(\tau)\right)\,{}_a\mathcal{J}^{\xi,\delta}\left(g(\tau)\right).$$

It follows that

$$\frac{{}_a\mathcal{J}^{\xi,\delta}(f(\tau))}{{}_a\mathcal{J}^{\xi,\delta}\left(g(\tau)\right)}\geq\frac{{}_a\mathcal{J}^{\xi,\delta}\left(\Phi(f(\tau))h(\tau)\right)}{{}_a\mathcal{J}^{\xi,\delta}\left(\frac{\Phi(f(\tau))}{f(\tau)}g(\tau)h(\tau)\right)}. \tag{19}$$

In addition, since $f\leq g$ on $[a,b]$ and $\frac{\Phi(\tau)}{\tau}$ is an increasing function, for $\eta,\theta\in[a,b]$, we have

$$\frac{\Phi(f(\eta))}{f(\eta)}\leq\frac{\Phi(g(\eta))}{g(\eta)}. \tag{20}$$

Multiplying both sides of Equation (20) by $\frac{1}{\delta^{\xi}\Gamma(\xi)}\exp[\frac{\delta-1}{\delta}(\tau-\eta)](\tau-\eta)^{\xi-1}g(\eta)h(\eta)$ and integrating the resultant inequality with respect to η over $[a,\tau]$, $a<\tau\leq b$, we get

$$\frac{1}{\delta^{\xi}\Gamma(\xi)}\int_a^{\tau}\exp[\frac{\delta-1}{\delta}(\tau-\eta)](\tau-\eta)^{\xi-1}\frac{\Phi(f(\eta))}{f(\eta)}g(\eta)h(\eta)d\eta$$
$$\leq\frac{1}{\delta^{\xi}\Gamma(\xi)}\int_a^{\tau}\exp[\frac{\delta-1}{\delta}(\tau-\eta)](\tau-\eta)^{\xi-1}\frac{\Phi(g(\eta))}{g(\eta)}g(\eta)h(\eta)d\eta,$$

which, in view of Equation (4), can be written as

$${}_a\mathcal{J}^{\xi,\delta}\left(\frac{\Phi(f(\tau))}{f(\tau)}g(\tau)h(\tau)\right)\leq{}_a\mathcal{J}^{\xi,\delta}\left(\Phi(g(\tau))h(\tau)\right). \tag{21}$$

Hence, from Equations (21) and (19), we obtain the required result. □

Remark 6. *Applying Theorem 5 for $\delta=1$, we get Theorem 3.5 proved by Dahmani [59].*

Remark 7. *Applying Theorem 5 for $\delta=\xi=1$ and $x=b$, we get Theorem 2.*

Theorem 6. *Suppose that the functions f, h, and g are positive and continuous on $[a,b]$, $(a<b)$ and $f\leq g$ on $[a,b]$. If the function $\frac{f}{g}$ is decreasing and the functions f and h are increasing on $[a,b]$, then, for any convex function Φ with $\Phi(0)=0$, the following inequality satisfies the proportional fractional integral operator given by Equation* (4)

$$\frac{{}_a\mathcal{J}^{\xi,\delta}\left[f(\tau)\right]\,{}_a\mathcal{J}^{\lambda,\delta}\left[\Phi(g(\tau))h(\tau)\right]+{}_a\mathcal{J}^{\lambda,\delta}\left[f(\tau)\right]\,{}_a\mathcal{J}^{\xi,\delta}\left[\Phi(g(\tau))h(\tau)\right]}{{}_a\mathcal{J}^{\xi,\delta}\left[g(\tau)\right]\,{}_a\mathcal{J}^{\lambda,\delta}\left[\Phi(f(\tau))h(\tau)\right]+{}_a\mathcal{J}^{\lambda,\delta}\left[g(\tau)\right]\,{}_a\mathcal{J}^{\xi,\delta}\left[\Phi(f(\tau))h(\tau)\right]}\geq 1, \tag{22}$$

where $\delta\in(0,1]$, $\xi,\lambda\in\mathbb{C}$ with $Re(\xi)>0$ and $Re(\lambda)>0$.

Proof. Multiplying both sides of Equation (18) by $\frac{1}{\delta^{\lambda}\Gamma(\lambda)}\exp[\frac{\delta-1}{\delta}(\tau-\theta)](\tau-\theta)^{\lambda-1}$ and integrating the resultant inequality with respect to θ over $[a,\tau], a<\tau\leq b$, we get

$$\begin{aligned}&{}_a\mathcal{J}^{\lambda,\delta}(f(\tau))\,{}_a\mathcal{J}^{\xi,\delta}\left(\frac{\Phi(f(\tau))}{f(\tau)}g(\tau)h(\tau)\right)+{}_a\mathcal{J}^{\lambda,\delta}\left(\frac{\Phi(f(\tau))}{f(\tau)}g(\tau)h(\tau)\right){}_a\mathcal{J}^{\xi,\delta}\left(f(\tau)\right)\\ &\geq {}_a\mathcal{J}^{\xi,\delta}\left(g(\tau)\right)\,{}_a\mathcal{J}^{\lambda,\delta}\left(\frac{\Phi(f(\tau))}{f(\tau)}f(\tau)h(\tau)\right)+{}_a\mathcal{J}^{\xi,\delta}\left(\frac{\Phi(f(\tau))}{f(\tau)}f(\tau)h(\tau)\right){}_a\mathcal{J}^{\lambda,\delta}\left(g(\tau)\right).\end{aligned}\tag{23}$$

Since $f\leq g$ on $[a,b]$ and $\frac{\Phi(\tau)}{\tau}$ is an increasing function, for $\eta,\theta\in[1,x], a<\tau\leq b$, we have

$$\frac{\Phi(f(\eta))}{f(\eta)}\leq\frac{\Phi(g(\eta))}{g(\eta)}.\tag{24}$$

Multiplying both sides of Equation (24) by $\frac{1}{\delta^{\xi}\Gamma(\xi)}\exp[\frac{\delta-1}{\delta}(\tau-\eta)](\tau-\eta)^{\xi-1}g(\eta)h(\eta),\ \eta\in[a,x],\ a<\tau\leq b$ and integrating the resultant inequality with respect to η over $[a,\tau],\ a<\tau\leq b$, we get

$${}_a\mathcal{J}^{\xi,\delta}\left(\frac{\Phi(f(\tau))}{f(\tau)}g(\tau)h(\tau)\right)\leq {}_a\mathcal{J}^{\xi,\delta}\left(\Phi(g(\tau))h(\tau)\right).\tag{25}$$

Similarly, one can obtain

$${}_a\mathcal{J}^{\eta,\delta}\left(\frac{\Phi(f(\tau))}{f(\tau)}g(\tau)h(\tau)\right)\leq {}_a\mathcal{J}^{\eta,\delta}\left(\Phi(g(\tau))h(\tau)\right).\tag{26}$$

Hence, from Equations (23), (25), and (26), we obtain the required inequality in Equation (22). □

Remark 8. *If we consider $\xi=\lambda$, then Theorem 6 leads to Theorem 5.*

Remark 9. *Applying Theorem 6 for $\delta=1$, we get Theorem 3.7 of Dahmani [59].*

4. Concluding Remarks

Some interesting integral inequalities for convex functions were presented by Liu et al. ([7] Theorems 9 and 10). Later, Dahmani [59] improved these integral inequalities by utilizing the R-L fractional integral operator. Here, we present some new fractional proportional integral inequalities for convex functions by utilizing the proportional fractional integrals. In fact, we established the inequalities presented in Theorem 1 and Theorem 2 using the fractional proportional integrals, which are nonlocal and their orders depend on two indices: δ, which is the proportional index, and ξ, which is the iterated index.

Author Contributions: All the authors contributed equally. All authors have read and agreed to the published version of the manuscript.

References

1. Mitrinovic, D.S.; Pecaric, J.E.; Fink, A.M. *Classical and New Inequalities in Analysis*; Kluwer Academic Publishers: Dordrecht, The Netherlands, 1993.
2. Pachpatte, B.G. *Mathematical Inequalities*, 1st ed.; North-Holland Mathematical Library (Volume 67) (Book 67); Elsevier Science: Amsterdam, The Netherlands, 2005.
3. Qi, F. Several integral inequalities. *JIPAM* **2000**, *1*, 19.
4. Sarikaya, M.Z.; Yildirim, H.; Saglam, A. On Hardy type integral inequality associated with the generalized translation. *Int. J. Contemp. Math. Sci.* **2006** *1*, 333–340. [CrossRef]

5. Ngo, Q.A.; Thang, D.D.; Dat, T.T.; Tuan, D.A. Notes on an integral inequality. *J. Inequal. Pure Appl. Math.* **2006**, *7*, 120.
6. Liu, W.J.; Cheng, G.S.; Li, C.C. Further development of an open problem concerning an integral inequality. *JIPAM* **2008**, *9*, 14.
7. Liu, W.J.; Ngô, Q.A.; Huy, V.N. Several interesting integral inequalities. *J. Math. Inequal.* **2009**, *3*, 201–212. [CrossRef]
8. Bougoufa, L. An integral inequality similar to Qi inequality. *JIPAM* **2005**, *6*, 27.
9. Boukerrioua, K.; Guezane Lakoud, A. On an open question regarding an integral inequality. *JIPAM* **2007**, *8*, 77.
10. Dahmani, Z.; Bedjaoui, N. Some generalized integral inequalities. *J. Adv. Res. Appl. Math.* **2011**, *3*, 58–66. [CrossRef]
11. Dahmani, Z.; Metakkel Elard, H. Generalizations of some integral inequalities using Riemann-Liouville operator. *Int. J. Open Probl. Compt. Math.* **2011**, *4*, 40–46.
12. Liu, W.J.; Li, C.C.; Dong, J.W. On an open problem concerning an integral inequality. *JIPAM* **2007**, *8*, 74.
13. Dahmani, Z.; Tabharit, L. On weighted Gruss type inequalities via fractional integration. *J. Adv. Res. Pure Math.* **2010**, *2*, 31–38. [CrossRef]
14. Dahmani, Z. New inequalities in fractional integrals. *Int. J. Nonlinear Sci.* **2010**, *9*, 493–497.
15. Nisar, K.S.; Qi, F.; Rahman, G.; Mubeen, S.; Arshad, M. Some inequalities involving the extended gamma function and the Kummer confluent hypergeometric k-function. *J. Inequal. Appl.* **2018**, *2018*, 135. [CrossRef] [PubMed]
16. Nisar, K.S.; Rahman, G.; Choi, J.; Mubeen, S.; Arshad, M. Certain Gronwall type inequalities associated with Riemann-Liouville k- and Hadamard k-fractional derivatives and their applications. *East Asian Math. J.* **2018**, *34*, 249–263.
17. Rahman, G.; Nisar, K.S.; Mubeen, S.; Choi, J. Certain Inequalities involving the (k, η)-fractional integral operator. *Far East J. Math. Sci. (FJMS)* **2018**, *103*, 1879–1888. [CrossRef]
18. Sarikaya, M.Z.; Dahmani, Z.; Kiris, M.E.; Ahmad, F. (k, s)-Riemann-Liouville fractional integral and applications. *Hacet. J. Math. Stat.* **2016**, *45*, 77–89. [CrossRef]
19. Set, E.; Tomar, M.; Sarikaya, M.Z. On generalized Grüss type inequalities for k-fractional integrals. *Appl. Math. Comput.* **2015**, *269*, 29–34. [CrossRef]
20. Sarikaya, M.Z.; Budak, H. Generalized Ostrowski type inequalities for local fractional integrals. *Proc. Am. Math. Soc.* **2017**, *145*, 1527–1538. [CrossRef]
21. Set, E.; Noor, M.A.; Awan, M.U.; Gözpinar, A. Generalized Hermite-Hadamard type inequalities involving fractional integral operators. *J. Inequal. Appl.* **2017**, *169*, 10. [CrossRef]
22. Dahmani, Z. New classes of integral inequalities of fractional order. *LE MATEMATICHE* **2014**, *LXIX*, 237–247. [CrossRef]
23. Podlubny, I. *Fractional Differential Equations*; Academic Press: London, UK, 1999.
24. Samko, S.G.; Kilbas, A.A.; Marichev, O.I. *Fractional Integrals and Derivatives, Theory and Applications*; Taylor & Francis: Abingdon, UK, 1993.
25. Abdeljawad, T. On Conformable Fractional Calculus. *J. Comput. Appl. Math.* **2015**, *279*, 57–66. [CrossRef]
26. Abdeljawad, T.; Baleanu, D. Monotonicity results for fractional difference operators with discrete exponential kernels. *Adv. Differ. Equ.* **2017**, *2017*, 78. [CrossRef]
27. Abdeljawad, T.; Baleanu, D. On Fractional Derivatives with Exponential Kernel and their Discrete Versions. *Rep. Math. Phys.* **2017**, *80*, 11–27. [CrossRef]
28. Anderson, D.R.; Ulness, D.J. Newly defined conformable derivatives. *Adv. Dyn. Syst. Appl.* **2015**, *10*, 109–137.
29. Atangana, A.; Baleanu, D. New fractional derivatives with nonlocal and non-singular kernel. Theory and application to heat transfer model. *Therm. Sci.* **2016**, *20*, 763–769. [CrossRef]
30. Caputo, M.; Fabrizio, M. A new Definition of Fractional Derivative without Singular Kernel. *Progr. Fract. Differ. Appl.* **2015**, *1*, 73–85.
31. Jarad, F.; Ugurlu, E.; Abdeljawad, T.; Baleanu, D. On a new class of fractional operators. *Adv. Differ. Equ.* **2017**, *2017*, 247. [CrossRef]
32. Khalil, R.; Horani, M.A.; Yousef, A.; Sababheh, M. A new definition of fractional derivative. *J. Comput. Appl. Math.* **2014**, *264*, 65–70. [CrossRef]
33. Losada, J.; Nieto, J.J. Properties of a New Fractional Derivative without Singular Kernel. *Progr. Fract.*

Differ. Appl. **2015**, *1*, 87–92.

34. Rahimi, Z.; Sumelka, W.; Yang, X.J. A new fractional nonlocal model and its application in free vibration of Timoshenko and Euler-Bernoulli beams. *Eur. Phys. J. Plus* **2017**, *132*, 479. [CrossRef]
35. Khan, M.A.; Khurshid, Y.; Du, T.-S.; Chu, Y.-M. Generalization of Hermite-Hadamard type inequalities via conformable fractional integrals. *J. Funct. Spaces* **2018**, *2018*, 5357463.
36. Khan, M.A.; Iqbal, A.; Suleman, M.; Chu, Y.-M. Hermite-Hadamard type inequalities for fractional integrals via Green's function. *J. Inequal. Appl.* **2018**, *2018*, 161. [CrossRef] [PubMed]
37. Huang, C.J.; Rahman, G.; Nisar, K.S.; Ghaffar, A.; Qi, F. Some Inequalities of Hermite-Hadamard type for k-fractional conformable integrals. *Aust. J. Math. Anal. Appl.* **2019**, *16*, 1–9.
38. Khurshid, Y.; Khan, M.A.; Chu, Y.-M. Conformable integral inequalities of the Hermite-Hadamard type in terms of GG- and GA-2 convexities. *J. Funct. Spaces* **2019**, *2019*, 6926107. [CrossRef]
39. Khurshid, Y.; Khan, M.A.; Chu, Y.-M.; Khan, Z.A. Hermite-Hadamard-Fejer inequalities for conformable fractional integrals via preinvex functions. *J. Funct. Spaces* **2019**, *2019*, 3146210. [CrossRef]
40. Mubeen, S.; Habib, S.; Naeem, M.N. The Minkowski inequality involving generalized k-fractional conformable integral. *J. Inequal. Appl.* **2019**, *2019*, 81. [CrossRef]
41. Nisar, K.S.; Rahman, G.; Mehrez, K. Chebyshev type inequalities via generalized fractional conformable integrals. *J. Inequal. Appl.* **2019**, *2019*, 245. [CrossRef]
42. Niasr, K.S.; Tassadiq, A.; Rahman, G.; Khan, A. Some inequalities via fractional conformable integral operators. *J. Inequal. Appl.* **2019**, *2019*, 217. [CrossRef]
43. Qi, F.; Rahman, G.; Hussain, S.M.; Du, W.S.; Nisar, K.S. Some inequalities of Čebyšev type for conformable k-fractional integral operators. *Symmetry* **2018**, *10*, 614. [CrossRef]
44. Rahman, G.; Nisar, K.S.; Qi, F. Some new inequalities of the Gruss type for conformable fractional integrals. *Aims Math.* **2018**, *3*, 575–583. [CrossRef]
45. Rahman, G.; Ullah, Z.; Khan, A.; Set, E.; Nisar, K.S. Certain Chebyshev type inequalities involving fractional conformable integral operators. *Mathematics* **2019**, *7*, 364. [CrossRef]
46. Ortega, A.; Rosales, J.J. Newton's law of cooling with fractional conformable derivative. *Revista Mexicana de Física* **2018**, *64*, 172–175. [CrossRef]
47. Hammad, M.A.; Khalil, R. Abel's formula and Wronskian for conformable fractional differential equations. *Int. J. Differ. Equ. Appl.* **2014**, *13*, 177–183.
48. Ilie, M.; Biazar, J.; Ayati, Z. General solution of Bernoulli and Riccati fractional differential equations based on conformable fractional derivative. *Int. J. Appl. Math. Res.* **2017**, *6*, 49–51.
49. Meng, S.; Cui, Y. The extremal solution to conformable fractional differential equations involving integral boundary condition. *Mathematics* **2019**, *7*, 186; doi:10.3390/math7020186. [CrossRef]
50. Jarad, F.; Abdeljawad, T.; Alzabut, J. Generalized fractional derivatives generated by a class of local proportional derivatives. *Eur. Phys. J. Spec. Top.* **2017**, *226*, 3457–3471. [CrossRef]
51. Wang, M.-K.; Chu, H.-H.; Chu, Y.-M. Precise Bounds for the Weighted Hölder Mean of the Complete P-Elliptic Integrals. *J. Math. Anal. Appl.* **2019**, *480*. [CrossRef]
52. Yang, Z.-H.; Qian, W.-M.; Chu, Y.-M.; Zhang, W. On rational bounds for the gamma function. *J. Inequal. Appl.* **2017**, *2017*, 210. [CrossRef]
53. Yang, Z.-H.; Qian, W.-M.; Chu, Y.-M.; Zhang, W. Monotonicity rule for the quotient of two functions and its application. *J. Inequal. Appl.* **2017**, *2017*, 106. [CrossRef]
54. Alzabut, J.; Abdeljawad, T.; Jarad, F.; Sudsutad, W. A Gronwall inequality via the generalized proportional fractional derivative with applications. *J. Inequal. Appl.* **2019**, *2019*, 101. [CrossRef]
55. Rahman, G.; Khan, A.; Abdeljawad, T.; Nisar, K.S. The Minkowski inequalities via generalized proportional fractional integral operators. *Adv. Differ. Equ.* **2019**, *2019*, 287. [CrossRef]
56. Rahman, G.; Abdeljawad, T.; Khan, A.; Nisar, K.S. Some fractional proportional integral inequalities. *J. Inequal. Appl.* **2019**, *2019*, 244. [CrossRef]
57. Rahman, G.; Abdeljawad, T.; Jarad, F.; Khan, A.; Nisar, K.S. Certain inequalities via generalized proportional Hadamard fractional integral operators. *Adv. Differ. Equ.* **2019**, *2019*, 454. [CrossRef]
58. Rahman, G.; Abdeljawad, T.; Jarad, F.; Nisar, K.S. Bounds of generalized proportional fractional integrals in general form via convex functions and their applications. *Mathematics* **2020**, *8*, 113. [CrossRef]
59. Dahmani, Z. A note on some new fractional results involving convex functions. *Acta Math. Univ. Comen.* **2012**, *LXXXI*, 241–246.

14

Integral Representation for the Solutions of Autonomous Linear Neutral Fractional Systems with Distributed Delay

Ekaterina Madamlieva [1], Mihail Konstantinov [2,*], Marian Milev [3] and Milena Petkova [1]

1 Faculty of Mathematics and Informatics, University of Plovdiv, 4000 Plovdiv, Bulgaria; ekaterinaa.b.m@gmail.com (E.M.); milenapetkova@uni-plovdiv.bg (M.P.)
2 Department of Mathematics, University of Architecture, Civil Engineering and Geodesy, 1046 Sofia, Bulgaria
3 Department of Matematics and Physics, University Of Food Technology, 4000 Plovdiv, Bulgaria; marianmilev2002@gmail.com
* Correspondence: misho.konstantinov@gmail.com

Abstract: The aim of this work is to obtain an integral representation formula for the solutions of initial value problems for autonomous linear fractional neutral systems with Caputo type derivatives and distributed delays. The results obtained improve and extend the corresponding results in the particular case of fractional systems with constant delays and will be a useful tool for studying different kinds of stability properties. The proposed results coincide with the corresponding ones for first order neutral linear differential systems with integer order derivatives.

Keywords: fractional derivatives; neutral fractional systems; distributed delay; integral representation

---

1. Introduction and Notations

Fractional Calculus has a long history, but it has attracted considerable attention recently as an important tool for modeling of various real problems, such as viscoelastic systems, diffusion processes, signal and control processing, and seismic processes. Detailed information about the fractional calculus theory and its applications can be found in the monographs [1–4]. Some results for fractional linear systems with delays are in given in the book [5]. The monograph [6] is devoted to the impulsive differential and functional differential equations with fractional derivatives, as well as to some of their applications.

It is well known that the study of linear fractional equations (integral representation, several types of stability, etc.) is an evergreen theme for research. Concerning these fields of fundamental and qualitative investigations for linear fractional ordinary differential equations and systems we refer to [2,4,7] and the references therein. Using the Laplace transform method, several interesting results in this direction are obtained in [8,9] as well. Regarding works concerning fractional differential systems with constant delays, we point out [10–13]. Concerning the retarded differential systems with variable or distributed delays—fundamental theory and application (stability properties)—we refer to [11,14–18]. Neutral fractional systems with distributed delays are essentially studied less (see [19–21]). Stability properties of retarded fractional systems with derivatives of distributed order are studied in [22]. One of the existing best applications of fractional order equations with delays is modeling human manual control, in which perceptual and neuromuscular delays introduce a delay term. As interesting studies, we refer to [23,24].

The problem of establishing an integral representation for the solutions for neutral or delayed linear fractional differential equations and/or systems needs a theorem for the existence of a fundamental matrix, i.e., theorem for existence and uniqueness of the solution to the initial value problem (IVP) in the case of discontinuous initial functions. As far as we know, there are only a few results concerning the IVP for delayed and neutral systems with discontinuous initial function, for the delayed case [14,15,25–27] and for the neutral case [28].

The aim of the work is to prove an integral representation formula for the general solution of an autonomous linear fractional neutral system with Caputo type derivatives and distributed delays. Note that our results extend and improve the results obtained in [10,12,15]. The proposed results coincide with the corresponding ones for a first order neutral linear differential system with integer order derivatives.

The paper is organized as follows. In Section 2, we recall some necessary definitions of Riemann-Liouville and Caputo fractional derivatives, as well as part of their properties. In this section, we also present the linear neutral fractional system under consideration together with some conditions. In Section 3, as a main result, integral representations of the solutions of the IVP for autonomous linear fractional neutral system with Caputo type derivatives and distributed delays are obtained for the homogeneous and inhomogeneous case. In Section 4, we present an illustrative example. In Section 5 we explain the practical benefits and application options of the obtained theoretical results.

In what follows, we use the notations: $\mathbb{N}$, $\mathbb{R}$ and $\mathbb{C}$ – the sets of natural, real and complex numbers, respectively; $\langle m, n\rangle$ – the set of integers $m, m+1, \ldots, n$ $(m \le n)$; $\mathbb{R}^{n\times n}$ – the space of real $n \times n$ matrices A with elements A_{pq}; $\mathbb{R}^n = \mathbb{R}^{n\times 1}$; $A^\top$ – the transposed matrix A with elements $(A^\top)_{pq} = A_{qp}$. The elements of $\mathbb{R}^n$ are the real column n-vectors $x = [x_1; x_2; \ldots; x_n]$ with elements x_k. The row n-vectors are denoted as $\xi = [\xi_1, \xi_2, \ldots, \xi_n]$(note that the elements of a vector column and a vector row are separated by ";" and ",", respectively). The identity and the zero matrices are denoted by E and Θ, respectively.

We also denote $\mathbb{C}_+ = \{p \in \mathbb{C} | \mathrm{Re}(p) > 0\}$, $\overline{\mathbb{C}}_+ = \{p \in \mathbb{C} | \mathrm{Re}\, p \ge 0\}$, $\mathbb{C}_- = \mathbb{C} \setminus \overline{\mathbb{C}}_+$, $\mathbb{R}_+ = (0, \infty)$, and $J_s = [s, \infty)$. For $p \in \mathbb{C}$, $y = [y_1; y_2; \ldots; y_n] \in \mathbb{C}^n$ and $\beta = (\beta_1, \beta_2, \ldots, \beta_n)$, $\beta_k \in [-1, 1]$ we set $I_\beta(p) = \mathrm{diag}\,(p^{\beta_1}, p^{\beta_2}, \ldots, p^{\beta_n})$ and $I_\beta(y) = \mathrm{diag}\left(y_1^{\beta_1}, y_2^{\beta_2}, \ldots, y_n^{\beta_n}\right)$. The linear space of locally Lebesgue integrable functions $f : \mathbb{R} \to \mathbb{R}$ is denoted by $L_1^{\mathrm{loc}}(\mathbb{R}, \mathbb{R})$.

2. Preliminaries and Problem Statement

Below, the definitions of Riemann–Liouville and Caputo fractional derivatives and some of their properties necessary for our exposition are described in order to avoid possible misunderstandings. For more details and other properties, we refer to [2–4].

Let $\alpha \in (0, 1)$ be an arbitrary number. Then for $a \in \mathbb{R}$, each $t > a$ and $f \in L_1^{\mathrm{loc}}(\mathbb{R}, \mathbb{R})$ the left-sided fractional integral operator, the left side Riemann-Liouville and Caputo fractional derivatives of order α are defined by

$$\begin{aligned}
(D_{a+}^{-\alpha} f)(t) &= \frac{1}{\Gamma(\alpha)} \int_a^t (t-s)^{\alpha-1} f(s) \mathrm{d}s, \ {}_{RL}D_{a+}^{\alpha} f(t) = \frac{\mathrm{d}}{\mathrm{d}t}\left(D_{a+}^{-(1-\alpha)} f(t)\right), \\
{}_C D_{a+}^{\alpha} f(t) &= {}_{RL}D_{a+}^{\alpha}[f(s) - f(a)](t) = {}_{RL} D_{a+}^{\alpha} f(t) - \frac{f(a)}{\Gamma(1-\alpha)}(t-a)^{-\alpha},
\end{aligned}$$

respectively. The following relations [4] involving fractional derivatives will be used

$$(D_{a+}^{0} f)(t) = f(t),\ {}_C D_{a+}^{\alpha} D_{a+}^{-\alpha} f(t) = f(t),\ D_{a+}^{-\alpha}\, {}_C D_{a+}^{\alpha} f(t) = f(t) - f(a).$$

Concerning the Laplace transform $\mathfrak{L}$,

$$\mathfrak{L} f(p) = \int_0^\infty \exp(-pt) f(t) \mathrm{d}t,\ p \in \mathbb{C},$$

we shall need the relations

$$\mathfrak{L}D_{0+}^{-\alpha}f(p) = p^{-\alpha}(\mathfrak{L}f)(p),\ \mathfrak{L}_{RL}D_{0+}^{\alpha}f(p) = p^{\alpha}(\mathfrak{L}f)(p) - [{}_{RL}D_{0+}^{\alpha-1}f(t)]_{t=0},$$
$$\mathfrak{L}_C D_{0+}^{\alpha}f(p) = p^{\alpha}(\mathfrak{L}f)(p) - p^{\alpha-1}f(0).$$

In what follows, we consider the autonomous linear neutral fractional system with distributed delay

$$D^{\alpha}\Big(X(t) - \sum_{l=1}^{r}\int_{-\tau}^{0}\Big[\mathrm{d}_{\theta}V^{l}(\theta)\Big]X(t+\theta)\Big) = \sum_{i=0}^{m}\int_{-\sigma}^{0}\Big[\mathrm{d}_{\theta}U^{i}(\theta)\Big]X(t+\theta) + F(t), \tag{1}$$

as well as the corresponding homogeneous system

$$D^{\alpha}\Big(X(t) - \sum_{l=1}^{r}\int_{-\tau}^{0}\Big[\mathrm{d}_{\theta}V^{l}(\theta)\Big]X(t+\theta)\Big) = \sum_{i=0}^{m}\int_{-\sigma}^{0}\Big[\mathrm{d}_{\theta}U^{i}(\theta)\Big]X(t+\theta), \tag{2}$$

where

$$X, F : J_0 \to \mathbb{R}^n,\ U^i, V^l : \mathbb{R} \to \mathbb{R}^{n\times n},\ U^i(\theta) = \Big[u_{kj}^{i}(\theta)\Big],\ V^l(\theta) = \Big[v_{kj}^{l}(\theta)\Big],$$
$$\tau, \sigma > 0,\ \tau_r \in (0,\tau],\ l \in \langle 1, r\rangle,\ \sigma_i \in (0,\sigma],\ i \in \langle 1, m\rangle,\ h = \max(\sigma,\tau),\ \sigma_0 = 0,$$
$$\alpha = (\alpha_1, \alpha_2, \ldots, \alpha_n),\ \alpha_k \in (0,1),\ k \in \langle 1, n\rangle,\ J_s = [s,\infty).$$

For simplicity, D^{α_k} denotes the left side Caputo fractional derivative ${}_C D_{0+}^{\alpha_k}$ in (1) and (2), and we use the notations

$$D^{\alpha}X(t) = [D^{\alpha_1}x_1(t); D^{\alpha_2}x_2(t); \ldots; D^{\alpha_n}x_n(t)],\ D^{\alpha} = \mathrm{diag}(D^{\alpha_1}, D^{\alpha_2}, \ldots, D^{\alpha_n}),$$
$$X(t) = [x_1(t); x_2(t); \ldots; x_n(t)],\ F(t) = [f_1(t); f_2(t); \ldots; f_n(t)].$$

Denote by $BV[-h,0]$ the linear space of matrix valued functions

$$W : \mathbb{R} \to \mathbb{R}^{n\times n},\ W(\theta) = \Big[\omega_{kj}(\theta)\Big]$$

with bounded variation in θ on $[-h,0]$,

$$\mathrm{Var}_{[-h,0]}W(.) = \sum_{k,j=1}^{n}\mathrm{Var}_{[-h,0]}w_{kj}(.),\ |W(\theta)| = \sum_{k,j=1}^{n}|w_{kj}(\theta)|.$$

As a space of initial functions, we use the Banach space $\widetilde{C} = PC([-h,0],\mathbb{R}^n)$ of the piecewise continuous on $[-h,0]$ vector functions $\Phi = [\phi_1; \phi_2; \ldots; \phi_n] : [-h,0] \to \mathbb{R}^n$ with norm

$$||\Phi|| = \sum_{k=1}^{n}\sup_{s\in[-h,0]}|\phi_k(s)| < \infty.$$

The initial condition for the system (1) or (2) is

$$X(t) = \Phi(t),\ t \in [-h,0]. \tag{3}$$

Definition 1. *The vector function X is a solution of the IVP* (1), (3) *in the interval J_{-h} if $X|_{J_0} \in C(J_0,\mathbb{R}^n)$ and if it satisfies the system* (1) *for $t \in \mathbb{R}_+$ and the initial condition* (3) *for $t \in [-h,0]$.*

We say that for the kernels $U^i : \mathbb{R} \to \mathbb{R}^{n\times n}$, $V^l : \mathbb{R} \to \mathbb{R}^{n\times n}$ the assumptions (SA) are fulfilled, if for each $i \in \langle 0, m\rangle$ and $l \in \langle 1, r\rangle$ the following conditions hold.

(SA1) The matrix valued functions $\theta \mapsto U^i(\theta)$ and $\theta \mapsto V^l(\theta)$ are measurable in $\theta \in \mathbb{R}$ and normalized so that $U^i(\theta) = 0$ and $V^l(\theta) = 0$ for $\theta \geq 0$, $U^i(\theta) = U^i(-\sigma_i)$ for $\theta \leq -\sigma_i$ and $V^l(\theta) = V^l(-\tau_l)$ for $\theta \leq -\tau_l$.

(SA2) The kernels $U^i(\theta)$ and $V^l(\theta)$ are left continuous for $\theta \in (-\sigma, 0)$ and $\theta \in (-\tau, 0]$ and $U^i(\cdot)$, $V^l(\cdot) \in BV[-h, 0]$.

(SA3) The Lebesgue decomposition of the kernels $U^i(\theta)$ and $V^l(\theta)$ for $\theta \in [-h, 0]$ is

$$\begin{aligned} U^i(\theta) &= \aleph^i(\theta) + \int_{-h}^{\theta} B^i(s)\mathrm{d}s + \Upsilon^i(\theta), \\ V^l(\theta) &= \widetilde{\aleph}^l(\theta) + \int_{-h}^{\theta} \widetilde{B}^l(s)\mathrm{d}s + \widetilde{\Upsilon}^l(\theta), \end{aligned}$$

where $A^i = \left[a^i_{kj}\right]$, $\widetilde{A}^l = \left[\widetilde{a}^l_{kj}\right] \in \mathbb{R}^{n\times n}$ and

$$\begin{aligned} &\aleph^i(\theta) = \left[a^i_{kj} H(\theta + \sigma_i)\right], \ \widetilde{\aleph}^l(\theta) = \left[\widetilde{a}^l_{kj} H(\theta + \tau_l)\right], \\ &\Upsilon^i(\theta) = \left[g^i_{kj}(\theta)\right], \ \widetilde{\Upsilon}^l(\theta) = \left[\widetilde{g}^l_{kj}(\theta)\right] \in C(\mathbb{R}, \mathbb{R}^{n\times n}), \\ &B^i(\theta) = \left[b^i_{kj}(\theta)\right], \ \widetilde{B}^l(\theta) = \left[\widetilde{b}^l_{kj}(\theta)\right] \in L^{\mathrm{loc}}_1(\mathbb{R}^{n\times n}, \mathbb{R}^{n\times n}). \end{aligned}$$

Remark 1. *The conditions (SA) are used essentially in the work [21] to establish an apriory estimate of all solutions of the IP* (1), (3), *which estimate guaranties that the Laplace transform can be correct applied to System* (2) *and to System* (1) *too, when the function F is exponentially bounded.*

Let $s \geq 0$ be an arbitrary number, $J_s = [s, \infty)$ and consider the matrix IVP

$$D^\alpha\left(Q(t,s) - \sum_{l=1}^{r}\int_{-\tau}^{0}\left[\mathrm{d}_\theta V^l(\theta)\right]Q(t+\theta,s)\right) = \sum_{i=0}^{m}\int_{-\sigma}^{0}\left[\mathrm{d}_\theta U^i(t,\theta)\right]Q(t+\theta,s) \tag{4}$$

with initial condition

$$Q(t,t) = I;\ Q(t,s) = 0,\ t < s. \tag{5}$$

Definition 2. *For each $s \geq 0$ the matrix valued function*

$$t \mapsto Q(t,s) = \left[\gamma_{kj}(t,s)\right],\ Q(\cdot,s) : J_s \to \mathbb{R}^{n\times n},$$

is called a solution of the IVP (4), (5) *for $t \in J_s$, if $Q(\cdot,s)$ is continuous in t on J_s and satisfies the matrix Equation* (4) *for $t \in (s,\infty)$ and the initial condition* (5).

It is well known that the problem of existence of a fundamental matrix for a linear homogeneous fractional system (delayed or neutral) leads to establishing that the corresponding IVP (4), (5) with discontinuous initial function has a unique solution. In the case when $s = 0$, the matrix $Q(t) = Q(t,0)$ will be called fundamental (or Cauchy) matrix of system (2).

Following [20,21], we introduce the characteristic matrix of System (2)

$$G(p) = I_\alpha(p) - W(p), \tag{6}$$

where

$$\begin{aligned} W(p) &= \sum_{i=0}^{m} U_i(p) + I_\alpha(p) \sum_{l=1}^{r} V_l(p),\ i \in \langle 0, m \rangle,\ l \in \langle 1, r \rangle, \\ U_i(p) &= \left[\int_{-h}^{0} \exp(p\theta) \mathrm{d}u^i_{kj}(\theta) \right],\ V_l(p) = \left[\int_{-h}^{0} \exp(p\theta) \mathrm{d}v^l_{kj}(\theta) \right]. \end{aligned}$$

3. Main Results

The results in this section are a generalization of the results concerning the autonomous case obtained in [10,15,16,25].

Theorem 1. *Let us assume the conditions (SA) are satisfied. Then the IVP* (4), (5) *has a unique solution* $Q(t,s)$ *in* J_s *for every* $s \geq 0$ *and the fundamental matrix* $Q(t,0) = Q(t)$ *of Equation* (2) *is*

$$Q(t) = \mathfrak{L}^{-1}\left(I_{\alpha-1}(p) G^{-1}(p) \right)(t). \tag{7}$$

Proof. Using the results from [28], we obtain that the IVP (4), (5) has a unique solution $Q(t,s)$ in J_s for every $s \geq 0$, and hence, a fundamental matrix $Q(t,0) = Q(t)$. In virtue of Theorem 3 [21], we can conclude that the Laplace transform can be applied to both sides of Equation (4). Substituting $t + \theta = \eta$ we obtain

$$\begin{aligned} &\int_0^\infty \exp(-pt) \sum_{i=0}^{m} \int_{-h}^{0} \left[\mathrm{d}_\theta U^i(\theta) \right] Q(t+\theta)\mathrm{d}t = \sum_{i=0}^{m} \int_{-h}^{0} \left[\mathrm{d}_\theta U^i(\theta) \right] \left(\exp(p\theta) \int_\theta^0 \exp(-p\eta) Q(\eta) \mathrm{d}\eta \right) \\ &+ \int_0^\infty \exp(-p\eta) Q(\eta) \mathrm{d}\eta \sum_{i=0}^{m} \int_{-h}^{0} \exp(p\theta) \mathrm{d}_\theta U^i(\theta) = \mathfrak{L}Q(p) \sum_{i=0}^{m} \int_{-h}^{0} \exp(p\theta) \mathrm{d}_\theta U^i(\theta). \end{aligned}$$

In a similar way for the left-hand side of Equation (4), we have that

$$\begin{aligned} \mathfrak{L}D^\alpha &\left(Q(t) - \sum_{l=1}^{r} \int_{-\tau}^{0} \left[\mathrm{d}_\theta V^l(\theta) \right] Q(t+\theta) \right)(p) = I_\alpha(p) \mathfrak{L}Q(p) \left[E - \sum_{l=1}^{r} \int_{-h}^{0} \exp(p\theta) \mathrm{d}_\theta V^l(\theta) \right] \\ &- I_{\alpha-1}(p) \left[E - \sum_{l=1}^{r} \int_{-\tau}^{0} \left[\mathrm{d}_\theta V^l(\theta) \right] Q(\theta) \right]. \end{aligned} \tag{8}$$

From Equation (8), it follows that

$$\mathfrak{L}Q(p) \left[I_\alpha - I_\alpha(p) \sum_{l=1}^{r} \int_{-h}^{0} \exp(p\theta) \mathrm{d}_\theta V^l(\theta) - \sum_{l=1}^{m} \int_{-h}^{0} \exp(p\theta) \mathrm{d}_\theta U^i(\theta) \right] = I_{\alpha-1}(p)$$

and hence, $\mathfrak{L}Q(p) = I_{\alpha-1}(p) G^{-1}(p)$, which completes the proof. □

Let us introduce the following functions:

$$\Phi_l(t) = \Phi(t),\ t \in [-\tau_l, 0],\ \ \Phi_l(t) = 0,\ t \in \mathbb{R} \setminus [-\tau_l, 0],\ \ l \in \langle 1, r \rangle,$$

and

$$\Phi_i(t) = \Phi(t),\ t \in [-\sigma_i, 0],\ \ \Phi_i(t) = 0,\ t \in \mathbb{R} \setminus [-\sigma_i, 0],\ \ i \in \langle 1, m \rangle.$$

Then, applying the Laplace transform second shifting theorem, we obtain

$$\begin{aligned} \int_\theta^0 \exp(-p(\eta-\theta))\Phi_l(\eta)\mathrm{d}\eta &= \exp(p\theta)\mathfrak{L}\Phi_l(t)(p), \\ \int_\theta^0 \exp(-p(\eta-\theta))\Phi_i(\eta)\mathrm{d}\eta &= \exp(p\theta)\mathfrak{L}\Phi_i(t)(p). \end{aligned} \tag{9}$$

Now we are in position to prove the following theorem.

Theorem 2. *Let us assume the conditions (SA) are satisfied. Then for each $\Phi \in \widetilde{C}$ the IVP* (2), (3) *has a unique solution $X_\Phi(t)$ with the integral representation:*

$$\begin{aligned} X_\Phi(t) &= Q(t)\left(\Phi(0) - \sum_{l=1}^{r}\int_{-h}^{0}\left[\mathrm{d}_\theta V^l(\theta)\right]\Phi_l(\theta)\right) + \sum_{l=1}^{r}\int_{-h}^{0}\left[\mathrm{d}_\theta V^l(\theta)\right]\Phi_l(t+\theta) \\ &+ \sum_{l=1}^{r}\int_{-h}^{0}\left[\mathrm{d}_\theta V^l(\theta)\right] D^{\frac{1}{2}}Q(t) * D^{\frac{1}{2}}\Phi_l(t+\theta) + \sum_{i=0}^{m}\int_{-h}^{0}[\mathrm{d}_\theta U^i(\theta)]D^{1-\alpha}Q(t) * \Phi_i(t+\theta) \\ &+ \sum_{i=0}^{m}\int_{-h}^{0}[\mathrm{d}_\theta U^i(\theta)]D^{-\alpha}\Phi_i(t+\theta). \end{aligned} \tag{10}$$

Proof. Let $\Phi \in \widetilde{C}$. Then using the results from [26], we can conclude that the IVP (2), (3) has a unique solution $X_\Phi(t)$. In virtue of Theorem 3 from [21], we can conclude that the Laplace transform can be applied to both sides of Equation (2). Then, substituting $X_\Phi(t)$ in Equation (2), applying the Laplace transform to Equation (2) and substituting $t + \theta = \eta$, we obtain for the right-hand side of Equation (2)

$$\begin{aligned} \mathfrak{L}\left(\sum_{i=0}^{m}\int_{-h}^{0}[\mathrm{d}_\theta U^i(\theta)]X_\Phi(t+\theta)\right)(p) &= \int_0^\infty \exp(-p\eta)X_\Phi(\eta)\mathrm{d}\eta \sum_{i=0}^{m}\int_{-h}^{0}\exp(p\theta)\mathrm{d}_\theta U^i(\theta) \\ + \sum_{i=0}^{m}\int_{-h}^{0}[\mathrm{d}_\theta U^i(\theta)]\left(\int_\theta^0 \exp(p(\theta-\eta))X_\Phi(\eta)\mathrm{d}\eta\right) &= \mathfrak{L}X_\Phi(t)(p)\sum_{i=0}^{m}\int_{-h}^{0}\exp(p\theta)\mathrm{d}_\theta U^i(\theta) \\ &+ \sum_{i=0}^{m}\int_{-h}^{0}[\mathrm{d}_\theta U^i(\theta)]\int_\theta^0 \exp(p(\theta-\eta))\Phi_i(\eta)\mathrm{d}\eta. \end{aligned} \tag{11}$$

Similarly, for the left-hand side of Equation (2), one obtains that

$$\begin{aligned} \mathfrak{L}D^\alpha\left(X_\Phi(t) - \sum_{l=1}^{r}\int_{-h}^{0}\left[\mathrm{d}_\theta V^l(\theta)\right]X_\Phi(t+\theta)\right)(p) &= -I_{\alpha-1}(p)\left(\Phi(0) - \sum_{l=1}^{r}\int_{-h}^{0}\left[\mathrm{d}_\theta V^l(\theta)\right]\Phi_l(\theta)\right) \\ &+ \mathfrak{L}X_\Phi(t)(p)\left(I_\alpha(p) - I_\alpha(p)\sum_{l=1}^{r}\int_{-h}^{0}\exp(p\theta)\mathrm{d}_\theta V^l(\theta)\right) \\ &- I_\alpha(p)\sum_{l=1}^{r}\int_{-h}^{0}[\mathrm{d}_\theta V^l(\theta)]\int_\theta^0 \exp(p(\theta-\eta))\Phi_l(\eta)\mathrm{d}\eta. \end{aligned} \tag{12}$$

From Equations (11) and (12), it follows

$$\begin{aligned} &\mathfrak{L}X_\Phi(p)\left(I_\alpha(p) - I_\alpha(p)\sum_{l=1}^{r}\int_{-h}^{0}\exp(p\theta)d_\theta V^l(\theta) - \sum_{i=0}^{m}\int_{-h}^{0}\exp(p\theta)d_\theta U^i(\theta)\right) \\ &= I_{\alpha-1}(p)\left(\Phi(0) - \sum_{l=1}^{r}\int_{-h}^{0}\left[\mathrm{d}_\theta V^l(\theta)\right]\Phi_l(\theta)\right) + I_\alpha(p)\sum_{l=1}^{r}\int_{-h}^{0}\left[\mathrm{d}_\theta V^l(\theta)\right]\int_\theta^0 \exp(p(\theta-\eta))\Phi_l(\eta)\mathrm{d}\eta \\ &+ \sum_{i=0}^{m}\int_{-h}^{0}\left[\mathrm{d}_\theta U^i(\theta)\right]\int_\theta^0 \exp(p(\theta-\eta))\Phi_i(\eta)\mathrm{d}\eta \end{aligned}$$

and hence,

$$\begin{aligned}\mathfrak{L}X_{\Phi}(p) &= G^{-1}(p)I_{\alpha-1}(p)\left(\Phi(0)-\sum_{l=1}^{r}\int_{-h}^{0}\left[\mathrm{d}_{\theta}V^{l}(\theta)\right]\Phi_{l}(\theta)\right)\\ &+G^{-1}(p)I_{\alpha}(p)+\sum_{l=1}^{r}\int_{-h}^{0}\left[\mathrm{d}_{\theta}V^{l}(\theta)\right]\int_{\theta}^{0}\exp(p(\theta-\eta))\Phi_{l}(\eta)\mathrm{d}\eta\\ &+G^{-1}(p)\sum_{i=0}^{m}\int_{-h}^{0}\left[\mathrm{d}_{\theta}U^{i}(\theta)\right]\int_{\theta}^{0}\exp(p(\theta-\eta))\Phi_{i}(\eta)\mathrm{d}\eta.\end{aligned} \tag{13}$$

The representations of Equations (7) and (13) imply that

$$\begin{aligned}\mathfrak{L}X_{\Phi}(p) &= \mathfrak{L}Q(t)(p)\left(\Phi(0)-\sum_{l=1}^{r}\int_{-h}^{0}[\mathrm{d}_{\theta}V^{l}(\theta)]\Phi_{l}(\theta)\right)\\ &+I_{1}(p)\mathfrak{L}Q(t)(p)\sum_{l=1}^{r}\int_{-h}^{0}\left[\mathrm{d}_{\theta}V^{l}(\theta)\right]\int_{\theta}^{0}\exp(p(\theta-\eta))\Phi_{l}(\eta)\mathrm{d}\eta\\ &+I_{1-\alpha}(p)\mathfrak{L}Q(t)(p)\sum_{i=0}^{m}\int_{-h}^{0}\left[\mathrm{d}_{\theta}U^{i}(\theta)\right]\int_{\theta}^{0}\exp(p(\theta-\eta))\Phi_{i}(\eta)\mathrm{d}\eta.\end{aligned} \tag{14}$$

In view of Equation (9), we obtain for the second term in the right-hand side of Equation (14) that

$$\begin{aligned}&I_{1}(p)\mathfrak{L}Q(t)(p)\sum_{l=1}^{r}\int_{-h}^{0}\left[\mathrm{d}_{\theta}V^{l}(\theta)\right]\int_{\theta}^{0}\exp(p(\theta-\eta))\Phi_{l}(\eta)\mathrm{d}\eta\\ &=I_{\frac{1}{2}}(p)\mathfrak{L}Q(t)(p)I_{\frac{1}{2}}(p)\sum_{l=1}^{r}\int_{-h}^{0}\left[\mathrm{d}_{\theta}V^{l}(\theta)\right]\mathfrak{L}\Phi_{l}(t+\theta)(p)\\ &=\mathfrak{L}D^{\frac{1}{2}}Q(t)(p)+I_{-\frac{1}{2}}(p)I_{\frac{1}{2}}(p)\sum_{l=1}^{r}\int_{-h}^{0}\left[\mathrm{d}_{\theta}V^{l}(\theta)\right]\mathfrak{L}\Phi_{l}(t+\theta)(p)\\ &=\mathfrak{L}D^{\frac{1}{2}}Q(t)(p)\sum_{l=1}^{r}\int_{-h}^{0}\left[\mathrm{d}_{\theta}V^{l}(\theta)\right]I_{\frac{1}{2}}(p)\mathfrak{L}\Phi_{l}(t+\theta)(p)\\ &+\sum_{l=1}^{r}\int_{-h}^{0}\left[\mathrm{d}_{\theta}V^{l}(\theta)\right]\mathfrak{L}\Phi_{l}(t+\theta)(p).\end{aligned} \tag{15}$$

For the first term in the right-hand side of Equation (15) we have

$$\begin{aligned}&\mathfrak{L}D^{\frac{1}{2}}Q(t)(p)\sum_{l=1}^{r}\int_{-h}^{0}\left[\mathrm{d}_{\theta}V^{l}(\theta)\right]I_{\frac{1}{2}}(p)\mathfrak{L}\Phi_{l}(t+\theta)(p)\\ &=\sum_{l=1}^{r}\int_{-h}^{0}\left[\mathrm{d}_{\theta}V^{l}(\theta)\right]\mathfrak{L}D^{\frac{1}{2}}Q(t)(p)\mathfrak{L}D^{\frac{1}{2}}\Phi_{l}(t+\theta)(p)\end{aligned}$$

and hence, from Equation (15) it follows

$$\begin{aligned}&I_{1}(p)\mathfrak{L}Q(t)(p)\sum_{l=1}^{r}\int_{-h}^{0}\left[\mathrm{d}_{\theta}V^{l}(\theta)\right]\int_{\theta}^{0}\exp(p(\theta-\eta))\Phi_{l}(\eta)\mathrm{d}\eta\\ &=\sum_{l=1}^{r}\int_{-h}^{0}\left[\mathrm{d}_{\theta}V^{l}(\theta)\right](\mathfrak{L}D^{\frac{1}{2}}Q(t)(p)\mathfrak{L}D^{\frac{1}{2}}\Phi_{l}(t+\theta)(p)\\ &+\sum_{l=1}^{r}\int_{-h}^{0}[\mathrm{d}_{\theta}V^{l}(\theta)]\mathfrak{L}\Phi_{l}(t+\theta)(p).\end{aligned} \tag{16}$$

Analogously, for the third term in the right-hand side of Equation (14), we have

$$\begin{aligned}
&I_{1-\alpha}(p)\mathfrak{L}Q(t)(p)\sum_{i=0}^{m}\int_{-h}^{0}\left[\mathrm{d}_\theta U^i(\theta)\right]\int_{\theta}^{0}\exp(p(\theta-\eta))\Phi_i(\eta)\mathrm{d}\eta \\
&= I_{1-\alpha}(p)\mathfrak{L}Q(t)(p)\sum_{i=0}^{m}\int_{-h}^{0}\left[\mathrm{d}_\theta U^i(\theta)\right]\mathfrak{L}\Phi_i(t+\theta)(p) \\
&= \sum_{i=0}^{m}\int_{-h}^{0}\left[\mathrm{d}_\theta U^i(\theta)\right]\mathfrak{L}D^{1-\alpha}Q(t)(p)\mathfrak{L}\Phi_i(t+\theta)(p) \\
&+ \sum_{i=0}^{m}\int_{-h}^{0}[\mathrm{d}_\theta U^i(\theta)]\mathfrak{L}D^{-\alpha}\Phi_i(t+\theta)(p)
\end{aligned} \tag{17}$$

From Equations (14), (16) and (17), it follows that

$$\begin{aligned}
\mathfrak{L}X_\Phi(p) &= \mathfrak{L}Q(t)(p)\left(\Phi(0)-\sum_{l=1}^{r}\int_{-h}^{0}\left[\mathrm{d}_\theta V^l(\theta)\right]\Phi(\theta)\right)+\sum_{l=1}^{r}\int_{-h}^{0}\left[\mathrm{d}_\theta V^l(\theta)\right]\mathfrak{L}\Phi_l(t+\theta)(p) \\
&+\sum_{l=1}^{r}\int_{-h}^{0}\left[\mathrm{d}_\theta V^l(\theta)\right]\mathfrak{L}D^{\frac{1}{2}}Q(t)(p)\mathfrak{L}D^{\frac{1}{2}}\Phi_l(t+\theta)(p) \\
&+\sum_{i=0}^{m}\int_{-h}^{0}\left[\mathrm{d}_\theta U^i(\theta)\right]\mathfrak{L}D^{1-\alpha}Q(t)(p)\mathfrak{L}\Phi_i(t+\theta)(p)+\sum_{i=0}^{m}\int_{-h}^{0}\left[\mathrm{d}_\theta U^i(\theta)\right]\mathfrak{L}D^{-\alpha}\Phi_i(t+\theta)(p)
\end{aligned} \tag{18}$$

Applying the inverse Laplace transform to both sides of Equation (18), we obtain Equation (10). □

Theorem 3. *Let the following conditions be satisfied:*

(i) *The conditions (SA) hold.*
(ii) *The function $F \in L_1^{\mathrm{loc}}(\bar{\mathbb{R}}_+, \mathbb{R}^n)$ is exponentially bounded.*

Then the solution $X^F(t)$ of the IVP (1), (3) *with initial function $\Phi(t) \equiv 0$, $t \in [-h, 0]$ has the following representation:*

$$X^F(t) = \int_0^t D^{1-\alpha}Q(t-s)F(s)\mathrm{d}s + D^{-\alpha}F(t), \tag{19}$$

where $Q(t)$ is the fundamental matrix of System (2).

Proof. First we substitute $X_\Phi(t)$ in Equation (1) and use the fact that $X^F(t) = 0$, $t \in [-h, 0]$. Since the function F is exponentially bounded, then we can apply to both sides the Laplace transform in order to get

$$\begin{aligned}
&\mathfrak{L}X^F(t)(p)\left(I_\alpha(p) - I_\alpha(p)\sum_{l=1}^{r}\int_{-h}^{0}\exp(p\theta)\mathrm{d}_\theta V^l(\theta) - \sum_{i=0}^{m}\int_{-h}^{0}\exp(p\theta)\mathrm{d}_\theta U^i(\theta)\right) \\
&= \mathfrak{L}X^F(t)(p)G(p) = \mathfrak{L}F(t)(p).
\end{aligned} \tag{20}$$

Now it follows from the equality $G^{-1}(p) = I_{1-\alpha}(p)\mathfrak{L}Q(t)(p)$ that

$$\begin{aligned}
&\mathfrak{L}X^F(t)(p) = I_{1-\alpha}(p)I_{\alpha-1}(p)G^{-1}(p)\mathfrak{L}F(t)(p) \\
&= I_{1-\alpha}(p)\mathfrak{L}Q(t)(p)\mathfrak{L}F(t)(p) = \left(\mathfrak{L}D^{1-\alpha}Q(t)(p) + I_{-\alpha}(p)\right)\mathfrak{L}F(t)(p) \\
&= \mathfrak{L}D^{1-\alpha}Q(t)(p)\mathfrak{L}F(t)(p) + \mathfrak{L}D^{-\alpha}F(t)(p).
\end{aligned} \tag{21}$$

Finally, we apply the inverse Laplace transform to Equation (21) and the representation Equation (19) follows. □

Corollary 1. *Let the conditions of Theorem 3 hold. Then for every initial function $\Phi \in \widetilde{C}$, the corresponding unique solution $X^F_\Phi(t)$ of the IVP (1), (3) has the integral representation*

$$\begin{aligned} X^F_\phi(t) = &\int_0^t D^{1-\alpha}Q(t-s)F(s)\mathrm{d}s + D^{-\alpha}F(t) + Q(t)\left(\Phi(0) - \sum_{l=1}^{r}\int_{-h}^{0}\left[\mathrm{d}_\theta V^l(\theta)\right]\Phi(\theta)\right) \\ &+\sum_{l=1}^{r}\int_{-h}^{0}\left[\mathrm{d}_\theta V^l(\theta)\right]\Phi(t+\theta) + \sum_{l=1}^{r}\int_{-h}^{0}\left[\mathrm{d}_\theta V^l(\theta)\right] D^{\frac{1}{2}}Q(t) * D^{\frac{1}{2}}\Phi_l(t+\theta) \\ &+\sum_{i=0}^{m}\int_{-h}^{0}\left[\mathrm{d}_\theta U^i(\theta)\right] D^{1-\alpha}Q(t) * \Phi_i(t+\theta) + \sum_{i=0}^{m}\int_{-h}^{0}\left[\mathrm{d}_\theta U^i(\theta)\right] D^{-\alpha}\Phi_i(t+\theta), \end{aligned}$$

where $Q(t)$ is the fundamental matrix of System (2).

Proof. Let $\Phi \in \widetilde{C}$ be an arbitrary initial function and let the functions $X_\Phi(t)$ and $X^F(t)$ be defined by the Equalities (9) and (19), respectively. Then, according to the superposition principle, the function $X_\Phi(t) + X^F(t)$ is the unique solution of the IVP (1), (3). Now the statement of Corollary 1 follows immediately from Theorems 2 and 3. □

4. Example

First, we give some results needed for the illustrative example presented below:

The delayed Mittag-Leffler type matrix function $\mathbf{E}^{B,\tau}_{\alpha,1} : \mathbb{R} \to \mathbb{R}^{n\times n}$ for every matrix $B \in \mathbb{R}^{n\times n}$ and for $\tau \in \mathbb{R}_+$ is defined by

$$\mathbf{E}^{Bt^\alpha}_\tau(t) := I + \sum_{k=1}^{\infty}\frac{B^k(t-(k-1)\tau)^{\alpha k}}{\Gamma(\alpha k+1)}H(k\tau - t), \quad t \geq 0 \tag{22}$$

with $\mathbf{E}^{Bt^\alpha}_\tau(0) := I$, $\mathbf{E}^{Bt^\alpha}_\tau(t) := \Theta$ for $t < 0$ and $H(t)$ is the Heavyside function with $H(0) = 1$. This is a slight modification of the original definition in [29], and note that for each $t \geq 0$, the sum in Equation (22) is finite and for $\tau = 0$ we have

$$\mathbf{E}^{Bt^\alpha}_0(t) := E_\alpha(Bt^\alpha) = \sum_{k=0}^{\infty}\frac{B^k t^{\alpha k}}{\Gamma(\alpha k+1)}, \quad t \geq 0, \tag{23}$$

where the right side is the standard Mittag-Leffler type matrix function.

Example 1. *Consider the nonhomogeneous system for $t > 0$:*

$$\begin{aligned} &D^{0.5}_{0+}x_1(t) = x_1(t-1) + 1 \\ &D^{0.5}_{0+}(x_2(t) + x_1(t-1) + x_2(t-1)) = x_2(t) + x_2(t-1) + x_1(t-2) \end{aligned} \tag{24}$$

with the initial conditions

$$\Phi(t) = (0,2)^T, t \in [-2,0] \quad \textit{i.e.} \quad x_1(t) = 0,\ x_2(t) = 2 \quad \textit{for} \quad t \in [-2,0]. \tag{25}$$

The homogenious system has the form

$$\begin{aligned} &D^{0.5}_{0+}\overline{x}_1(t) = \overline{x}_1(t-1) \\ &D^{0.5}_{0+}(\overline{x}_2(t) + \overline{x}_1(t-1) + \overline{x}_2(t-1)) = \overline{x}_2(t) + \overline{x}_2(t-1) + \overline{x}_1(t-2) \end{aligned} \tag{26}$$

and introduce the following initial conditions necessary for the calculating the fundamental matrix $Q(t)$:

$$1.\ x_1(0) = 1, x_2(0) = 0 \quad \textit{and} \quad x_1(t) = 0, x_2(t) = 0 \quad \textit{for} \quad t \in [-2,0); \tag{27}$$

$$2.\ x_1(0) = 0, x_2(0) = 1 \quad and \quad x_1(t) = 0, x_2(t) = 0 \quad for \quad t \in [-2, 0). \tag{28}$$

Let consider the IP (26), (27). Then the first Equation of (26) in virtue of Theorem 3.1 in [29] has the solution $\overline{x}_1^{\tau}(t) = \mathbf{E}_1^{t^{0.5}}$ $(\tau = 1, \alpha = 0.5)$. *Taking into account Equation (27), it is simple to check that* $(D_{0+}^{0.5}\overline{x}_1^1(s-1))(t) = (D_{0+}^{0.5}\overline{x}_1^1(s))(t-1)$, *and then in virtue of Theorem 3.1 in [29] we have that* $(D_{0+}^{0.5}\overline{x}_1^1)(t-1) = \overline{x}_1^1(t-2)$, *and hense, from the second equation and Equation (27), we obtain that* $\overline{x}_2^1(t) \equiv 0$ *for* $t \in [-2, \infty)$. *Thus the IP (26), (27) have the following solution* $\overline{x}_1^1(t) = \mathbf{E}_1^{t^{0.5}}$, $\overline{x}_2^1(t) \equiv 0$ *for* $t \in [-1, \infty)$.

Consider the IP (26), (28) . Then obviously $\overline{x}_1^2(t) \equiv 0$ *for* $t \in [-2, \infty)$ *and the second equation become the form:* $D_{0+}^{0.5}(\overline{x}_2(t) + \overline{x}_2(t-1)) = \overline{x}_2(t) + \overline{x}_2(t-1)$ *and by making the substitutuon* $y(t) = \overline{x}_2(t) + \overline{x}_2(t-1)$ *we obtain the equations* $D_{0+}^{0.5}y(t) = y(t)$ *with initial codition* $y(0) = 1$, *i.e., the following IP*

$$D_{0+}^{0.5}y(t) = y(t), \quad t > 0; \quad y(0) = 1. \tag{29}$$

Applying Lemma 2.23 in [2] for the case when $\lambda = 1$, $\tau = 1$, $\alpha = 0.5$ *we obtain that the solution of the IP (29) is the fuction* $y(t) = \mathbf{E}_1^{t^{0.5}}(t) = \sum_{k=0}^{\infty} \frac{t^{\alpha k}}{\Gamma(\alpha k + 1)}$. *Then, using the step method, we obtain for each* $k \in \mathbb{N}$ *and* $t \in [k-1, k)]$ *that* $\overline{x}_2^2(t) = \sum_{k=1}^{\infty} (-1)^{k-1} \mathbf{E}_1^{t^{0.5}}(t-(k-1))H(k-t)$ *for* $t > 0$. *Thus, we obtain that the fundamental matrix have the form:*

$$Q(t) = \begin{pmatrix} \mathbf{E}_1^{t^{0.5}}(t) & 0 \\ 0 & \sum_{k=1}^{\infty}(-1)^{k-1}\mathbf{E}_1^{t^{0.5}}(t-(k-1))H(k-t) \end{pmatrix}. \tag{30}$$

In the IP (24), (25) we have that: $\Phi(t) = (0, 2)^T, t \in [-2, 0]$; $F(t) = (1, 0)^T$. *Then from Equation (19), we have*

$$x_1^F(t) = \frac{1}{\Gamma(0.5)} \int_0^t \left(\int_0^{t-s} (t-s-\eta)^{-0.5} \left(\mathbf{E}_1^{t^{0.5}}\right)'(\eta) \mathrm{d}\eta \right) \mathrm{d}s + \frac{\sqrt{t}}{\Gamma(1.5)},$$
$$x_2^F(t) = 0.$$

From Equation (10), it follows

$$x_1^{\Phi}(t) = 0,$$
$$x_2^{\Phi}(t) = 2 + 2\int_0^t \left(\int_0^s (s-\eta)^{-0.5} \left(\sum_{k=1}^{\infty} (-1)^{k-1}\mathbf{E}_1^{t^{0.5}}(\eta-(k-1))H(k-\eta) \right)' \mathrm{d}\eta \right) \mathrm{d}s + \frac{2\sqrt{t}}{\Gamma(1.5)}.$$

Then, the solution of the IP (24), (25), according Corollary 1, is

$$x_1(t) = x_1^{\Phi}(t) + x_1^F(t) = \frac{1}{\Gamma(0.5)} \int_0^t \left(\int_0^{t-s} (t-s-\eta)^{-0.5} \left(\mathbf{E}_1^{t^{0.5}}\right)'(\eta) \mathrm{d}\eta \right) \mathrm{d}s + \frac{\sqrt{t}}{\Gamma(1.5)},$$

$$x_2(t) = x_2^{\Phi}(t) + x_2^F(t) = 2 + 2\int_0^t \left(\int_0^s (s-\eta)^{-0.5} \left(\sum_{k=1}^{\infty} (-1)^{k-1}\mathbf{E}_1^{t^{0.5}}(\eta-(k-1))H(k-\eta) \right)' \mathrm{d}\eta \right) \mathrm{d}s$$
$$+ \frac{2\sqrt{t}}{\Gamma(1.5)}.$$

5. Conclusions

Following the investigations way in the case of functional differential systems with integer order derivatives, we proved a formula for integral representation of the solutions of Cauchy problem for fractional neutral systems, which improves and extends the corresponding former results obtained in the particular case of fractional systems with constant delays. However, the main idea is not only to make a standard generalization of existing results, but as in the case of systems with integer derivatives, the proved formula to be an useful tool for further study of different kinds stability properties of linear neutral fractional systems, which have a lot of practical applications.

As examples in this direction, we refer to the works [29,30], where finite time stability is studied by this approach, i.e., in the partial case of one constant delay. In the mentioned articles, first a formula for integral representation of the solutions of Cauchy problem is proved, and then, using the obtained result, sufficient conditions for finite time stability of the considered fractional delayed system are established. Furthermore, applying the same approach, in [16], the asymptotic stability properties of nonlinear perturbed linear fractional delayed systems are studied .

Author Contributions: The authors contribution in the article are equal. All authors have read and agreed to the published version of the manuscript.

Acknowledgments: The authors are grateful to the anonymous reviewers for their very helpful comments.

References

1. Diethelm, K. *The Analysis of Fractional Differential Equations, an Application–Oriented Exposition Using Differential Operators of Caputo Type*; Lecture Notes in Mathematics; Springer: Berlin, Germany, 2010; Volume 2004.
2. Kilbas, A.A.; Srivastava, H.M.; Trujillo, J.J. *Theory and Applications of Fractional Differential Equations*; Elsevier Science B.V: Amsterdam, The Netherlands, 2006.
3. Kiryakova, V. *Generalized Fractional Calculus and Applications*; Longman Scientific & Technical: Harlow, UK; John Wiley & Sons, Inc.: New York, NY, USA, 1994.
4. Podlubny, I. *Fractional Differential Equation*; Academic Press: San Diego, CA, USA, 1999.
5. Kaczorek, T. *Selected Problems of Fractional Systems Theory*; Lecture Notes in Control and Information Sciences; Springer: Berlin/Heidelberg, Germany, 2011.
6. Stamova, I.; Stamov, G. *Functional and Impulsive Differential Equations of Fractional Order*; Qualitative analysis and applications; CRC Press: Boca Raton, FL, USA, 2017.
7. Bonilla, B.; Rivero, M.; Trujillo, J. On systems of linear fractional differential equations with constant coefficients. *Appl. Comput. Math.* **2007**, *187*, 68–78. [CrossRef]
8. Li, K.; Peng, J. Laplace transform and fractional differential equations, *Appl. Math. Let.* **2011**, *24*, 2019–2023. [CrossRef]
9. Lin, S.D.; Lu, C.H. Laplace transform for solving some families of fractional differential equations and its applications. *Adv. Differ. Equ.* **2013**, *137*, 1–9. [CrossRef]
10. Golev, A.; Milev, M. Integral representation of the solution of the Cauchy problem for autonomous linear neutral fractional system. *Int. J. Pure Appl. Math.* **2018**, *119*, 235–247.
11. Kiskinov H.; Zahariev A. Asymptotic stability of delayed fractional systems with nonlinear perturbation. *AIP Conf. Proc.* **2018**, *2048*, 050014.
12. Zhang, H.; Cao, J.; Jiang, W. General solution of linear fractional neutral differential difference equations. *Discret. Dyn. Nat. Soc.* **2013**. [CrossRef]
13. Zhang, H.; Wu, D. Variation of constant formulae for time invariant and time varying Caputo fractional delay differential systems. *J. Math. Res. Appl.* **2014**, *34*, 549-560.
14. Kiskinov, H.; Zahariev A. On fractional systems with Riemann-Liouville derivatives and distributed delays—Choice of initial conditions, existence and uniqueness of the solutions. *Eur. Phys. Jo. Spec. Top.* **2017**, *26*, 3473-3487. [CrossRef]
15. Krol, K. Asymptotic properties of fractional delay differential equations, *Appl. Math. Comput.* **2011**, *218*, 1515–1532.
16. Milev, M.; Zlatev, S. A note about stability of fractional retarded linear systems with distributed delays. *Int. J. Pure Appl. Math.* **2017**, *115*, 873–881. [CrossRef]

17. Veselinova, M.; Kiskinov, H.; Zahariev, A. Stability analysis of linear fractional differential system with distributed delays. *AIP Conf. Proc.* **2015**, *1690*, 040013.
18. Veselinova, M.; Kiskinov, H.; Zahariev, A. About stability conditions for retarded fractional differential systems with distributed delays. *Commun. Appl. Anal.* **2016**, *20*, 325–334.
19. Kiskinov H.; Milev N.; Zahariev A. A comparison type theorem for linear neutral fractional systems with distributed delays. *AIP Conf. Proc.* **2017**, *1910*, 050009.
20. Veselinova, M.; Kiskinov, H.; Zahariev, A. Stability analysis of neutral linear fractional system with distributed delays. *Filomat* **2016**, *30*, 841–851. [CrossRef]
21. Veselinova, M.; Kiskinov, H.; Zahariev, A. Explicit conditions for stability of neutral linear fractional system with distributed delays. *AIP Conf. Proc.* **2016**, *1789*, 040005.
22. Boyadzhiev, D.; Kiskinov, H.; Veselinova, M.; Zahariev, A. Stability analysis of linear distributed order fractional systems with distributed delays. *Fract. Calc. Appl. Anal.* **2017**, *20*, 914–935. [CrossRef]
23. Martinez-Garcia, M.; Gordon, T.; Shu, L. Extended crossover model for human-control of fractional order plants. *IEEE Access* **2017**, 5, 27622-27635. [CrossRef]
24. Martinez-Garcia, M.; Zhang, Yu; Gordon T. Memory pattern identification for feedback tracking control in human-machine systems. *Hum. Factors* **2019**. [CrossRef]
25. Boyadzhiev, D.; Kiskinov, H.; Zahariev, A. Integral representation of solutions of fractional system with distributed delays. *Integral Transforms Spec. Funct.* **2018**, *29*, 725–744. [CrossRef]
26. Zahariev, A.; Kiskinov, H.; Angelova E. Linear fractional system of incommensurate type with distributed delay and bounded Lebesgue measurable initial conditions. *Dyn. Syst. Appl.* **2019**, *28*, 491–506.
27. Zahariev, A.; Kiskinov, H.; Angelova, E. Smoothness of the fundamental matrix of linear fractional system with variable delays. *Neural Parallel Sci. Comput.* **2019**, *27*, 71–83.
28. Zahariev, A.; Kiskinov, H. Existence of fundamental matrix for neutral linear fractional system with distributed delays. *Int. J. Pure and Appl. Math.* **2018**, *119*, 31–51.
29. Li, M.; Wang, J. Finite time stability of fractional delay differential equations. *Appl. Math. Lett.* **2017**, *64*, 170–176. [CrossRef]
30. Li, M.; Wang, J. Exploring delayed Mittag-Leffler type matrix functions to study finite time stability of fractional delay differential equations. *Appl. Mathe. Comput.* **2018**, *324*, 254–265. [CrossRef]

15

On the Nonlocal Fractional Delta-Nabla Sum Boundary Value Problem for Sequential Fractional Delta-Nabla Sum-Difference Equations

Jiraporn Reunsumrit [1,†] **and Thanin Sitthiwirattham** [2,*,†]

1 Department of Mathematics, Faculty of Applied Science, King Mongkut's University of Technology North Bangkok, Bangkok 10800, Thailand; jiraporn.r@sci.kmutnb.ac.th

2 Mathematics Department, Faculty of Science and Technology, Suan Dusit University, Bangkok 10300, Thailand

* Correspondence: thanin_sit@dusit.ac.th

† These authors contributed equally to this work.

Abstract: In this paper, we propose sequential fractional delta-nabla sum-difference equations with nonlocal fractional delta-nabla sum boundary conditions. The Banach contraction principle and the Schauder's fixed point theorem are used to prove the existence and uniqueness results of the problem. The different orders in one fractional delta differences, one fractional nabla differences, two fractional delta sum, and two fractional nabla sum are considered. Finally, we present an illustrative example.

Keywords: sequential fractional delta-nabla sum-difference equations; nonlocal fractional delta-nabla sum boundary value problem; existence; uniqueness

1. Introduction

Nowaday, fractional calculus is attractive knowledge for many reseachers in many fields. In particular, the fractional calculus has been used in many research works related to biological, biomechanics, magnetic fields, echanics of micro/nano structures, and physical problems (see [1–7]). We can find fractional delta difference calculus and fractional nabla difference calculus in [8–24] and [25–36], respectively. Definitions and properties of fractional difference calculus are presented in the book [37].

We note that there are a few papers using the delta-nabla calculus as a tool. For example, Malinowska and Torres [38] presented the delta-nabla calculus of variations. Dryl and Torres [39,40] studied the delta-nabla calculus of variations for composition functionals on time scales, and a general delta-nabla calculus of variations on time scales with application to economics. Ghorbanian and Rezapour [41] proposed a two-dimensional system of delta-nabla fractional difference inclusions. Liu, Jin and Hou [42] investigated existence of positive solutions for discrete delta-nabla fractional boundary value problems with p-Laplacian.

In this paper, we aim to extend the study of delta-nabla calculus that has appeared in discrete fractional boundary value problems. We have found that the research works related to delta-nabla calculus were presented as above. However, the boundary value problem for sequential fractional delta-nabla difference equation has not been studied before. Our problem is sequential fractional delta-nabla sum-difference equations with nonlocal fractional delta-nabla sum boundary conditions as given by

$$\begin{aligned}
\Delta^{\alpha}\nabla^{\beta}u(t) &= F\left[t+\alpha-1, u(t+\alpha-1), \left(\mathcal{S}^{\theta}u\right)(t+\alpha-1), \left(\mathcal{T}^{\phi}u\right)(t+\alpha-1)\right] \\
\Delta^{-\omega}u(\alpha+\omega-2) &= \kappa\nabla^{-\gamma}u(\alpha-2) \\
u(T+\alpha) &= \lambda u(\eta), \quad \eta \in \mathbb{N}_{\alpha-1,T+\alpha-1}
\end{aligned} \tag{1}$$

where $t \in \mathbb{N}_{0,T} := \{0,1,\ldots,T\}$; $\alpha \in (1,2]$; $\beta,\theta,\phi,\omega,\gamma \in (0,1]$; $\alpha+\beta \in (2,3]$; $T > 0$; κ,λ are given constants; $F \in C(\mathbb{N}_{\alpha-2,T+\alpha} \times \mathbb{R}^3, \mathbb{R})$; and for $\varphi, \psi \in C(\mathbb{N}_{\alpha-2,T+\alpha} \times \mathbb{N}_{\alpha-2,T+\alpha}, [0,\infty))$, we define

$$\left(\mathcal{S}^{\theta}u\right)(t) := \left[\nabla^{-\theta}\varphi u\right](t) = \frac{1}{\Gamma(\theta)}\sum_{s=\alpha-2}^{t}(t-\rho(s))^{\overline{\theta-1}}\varphi(t,s)\,u(s),$$

$$\left(\mathcal{T}^{\phi}u\right)(t) := \left[\Delta^{-\phi}\psi u\right](t+\phi) = \frac{1}{\Gamma(\phi)}\sum_{s=\alpha-\phi-2}^{t-\phi}(t-\sigma(s))^{\underline{\phi-1}}\psi(t,s+\phi)\,u(s+\phi).$$

The objective of this research is to investigate the solution of the boundary value problem (1). The basic knowledge is disscussed in Section 2, the existence results are presented in Section 3, and an example is provided in Section 4.

2. Preliminaries

We give the notations, definitions, and lemmas as follows. The forward operator and the backward operator are defined as $\sigma(t) := t+1$, and $\rho(t) := t-1$, respectively.

For $t, \alpha \in \mathbb{R}$, we define the generalized falling and rising functions as follows:

- The generalized falling function
$$t^{\underline{\alpha}} := \frac{\Gamma(t+1)}{\Gamma(t+1-\alpha)}$$
for any $t+1-\alpha$ is not a pole of the Gamma function. If $t+1-\alpha$ is a pole and $t+1$ is not a pole, then $t^{\underline{\alpha}} = 0$.
- The generalized rising function
$$t^{\overline{\alpha}} := \frac{\Gamma(t+\alpha)}{\Gamma(t)}$$
for any t is not a pole of the Gamma function. If t is a pole and $t+\alpha$ is not a pole, then $t^{\overline{\alpha}} = 0$.

Definition 1. *For $\alpha > 0$ and f defined on $\mathbb{N}_a := \{a, a+1, \ldots\}$, the α-order fractional delta sum of f is defined by*

$$\Delta^{-\alpha}f(t) := \frac{1}{\Gamma(\alpha)}\sum_{s=a}^{t-\alpha}(t-\sigma(s))^{\underline{\alpha-1}}f(s), \quad t \in \mathbb{N}_{a+\alpha}.$$

The α-order fractional nabla sum of f is defined by

$$\nabla^{-\alpha}f(t) := \frac{1}{\Gamma(\alpha)}\sum_{s=a}^{t}(t-\rho(s))^{\overline{\alpha-1}}f(s), \quad t \in \mathbb{N}_{a}.$$

Definition 2. *For $\alpha > 0$, $N \in \mathbb{N}$ where $0 \leq N-1 < \alpha < N$ and f defined on $\mathbb{N}_a$, the α-order Riemann-Liouville fractional delta difference of f is defined by*

$$\Delta^{\alpha}f(t) := \Delta^{N}\Delta^{-(N-\alpha)}f(t) = \frac{1}{\Gamma(-\alpha)}\sum_{s=a}^{t+\alpha}(t-\sigma(s))^{\underline{-\alpha-1}}f(s), \quad t \in \mathbb{N}_{a+N-\alpha}.$$

The α-order Riemann-Liouville fractional nabla difference of f is defined by

$$\nabla^{\alpha} f(t) := \nabla^{N}\nabla^{-(N-\alpha)} f(t) = \frac{1}{\Gamma(-\alpha)}\sum_{s=a}^{t}(t-\rho(s))^{\overline{-\alpha-1}} f(s), \quad t \in \mathbb{N}_{a+N}.$$

Lemma 1 ([15]). *Let $0 \le N-1 < \alpha \le N$, $N \in \mathbb{N}$ and $y : \mathbb{N}_a \to \mathbb{R}$. Then,*

$$\Delta^{-\alpha}\Delta^{\alpha} y(t) = y(t) + C_1(t-a)^{\underline{\alpha-1}} + C_2(t-a)^{\underline{\alpha-2}} + \ldots + C_N(t-a)^{\underline{\alpha-N}},$$

for some $C_i \in \mathbb{R}$, with $1 \le i \le N$.

Lemma 2 ([28]). *Let $0 \le N-1 < \alpha \le N$, $N \in \mathbb{N}$ and $y : \mathbb{N}_{a+1} \to \mathbb{R}$. Then,*

$$\nabla^{-\alpha}\nabla^{\alpha} y(t) = \begin{cases} y(t), & \alpha \notin \mathbb{N} \\ y(t) - \sum_{k=0}^{N-1} \frac{(t-a)^{\overline{k}}}{k!}\nabla^{k} f(a), & \alpha = N, \end{cases}$$

for some $t \in \mathbb{N}_{a+N}$.

We next provide a linear variant of our problem (1).

Lemma 3. *Let $\Lambda \neq 0$; $\alpha \in (1,2]$; $\beta,\omega,\gamma \in (0,1)$; $\alpha+\beta \in (2,3]$; $T > 0$; κ,λ are given constants; and $h \in C(\mathbb{N}_{\alpha-2,T+\alpha},\mathbb{R})$. Then the problem*

$$\Delta^{\alpha}\nabla^{\beta} u(t) = h(t+\alpha-1), \quad t \in \mathbb{N}_{0,T} \tag{2}$$
$$\Delta^{-\omega} u(\alpha+\omega-2) = \kappa\nabla^{-\gamma} u(\alpha-2) \tag{3}$$
$$u(T+\alpha) = \lambda u(\eta), \quad \eta \in \mathbb{N}_{\alpha-1,T+\alpha-1} \tag{4}$$

has the unique solution

$$\begin{aligned} u(t) &= \frac{\mathcal{O}[h]}{\Lambda\Gamma(\beta)}\sum_{s=\alpha-1}^{t}(t-\rho(s))^{\overline{\beta-1}} s^{\underline{\alpha-1}} \\ &\quad + \frac{1}{\Gamma(\beta)\Gamma(\alpha)}\sum_{s=\alpha}^{t}\sum_{r=0}^{s-\alpha}(t-\rho(s))^{\overline{\beta-1}}(s-\sigma(r))^{\underline{\alpha-1}} h(r+\alpha-1) \end{aligned} \tag{5}$$

where the functional $\mathcal{O}[h]$ and the constant Λ are defined by

$$\begin{aligned} \mathcal{O}[h] &= \frac{\lambda}{\Gamma(\beta)\Gamma(\alpha)}\sum_{s=\alpha}^{\eta}\sum_{r=0}^{s-\alpha}(\eta-\rho(s))^{\overline{\beta-1}}(s-\sigma(r))^{\underline{\alpha-1}} h(r+\alpha-1) \\ &\quad - \frac{1}{\Gamma(\beta)\Gamma(\alpha)}\sum_{s=\alpha}^{T+\alpha}\sum_{r=0}^{s-\alpha}(T+\alpha-\rho(s))^{\overline{\beta-1}}(s-\sigma(r))^{\underline{\alpha-1}} h(r+\alpha-1), \end{aligned} \tag{6}$$

$$\Lambda = \frac{1}{\Gamma(\beta)}\sum_{s=\alpha-1}^{T+\alpha}(T+\alpha-\rho(s))^{\overline{\beta-1}} s^{\underline{\alpha-1}} - \frac{\lambda}{\Gamma(\beta)}\sum_{s=\alpha-1}^{\eta}(\eta-\rho(s))^{\overline{\beta-1}} s^{\underline{\alpha-1}}. \tag{7}$$

Proof. Using the fractional delta sum of order α for (2), we obtain

$$\nabla^{\beta} u(t) = C_1 t^{\underline{\alpha-1}} + C_2 t^{\underline{\alpha-2}} + \frac{1}{\Gamma(\alpha)}\sum_{s=0}^{t-\alpha}(t-\sigma(s))^{\underline{\alpha-1}} h(s+\alpha-1), \tag{8}$$

for $t \in \mathbb{N}_{\alpha-2,T+\alpha}$.

Taking the fractional nabla sum of order β for (8), we get

$$\begin{aligned} u(t) &= \frac{1}{\Gamma(\beta)}\sum_{s=\alpha-2}^{t}(t-\rho(s))^{\overline{\beta-1}}\left[C_1 s^{\underline{\alpha-1}}+C_2 s^{\underline{\alpha-2}}\right] \\ &+\frac{1}{\Gamma(\beta)\Gamma(\alpha)}\sum_{s=\alpha}^{t}\sum_{r=0}^{s-\alpha}(t-\rho(s))^{\overline{\beta-1}}(s-\sigma(r))^{\underline{\alpha-1}}h(r+\alpha-1), \end{aligned} \tag{9}$$

for $t\in\mathbb{N}_{\alpha-2,T+\alpha}$.

Using the fractional delta sum of order ω for (9), we have

$$\begin{aligned} \Delta^{-\omega}u(t) &= \sum_{s=\alpha}^{t-\omega}\sum_{r=\alpha}^{s}\frac{(t-\sigma(s))^{\underline{\omega-1}}(s-\rho(r))^{\overline{\beta-1}}}{\Gamma(\omega)\Gamma(\beta)}\left[C_1 r^{\underline{\alpha-1}}+C_2 r^{\underline{\alpha-2}}\right] \\ &+\sum_{s=\alpha}^{t-\omega}\sum_{r=\alpha}^{s}\sum_{\xi=0}^{r-\alpha}\frac{(t-\sigma(s))^{\underline{\omega-1}}(s-\rho(r))^{\overline{\beta-1}}(r-\sigma(\xi))^{\underline{\alpha-1}}}{\Gamma(\omega)\Gamma(\beta)\Gamma(\alpha)}h(\xi+\alpha-1), \end{aligned} \tag{10}$$

for $t\in\mathbb{N}_{\alpha+\omega-1,T+\alpha+\omega}$.

Taking the fractional nabla sum of order γ for (9), we have

$$\begin{aligned} \nabla^{-\gamma}u(t) &= \sum_{s=\alpha}^{t}\sum_{r=\alpha-2}^{s}\frac{(t-\rho(s))^{\overline{\gamma-1}}(s-\rho(r))^{\overline{\beta-1}}}{\Gamma(\gamma)\Gamma(\beta)}\left[C_1 r^{\underline{\alpha-1}}+C_2 r^{\underline{\alpha-2}}\right] \\ &+\sum_{s=\alpha}^{t}\sum_{r=\alpha}^{s}\sum_{\xi=0}^{r-\alpha}\frac{(t-\rho(s))^{\overline{\gamma-1}}(s-\rho(r))^{\overline{\beta-1}}(r-\sigma(\xi))^{\underline{\alpha-1}}}{\Gamma(\omega)\Gamma(\beta)\Gamma(\alpha)}h(\xi+\alpha-1), \end{aligned} \tag{11}$$

for $t\in\mathbb{N}_{\alpha-2,T+\alpha}$.

By substituting $t=\alpha+\omega-2$ and $t=\alpha-2$ into (10) and (11), respectively; and using the condition (3), we obtain

$$C_2=0$$

Substitute $C_2=0$ and apply the condition (4). Then, we obtain

$$C_1=\frac{\mathcal{O}[h]}{\Lambda}$$

where $\mathcal{O}[h]$ and Λ are defined by (6)–(7), respectively. Substituting the constants C_1 and C_2 into (9), we obtain (5). □

3. Existence and Uniqueness Result

Define $\mathcal{C}=C(\mathbb{N}_{\alpha-2,T+\alpha},\mathbb{R})$ is the Banach space of all function u and define the norm as

$$\|u\|_{\mathcal{C}}=\|u\|+\|\nabla^{-\theta}u\|+\|\Delta^{-\phi}u\|$$

where $\|u\|=\max_{t\in\mathbb{N}_{\alpha-2,T+\alpha}}|u(t)|$, $\|\nabla^{-\theta}u\|=\max_{t\in\mathbb{N}_{\alpha-2,T+\alpha}}|\nabla^{-\theta}u(t)|$ and $\|\Delta^{-\phi}u\|=\max_{t\in\mathbb{N}_{\alpha-2,T+\alpha}}|\Delta^{-\phi}u(t+\phi)|$.

In addition, we define the operator $\mathcal{F}:\mathcal{C}\to\mathcal{C}$ by

$$\begin{aligned} (\mathcal{F}u)(t) &= \frac{\mathcal{O}[h]}{\Lambda\Gamma(\beta)}\sum_{s=\alpha-1}^{t}(t-\rho(s))^{\overline{\beta-1}}s^{\underline{\alpha-1}}+\frac{1}{\Gamma(\beta)\Gamma(\alpha)}\sum_{s=\alpha}^{t}\sum_{r=0}^{s-\alpha}(t-\rho(s))^{\overline{\beta-1}}(s-\sigma(r))^{\underline{\alpha-1}}\times \\ & F\left[r+\alpha-1,u(r+\alpha-1),\left(\mathcal{S}^{\theta}u\right)(r+\alpha-1),\left(\mathcal{T}^{\phi}u\right)(r+\alpha-1)\right], \end{aligned} \tag{12}$$

where $\Lambda \neq 0$ is defined by (7) and the functional $\mathcal{O}[F(u)]$ is defined by

$$\begin{aligned}
\mathcal{O}[F(u)] &= \frac{\lambda}{\Gamma(\beta)\Gamma(\alpha)} \sum_{s=\alpha}^{\eta} \sum_{r=0}^{s-\alpha} (\eta-\rho(s))^{\overline{\beta-1}} (s-\sigma(r))^{\underline{\alpha-1}} \times \\
&\quad F\left[r+\alpha-1, u(r+\alpha-1), \left(\mathcal{S}^{\theta}u\right)(r+\alpha-1), \left(\mathcal{T}^{\phi}u\right)(r+\alpha-1)\right] \\
&\quad - \frac{1}{\Gamma(\beta)\Gamma(\alpha)} \sum_{s=\alpha}^{T+\alpha} \sum_{r=0}^{s-\alpha} (T+\alpha-\rho(s))^{\overline{\beta-1}} (s-\sigma(r))^{\underline{\alpha-1}} \times \\
&\quad F\left[r+\alpha-1, u(r+\alpha-1), \left(\mathcal{S}^{\theta}u\right)(r+\alpha-1), \left(\mathcal{T}^{\phi}u\right)(r+\alpha-1)\right]. \qquad (13)
\end{aligned}$$

Obviously, the operator $\mathcal{F}$ has the fixed points if and only if the boundary value problem (1) has solutions. Firstly, we show the existence and uniqueness result of the boundary value problem (1) by using the Banach contraction principle.

Theorem 1. *Assume that* $F : \mathbb{N}_{\alpha-2,T+\alpha} \times \mathbb{R}^3 \to \mathbb{R}$ *is continuous,* $\varphi, \psi : \mathbb{N}_{\alpha-2,T+\alpha} \times \mathbb{N}_{\alpha-2,T+\alpha} \to [0,\infty)$ *with* $\varphi_0 = \max\{\varphi(t-1,s) : (t,s) \in \mathbb{N}_{\alpha-2,T+\alpha} \times \mathbb{N}_{\alpha-2,T+\alpha}\}$ *and* $\psi_0 = \max\{\psi(t-1,s) : (t,s) \in \mathbb{N}_{\alpha-2,T+\alpha} \times \mathbb{N}_{\alpha-2,T+\alpha}\}$. *Suppose that the following conditions hold:*

(H_1) *there exist constants* $L_1, L_2, L_3 > 0$ *such that for each* $t \in \mathbb{N}_{\alpha-2,T+\alpha}$ *and* $u, v \in \mathbb{R}$

$$\begin{aligned}
&\left|F\left(t,u,\left(\mathcal{S}^{\theta}u\right),\left(\mathcal{T}^{\phi}u\right)\right) - F\left(t,v,\left(\mathcal{S}^{\theta}v\right),\left(\mathcal{T}^{\phi}v\right)\right)\right| \\
\leq\ & L_1|u-v| + L_2\left|\left(\mathcal{S}^{\theta}u\right) - \left(\mathcal{S}^{\theta}v\right)\right| + L_3\left|\left(\mathcal{T}^{\phi}u\right) - \left(\mathcal{T}^{\phi}v\right)\right|.
\end{aligned}$$

Then the problem (1) *has a unique solution on* $\mathbb{N}_{\alpha-2,T+\alpha}$ *provided that*

$$\chi := \left\{L_1 + L_2\varphi_0\frac{(T+3)^{\overline{\theta}}}{\Gamma(\theta+1)} + L_3\psi_0\frac{(T+2)^{\underline{\phi}}}{\Gamma(\phi+1)}\right\}[\Omega_1+\Omega_2+\Omega_3] < 1 \qquad (14)$$

where

$$\Theta = \frac{1}{\Gamma(\alpha+1)\Gamma(\beta+1)}\left[\lambda\eta^{\underline{\alpha}}(\eta-\alpha+1)^{\overline{\beta}} + (T+\alpha)^{\underline{\alpha}}(T+1)^{\overline{\beta}}\right], \qquad (15)$$

$$\Omega_1 = \frac{1}{\Gamma(\beta+1)}\left[\frac{\Theta}{|\Lambda|}(T+\alpha)^{\underline{\alpha-1}}(T+2)^{\overline{\beta-1}} + \frac{(T+\alpha)^{\underline{\alpha}}}{\Gamma(\alpha+1)}(T+1)^{\overline{\beta}}\right], \qquad (16)$$

$$\Omega_2 = \frac{1}{\Gamma(\theta+1)\Gamma(\beta+1)}\left[\frac{\Theta}{|\Lambda|}(T+\alpha)^{\underline{\alpha-1}}(T+2)^{\overline{\beta-1}}(T+2)^{\overline{\theta}} + \frac{(T+\alpha)^{\underline{\alpha}}}{\Gamma(\alpha+1)}(T+1)^{\overline{\beta}}(T+1)^{\overline{\theta}}\right], \qquad (17)$$

$$\Omega_3 = \frac{1}{\Gamma(\phi+1)\Gamma(\beta+1)}\left[\frac{\Theta}{|\Lambda|}(T+\alpha)^{\underline{\alpha-1}}(T+2)^{\overline{\beta-1}}(T+1)^{\underline{\phi}} + \frac{(T+\alpha)^{\underline{\alpha}}}{\Gamma(\alpha+1)}(T+1)^{\overline{\beta}}T^{\underline{\phi}}\right]. \qquad (18)$$

Proof. We shall show that $\mathcal{F}$ is a contraction. For any $u, v \in \mathcal{C}$ and for each $t \in \mathbb{N}_{\alpha-2,T+\alpha}$, we have

$$\begin{aligned}
\Big|\mathcal{O}[F(u)] - \mathcal{O}[F(v)]\Big| \le\ & \Bigg| \tfrac{\lambda}{\Gamma(\beta)\Gamma(\alpha)} \sum_{s=\alpha}^{\eta} \sum_{r=0}^{s-\alpha} (\eta - \rho(s))^{\overline{\beta-1}} (s-\sigma(r))^{\underline{\alpha-1}} \times \\
& \Big[L_1|u-v| + L_2\Big|(\mathcal{S}^{\theta}u) - (\mathcal{S}^{\theta}v)\Big| + L_3\Big|(\mathcal{T}^{\phi}u) - (\mathcal{T}^{\phi}v)\Big|\Big] \\
& - \tfrac{1}{\Gamma(\beta)\Gamma(\alpha)} \sum_{s=\alpha}^{T+\alpha} \sum_{r=0}^{s-\alpha} (T+\alpha-\rho(s))^{\overline{\beta-1}} (s-\sigma(r))^{\underline{\alpha-1}} \times \\
& \Big[L_1|u-v| + L_2\Big|(\mathcal{S}^{\theta}u) - (\mathcal{S}^{\theta}v)\Big| + L_3\Big|(\mathcal{T}^{\phi}u) - (\mathcal{T}^{\phi}v)\Big|\Big] \Bigg| \\
\le\ & \|u-v\|_{\mathcal{C}} \left\{ L_1 + L_2\varphi_0 \tfrac{(T+3)^{\overline{\theta}}}{\Gamma(\theta+1)} + L_3\psi_0 \tfrac{(T+2)^{\underline{\phi}}}{\Gamma(\phi+1)} \right\} \Bigg| \lambda \sum_{s=\alpha}^{\eta} \sum_{r=0}^{s-\alpha} \tfrac{(\eta-\rho(s))^{\overline{\beta-1}}}{\Gamma(\beta)} \times \\
& \tfrac{(s-\sigma(r))^{\underline{\alpha-1}}}{\Gamma(\alpha)} - \sum_{s=\alpha}^{T+\alpha} \sum_{r=0}^{s-\alpha} \tfrac{(T+\alpha-\rho(s))^{\overline{\beta-1}} (s-\sigma(r))^{\underline{\alpha-1}}}{\Gamma(\beta)\Gamma(\alpha)} \Bigg| \\
\le\ & \|u-v\|_{\mathcal{C}} \left\{ L_1 + L_2\varphi_0 \tfrac{(T+3)^{\overline{\theta}}}{\Gamma(\theta+1)} + L_3\psi_0 \tfrac{(T+2)^{\underline{\phi}}}{\Gamma(\phi+1)} \right\} \Theta,
\end{aligned} \tag{19}$$

and

$$\begin{aligned}
& \big|(\mathcal{F}u)(t) - (\mathcal{F}v)(t)\big| \\
\le\ & \left| \frac{\mathcal{O}[F(u)] - \mathcal{O}[F(v)]}{\Lambda} \right| \sum_{s=\alpha-1}^{T+\alpha} (T+\alpha-\rho(s))^{\overline{\beta-1}} s^{\underline{\alpha-1}} + \sum_{s=\alpha}^{T+\alpha} \sum_{r=0}^{s-\alpha} \frac{(T+\alpha-\rho(s))^{\overline{\beta-1}}}{\Gamma(\beta)} \times \\
& \frac{(s-\sigma(r))^{\underline{\alpha-1}}}{\Gamma(\alpha)} \Big[L_1|u-v| + L_2\Big|\Big(\mathcal{S}^{\theta}u\Big) - \Big(\mathcal{S}^{\theta}v\Big)\Big| + L_3\Big|(\mathcal{T}^{\phi}u) - (\mathcal{T}^{\phi}v)\Big|\Big] \\
\le\ & \|u-v\|_{\mathcal{C}} \left\{ L_1 + L_2\varphi_0 \frac{(T+3)^{\overline{\theta}}}{\Gamma(\theta+1)} + L_3\psi_0 \frac{(T+2)^{\underline{\phi}}}{\Gamma(\phi+1)} \right\} \left\{ \frac{\Theta(T+\alpha)^{\underline{\alpha-1}}}{|\Lambda|} \times \right. \\
& \left. \sum_{s=\alpha-1}^{T+\alpha} (T+\alpha-\rho(s))^{\overline{\beta-1}} + \sum_{s=\alpha}^{T+\alpha} \sum_{r=0}^{s-\alpha} \frac{(T+\alpha-\rho(s))^{\overline{\beta-1}} (s-\sigma(r))^{\underline{\alpha-1}}}{\Gamma(\beta)\Gamma(\alpha)} \right\} \\
\le\ & \|u-v\|_{\mathcal{C}} \left\{ L_1 + L_2\varphi_0 \frac{(T+3)^{\overline{\theta}}}{\Gamma(\theta+1)} + L_3\psi_0 \frac{(T+2)^{\underline{\phi}}}{\Gamma(\phi+1)} \right\} \Omega_1.
\end{aligned} \tag{20}$$

Next, we consider the following $(\nabla^{-\theta}\mathcal{F}u)$ and $(\Delta^{-\phi}\mathcal{F}u)$ as

$$\begin{aligned}
(\nabla^{-\theta}\mathcal{F}u)(t) =\ & \frac{\mathcal{O}[F(u)]}{\Lambda} \sum_{s=\alpha}^{t} \sum_{r=\alpha-1}^{s} \frac{(t-\rho(s))^{\overline{\theta-1}} (s-\rho(r))^{\overline{\beta-1}}}{\Gamma(\theta)\Gamma(\beta)} r^{\underline{\alpha-1}} \\
& + \sum_{s=\alpha}^{t} \sum_{r=\alpha}^{s} \sum_{\xi=0}^{r-\alpha} \frac{(t-\rho(s))^{\overline{\theta-1}} (s-\rho(r))^{\overline{\beta-1}} (r-\sigma(\xi))^{\underline{\alpha-1}}}{\Gamma(\theta)\Gamma(\beta)\Gamma(\alpha)} \times \\
& F\Big[\xi+\alpha-1, u(\xi+\alpha-1), \Delta^{\theta}u(\xi+\alpha-\theta+1), \nabla^{\gamma}u(\xi+\alpha+1)\Big]\Big],
\end{aligned} \tag{21}$$

and

$$\begin{aligned}
(\Delta^{-\phi}\mathcal{F}u)(t+\phi) =\ & \frac{\mathcal{O}[F(u)]}{\Lambda} \sum_{s=\alpha}^{t-\phi} \sum_{r=\alpha-1}^{s} \frac{(t+\phi-\sigma(s))^{\underline{\phi-1}} (s-\rho(r))^{\overline{\beta-1}}}{\Gamma(\phi)\Gamma(\beta)} r^{\underline{\alpha-1}} \\
& + \sum_{s=\alpha}^{t-\phi} \sum_{r=\alpha}^{s} \sum_{\xi=0}^{r-\alpha} \frac{(t+\phi-\sigma(s))^{\underline{\phi-1}} (s-\rho(r))^{\overline{\beta-1}} (r-\sigma(\xi))^{\underline{\alpha-1}}}{\Gamma(\phi)\Gamma(\beta)\Gamma(\alpha)} \times \\
& F\Big[\xi+\alpha-1, u(\xi+\alpha-1), \Delta^{\theta}u(\xi+\alpha-\theta+1), \nabla^{\gamma}u(\xi+\alpha+1)\Big]\Big].
\end{aligned} \tag{22}$$

Similarly to the above, we have

$$|(\nabla^{-\theta}\mathcal{F}u)(t)-(\nabla^{-\theta}\mathcal{F}v)(t)| \leq \|u-v\|_C\left\{L_1+L_2\varphi_0\frac{(T+3)^{\overline{\theta}}}{\Gamma(\theta+1)}+L_3\psi_0\frac{(T+2)^{\underline{\phi}}}{\Gamma(\phi+1)}\right\}\Omega_2, \tag{23}$$

$$|\Delta^{-\phi}\mathcal{F}u)(t+\phi)-\Delta^{-\phi}\mathcal{F}v)(t+\phi)| \leq \|u-v\|_C\left\{L_1+L_2\varphi_0\frac{(T+3)^{\overline{\theta}}}{\Gamma(\theta+1)}+L_3\psi_0\frac{(T+2)^{\underline{\phi}}}{\Gamma(\phi+1)}\right\}\Omega_3. \tag{24}$$

From (20), (23) and (24), we get

$$\begin{aligned}\|(\mathcal{F}u)-(\mathcal{F}v)\|_C &\leq \|u-v\|_C\left\{L_1+L_2\varphi_0\frac{(T+3)^{\overline{\theta}}}{\Gamma(\theta+1)}+L_3\psi_0\frac{(T+2)^{\underline{\phi}}}{\Gamma(\phi+1)}\right\}[\Omega_1+\Omega_2+\Omega_3] \\ &= \chi\|u-v\|_C. \end{aligned} \tag{25}$$

By (14), we have $\|(\mathcal{F}u)(t)-(\mathcal{F}v)(t)\|_C < \|u-v\|_C$.

Consequently, $\mathcal{F}$ is a contraction. Therefore, by the Banach fixed point theorem, we get that $\mathcal{F}$ has a fixed point which is a unique solution of the problem (1) on $t \in \mathbb{N}_{\alpha-2,T+\alpha}$. □

4. Existence of at Least One Solution

Next, we provide Arzelá-Ascoli theorem and Schauder's fixed point theorem that will be used to prove the existence of at least one solution of (1).

Lemma 4 ([43]). *(Arzelá-Ascoli theorem) A set of function in $C[a,b]$ with the sup norm is relatively compact if and only it is uniformly bounded and equicontinuous on $[a,b]$.*

Lemma 5 ([43]). *A set is compact if it is closed and relatively compact.*

Lemma 6 ([44]). *(Schauder's fixed point theorem) Let (D,d) be a complete metric space, U be a closed convex subset of D, and $T : D \to D$ be the map such that the set $Tu : u \in U$ is relatively compact in D. Then the operator T has at least one fixed point $u^* \in U$: $Tu^* = u^*$.*

Theorem 2. *Assuming that (H_1) holds, problem (1) has at least one solution on $\mathbb{N}_{\alpha-3,T+\alpha}$.*

Proof. The proof is organized as follows.
Step I. Verify $\mathcal{F}$ map bounded sets into bounded sets in $B_R = \{u \in C(\mathbb{N}_{\alpha-2,T+\alpha}) : \|u\|_C \leq R\}$.

Let $\max\limits_{t\in\mathbb{N}_{\alpha-2,T+\alpha}} |F(t,0,0,0)| = M$ and choose a constant

$$R \geq \frac{M[\Omega_1+\Omega_2+\Omega_3]}{1-[\Omega_1+\Omega_2+\Omega_3]\left\{L_1+L_2\varphi_0\frac{(T+3)^{\overline{\theta}}}{\Gamma(\theta+1)}+L_3\psi_0\frac{(T+2)^{\underline{\phi}}}{\Gamma(\phi+1)}\right\}}. \tag{26}$$

For each $u \in B_R$, we obtain

$$\begin{aligned}
&\left|\mathcal{O}[F(u)]\right| \\
\leq\ & \left|\lambda \sum_{s=\alpha}^{\eta}\sum_{r=0}^{s-\alpha} \frac{(\eta-\rho(s))^{\overline{\beta-1}}(s-\sigma(r))^{\underline{\alpha-1}}}{\Gamma(\beta)\Gamma(\alpha)} - \sum_{s=\alpha}^{T+\alpha}\sum_{r=0}^{s-\alpha} \frac{(T+\alpha-\rho(s))^{\overline{\beta-1}}(s-\sigma(r))^{\underline{\alpha-1}}}{\Gamma(\beta)\Gamma(\alpha)}\right| \\
&\left[\left|F\left[r+\alpha-1, u(r+\alpha-1), \left(\mathcal{S}^{\theta}u\right)(r+\alpha-\theta+1), \left(\mathcal{T}^{\phi}v\right)(r+\alpha+1)\right]\right.\right. \qquad (27) \\
&\left.\left. -F(r+\alpha-1,0,0,0)\right| + \left|F(r+\alpha-1,0,0,0)\right|\right] \\
\leq\ & \left\{\left[L_1 + L_2\varphi_0\frac{(T+3)^{\overline{\theta}}}{\Gamma(\theta+1)} + L_3\psi_0\frac{(T+2)^{\underline{\phi}}}{\Gamma(\phi+1)}\right]\|u\|_{\mathcal{C}} + M\right\}\Theta
\end{aligned}$$

and

$$\begin{aligned}
&\left|(\mathcal{F}u)(t)\right| \\
\leq\ & \left|\frac{\mathcal{O}[F(u)]}{\Lambda}\right| \sum_{s=\alpha-1}^{T+\alpha}(T+\alpha-\rho(s))^{\overline{\beta-1}}s^{\underline{\alpha-1}} + \sum_{s=\alpha}^{T+\alpha}\sum_{r=0}^{s-\alpha}\frac{(T+\alpha-\rho(s))^{\overline{\beta-1}}}{\Gamma(\beta)} \times \\
&\frac{(s-\sigma(r))^{\underline{\alpha-1}}}{\Gamma(\alpha)}\left[\left|F\left[r+\alpha-1, u(r+\alpha-1), \left(\mathcal{S}^{\theta}u\right)(r+\alpha-\theta+1),\right.\right.\right. \qquad (28) \\
&\left.\left.\left(\mathcal{T}^{\phi}v\right)(r+\alpha+1)\right] - F(r+\alpha-1,0,0,0)\right| + \left|F(r+\alpha-1,0,0,0)\right|\right] \\
\leq\ & \left\{\left[L_1 + L_2\varphi_0\frac{(T+3)^{\overline{\theta}}}{\Gamma(\theta+1)} + L_3\psi_0\frac{(T+2)^{\underline{\phi}}}{\Gamma(\phi+1)}\right]\|u\|_{\mathcal{C}} + M\right\}\Omega_1.
\end{aligned}$$

In addition, we have

$$\left|(\nabla^{-\theta}\mathcal{F}u)(t)\right| \leq \left\{\left[L_1 + L_2\varphi_0\frac{(T+3)^{\overline{\theta}}}{\Gamma(\theta+1)} + L_3\psi_0\frac{(T+2)^{\underline{\phi}}}{\Gamma(\phi+1)}\right]\|u\|_{\mathcal{C}} + M\right\}\Omega_2, \qquad (29)$$

$$\left|(\Delta^{-\phi}\mathcal{F}u)(t+\phi)\right| \leq \left\{\left[L_1 + L_2\varphi_0\frac{(T+3)^{\overline{\theta}}}{\Gamma(\theta+1)} + L_3\psi_0\frac{(T+2)^{\underline{\phi}}}{\Gamma(\phi+1)}\right]\|u\|_{\mathcal{C}} + M\right\}\Omega_3. \qquad (30)$$

From (28), (29) and (30), we have

$$\begin{aligned}
\|(\mathcal{F}u)(t)\|_{\mathcal{C}} &\leq \left\{\left[L_1 + L_2\varphi_0\frac{(T+3)^{\overline{\theta}}}{\Gamma(\theta+1)} + L_3\psi_0\frac{(T+2)^{\underline{\phi}}}{\Gamma(\phi+1)}\right]\|u\|_{\mathcal{C}} + M\right\}[\Omega_1+\Omega_2+\Omega_3] \\
&\leq R. \qquad (31)
\end{aligned}$$

So, $\|\mathcal{F}u\|_{\mathcal{C}} \leq R$. Therefore $\mathcal{F}$ is uniformly bounded.

Step II. Since F and H are continuous, the operator $\mathcal{F}$ is continuous on B_R.

Step III. Prove that $\mathcal{F}$ is equicontinuous on B_R. For any $\epsilon > 0$, there exists a positive constant $\rho^* = \max\{\delta_1, \delta_2, \delta_3, \delta_4, \delta_5, , \delta_6\}$ such that for $t_1, t_2 \in \mathbb{N}_{\alpha-2,T+\alpha}$
$\left|(t_2-\alpha+2)^{\overline{\beta}} - (t_1-\alpha+2)^{\overline{\beta}}\right| < \frac{\epsilon\Gamma(\beta+1)|\Lambda|}{6(T+\alpha)^{\underline{\alpha-1}}\Theta\|F\|}$, where $|t_2-t_1| < \delta_1$,
$\left|(t_2-\alpha+1)^{\overline{\beta}} - (t_1-\alpha+1)^{\overline{\beta}}\right| < \frac{\epsilon\Gamma(\alpha+1)\Gamma(\beta+1)}{6(T+\alpha)^{\underline{\alpha}}\|F\|}$, where $|t_2-t_1| < \delta_2$, $\left|(t_2-\alpha+2)^{\overline{\theta}} - (t_1-\alpha+2)^{\overline{\theta}}\right| < \frac{\epsilon\Gamma(\beta+1)\Gamma(\theta+1)|\Lambda|}{6(T+2)^{\overline{\beta}}(T+\alpha)^{\underline{\alpha-1}}\Theta\|F\|}$, where $|t_2-t_1| < \delta_3$,
$\left|(t_2-\alpha+1)^{\overline{\theta}} - (t_1-\alpha+1)^{\overline{\theta}}\right| < \frac{\epsilon\Gamma(\alpha+1)\Gamma(\beta+1)\Gamma(\theta+1)}{6(T+1)^{\overline{\beta}}(T+\alpha)^{\underline{\alpha}}\Theta\|F\|}$, where $|t_2-t_1| < \delta_4$,

$\left|(t_2-\alpha+1)^{\underline{\phi}}-(t_1-\alpha+1)^{\underline{\phi}}\right| < \frac{\epsilon\Gamma(\beta+1)\Gamma(\phi+1)|\Lambda|}{6(T+2)^{\overline{\beta}}(T+\alpha)^{\underline{\alpha-1}}\Theta\|F\|}$, where $|t_2-t_1|<\delta_5$,
$\left|(t_2-\alpha)^{\underline{\phi}}-(t_1-\alpha)^{\underline{\phi}}\right| < \frac{\epsilon\Gamma(\alpha+1)\Gamma(\beta+1)\Gamma(\phi+1)}{6(T+1)^{\overline{\beta}}(T+\alpha)^{\underline{\alpha}}\|F\|}$, where $|t_2-t_1|<\delta_6$.
Then we have

$$\begin{aligned}
&|(\mathcal{F}u)(t_2)-(\mathcal{F}u)(t_1)| \\
\le\ & \left|\frac{\mathcal{O}[F(u)]}{\Lambda}\right|\left|\frac{1}{\Gamma(\beta)}\sum_{s=\alpha-1}^{t_2}(t_2-\rho(s))^{\overline{\beta-1}}s^{\underline{\alpha-1}}-\frac{1}{\Gamma(\beta)}\sum_{s=\alpha-1}^{t_1}(t_1-\rho(s))^{\overline{\beta-1}}s^{\underline{\alpha-1}}\right| \\
&\left|\sum_{s=\alpha}^{t_2}\sum_{r=0}^{s-\alpha}\frac{(t_2-\rho(s))^{\overline{\beta-1}}(s-\sigma(r))^{\underline{\alpha-1}}}{\Gamma(\beta)\Gamma(\alpha)}\times\right. \\
&F\left[r+\alpha-1,u(r+\alpha-1),\left(\mathcal{S}^{\theta}u\right)(r+\alpha-1),\left(\mathcal{T}^{\phi}u\right)(r+\alpha-1)\right] \\
&-\sum_{s=\alpha}^{t_1}\sum_{r=0}^{s-\alpha}\frac{(t_1-\rho(s))^{\overline{\beta-1}}(s-\sigma(r))^{\underline{\alpha-1}}}{\Gamma(\beta)\Gamma(\alpha)}\times \\
&\left.F\left[r+\alpha-1,u(r+\alpha-1),\left(\mathcal{S}^{\theta}u\right)(r+\alpha-1),\left(\mathcal{T}^{\phi}u\right)(r+\alpha-1)\right]\right| \\
<\ & \frac{\Theta\|F\|(T+\alpha)^{\underline{\alpha-1}}}{|\Lambda|\Gamma(\beta+1)}\left|(t_2-\alpha+2)^{\overline{\beta}}-(t_1-\alpha+2)^{\overline{\beta}}\right|+\frac{\|F\|(T+\alpha)^{\underline{\alpha}}}{\Gamma(\alpha+1)\Gamma(\beta+1)}\times \\
&\left|(t_2-\alpha+1)^{\overline{\beta}}-(t_1-\alpha+1)^{\overline{\beta}}\right| \\
<\ & \frac{\epsilon}{6}+\frac{\epsilon}{6}=\frac{\epsilon}{3}.
\end{aligned} \tag{32}$$

Similarly to the above, we have

$$\begin{aligned}
\left|(\nabla^{-\theta}\mathcal{F}u)(t_2)-(\nabla^{-\theta}\mathcal{F}u)(t_1)\right| <\ & \frac{\Theta\|F\|(T+\alpha)^{\underline{\alpha-1}}(T+2)^{\overline{\beta}}}{|\Lambda|\Gamma(\beta+1)\Gamma(\theta+1)}\left|(t_2-\alpha+2)^{\overline{\theta}}-(t_1-\alpha+2)^{\overline{\theta}}\right| \\
&+\frac{\|F\|(T+\alpha)^{\underline{\alpha}}(T+1)^{\overline{\beta}}}{\Gamma(\alpha+1)\Gamma(\beta+1)\Gamma(\theta+1)}\left|(t_2-\alpha+1)^{\overline{\theta}}-(t_1-\alpha+1)^{\overline{\theta}}\right| \\
<\ & \tfrac{\epsilon}{6}+\tfrac{\epsilon}{6}=\tfrac{\epsilon}{3},
\end{aligned} \tag{33}$$

$$\begin{aligned}
\left|(\Delta^{-\phi}\mathcal{F}u)(t_2+\phi)-(\Delta^{-\phi}\mathcal{F}u)(t_1+\phi)\right| <\ & \frac{\Theta\|F\|(T+\alpha)^{\underline{\alpha-1}}(T+2)^{\overline{\beta}}}{|\Lambda|\Gamma(\beta+1)\Gamma(\phi+1)}\left|(t_2-\alpha+2)^{\overline{\phi}}-(t_1-\alpha+2)^{\overline{\phi}}\right| \\
&+\frac{\|F\|(T+\alpha)^{\underline{\alpha}}(T+1)^{\overline{\beta}}}{\Gamma(\alpha+1)\Gamma(\beta+1)\Gamma(\phi+1)}\left|(t_2-\alpha)^{\overline{\phi}}-(t_1-\alpha)^{\overline{\phi}}\right| \\
<\ & \tfrac{\epsilon}{6}+\tfrac{\epsilon}{6}=\tfrac{\epsilon}{3}.
\end{aligned} \tag{34}$$

Hence

$$\left\|(\mathcal{F}u)(t_2)-(\mathcal{F}u)(t_1)\right\|_{\mathcal{C}} < \frac{\epsilon}{3}+\frac{\epsilon}{3}+\frac{\epsilon}{3} = \epsilon. \tag{35}$$

It implies that the set $\mathcal{F}(B_R)$ is equicontinuous. By the results of Steps I to III and the Arzelá-Ascoli theorem, $\mathcal{F}:\mathcal{C}\to\mathcal{C}$ is completely continuous. By Schauder fixed point theorem, the boundary value problem (1) has at least one solution. □

5. An Example

Here, we provide a sequential fractional delta-nabla sum-difference equations with nonlocal fractional delta-nabla sum boundary conditions

$$\begin{aligned}
\Delta^{\frac{3}{2}}\nabla^{\frac{2}{3}}u(t) &= \frac{e^{-\sin^2\left(t+\frac{1}{2}\right)}}{\left(t+\frac{201}{2}\right)^2}\cdot\left[\frac{2\left|u\left(t+\frac{1}{2}\right)\right|}{\left|u\left(t+\frac{1}{2}\right)\right|+1}+\frac{\left|\left(\mathcal{S}^{\frac{1}{5}}u\right)\left(t+\frac{1}{2}\right)\right|}{\left|\left(\mathcal{S}^{\frac{1}{5}}u\right)\left(t+\frac{1}{2}\right)\right|+1}+\frac{3\left|\left(\mathcal{T}^{\frac{3}{4}}u\right)\left(t+\frac{1}{2}\right)\right|}{\left|\left(\mathcal{T}^{\frac{3}{4}}u\right)\left(t+\frac{1}{2}\right)\right|+1}\right] \\
\Delta^{-\frac{1}{3}}u\left(-\frac{11}{2}\right) &= 2\nabla^{-\frac{2}{5}}u\left(-\frac{1}{2}\right) \\
u\left(\frac{15}{2}\right) &= 3u\left(-\frac{7}{2}\right),
\end{aligned} \tag{36}$$

where

$$\left(\mathcal{S}^{\theta}u\right)\left(t+\frac{1}{2}\right)=\frac{1}{\Gamma\left(\frac{1}{5}\right)}\sum_{s=-\frac{1}{2}}^{t+\frac{1}{2}}\left(t+\frac{1}{2}-\rho(s)\right)^{\overline{-\frac{4}{5}}}\frac{e^{-(s+1)}}{\left(t+\frac{21}{2}\right)^4}u(s),$$

$$\left(\mathcal{T}^{\phi}u\right)\left(t+\frac{1}{2}\right)=\frac{1}{\Gamma\left(\frac{3}{4}\right)}\sum_{s=-\frac{5}{4}}^{t-\frac{1}{4}}\left(t+\frac{1}{2}-\sigma(s)\right)^{\underline{-\frac{1}{4}}}\frac{e^{-\left(s+\frac{3}{4}\right)}}{\left(t+\frac{201}{2}\right)^2}u\left(s+\frac{3}{4}\right).$$

Letting $\alpha = \frac{3}{2}$, $\beta = \frac{2}{3}$, $\theta = \frac{1}{5}$, $\phi = \frac{3}{4}$, $\omega = \frac{1}{3}$, $\gamma = \frac{2}{5}$, $\eta = \frac{7}{2}$, $\kappa = 2$, $\lambda = 3$, $T = 6$, $\varphi(t,s) = \frac{e^{-(s+1)}}{(t+10)^4}$, $\psi(t,s+\phi) = \frac{e^{-\left(s+\frac{3}{4}\right)}}{(t+100)^2}$, and $F\left[t,u(t),(\mathcal{S}^{\theta}u)(t),(\mathcal{T}^{\phi}u)(t)\right] = \frac{e^{-\sin^2 t}}{(t+100)^2}\cdot\left[\frac{2|u(t)|}{1+|u(t)|}+\frac{\left|(\mathcal{S}^{\frac{1}{5}}u)(t)\right|}{\left|(\mathcal{S}^{\frac{1}{5}}u)(t)\right|+1}+3\frac{\left|(\mathcal{T}^{\frac{3}{4}}u)(t)\right|}{\left|(\mathcal{T}^{\frac{3}{4}}u)(t)\right|+1}\right]$, we find that

$$|\Lambda| = 3.29500,\ \ \Theta = 29.34125,\ \ \Omega_1 = 57.88022,\ \ \Omega_2 = 246.53142,\ \ \Omega_3 = 699.50153.$$

where $|\Lambda|, \Theta, \Omega_1, \Omega_2, \Omega_3$ are defined by (7), (15)–(18), respectively. In addition, for $(t,s) \in \mathbb{N}_{-\frac{1}{2},\frac{15}{2}} \times \mathbb{N}_{-\frac{1}{2},\frac{15}{2}}$, we find that

$$\varphi_0 = \max\{\varphi(t-1,s)\} = 0.0000745 \quad \text{and} \quad \psi_0 = \max\{\psi(t-1,s)\} = 0.0000787.$$

For each $t \in \mathbb{N}_{-\frac{1}{2},\frac{15}{2}}$, we obtain

$$\begin{aligned}
&\left|F\left[t,u(t),\left(\mathcal{S}^{\theta}u\right)(t),\left(\mathcal{T}^{\phi}u\right)(t)\right]-F\left[t,v(t),\left(\mathcal{S}^{\theta}v\right)(t),\left(\mathcal{T}^{\phi}v\right)(t)\right]\right| \\
&< \frac{1}{\left(-\frac{1}{2}+100\right)^2}\cdot\left[2\frac{|u|-|v|}{[1+|u|]\,[1+|v|]}+\frac{|\left(\mathcal{S}^{\theta}u\right)|-|\left(\mathcal{S}^{\theta}v\right)|}{[|\left(\mathcal{S}^{\theta}u\right)|+1]\,[|\left(\mathcal{S}^{\theta}v\right)|+1]}\right. \\
&\quad \left.+3\frac{|\left(\mathcal{T}^{\phi}u\right)|-|\left(\mathcal{T}^{\phi}v\right)|}{[|\left(\mathcal{T}^{\phi}u\right)|+1]\,[|\left(\mathcal{T}^{\phi}v\right)|+1]}\right] \\
&\le \frac{8}{39601}|u-v|+\frac{4}{39601}\left|\left(\mathcal{S}^{\theta}u\right)-\left(\mathcal{S}^{\theta}v\right)\right|+\frac{12}{39601}\left|\left(\mathcal{T}^{\phi}u\right)-\left(\mathcal{T}^{\phi}v\right)\right|.
\end{aligned}$$

Therefore, (H_1) holds with $L_1 = \dfrac{8}{39601}$, $L_2 = \dfrac{4}{39601}$, and $L_3 = \frac{12}{39601}$.

Then, we can show that (25) is true as follows

$$\chi = \left\{L_1 + L_2\varphi_0\frac{(9)^{\overline{\frac{1}{5}}}}{\Gamma(\frac{6}{5})} + L_3\psi_0\frac{(8)^{\underline{\frac{3}{4}}}}{\Gamma(\frac{7}{4})}\right\}[\Omega_1+\Omega_2+\Omega_3] \approx 0.20294 < 1.$$

Hence, by Theorem 1, the problem (36) has a unique solution.

6. Conclusions

We estabish the conditions for the existence and unique results of the solution for a fractional delta-nabla difference equations with fractional delta-nabla sum-difference boundary value conditions by using Banach contraction principle and the conditions for result of at least one solution by using the Schauder's fixed point theorem.

Author Contributions: Conceptualization, J.R. and T.S.; methodology, J.R. and T.S.; formal Analysis, J.R. and T.S.; investigation, J.R. and T.S.; writing—original draft preparation, J.R. and T.S.; writing—review & editing, J.R. and T.S.; funding acquisition, J.R. All authors have read and agreed to the published version of the manuscript.

Acknowledgments: The last author of this research was supported by Suan Dusit University.

References

1. Wu, G.C.; Baleanu, D. Discrete fractional logistic map and its chaos. *Nonlinear Dyn.* **2014**, *75*, 283–287. [CrossRef]
2. Wu, G.C.; Baleanu, D. Chaos synchronization of the discrete fractional logistic map. *Signal Process.* **2014**, *102*, 96–99. [CrossRef]
3. Wu, G.C.; Baleanu, D.; Xie, H.P.; Chen, F.L. Chaos synchronization of fractional chaotic maps based on stability results. *Phys. A* **2016**, *460*, 374–383. [CrossRef]
4. Voyiadjis, G.Z.; Sumelka, W. Brain modelling in the framework of anisotropic hyperelasticity with time fractional damage evolution governed by the Caputo-Almeida fractional derivative. *J. Mech. Behav. Biomed.* **2019**, *89*, 209–216. [CrossRef]
5. Caputo, M.; Ciarletta, M.; Fabrizio, M.; Tibullo, V. Melting and solidification of pure metals by a phase-field model. *Rend Lincei-Mat. Appl.* **2017**, *28*, 463–478. [CrossRef]
6. Gómez-Aguilar, J.F. Fractional Meissner–Ochsenfeld effect in superconductors. *Phys Lett. B* **2019**, *33*, 1950316.
7. Rahimi, Z.; Rezazadeh, G.; Sumelka, W.; Yang, X.-J. A study of critical point instability of micro and nano beams under a distributed variable-pressure force in the framework of the inhomogeneous non-linear nonlocal theory. *Arch. Mech.* **2017**, *69*, 413–433.
8. Agarwal, R.P.; Baleanu, D.; Rezapour, S.; Salehi, S. The existence of solutions for some fractional finite difference equations via sum boundary conditions. *Adv. Differ. Equ.* **2014**, *2014*, 282. [CrossRef]
9. Goodrich, C.S. On discrete sequential fractional boundary value problems. *J. Math. Anal. Appl.* **2012**, *385*, 111–124. [CrossRef]
10. Goodrich, C.S. On a discrete fractional three-point boundary value problem. *J. Differ. Equ. Appl.* **2012**, *18*, 397–415. [CrossRef]
11. Lv, W. Existence of solutions for discrete fractional boundary value problems witha p-Laplacian operator. *Adv. Differ. Equ.* **2012**, *2012*, 163. [CrossRef]
12. Ferreira, R. Existence and uniqueness of solution to some discrete fractional boundary value problems of order less than one. *J. Differ. Equ. Appl.* **2013**, *19*, 712–718. [CrossRef]
13. Abdeljawad, T. On Riemann and Caputo fractional differences. *Comput. Math. Appl.* **2011**, *62*, 1602–1611. [CrossRef]
14. Atici, F.M.; Eloe, P.W. Two-point boundary value problems for finite fractional difference equations. *J. Differ. Equ. Appl.* **2011**, *17*, 445–456. [CrossRef]
15. Atici, F.M.; Eloe, P.W. A transform method in discrete fractional calculus. *Int. J. Differ. Equ.* **2007**, *2*, 165–176.
16. Atici, F.M.; Eloe, P.W. Initial value problems in discrete fractional calculus. *Proc. Am. Math. Soc.* **2009**, *137*, 981–989. [CrossRef]
17. Sitthiwirattham, T.; Tariboon, J.; Ntouyas, S.K. Existence results for fractional difference equations with three-point fractional sum boundary conditions. *Discret. Dyn. Nat. Soc.* **2013** *2013*, 104276. [CrossRef]
18. Sitthiwirattham, T.; Tariboon, J.; Ntouyas, S.K. Boundary value problems for fractional difference equations with three-point fractional sum boundary conditions. *Adv. Differ. Equ.* **2013**, *2013*, 296. [CrossRef]

19. Sitthiwirattham, T. Existence and uniqueness of solutions of sequential nonlinear fractional difference equations with three-point fractional sum boundary conditions. *Math. Methods Appl. Sci.* **2015**, *38*, 2809–2815. [CrossRef]
20. Sitthiwirattham, T. Boundary value problem for p-Laplacian Caputo fractional difference equations with fractional sum boundary conditions. *Math. Methods Appl. Sci.* **2016**, *39*, 1522–1534. [CrossRef]
21. Reunsumrit, J.; Sitthiwirattham, T. Positive solutions of three-point fractional sum boundary value problem for Caputo fractional difference equations via an argument with a shift. *Positivity* **2014**, *20*, 861–876. [CrossRef]
22. Reunsumrit, J.; Sitthiwirattham, T. On positive solutions to fractional sum boundary value problems for nonlinear fractional difference equations. *Math. Methods Appl. Sci.* **2016**, *39*, 2737–2751. [CrossRef]
23. Kaewwisetkul, B.; Sitthiwirattham, T. On nonlocal fractional sum-difference boundary value problems for Caputo fractional functional difference equations with delay. *Adv. Differ. Equ.* **2017**, *2017*, 219. [CrossRef]
24. Chasreechai, S.; Sitthiwirattham, T. On separate fractional sum-difference boundary value problems with n-point fractional sum-difference boundary conditions via arbitrary different fractional orders. *Mathematics* **2019**, *2019*, 471. [CrossRef]
25. Setniker, A. Sequntial Differences in Nabla Fractional Calculus. Ph.D. Thesis, University of Nebraska, Lincoln, NE, USA, 2019.
26. Anastassiou, G.A. Nabla discrete calculus and nabla inequalities. *Math. Comput. Model.* **2010**, *51*, 562–571. [CrossRef]
27. Anastassiou, G.A. Foundations of nabla fractional calculus on time scales and inequalities. *Comput. Math. Appl.* **2010**, *59*, 3750–3762. [CrossRef]
28. Abdeljawad, T.; Atici, F.M. On the definitions of nabla fractional operators. *Abstr. Appl. Anal.* **2012**, *2012*, 406757. [CrossRef]
29. Abdeljawad, T. On delta and nabla Caputo fractional differences and dual identities. *Discret. Dyn. Nat. Soc.* **2013**, *2013*, 406910. [CrossRef]
30. Abdeljawad, T.; Abdall, B. Monotonicity results for delta and nabla Caputo and Riemann fractional differences via dual identities. *Filomat* **2017**, *31*, 3671–3683. [CrossRef]
31. Ahrendt, K.; Castle, L.; Holm, M.; Yochman, K. Laplace transforms for the nabla-difference operator and a fractional variation of parameters formula. *Commun. Appl. Anals.* **2012**, *16*, 317–347.
32. Atici, F.M.; Eloe, P.W. Discrete fractional calculus with the nabla operator. *Electron. Qual. Theory* **2009**. [CrossRef]
33. Atici, F.M.; Eloe, P.W. Linear systems of fractional nabla difference equations. *Rocky Mt. Math.* **2011**, *241*, 353–370. [CrossRef]
34. Baoguoa, J.; Erbe, L.; Peterson, A. Convexity for nabla and delta fractional differences. *J. Differ. Equ. Appl.* **2015**, *21*, 360–373. [CrossRef]
35. Baoguoa, J.; Erbe, L.; Peterson, A. Two monotonicity results for nabla and delta fractional differences. *Arch. Math.* **2015**, *104*, 589–597.
36. Natália, M.; Torres, D.F.M. Calculus of variations on timescales with nabla derivatives. *Nonlinear Anal.* **2018**, *71*, 763–773.
37. Goodrich, C.S.; Peterson, A.C. *Discrete Fractional Calculus*; Springer: New York, NY, USA, 2015.
38. Malinowska, A.B.; Torres, D.F.M. The delta-nabla calculus of variations. *Fasc. Math.* **2009**, *44*, 75–83.
39. Dryl, M.; Torres, D.F.M. The Delta-nabla calculus of variations for composition functionals on time scales. *Int. J. Differ. Equ.* **2003**, *8*, 27–47.
40. Dryl, M.; Torres, D.F.M. A general delta-nabla calculus of variations on time scales with application to economics. *Int. J. Dyn. Syst. Differ. Equ.* **2014**, *5*, 42–71. [CrossRef]
41. Ghorbanian, V.; Rezapour, S. A two-dimensional system of delta-nabla fractional difference inclusions. *Novi. Sad. J. Math.* **2017**, *47*, 143–163.
42. Liu, H.; Jin, Y.; Hou, C. Existence of positive solutions for discrete delta-nabla fractional boundary value problems with p-Laplacian. *Bound. Value Probl.* **2017**, *2017*, 60. [CrossRef]
43. Griffel, D.H. *Applied Functional Analysis*; Ellis Horwood Publishers: Chichester, UK, 1981.
44. Guo, D.; Lakshmikantham, V. *Nonlinear Problems in Abstract Cone*; Academic Press: Orlando, FL, USA, 1988.

16

Some Fractional Hermite–Hadamard type Inequalities for Interval-Valued Functions

Fangfang Shi [1], Guoju Ye [1,*], Dafang Zhao [2] and Wei Liu [1]

1 College of Science, Hohai University, Nanjing 210098, China; 15298387799@163.com (F.S.); liuw626@hhu.edu.cn (W.L.)

2 School of Mathematics and Statistics, Hubei Normal University, Huangshi 435002, China; dafangzhao@163.com

* Correspondence: yegj@hhu.edu.cn;

Abstract: In this paper, firstly we prove the relationship between interval h-convex functions and interval harmonically h-convex functions. Secondly, several new Hermite–Hadamard type inequalities for interval h-convex functions via interval Riemann–Liouville type fractional integrals are established. Finally, we obtain some new fractional Hadamard–Hermite type inequalities for interval harmonically h-convex functions by using the above relationship. Also we discuss the importance of our results and some special cases. Our results extend and improve some previously known results.

Keywords: Hermite–Hadamard type inequalities; interval-valued functions; fractional integrals

---

1. Introduction

Hermite–Hadamard inequality was firstly discovered by Hermite and Hadamard for convex functions are considerable significant in the literature. Since Hermite–Hadamard inequality has been regarded as one of the most useful inequalities in mathematical analysis and optimization, many papers have provided generalizations, refinements and extensions, see [1–4]. Due to the fractional integral has played a irreplaceable part in various scientific fields and importance of Hermite–Hadamard type inequalities, Sarikaya et al. [5] presented Hermite–Hadamard type inequalities via fractional integrals. Moreover, many papers relating to fractional integral inequalities have been obtained for different classes of functions, see [6–8].

On the other hand, interval analysis was initially developed as an attempt to deal with interval uncertainty that appears in computer graphics [9], automatic error analysis [10], and many others. Recently, several authors have extended their research by combining integral inequalities with interval-valued functions (IVFs), one can see Chalco-Cano et al. [11], Román-Flores et al. [12], Flores-Franulič et al. [13], Zhao et al. [14,15], An et al. [16]. As a further extension, more and more Hermite–Hadamard type inequalities involving interval Riemann–Liouville type fractional integral have been obtained for different classes of IVFs, see for interval convex functions [17], for interval harmonically convex functions [18] and the references therein.

Motivated by the ongoing research, We proved the relationship between interval h-convex functions and interval harmonically h-convex functions, then we establish some new Hermite–Hadamard type inequalities for interval h-convex functions and interval harmonically h-convex functions via interval Riemann–Liouville type fractional integrals. Our results extend and improve some known results. Also we discuss the importance of our results and some special cases. In addition, results obtained in this paper may be extended for other classes of convex functions including interval (h_1, h_2)-convex functions and interval Log-h-convex functions and used as a tool to investigate the research of optimization and probability, among others.

2. Preliminaries and Result

Let us denote by $\mathbb{R}_{\mathcal{I}}$ the collection of all nonempty closed intervals of the real line $\mathbb{R}$. We call $[z] = [\underline{z}, \overline{z}]$ positive if $\underline{z} > 0$. We denote by $\mathbb{R}_{\mathcal{I}}^{+}$ and $\mathbb{R}^{+}$ the set of all positive intervals and the set of all positive numbers of $\mathbb{R}$, respectively. For $[z] = [\underline{z}, \overline{z}], [s] = [\underline{s}, \overline{s}] \in \mathbb{R}_{\mathcal{I}}$, the inclusion "$\subseteq$" is defined by

$$[\underline{z}, \overline{z}] \subseteq [\underline{s}, \overline{s}] \Leftrightarrow \underline{s} \leq \underline{z},\ \overline{z} \leq \overline{s}.$$

For $\lambda \in \mathbb{R}$, the Minkowski addition and scalar multiplication are defined by

$$[z] + [s] = [\underline{z}, \overline{z}] + [\underline{s}, \overline{s}] = [\underline{z} + \underline{s}, \overline{z} + \overline{s}];$$

$$\lambda[z] = \lambda[\underline{z}, \overline{z}] = \begin{cases} [\lambda\underline{z}, \lambda\overline{z}], & \lambda > 0, \\ \{0\}, & \lambda = 0, \\ [\lambda\overline{z}, \lambda\underline{z}], & \lambda < 0 \end{cases}$$

respectively. The conception of Riemann integral for interval-valued function is introduced in [19]. Moreover, we have

Definition 1. *[19] Let $f : [a,b] \to \mathbb{R}_{\mathcal{I}}$ be an interval-valued function such that $f = [\underline{f}, \overline{f}]$. Then the f is Riemann integrable on $[a,b]$ iff $\underline{f}$ and $\overline{f}$ are Riemann integrable on $[a,b]$ and*

$$\int_a^b f(t)dt = \left[\int_a^b \underline{f}(t)dt, \int_a^b \overline{f}(t)dt\right].$$

The set of all Riemann integrable IVFs on $[a,b]$ will be denoted by $\mathcal{IR}_{([a,b])}$. For more basic notations with interval analysis, see [19,20]. Furthermore, we recall the following results in [17].

Definition 2. *Let $f : [a,b] \to \mathbb{R}_{\mathcal{I}}$ be an interval-valued function and $f \in \mathcal{IR}_{([a,b])}$. Then the interval Riemann–Liouville type fractional integrals of f are defined by*

$$\mathcal{J}_{a^+}^{\alpha} f(t) = \frac{1}{\Gamma(\alpha)} \int_a^t (t-\nu)^{\alpha-1} f(\nu) d\nu, \quad t > a.$$

and

$$\mathcal{J}_{b^-}^{\alpha} f(t) = \frac{1}{\Gamma(\alpha)} \int_t^b (\nu-t)^{\alpha-1} f(\nu) d\nu, \quad t < b.$$

where $\alpha > 0$ and Γ is the Gamma function.

Definition 3. *[14] Let $h : [0,1] \to \mathbb{R}^{+}$ be a non-negative function. We say that $f : [a,b] \to \mathbb{R}_{\mathcal{I}}^{+}$ is interval h-convex function or that $f \in SX(h, [a,b], \mathbb{R}_{\mathcal{I}}^{+})$, if for all $x, y \in [a,b]$ and $\nu \in [0,1]$, we have*

$$f(\nu x + (1-\nu)y) \supseteq h(\nu)f(x) + h(1-\nu)f(y).$$

Definition 4. *[15] Let $h : [0,1] \to \mathbb{R}^{+}$ be a non-negative function. We say that $f : [a,b] \to \mathbb{R}_{\mathcal{I}}^{+}$ is interval harmonically h-convex function or that $f \in SHX(h, [a,b], \mathbb{R}_{\mathcal{I}}^{+})$, if for all $x, y \in [a,b]$ and $\nu \in [0,1]$, we have*

$$f\left(\frac{xy}{\nu x + (1-\nu)y}\right) \supseteq h(1-\nu)f(x) + h(\nu)f(y).$$

Next, we will present the relationship between interval h-convex functions and interval harmonically h-convex functions which will be used in Section 4.

Theorem 1. $f(x) \in SHX(h, [a,b], \mathbb{R}_{\mathcal{I}}^{+})$ *iff* $f(\frac{1}{x}) \in SX(h, [a,b], \mathbb{R}_{\mathcal{I}}^{+})$.

Proof. Let $\psi(x) = f(\frac{1}{x})$. Since $f \in SHX(h, [a,b], \mathbb{R}_{\mathcal{I}}^{+})$,

$$f\left(\frac{xy}{\nu x + (1-\nu)y}\right) \supseteq h(1-\nu)f(x) + h(\nu)f(y). \tag{1}$$

By using $A = \frac{1}{x}$ and $B = \frac{1}{y}$ to replace x and y, respectively, applying (1)

$$\begin{aligned} f\left(\frac{\frac{1}{xy}}{\nu\frac{1}{x} + (1-\nu)\frac{1}{y}}\right) &= f\left(\frac{1}{(1-\nu)x + \nu y}\right) \\ &= \psi\left((1-\nu)x + \nu y\right) \\ &\supseteq h(\nu)f\left(\frac{1}{y}\right) + h(1-\nu)f\left(\frac{1}{x}\right) \\ &= h(\nu)\psi(y) + h(1-\nu)\psi(x), \end{aligned}$$

which gives that $\psi \in SX(h, [a,b], \mathbb{R}_{\mathcal{I}}^{+})$.

On the other hand, if $\psi \in SX(h, [a,b], \mathbb{R}_{\mathcal{I}}^{+})$, then

$$\psi\left(\nu x + (1-\nu)y\right) \supseteq h(\nu)\psi(x) + h(1-\nu)\psi(y).$$

In the same way as above, we have

$$\begin{aligned} \psi\left(\nu\frac{1}{x} + (1-\nu)\frac{1}{y}\right) = f\left(\frac{1}{\nu\frac{1}{x} + (1-\nu)\frac{1}{y}}\right) &= f\left(\frac{xy}{(1-\nu)x + \nu y}\right) \\ \supseteq h(\nu)\psi\left(\frac{1}{x}\right) + h(1-\nu)\psi\left(\frac{1}{y}\right)& \\ = h(\nu)f(x) + h(1-\nu)f(y),& \end{aligned}$$

which gives that $f \in SHX(h, [a,b], \mathbb{R}_{\mathcal{I}}^{+})$. We have completed the proof. □

Remark 1. *If $h(\nu) = \nu$ and $\underline{f} = \overline{f}$, then we get the Lemma 2.1 of [6].*

3. Fractional Hermite–Hadamard Type Inequalities of Interval h-Convex Functions

In this section, we will prove some new Hermite–Hadamard type inequalities for interval h-convex functions via interval Riemann–Liouville type integrals.

Theorem 2. *Let $f : [a,b] \to \mathbb{R}_{\mathcal{I}}^{+}$ be an interval-valued function such that $f = [\underline{f}, \overline{f}]$ and $f \in \mathcal{IR}_{([a,b])}$, $h : [0,1] \to \mathbb{R}^{+}$ be a non-negative function and $h\left(\frac{1}{2}\right) \neq 0$. If $f \in SX(h, [a,b], \mathbb{R}_{\mathcal{I}}^{+})$, then*

$$\begin{aligned} \frac{1}{\alpha h(\frac{1}{2})} f\left(\frac{a+b}{2}\right) &\supseteq \frac{\Gamma(\alpha)}{(b-a)^{\alpha}}\left[\mathcal{J}_{a^+}^{\alpha} f(b) + \mathcal{J}_{b^-}^{\alpha} f(a)\right] \\ &\supseteq [f(a) + f(b)]\int_0^1 \nu^{\alpha-1}\left[h(\nu) + h(1-\nu)\right] d\nu \end{aligned} \tag{2}$$

with $\alpha > 0$.

Proof. Since $f \in SX(h, [a,b], \mathbb{R}_{\mathcal{I}}^{+})$, we have

$$\frac{1}{h(\frac{1}{2})} f\left(\frac{x+y}{2}\right) \supseteq f(x) + f(y).$$

Let $x = \nu a + (1-\nu)b$, $y = (1-\nu)a + \nu b$, $\nu \in [0,1]$, then

$$\frac{1}{h(\frac{1}{2})} f\Big(\frac{a+b}{2}\Big) \supseteq f(\nu a + (1-\nu)b) + f((1-\nu)a + \nu b). \tag{3}$$

Multiplying both sides (3) by $\nu^{\alpha-1}$ and integrating on $[0,1]$, we get

$$\begin{aligned}
&\frac{1}{\alpha h(\frac{1}{2})} f\left(\frac{a+b}{2}\right) \\
=&\frac{1}{h(\frac{1}{2})} f\left(\frac{a+b}{2}\right) \int_0^1 \nu^{\alpha-1} d\nu \\
\supseteq&\left[\int_0^1 \nu^{\alpha-1} f(\nu a + (1-\nu)b) d\nu + \int_0^1 \nu^{\alpha-1} f((1-\nu)a + \nu b) d\nu\right] \\
=&\Big[\int_0^1 \nu^{\alpha-1} \left(\underline{f}(\nu a + (1-\nu)b) + \underline{f}((1-\nu)a + \nu b)\right) d\nu, \\
&\int_0^1 \nu^{\alpha-1} \left(\overline{f}(\nu a + (1-\nu)b) + \overline{f}((1-\nu)a + \nu b)\right) d\nu\Big] \\
=&\Big[\int_b^a (\tau(\mu))^{\alpha-1} \underline{f}(\mu) \frac{d\mu}{a-b} + \int_a^b (1-\tau(\mu))^{\alpha-1} \underline{f}(\mu) \frac{d\mu}{b-a}, \\
&\int_b^a (\tau(\mu))^{\alpha-1} \overline{f}(\mu) \frac{d\mu}{a-b} + \int_a^b (1-\tau(\mu))^{\alpha-1} \overline{f}(\mu) \frac{d\mu}{b-a}\Big] \\
=&\frac{\Gamma(\alpha)}{(b-a)^\alpha} \left[J_{a^+}^\alpha \underline{f}(b) + J_{b^-}^\alpha \underline{f}(a), J_{a^+}^\alpha \overline{f}(b) + J_{b^-}^\alpha \overline{f}(a)\right] \\
=&\frac{\Gamma(\alpha)}{(b-a)^\alpha} \left[\mathcal{J}_{a^+}^\alpha f(b) + \mathcal{J}_{b^-}^\alpha f(a)\right].
\end{aligned} \tag{4}$$

where $\tau(\mu) = \frac{b-\mu}{b-a}$.

Similarly, since $f \in SX(h, [a,b], \mathbb{R}_\mathcal{I}^+)$,

$$f(\nu a + (1-\nu)b) + f((1-\nu)a + \nu b) \supseteq [h(\nu) + h(1-\nu)] [f(a) + f(b)]. \tag{5}$$

Multiplying both sides (5) by $\nu^{\alpha-1}$ and integrating on $[0,1]$, we have

$$\frac{\Gamma(\alpha)}{(b-a)^\alpha} [\mathcal{J}_{a^+}^\alpha f(b) + \mathcal{J}_{b^-}^\alpha f(a)] \supseteq [f(a) + f(b)] \int_0^1 \nu^{\alpha-1} [h(\nu) + h(1-\nu)] d\nu. \tag{6}$$

By combining (4) with (6), and the result follows. □

Example 1. *Suppose that $[a,b] = [1,2]$. Let $h(\nu) = \nu$ for all $\nu \in [0,1]$ and $\alpha = \frac{1}{2}$, $f : [a,b] \to \mathbb{R}_\mathcal{I}^+$ be defined by*

$$f(t) = \left[-\sqrt{t} + 2, \sqrt{t} + 2\right].$$

We obtain

$$\frac{1}{\alpha h(\frac{1}{2})} f\left(\frac{a+b}{2}\right) = [8 - 2\sqrt{6}, 8 + 2\sqrt{6}],$$

$$\frac{\Gamma(\alpha)}{(b-a)^\alpha} \left[\mathcal{J}_{a^+}^\alpha f(b) + \mathcal{J}_{b^-}^\alpha f(a)\right] = \left[7 - \sqrt{2} - \frac{\pi}{2} - log(\sqrt{2}+1), 9 + \sqrt{2} + \frac{\pi}{2} + log(\sqrt{2}+1)\right],$$

$$[f(a) + f(b)] \int_0^1 \nu^{\alpha-1} \left[h(\nu) + h(1-\nu)\right] d\nu = [6 - 2\sqrt{2}, 6 + 2\sqrt{2}].$$

Then we get

$$\left[8-2\sqrt{6},8+2\sqrt{6}\right] \supseteq \left[7-\sqrt{2}-\frac{\pi}{2}-log(\sqrt{2}+1),9+\sqrt{2}+\frac{\pi}{2}+log(\sqrt{2}+1)\right] \supseteq \left[6-2\sqrt{2},6+2\sqrt{2}\right].$$

Consequently, Theorem 2 is verified.

Remark 2. *If $\alpha = 1$, then we get Theorem 4.1 of [14]. If $h(\nu) = \nu$, then we get Theorem 2.5 of [17]. If $\underline{f} = \overline{f}$ and $\alpha = 1$, then we get Theorem 6 of [4]. If $\underline{f} = \overline{f}$ and $h(\nu) = \nu$, then we get Theorem 2 of [5].*

Theorem 3. *Let $f,g : [a,b] \to \mathbb{R}_{\mathcal{I}}^{+}$ be two interval-valued functions such that $f = [\underline{f},\overline{f}]$, $g = [\underline{g},\overline{g}]$ and $fg \in \mathcal{IR}_{([a,b])}$, $h_1,h_2 : [0,1] \to \mathbb{R}^{+}$ be non-negative functions. If $f \in SX(h_1,[a,b],\mathbb{R}_{\mathcal{I}}^{+})$, $g \in SX(h_2,[a,b],\mathbb{R}_{\mathcal{I}}^{+})$, then*

$$\begin{aligned}&\frac{\Gamma(\alpha)}{(b-a)^{\alpha}}\left[\mathcal{J}_{a^+}^{\alpha} f(b)g(b) + \mathcal{J}_{b^-}^{\alpha} f(a)g(a)\right]\\ \supseteq&\mathcal{M}(a,b)\int_0^1 \nu^{\alpha-1}\left[h_1(\nu)h_2(\nu)+h_1(1-\nu)h_2(1-\nu)\right]d\nu\\ &+\mathcal{N}(a,b)\int_0^1 \nu^{\alpha-1}\left[h_1(\nu)h_2(1-\nu)+h_1(1-\nu)h_2(\nu)\right]d\nu\end{aligned} \tag{7}$$

where

$$\mathcal{M}(a,b) = f(a)g(a)+f(b)g(b),\ \mathcal{N}(a,b) = f(a)g(b)+f(b)g(a).$$

Proof. By hypothesis, one has

$$f(\nu a+(1-\nu)b) \supseteq h_1(\nu)f(a)+h_1(1-\nu)f(b),$$

$$g(\nu a+(1-\nu)b) \supseteq h_2(\nu)g(a)+h_2(1-\nu)g(b).$$

Since $f,g \in \mathbb{R}_{\mathcal{I}}^{+}$, we obtain

$$\begin{aligned}&f(\nu a+(1-\nu)b)g(\nu a+(1-\nu)b)\\ \supseteq&h_1(\nu)h_2(\nu)f(a)g(a)+h_1(1-\nu)h_2(1-\nu)f(b)g(b)\\ &+h_1(\nu)h_2(1-\nu)f(a)g(b)+h_1(1-\nu)h_2(\nu)f(b)g(a).\end{aligned} \tag{8}$$

In the same way as above, we have

$$\begin{aligned}&f((1-\nu)a+\nu b)g((1-\nu)a+\nu b)\\ \supseteq&h_1(1-\nu)h_2(1-\nu)f(a)g(a)+h_1(\nu)h_2(\nu)f(b)g(b)\\ &+h_1(1-\nu)h_2(\nu)f(a)g(b)+h_1(\nu)h_2(1-\nu)f(b)g(a).\end{aligned} \tag{9}$$

By adding (8) and (9), we obtain

$$\begin{aligned}&f(\nu a+(1-\nu)b)g(\nu a+(1-\nu)b)+f((1-\nu)a+\nu b)g((1-\nu)a+\nu b)\\ \supseteq&\left[h_1(\nu)f(a)+h_1(1-\nu)f(b)\right]\left[h_2(\nu)g(a)+h_2(1-\nu)g(b)\right]\\ &+\left[h_1(1-\nu)f(a)+h_1(\nu)f(b)\right]\left[h_2(1-\nu)g(a)+h_2(\nu)g(b)\right]\\ =&\mathcal{M}(a,b)\left[h_1(\nu)h_2(\nu)+h_1(1-\nu)h_2(1-\nu)\right]\\ &+\mathcal{N}(a,b)\left[h_1(1-\nu)h_2(\nu)+h_1(\nu)h_2(1-\nu)\right].\end{aligned} \tag{10}$$

Multiplying both sides (10) by $\nu^{\alpha-1}$ and integrating on $[0,1]$, we have

$$\begin{aligned}&\int_0^1 \nu^{\alpha-1} f(\nu a+(1-\nu)b)g(\nu a+(1-\nu)b)d\nu\\&+\int_0^1 \nu^{\alpha-1} f((1-\nu)a+\nu b)g((1-\nu)a+\nu b)d\nu\\ \supseteq&\mathcal{M}(a,b)\int_0^1 \nu^{\alpha-1}[h_1(\nu)h_2(\nu)+h_1(1-\nu)h_2(1-\nu)]d\nu\\&+\mathcal{N}(a,b)\int_0^1 \nu^{\alpha-1}[h_1(1-\nu)h_2(\nu)+h_1(\nu)h_2(1-\nu)]d\nu.\end{aligned} \tag{11}$$

By Definition 2, we obtain

$$\int_0^1 \nu^{\alpha-1} f(\nu a+(1-\nu)b)g(\nu a+(1-\nu)b)d\nu=\frac{\Gamma(\alpha)}{(b-a)^\alpha}\mathcal{J}_{a^+}^{\alpha} f(b)g(b), \tag{12}$$

$$\int_0^1 \nu^{\alpha-1} f((1-\nu)a+\nu b)g((1-\nu)a+\nu b)d\nu=\frac{\Gamma(\alpha)}{(b-a)^\alpha}\mathcal{J}_{b^-}^{\alpha} f(a)g(a). \tag{13}$$

By substituting the equalities (12) and (13) in (11), then we have inequality (7). □

Remark 3. *If $h(\nu)=\nu$, then we get Theorem 3.5 of [17]. If $\alpha=1$, then we get Theorem 4.5 of [14]. If $\underline{f}=\overline{f}$ and $\alpha=1$, then we get Theorem 7 of [4].*

Theorem 4. *Let $f,g:[a,b]\to\mathbb{R}_{\mathcal{I}}^+$ be two interval-valued functions such that $f=[\underline{f},\overline{f}]$, $g=[\underline{g},\overline{g}]$ and $fg\in\mathcal{IR}_{([a,b])}$, $h_1,h_2:[0,1]\to\mathbb{R}^+$ be non-negative functions and $h_1(\frac{1}{2})h_2(\frac{1}{2})\neq 0$. If $f\in SX(h_1,[a,b],\mathbb{R}_{\mathcal{I}}^+)$, $g\in SX(h_2,[a,b],\mathbb{R}_{\mathcal{I}}^+)$, then*

$$\begin{aligned}&\frac{1}{\alpha h_1(\frac{1}{2})h_2(\frac{1}{2})}f\Big(\frac{a+b}{2}\Big)g\Big(\frac{a+b}{2}\Big)\\ \supseteq&\frac{\Gamma(\alpha)}{(b-a)^\alpha}\left[\mathcal{J}_{a^+}^{\alpha} f(b)g(b)+\mathcal{J}_{b^-}^{\alpha} f(a)g(a)\right]\\&+\mathcal{M}(a,b)\int_0^1 [\nu^{\alpha-1}+(1-\nu)^{\alpha-1}]h_1(\nu)h_2(1-\nu)d\nu\\&+\mathcal{N}(a,b)\int_0^1 [\nu^{\alpha-1}+(1-\nu)^{\alpha-1}]h_1(1-\nu)h_2(1-\nu)d\nu.\end{aligned} \tag{14}$$

Proof. Since $f\in SX(h_1,[a,b],\mathbb{R}_{\mathcal{I}}^+)$, $g\in SX(h_2,[a,b],\mathbb{R}_{\mathcal{I}}^+)$, we get

$$\begin{aligned}&f\Big(\frac{a+b}{2}\Big)g\Big(\frac{a+b}{2}\Big)\\=&f\Big(\frac{\nu a+(1-\nu)b}{2}+\frac{(1-\nu)a+\nu b}{2}\Big)g\Big(\frac{\nu a+(1-\nu)b}{2}+\frac{(1-\nu)a+\nu b}{2}\Big)\\ \supseteq&h_1(\frac{1}{2})h_2(\frac{1}{2})\Big[f(\nu a+(1-\nu)b)+f((1-\nu)a+\nu b)\Big]\Big[q(\nu a+(1-\nu)b)+q((1-\nu)a+\nu b)\Big]\\ \supseteq&h_1(\frac{1}{2})h_2(\frac{1}{2})\Big[f(\nu a+(1-\nu)b)g(\nu a+(1-\nu)b)+f((1-\nu)a+\nu b)g((1-\nu)a+\nu b)\Big]\\&+h_1(\frac{1}{2})h_2(\frac{1}{2})\Big[\Big(h_1(\nu)h_2(1-\nu)+h_1(1-\nu)h_2(\nu)\Big)\mathcal{M}(a,b)\\&+\Big(h_1(\nu)h_2(\nu)+h_1(1-\nu)h_2(1-\nu)\Big)\mathcal{N}(a,b)\Big].\end{aligned} \tag{15}$$

Multiplying both sides (15) by $\nu^{\alpha-1}$ and integrating on $[0,1]$, we have inequality (14). □

Remark 4. *If $h(v) = v$, then we get Theorem 3.6 of [17]. If $\alpha = 1$, then we get Theorem 4.6 of [14]. If $\underline{f} = \overline{f}$ and $\alpha = 1$, then we get Theorem 8 of [4].*

4. Fractional Hermite–Hadamard Type Inequalities of Interval Harmonically h-Convex Functions

In this section, we will use the above results to get some Hermite–Hadamard type inequalities for interval harmonically h-convex functions via interval Riemann–Liouville type integrals and some special cases are also discussed.

Theorem 5. *Let $f : [a,b] \to \mathbb{R}_{\mathcal{I}}^{+}$ be an interval-valued function such that $f = [\underline{f}, \overline{f}]$ and $f \in \mathcal{IR}_{([a,b])}$, $h : [0,1] \to \mathbb{R}^{+}$ be a non-negative function and $h\left(\frac{1}{2}\right) \neq 0$. If $f \in SHX(h, [a,b], \mathbb{R}_{\mathcal{I}}^{+})$, then*

$$\begin{aligned}&\frac{1}{\alpha h\left(\frac{1}{2}\right)} f\left(\frac{2ab}{a+b}\right)\\ \supseteq&\Gamma(\alpha)\left(\frac{ab}{b-a}\right)^{\alpha}\left[\mathcal{J}^{\alpha}_{\left(\frac{1}{a}\right)-}(f\circ\xi)\left(\frac{1}{b}\right)+\mathcal{J}^{\alpha}_{\left(\frac{1}{b}\right)+}(f\circ\xi)\left(\frac{1}{a}\right)\right]\\ \supseteq&\left[f(a)+f(b)\right]\int_0^1 v^{\alpha-1}\left[h(v)+h(1-v)\right]dv\end{aligned} \tag{16}$$

where $\xi(x) = \frac{1}{x}$.

Proof. Let $\psi(x) = f\left(\frac{1}{x}\right)$. By Theorem 1, we have $\psi \in SX\left(h, [a,b], \mathbb{R}_{\mathcal{I}}^{+}\right)$ and

$$\frac{1}{h\left(\frac{1}{2}\right)}\psi\left(\frac{x+y}{2xy}\right) \supseteq \psi\left(\frac{1}{x}\right) + \psi\left(\frac{1}{y}\right).$$

Let $x = \frac{ab}{va+(1-v)b}$, $y = \frac{ab}{(1-v)a+vb}$, $v \in [0,1]$. Then

$$\frac{1}{h\left(\frac{1}{2}\right)}\psi\left(\frac{a+b}{2ab}\right) \supseteq \psi\left(\frac{va+(1-v)b}{ab}\right) + \psi\left(\frac{(1-v)a+vb}{ab}\right), \tag{17}$$

Multiplying both sides (17) by $v^{\alpha-1}$ and integrating on $[0,1]$, we have

$$\begin{aligned}&\frac{1}{\alpha h\left(\frac{1}{2}\right)}\psi\left(\frac{a+b}{2ab}\right)\\ \supseteq&\left[\int_0^1 v^{\alpha-1}\psi\left(\frac{va+(1-v)b}{ab}\right)dv+\int_0^1 v^{\alpha-1}\psi\left(\frac{(1-v)a+vb}{ab}\right)dv\right]\\ =&\Gamma(\alpha)\left(\frac{ab}{b-a}\right)^{\alpha}\left[\mathcal{J}^{\alpha}_{\left(\frac{1}{a}\right)-}\psi\left(\frac{1}{b}\right)+\mathcal{J}^{\alpha}_{\left(\frac{1}{b}\right)+}\psi\left(\frac{1}{a}\right)\right]\end{aligned} \tag{18}$$

Similarly, we have

$$\psi\left(\frac{va+(1-v)b}{ab}\right) \supseteq h(v)\psi\left(\frac{1}{b}\right) + h(1-v)\psi\left(\frac{1}{a}\right),$$

$$\psi\left(\frac{(1-v)a+vb}{ab}\right) \supseteq h(v)\psi\left(\frac{1}{a}\right) + h(1-v)\psi\left(\frac{1}{b}\right).$$

Then,

$$\psi\Big(\frac{\nu a+(1-\nu)b}{ab}\Big)+\psi\Big(\frac{(1-\nu)a+\nu b}{ab}\Big)\supseteq[h(\nu)+h(1-\nu)]\left[\psi\left(\frac{1}{a}\right)+\psi\left(\frac{1}{b}\right)\right]. \tag{19}$$

Multiplying both sides (19) by $\nu^{\alpha-1}$ and integrating on $[0,1]$, we have

$$\begin{aligned}&\Gamma(\alpha)\left(\frac{ab}{b-a}\right)^{\alpha}\left[\mathcal{J}^{\alpha}_{(\frac{1}{a})^-}\psi\left(\frac{1}{b}\right)+\mathcal{J}^{\alpha}_{(\frac{1}{b})^+}\psi\left(\frac{1}{a}\right)\right]\\&\supseteq\left[\psi\left(\frac{1}{a}\right)+\psi\left(\frac{1}{b}\right)\right]\int_0^1\nu^{\alpha-1}[h(\nu)+h(1-\nu)]\,dt.\end{aligned} \tag{20}$$

By (18) and (20), we have inequality (16). □

Remark 5. *If $h(\nu)=\nu$, then we get Theorem 3.6 of [18]. If $\alpha=1$, then we get Theorem 2 of [15]. If $\underline{f}=\overline{f}$ and $\alpha=1$, then we get Theorem 3.2 of [3]. If $\underline{f}=\overline{f}$ and $h(\nu)=\nu$, then we get Theorem 4 of [7].*

Theorem 6. *Let $f,g:[a,b]\to\mathbb{R}^+_{\mathcal{I}}$ be two interval-valued functions such that $f=[\underline{f},\overline{f}]$, $g=[\underline{g},\overline{g}]$ and $fg\in\mathcal{IR}_{([a,b])}$, $h_1,h_2:[0,1]\to\mathbb{R}^+$ be non-negative functions. If $f\in SHX(h_1,[a,b],\mathbb{R}^+_{\mathcal{I}})$, $g\in SHX(h_2,[a,b],\mathbb{R}^+_{\mathcal{I}})$, then*

$$\begin{aligned}&\Gamma(\alpha)\Big(\frac{ab}{b-a}\Big)^{\alpha}\left[\mathcal{J}^{\alpha}_{(\frac{1}{b})^+}(f\circ\xi)\left(\frac{1}{a}\right)(g\circ\xi)\left(\frac{1}{a}\right)+\mathcal{J}^{\alpha}_{(\frac{1}{a})^-}(f\circ\xi)\left(\frac{1}{b}\right)(g\circ\xi)\left(\frac{1}{b}\right)\right]\\&\supseteq\mathcal{M}(a,b)\int_0^1\left[\nu^{\alpha-1}+(1-\nu)^{\alpha-1}\right]h_1(\nu)h_2(\nu)d\nu\\&\quad+\mathcal{N}(a,b)\int_0^1\left[\nu^{\alpha-1}+(1-\nu)^{\alpha-1}\right]h_1(\nu)h_2(1-\nu)d\nu\end{aligned} \tag{21}$$

where $\xi(x)=\frac{1}{x}$.

Proof. The proof is completed by combining Theorems 1, 3 and 5. □

Remark 6. *If $h_1(\nu)=h_2(\nu)=\nu$, then*

$$\begin{aligned}&\Gamma(\alpha)\Big(\frac{ab}{b-a}\Big)^{\alpha}\left[\mathcal{J}^{\alpha}_{(\frac{1}{b})^+}(f\circ\xi)\left(\frac{1}{a}\right)(g\circ\xi)\left(\frac{1}{a}\right)+\mathcal{J}^{\alpha}_{(\frac{1}{a})^-}(f\circ\xi)\left(\frac{1}{b}\right)(g\circ\xi)\left(\frac{1}{b}\right)\right]\\&\supseteq\mathcal{M}(a,b)\int_0^1\nu^2\left[\nu^{\alpha-1}+(1-\nu)^{\alpha-1}\right]d\nu+\mathcal{N}(a,b)\int_0^1\nu(1-\nu)\left[\nu^{\alpha-1}+(1-\nu)^{\alpha-1}\right]d\nu.\end{aligned}$$

If $\alpha=1$, then we get Theorem 4 of [15]. If $\underline{f}=\overline{f}$ and $\alpha=1$, then we get Theorem 3.6 of [3].

Theorem 7. *Let $f,g:[a,b]\to\mathbb{R}^+_{\mathcal{I}}$ be two interval-valued functions such that $f=[\underline{f},\overline{f}]$, $g=[\underline{g},\overline{g}]$ and $fg\in\mathcal{IR}_{([a,b])}$, $h_1,h_2:[0,1]\to\mathbb{R}^+$ be non-negative functions and $h_1(\frac{1}{2})h_2(\frac{1}{2})\neq 0$. If $f\in SHX(h_1,[a,b],\mathbb{R}^+_{\mathcal{I}})$, $g\in SHX(h_2,[a,b],\mathbb{R}^+_{\mathcal{I}})$, then*

$$\begin{aligned}&\frac{1}{\alpha h_1(\frac{1}{2})h_2\left(\frac{1}{2}\right)}f\Big(\frac{2ab}{a+b}\Big)g\Big(\frac{2ab}{a+b}\Big)\\&\supseteq\Gamma(\alpha)\Big(\frac{ab}{b-a}\Big)^{\alpha}\left[\mathcal{J}^{\alpha}_{(\frac{1}{b})^+}(f\circ\xi)\left(\frac{1}{a}\right)(g\circ\xi)\left(\frac{1}{a}\right)+\mathcal{J}^{\alpha}_{(\frac{1}{a})^-}(f\circ\xi)\left(\frac{1}{b}\right)(g\circ\xi)\left(\frac{1}{b}\right)\right]\\&\quad+\int_0^1[\nu^{\alpha-1}+(1-\nu)^{\alpha-1}][h_1(\nu)h_2(1-\nu)\mathcal{M}(a,b)+h_1(\nu)h_2(\nu)\mathcal{N}(a,b)]d\nu\end{aligned}$$

where $\xi(x) = \frac{1}{x}$.

Proof. The proof is completed by combining Theorem 1, 4 and 5. □

Remark 7. *If* $h_1(\nu) = h_2(\nu) = \nu$, *then*

$$\begin{aligned}
&4f\left(\frac{2ab}{a+b}\right)g\left(\frac{2ab}{a+b}\right)\\
&\supseteq \Gamma(\alpha+1)\left(\frac{ab}{b-a}\right)^{\alpha}\left[\mathcal{J}^{\alpha}_{\left(\frac{1}{b}\right)^+}(f\circ\xi)\left(\frac{1}{a}\right)(g\circ\xi)\left(\frac{1}{a}\right)+\mathcal{J}^{\alpha}_{\left(\frac{1}{a}\right)^-}(f\circ\xi)\left(\frac{1}{b}\right)(g\circ\xi)\left(\frac{1}{b}\right)\right]\\
&\quad+\int_0^1\left[\nu^{\alpha-1}+(1-\nu)^{\alpha-1}\right]\left[\nu(1-\nu)\mathcal{M}(a,b)+\nu^2\mathcal{N}(a,b)\right]d\nu.
\end{aligned}$$

If $\alpha = 1$, *then we get Theorem 4 of [15].*

5. Conclusions

This paper proved the relationship between interval h-convex functions and interval harmonically h-convex functions. Further, we obtained some Hermite–Hadamard type inequalities for IVFs via interval Riemann–Liouville type fractional integrals. The results obtained in this article are the generalizations and refinements of the earlier works. Moreover, these results may be extended for other kinds of convex functions including interval (h_1, h_2)-convex functions and interval Log-h-convex functions and used as a method to establish the Hermite–Hadamard type inequalities for other types of interval harmonically convex functions. As a future research direction, we intend to investigate Hermite–Hadamard type inequalities for IVFs on time scales and some applications in interval optimization, probability, among others.

Author Contributions: Conceptualization, F.S., G.Y., D.Z., and W.L.; methodology, F.S., G.Y., D.Z., and W.L.; writing—original draft preparation, F.S.; writing—review and editing, F.S.; supervision, D.Z.; and W.L.; project administration, G.Y.; and funding acquisition, G.Y.; All authors have read and agreed to the published version of the manuscript.

Acknowledgments: The authors are very grateful to three anonymous referees, for several valuable and helpful comments, suggestions and questions, which helped them to improve the paper into present form.

References

1. Bombardelli, M.; Varošanec, S. Properties of h-convex functions related to the Hermite–Hadamard–Fejér inequalities. *Comput. Math. Appl.* **2009**, *58*, 1869–1877. [CrossRef]
2. Dragomir, S.S. Inequalities of Hermite–Hadamard type for h-convex functions on linear spaces. *Proyecciones* **2015**, *34*, 323–341. [CrossRef]
3. Noor, M.A.; Noor, K.I.; Awan, M.U.; Costache, S. Some integral inequalities for harmonically h-convex functions. *Politehn. Univ. Bucharest Sci. Bull. Ser. A Appl. Math. Phys.* **2015**, *77*, 5–16.
4. Sarikaya, M.Z.; Saglam, A.; Yildirim, H. On some Hadamard-type inequalities for h-convex functions. *J. Math. Inequal* **2008**, *2*, 335–341. [CrossRef]
5. Sarikaya, M.Z.; Set, E.; Yaldiz, H.; Başak, N. Hermite–Hadamard's inequalities for fractional integrals and related fractional inequalities. *Math. Comput. Model.* **2013**, *57*, 2403–2407. [CrossRef]
6. Chen, F. Extensions of the Hermite–Hadamard inequality for harmonically convex functions via fractional integrals. *Appl. Math. Comput.* **2015**, *268*, 121–128. [CrossRef]
7. Işcan, İ.; Wu, S.H. Hermite-Hadamard type inequalities for harmonically convex functions via fractional integrals. *Appl. Math. Comput.* **2014** *238*, 237–244. [CrossRef]

8. Noor, M.A.; Noor, K.I.; Mihai, M.V.; Awan, M.U. Fractional Hermite–Hadamard inequalities for some classes of differentiable preinvex functions. *Sci. Bull. Politeh. Univ. Buchar. Ser. A Appl. Math. Phys.* **2016**, *78*, 163–174.
9. Snyder, J.M. Interval analysis for computer graphics. *SIGGRAPH Comput. Graph* **1992**, *26*, 121–130. [CrossRef]
10. Rothwell, E.J.; Cloud, M.J. Automatic error analysis using intervals. *IEEE Trans. Ed.* **2011**, *55*, 9–15. [CrossRef]
11. Chalco-Cano, Y.; Lodwick, W.A.; Condori-Equice, W. Ostrowski type inequalities for interval-valued functions using generalized Hukuhara derivative. *Comput. Appl. Math.* **2012**, *31*, 457–472.
12. Flores-Franulič, A.; Chalco-Cano, Y.; Román-Flores, H. An Ostrowski type inequality for interval-valued functions. In Proceedings of the 2013 Joint IFSA World Congress and NAFIPS Annual Meeting (IFSA/NAFIPS), Edmonton, AB, Canada, 24–28 June 2013; Volume 35, pp. 1459–1462.
13. Román-Flores, H.; Chalco-Cano, Y.; Lodwick, W.A. Some integral inequalities for interval-valued functions. *Comput. Appl. Math.* **2016**, *35*, 1–13. [CrossRef]
14. Zhao, D.F.; An, T.Q.; Ye, G.J.; Liu, W. New Jensen and Hermite–Hadamard type inequalities for h-convex interval-valued functions. *J. Inequal. Appl.* **2018**, *2018*, 302. [CrossRef]
15. Zhao, D.F.; An, T.Q.; Ye, G.J.; Torres, D.F.M. On Hermite–Hadamard type inequalities for harmonically h-convex interval-valued functions. *Math. Inequal. Appl.* **2020**, *23*, 95–105.
16. An, Y.R.; Ye, G.J.; Zhao, D.F.; Liu, W. Hermite–Hadamard Type Inequalities for Interval $(h1, h2)$-Convex Functions. *Mathematics* **2019**, *7*, 436. [CrossRef]
17. Büdak, H.; Tunc, T.; Sarikaya, M.Z. Fractional Hermite–Hadamard type inequalities for interval-valued functions. *Proc. Amer. Math. Soc.* **2020**, *148*, 705–718. [CrossRef]
18. Liu, X.L.; Ye, G.J.; Zhao, D.F.; Liu, W. Fractional Hermite–Hadamard type inequalities for interval-valued functions. *J. Inequal. Appl.* **2019**, *2019*, 266. [CrossRef]
19. Markov, S. Calculus for interval functions of a real variable. *Computing* **1979**, *22*, 325–337. [CrossRef]
20. Moore, R.E. *Interval Analysis*; Prentice-Hall: Englewood Cliffs, NJ, USA, 1966.

Permissions

All chapters in this book were first published by MDPI; hereby published with permission under the Creative Commons Attribution License or equivalent. Every chapter published in this book has been scrutinized by our experts. Their significance has been extensively debated. The topics covered herein carry significant findings which will fuel the growth of the discipline. They may even be implemented as practical applications or may be referred to as a beginning point for another development.

The contributors of this book come from diverse backgrounds, making this book a truly international effort. This book will bring forth new frontiers with its revolutionizing research information and detailed analysis of the nascent developments around the world.

We would like to thank all the contributing authors for lending their expertise to make the book truly unique. They have played a crucial role in the development of this book. Without their invaluable contributions this book wouldn't have been possible. They have made vital efforts to compile up to date information on the varied aspects of this subject to make this book a valuable addition to the collection of many professionals and students.

This book was conceptualized with the vision of imparting up-to-date information and advanced data in this field. To ensure the same, a matchless editorial board was set up. Every individual on the board went through rigorous rounds of assessment to prove their worth. After which they invested a large part of their time researching and compiling the most relevant data for our readers.

The editorial board has been involved in producing this book since its inception. They have spent rigorous hours researching and exploring the diverse topics which have resulted in the successful publishing of this book. They have passed on their knowledge of decades through this book. To expedite this challenging task, the publisher supported the team at every step. A small team of assistant editors was also appointed to further simplify the editing procedure and attain best results for the readers.

Apart from the editorial board, the designing team has also invested a significant amount of their time in understanding the subject and creating the most relevant covers. They scrutinized every image to scout for the most suitable representation of the subject and create an appropriate cover for the book.

The publishing team has been an ardent support to the editorial, designing and production team. Their endless efforts to recruit the best for this project, has resulted in the accomplishment of this book. They are a veteran in the field of academics and their pool of knowledge is as vast as their experience in printing. Their expertise and guidance has proved useful at every step. Their uncompromising quality standards have made this book an exceptional effort. Their encouragement from time to time has been an inspiration for everyone.

The publisher and the editorial board hope that this book will prove to be a valuable piece of knowledge for researchers, students, practitioners and scholars across the globe.

List of Contributors

Jarunee Soontharanon and Saowaluck Chasreechai
Department of Mathematics, Faculty of Applied Science, King Mongkut's University of Technology North Bangkok, Bangkok 10800, Thailand

Thanin Sitthiwirattham
Mathematics Department, Faculty of Science and Technology, Suan Dusit University, Bangkok 10700, Thailand

Kui Liu
Department of Mathematics, Guizhou University, Guiyang 550025, China

Michal Fečkan
Department of Mathematical Analysis and Numerical Mathematics, Faculty of Mathematics, Physics and Informatics, Comenius University in Bratislava, Mlynská dolina, 842 48 Bratislava, Slovakia
Mathematical Institute, Slovak Academy of Sciences, Štefánikova 49, 814 73 Bratislava, Slovakia

D. O'Regan
School of Mathematics, Statistics and Applied Mathematics, National University of Ireland, H91 TK33 Galway, Ireland

JinRong Wang
Department of Mathematics, Guizhou University, Guiyang 550025, China
School of Mathematical Sciences, Qufu Normal University, Qufu 273165, China

Saowaluck Chasreechai
Department of Mathematics, Faculty of Applied Science, King Mongkut's University of Technology North Bangkok, Bangkok 10800, Thailand

Ehsan Pourhadi
International Center for Mathematical Modelling in Physics and Cognitive Sciences, Department of Mathematics, Linnaeus University, SE-351 95 Växjö, Sweden
Department of Mathematics, Iran University of Science and Technology, Narmak, Tehran 16846-13114, Iran

Reza Saadati
Department of Mathematics, Iran University of Science and Technology, Narmak, Tehran 16846-13114, Iran

Sotiris K. Ntouyas
Department of Mathematics, University of Ioannina, 451 10 Ioannina, Greece
Nonlinear Analysis and Applied Mathematics (NAAM)-Research Group, Department of Mathematics, Faculty of Science, King Abdulaziz University, Jeddah 21589, Saudi Arabia

Dandan Yang and Chuanzhi Bai
School of Mathematical Science, Huaiyin Normal University, Huaian 223300, Jiangsu, China

Yasemin Basci
Department of Mathematics, Faculty of Arts and Sciences, Bolu Abant Izzet Baysal University, 14280 Golkoy, Bolu, Turkey

Dumitru Baleanu
Department of Mathematics, Faculty of Arts and Sciences, Çankaya University, 06530 Ankara, Turkey
Institute of Space Sciences, 077125 Bucharest-Magurele, Romania

XiaWu and Jialu Zhang
School of Mathematics and Finance, Xiangnan University, Chenzhou 411105, China

Kamal Shah and Inam Ullah
Department of Mathematics, University of Malakand, Chakdara Dir(L), Khyber Pakhtunkhwa 18800, Pakistan

Poom Kumam
KMUTT Fixed Point Research Laboratory, KMUTT-Fixed Point Theory and Applications Research Group, Theoretical and Computational Science Center (TaCS), Science Laboratory Building, Faculty of Science, King Mongkut's University of Technology Thonburi (KMUTT), 126 Pracha-Uthit Road, Bang Mod, Thrung Khru, Bangkok 10140, Thailand

Hassan Eltayeb and Said Mesloub
Mathematics Department, College of Science, King Saud University, Riyadh 11451, Saudi Arabia

Yahya T. Abdalla
Department of Mathematics, College of Science, Sudan University of Science and Technology, Khartoum 11111, Sudan

Adem Kılıçman
Department of Mathematics and Institute for Mathematical Research, Universiti Putra Malaysia, Serdang 43400 UPM, Selangor, Malaysia

Youzheng Ding
School of Science, Shandong Jianzhu University, Jinan 250101, China

Jiafa Xu
School of Mathematical Sciences, Chongqing Normal University, Chongqing 401331, China

Zhengqing Fu
College of Mathematics and System Sciences, Shandong University of Science and Technology, Qingdao 266590, China

Le Dinh Long
Faculty of Natural Sciences, Thu Dau Mot University, Thu Dau Mot City 820000, Binh Duong Province, Vietnam

Yong Zhou
Faculty of Information Technology, Macau University of Science and Technology, Macau 999078, China
Faculty of Mathematics and Computational Science, Xiangtan University, Xiangtan 411105, China

Tran Thanh Binh
Institute of Research and Development, Duy Tan University, Da Nang 550000, Vietnam

Nguyen Can
Applied Analysis Research Group, Faculty of Mathematics and Statistics, Ton Duc Thang University, Ho Chi Minh City 700000, Vietnam

Bashir Ahmad, Ahmed Alsaedi and Hamed H. Al-Sulami
Nonlinear Analysis and Applied Mathematics (NAAM)-Research Group, Department of Mathematics, Faculty of Science, King Abdulaziz University, Jeddah 21589, Saudi Arabia

Gauhar Rahman
Department of Mathematics, Shaheed Benazir Bhutto University, Sheringal, Upper Dir 18000, Pakistan

Kottakkaran Sooppy Nisar
Department of Mathematics, College of Arts and Sciences, Prince Sattam bin Abdulaziz University, Wadi Aldawaser 11991, Saudi Arabia

Thabet Abdeljawad
Department of Mathematics and General Sciences, Prince Sultan University, Riyadh 12345, Saudi Arabia
Department of Medical Research, China Medical University, Taichung 40402, Taiwan
Department of Computer Science and Information Engineering, Asia University, Taichung 40402, Taiwan

Samee Ullah
Department of Mathematics, University of Malakand, Lower Dir 18800, Chakdara, Pakistan

Ekaterina Madamlieva and Milena Petkova
Faculty of Mathematics and Informatics, University of Plovdiv, 4000 Plovdiv, Bulgaria

Mihail Konstantinov
Department of Mathematics, University of Architecture, Civil Engineering and Geodesy, 1046 Sofia, Bulgaria

Marian Milev
Department of Matematics and Physics, University Of Food Technology, 4000 Plovdiv, Bulgaria

Jiraporn Reunsumrit
Department of Mathematics, Faculty of Applied Science, King Mongkut's University of Technology North Bangkok, Bangkok 10800, Thailand

Fangfang Shi, Guoju Ye and Wei Liu
College of Science, Hohai University, Nanjing 210098, China

Dafang Zhao
School of Mathematics and Statistics, Hubei Normal University, Huangshi 435002, China

Index

www.ingramcontent.com/pod-product-compliance
Lightning Source LLC
LaVergne TN
LVHW081031210926
848485LV00004B/998
* 9 7 8 1 6 4 7 2 5 4 4 0 7 *